REPRESENTATIONS AND CHARACTERS

Revisiting the Works of Harish-Chandra and André Weil

LECTURE NOTES SERIES
Institute for Mathematical Sciences, National University of Singapore

Series Editors: Zuowei Shen and Chen-Bo Zhu
Institute for Mathematical Sciences
National University of Singapore

ISSN: 1793-0758

Published

Vol. 45 *Representations and Characters: Revisiting the Works of Harish-Chandra and André Weil*
edited by Hung Yean Loke, Angela Pasquale, Tomasz Przebinda & Binyong Sun

Vol. 44 *Higher Recursion Theory and Set Theory*
edited by James Cummings, Andrew Marks, Yue Yang & Liang Yu

Vol. 43 *On the Langlands Program: Endoscopy and Beyond*
edited by Wee Teck Gan, Dihua Jiang, Lei Zhang & Chen-Bo Zhu

Vol. 42 *Aspects of Computation and Automata Theory with Applications*
edited by Noam Greenberg, Sanjay Jain, Keng Meng Ng, Sven Schewe, Frank Stephan, Guohua Wu & Yue Yang

Vol. 41 *Density Functionals for Many-Particle Systems: Mathematical Theory and Physical Applications of Effective Equations*
edited by Berthold-Georg Englert, Heinz Siedentop & Martin-Isbjörn Trappe

Vol. 40 *Modeling and Simulation for Collective Dynamics*
edited by Weizhu Bao, Peter A Markowich, Benoit Perthame & Eitan Tadmor

Vol. 39 *Models and Methods for Quantum Condensation and Fluids*
edited by Weizhu Bao, Yongyong Cai, Ionut Danaila & Peter A Markowich

Vol. 38 *Genealogies of Interacting Particle Systems*
edited by Matthias Birkner, Rongfeng Sun & Jan M Swart

Vol. 37 *Mathematics of Shapes and Applications*
edited by Sergey Kushnarev, Anqi Qiu & Laurent Younes

Vol. 36 *The Geometry, Topology and Physics of Moduli Spaces of Higgs Bundles*
edited by Richard Wentworth & Graeme Wilkin

Vol. 35 *Combinatorial and Toric Homotopy: Introductory Lectures*
edited by Alastair Darby, Jelena Grbić, Zhi Lü & Jie Wu

Vol. 34 *White Noise Analysis and Quantum Information*
edited by Luigi Accardi, Louis H Y Chen, Takeyuki Hida, Masanori Ohya, Si Si & Noboru Watanabe

For the complete list of titles in this series, please go to
https://www.worldscientific.com/series/LNIMSNUS

Lecture Notes Series, Institute for Mathematical Sciences,
National University of Singapore
Vol. 45

REPRESENTATIONS AND CHARACTERS

Revisiting the Works of Harish-Chandra and André Weil

Editors

Hung Yean Loke
National University of Singapore, Singapore

Angela Pasquale
Université de Lorraine, France

Tomasz Przebinda
University of Oklahoma, USA

Binyong Sun
Zhejiang University, China

World Scientific

JERSEY • LONDON • SINGAPORE • BEIJING • SHANGHAI • HONG KONG • TAIPEI • CHENNAI • TOKYO

Published by

World Scientific Publishing Co. Pte. Ltd.
5 Toh Tuck Link, Singapore 596224
USA office: 27 Warren Street, Suite 401-402, Hackensack, NJ 07601
UK office: 57 Shelton Street, Covent Garden, London WC2H 9HE

British Library Cataloguing-in-Publication Data
A catalogue record for this book is available from the British Library.

Lecture Notes Series, Institute for Mathematical Sciences,
National University of Singapore — Vol. 45
REPRESENTATIONS AND CHARACTERS
Revisiting the Works of Harish-Chandra and André Weil

ISBN 978-981-98-2543-1 (hardcover)
ISBN 978-981-98-2544-8 (ebook for institutions)
ISBN 978-981-98-2545-5 (ebook for individuals)

For any available supplementary material, please visit
https://www.worldscienti ic.com/worldscibooks/10.1142/14652#t=suppl

https://doi.org/10.1142/9789819825448_fmatter

Contents

https://doi.org/10.1142/9789819825448_fmatter

Foreword

The Institute for Mathematical Sciences (IMS) organizes thematic programs of longer duration as well as shorter workshops and public lectures. The themes are selected from among areas at the forefront of current research in the mathematical sciences and their applications. Each volume of the *IMS Lecture Notes Series* is a compendium of papers based on lectures and tutorials delivered at the IMS. The aim is to make original papers and expository articles on a subject of current interest available to the international research community. These volumes also serve as a record of activities that took place at the IMS. We hope that through regular publication of these Lecture Notes the Institute will achieve, in part, its objective of reaching out to the community of scholars in the promotion of research in the mathematical sciences.

Zuowei Shen
Chen-Bo Zhu
Series Editors

https://doi.org/10.1142/9789819825448_fmatter

Preface

The theory of representations of reductive groups over local fields is a profound and active area of research. It has found important applications to other areas of mathematics such as number theory, harmonic analysis and mathematical physics. It is also a crucial part of the Langlands program.

Two central questions in representation theory are to classify all equivalence classes of irreducible (unitary) representations and to decompose a representation into irreducible ones. Harish-Chandra and the Gelfand school made fundamental progress in this area in the 1960s and 1970s. Characters and matrix coefficients are among the most important analytic invariants of representations. Harish-Chandra's description of discrete series representations is based on the construction of their characters. Another useful method in understanding representations is to study dual pair correspondences via the Segal-Shale-Weil representations and the minimal representations. Despite the immense progress, there are still many interesting open questions.

A two week long conference on representation theory was held at Institute for Mathematical Sciences (IMS) from 1 to 15 July 2022. The date of the conference overlapped with the International Congress for Mathematicians (ICM) 2022. Through the efforts of the management of IMS, the conference became an overlay satellite conference of the virtual 2022 ICM. As one of the six events allowing ICM speakers to give their talks in front of a live audience, the conference received greater prominence and attracted in-person and online participants. The first three days were dedicated to a series of two- to three-hour tutorial lectures by Professor Roger Howe (Yale University and Texas A&M University), Professor Toshiyuki Kobayashi (University of Tokyo) and Professor Wee Teck Gan (National University of Singapore). Each lecturer presented an overview of his area of expertise targeted at postdoctoral fellows and graduate students. Both Howe's and Kobayashi's lectures were named as IMS Distinguished Visitor Lecture Series. Roger Howe gave a survey of the history and major results in the Theta Correspondences. This is the celebrated research program which he pioneered in the 1970s. Kobayashi's lectures touched on recent topics on proper group actions and their relation to representation theory. In his lectures, Wee Teck Gan interpreted the theta correspondence from the view point of quantization and the relative Langlands program. Then he proposed a framework to relate the characters of two representations in theta correspondence.

The highlight of the conference were the ICM lectures held in the afternoons from 4 to 7 July. These included the ICM Emmy Noether Lecture by Professor Marie-France Vignéras (Institut de mathématiques de Jussieu — Paris Rive Gauche) on the modular representations of p-adic groups and on the local Langlands correspondences over fields of nonzero characteristics. Most ICM invited lecturers of Section 3 (Number Theory) and Section 7 (Lie Theory and Generalizations) gave their ICM lectures at this conference. They were Raphaël Beuzart–Plessis (Aix–Marseille University), Atsushi Ichino (Kyoto University), Tasho Kaletha (University of Michigan), Sug Woo Shin (The University of California, Berkeley), Yiannis Sakellaridis (Johns Hopkins University), Chen-Bo Zhu (National University of Singapore) and Binyong Sun (Zhejiang University), and Weiqiang Wang (University of Virginia). Evgeny Feigin (HSE University) delivered an online lecture.

In addition to the local participants from the National University of Singapore, the institute hosted 37 overseas visitors. Besides the tutorial lectures and ICM talks, there were 31 research talks delivered at IMS and 11 online talks. While many talks were given by established scholars, it is heartening to note that more than half of the speakers were young and promising researchers. The lectures spanned a wide range of topics: the orbit method and nilpotent orbits, character formulas of supercuspidal representations, character expansions of representations of p-adic groups, higher branching laws, conformally invariant differential operations, Fourier transform for tempered unipotent representations of p-adic groups, harmonic analysis on GL_n over finite fields, Hecke algebras and theta correspondence over local and finite fields. It was a wonderful experience for the participants to meet each other, to learn recent results and to exchange ideas. All the lectures were broadcasted live over the internet and there were more than 200 registered online participants to view them.

The nine chapters of this volume are for the most part expanded versions of the lectures delivered at the conference by their authors and all of them were refereed. In the following, we give a brief summary of their content.

Anne-Marie Aubert reviews the construction of generalized affine Hecke algebras associated with Bernstein series for both smooth irreducible and enhanced L-parameters of p-adic reductive groups, and applies this framework to study Howe's correspondence. She also reviews the Deligne-Lusztig classification of representations of finite groups of Lie type and the eta correspondence. Specific results of this chapter include proving a correspondence between simple modules of Hecke algebras attached to dual pairs of p-adic symplectic and orthogonal groups, and deriving a correspondence for C^*-algebras in the case of tempered representations.

Cheng Chen surveys an approach to the local Gan-Gross-Prasad conjecture based on Mœglin and Waldspurger's works. The chapter outlines the statement of the conjecture and some works proving the basic tempered cases of the conjecture, including the trace formula approach for tempered Bessel cases and the theta correspondence for connecting the Bessel and Fourier-Jacobi cases. In this chapter, he modifies the step showing that, in all situations, the local Gan-Gross-Prasad conjecture can be proved from the basic tempered cases. For this, he generalizes a multiplicity formula of Mœglin and Waldspurger, which he recently proved for any local fields of characteristic zero.

Corina Ciobotaru and Jan Frahm study symmetry breaking operators for small groups defined on non-archimedean local fields. Let $\mathbb{E}/\mathbb{F}$ be quadratic extension of non-archimedean local fields. Fix a quasicharacter χ of $\mathbb{E}^\times$ and a quasicharacter χ' of $\mathbb{F}^\times$. Denote by π_χ and by $\tau_{\chi'}$ the corresponding principal series representations of $\mathrm{PGL}_2(\mathbb{E})$ and $\mathrm{PGL}_2(\mathbb{F})$ respectively. The main object of study in this chapter is the space $\mathrm{Hom}_{\mathrm{PGL}_2(\mathbb{F})}(\pi_\chi, \tau_{\chi'})$, known as the space of symmetry breaking operators. The authors construct and classify such operators explicitly in terms of distributions on $\mathbb{E}$. In particular, they show that this space has dimension one or two, depending on the characters χ and χ', which is consistent with previous results of D. Prasad.

Jonathan Cohen writes about the depth zero, irreducible, supercuspidal representations of $\mathrm{GSp}(4, \mathbb{F})$ over a nonarchimedean local field $\mathbb{F}$. He calculates the dimensions of the spaces of Klingen-invariant and Siegel-invariant vectors in the representation space of various levels.

Lucas Fresse and Kyo Nishiyama present a comprehensive survey on recent progress in the study of double flag varieties for a symmetric pair (G, K), where G is a classical connected reductive algebraic group over the complex numbers, together with several new results and theorems. The main focus is on double flag varieties with finitely many K-orbits and their classification, including a complete classification of finite-type double flag varieties for the symmetric pair of type AIII. The chapter also develops a Steinberg theory for double flag varieties, generalizing the classical Steinberg map and Robinson–Schensted correspondence, and introduces new combinatorial procedures for Steinberg maps in the type AIII setting. Furthermore, an application of embedding theory yields new examples of finite-type double flag varieties. The chapter addresses both experts and non-specialists, and concludes with open questions and directions for future research.

Max Gurevich studies the Harish-Chandra–Howe local character expansion for representations of a general linear group over a p-adic field. He provides formulas for the coefficients of that expansion in terms of the degenerate Whittaker models, going back to the works of Moeglin and Waldspurger, and in terms of Zelevinsky's parameters via Kazhdan–Lusztig polynomials. Throughout the chapter he guides the reader through the relevant literature making the reading easy and enjoyable.

Hongyu He discusses the orbit method and its applications to the branching problem. The main point is that the study of the orbit projections can often provide new insights and perspectives to the study of the branching problem and vice versa. The chapter specifically focuses on the noncompact case of the unitary groups $\mathrm{U}(p,q)$ and $\mathrm{U}(p,q-1)$. It shows that the Gan-Gross-Prasad (GGP) interlacing relation can be used to describe both the projection of elliptic coadjoint orbits and the branching laws for discrete series representations of these groups. This illustrates that the orbit method can be a useful tool for suggesting branching laws, although in the most general cases, it will not tell exactly which representations occur in the branching. The chapter also touches on the use of the wave front sets as a way to gain information about branching laws when the orbit method is not directly applicable.

Roger Howe presents his construction of the oscillator representation of the metaplectic group over the field of real numbers, its interactions with the Classical Invariant Theory via his theory of reductive dual pairs, as well as applications to representation theory, automorphic forms, and harmonic analysis. The harmonic analysis part focuses on the oscillator semigroup and its connections to PDEs. This chapter serves as a sequel to Roger Howe's 1980 paper on the role of the Heisenberg group in harmonic analysis and expands on topics which have been understood with much greater clarity since the publication of that article. It outlines the historical development of the theta correspondence (or Howe's correspondence) with the unique perspective from its founder. It concludes with a comprehensive and useful list of references.

Toshiyuki Kobayashi presents his recent developments on proper group actions, highlighting their connections with representation theory. He explores the geometric aspects of proper actions, including criteria for the properness of homogeneous spaces for reductive groups G, and the interplay between the properness of group actions and the discrete decomposability of unitary representations realized on function spaces. It also examines two contrasting new approaches to quantify proper actions: one called sharpness, which measures how strongly an action satisfies the properness condition, and another that uses dynamical volume estimates to measure how much an action deviates from being proper. These volume estimates have been particularly useful for establishing temperedness criteria for regular unitary representations on G-spaces. The chapter illustrates the key concepts with concrete examples from both geometry and representation theory.

The organizers are pleased and grateful for the enthusiastic participation of colleagues from around the world for their participation in the conference. Special thanks go to those who accepted to contribute to this proceedings volume. We would like to take this opportunity to thank IMS for the financial and administrative support. The success of the conference wouldn't have been possible without the essential expertise and dedication of all IMS staff. This event promoted greater sharing of views and ideas, inspired new research and strengthened collaborations.

We are grateful to Dr. Premala Chandra, the daughter of Harish-Chandra, and to Mrs. Caitlin Rizzo, Archivist at the Shelby White and Leon Levy Archives Center of the Institute for Advanced Study, Princeton, for granting us the permissions to display the pictures of Harish-Chandra and of André Weil on the cover of this book.

November 2025

Hung Yean Loke
National University of Singapore,
Singapore

Angela Pasquale
Université de Lorraine,
France

Tomasz Przebinda
The University of Oklahoma,
USA

Binyong Sun
Zhejiang University,
China

Institute for Mathematical Sciences

https://doi.org/10.1142/9789819825448_0001

Correspondences between Affine Hecke Algebras and Applications

Anne-Marie Aubert
Sorbonne Université and Université Paris Cité, CNRS, IMJ-PRG, F-75005 Paris, France
anne-marie.aubert@imj-prg.fr

We review the construction of generalized affine Hecke algebras attached to Bernstein series of both smooth irreducible and enhanced L-parameters of p-adic reductive groups and apply it to the study of the Howe correspondence.

Contents

1. Introduction

Affine Hecke algebras are known to be very useful to describe the smooth complex representations of reductive p-adic groups, in terms of the supercuspidal representations of their Levi subgroups. Their use in the classification of smooth irreducible representations started with the seminal papers of Kazhdan and Lusztig [34] for Iwahori-spherical representations and of Lusztig [43], [44] for unipotent representations.

For G a p-adic group, the theory of the Bernstein Center of [15] provides a decomposition of the category $\mathfrak{R}(G)$ of smooth representations of G as the direct product of a family of full subcategories $\mathfrak{R}^{\mathfrak{s}}(G)$, that are called Bernstein blocks. Here $\mathfrak{s}$ is the equivalence class of a pair (L, σ), where L is a Levi subgroup of G and σ is an irreducible supercuspidal representation of L (see §3 for more details). By [31], if G is symplectic group or an orthogonal group, each Bernstein block is Morita equivalent to the category of modules of an *extended affine Hecke algebra* $\mathcal{H}(G, \mathfrak{s})$, that is, the crossed product of an affine Hecke algebra by the group algebra of a finite group.

For G an arbitrary p-adic group, and for every inertial class $\mathfrak{s} = [L, \sigma]_G$ such that the restriction of σ to L^1 is multiplicity free, where L^1 is the subgroup of L generated by all its compact its subgroups, the Bernstein block $\mathfrak{R}^{\mathfrak{s}}(G)$ is Morita equivalent to the category of modules of an algebra $\mathcal{H}(G, \mathfrak{s})$, which is the crossed product of an affine Hecke algebra by a (2-cocycle) twisted group algebra of a finite group (see [59]).

In recent works, we showed that affine Hecke algebras play a similar role on the Galois side of the local Langlands correspondence. In [8], we introduced a notion of *cuspidality* for enhanced Langlands parameters for G, which conjecturally puts irreducible supercuspidal representations of G in bijection with such enhanced L-parameters. We also define *cuspidal support maps* and *Bernstein series* for enhanced L-parameters, in analogy with Bernstein's theory of representations of p-adic groups. In [9], to every Bernstein series $\Phi_{\mathrm{e}}^{\mathfrak{s}^\vee}(G)$ of enhanced Langlands parameters for G we canonically associated an affine Hecke algebra $\mathcal{H}(G^\vee, \mathfrak{s}^\vee)$, again possibly extended by the twisted group algebra of a finite group. We proved that the simple modules of this algebra are naturally in bijection with the members of the Bernstein series, and that the set of central characters of the algebra is naturally in bijection with the collection of cuspidal supports of these enhanced Langlands parameters. These bijections send tempered or (essentially) square-integrable representations to the expected kind of Langlands parameters.

Furthermore, in [9] and [10], we showed that for many reductive p-adic groups (in particular for classical groups), if a Bernstein series $\mathrm{Irr}(\mathfrak{R}^{\mathfrak{s}}(G))$ for G corresponds to a Bernstein series $\Phi_{\mathrm{e}}^{\mathfrak{s}^\vee}(G)$ of enhanced Langlands parameters for G via the local Langlands correspondence, then the algebras $\mathcal{H}(G,\mathfrak{s})$ and $\mathcal{H}(G^\vee,\mathfrak{s}^\vee)$ are Morita equivalent.

We will review the construction of these algebras, and explain how these results can be applied to describe the theta correspondence for p-adic groups. The theory of this correspondence, also known as the Howe correspondence, was initiated by Roger Howe and builds upon the work of Weil in a representation theoretic treatment: a key role in the theory is played by the Weil representation (so-called, oscillator representation), a representation of a non-linear-algebraic group (the metaplectic group), described notably in [11]. It turned out that this correspondence has many important uses in local representation theory. Furthermore, its global counterpart (which agrees with theta correspondence locally) provides one of a very few direct ways to explicitly construct automorphic forms.

We will also describe some of the recent results on the description of the Howe correspondence over both finite fields and non-archimedean local fields. For $(G_n, G'_{n'})$ an irreducible reductive dual pair, formed of a p-adic symplectic group $\mathrm{Sp}_{2n}(F)$ and a p-adic orthogonal group $\mathrm{O}_{2n'}(F)$, we will prove in Theorem 5.2 that the compatibility the Howe correspondence with parabolic induction leads to a collection of correspondences between simple modules of algebras $\mathcal{H}(G_n,\mathfrak{s})$ and $\mathcal{H}(G'_{n'},\mathfrak{s}')$, where $\mathfrak{s}'$ is explicitly described in term of $\mathfrak{s}$. We will in particular consider the case of tempered representations and derive in Theorem 5.3 a correspondence at the level of C^*-algebras when $n' = n$ or $n' = n+1$.

The study of the Howe correspondence has also been heavily influenced by the recent developments in the framework of the Langlands program, so that new results are formulated in this spirit (see notably [3], [25], and [24]). As a consequence of our results, we will show that the Howe correspondence induces a collection of correspondences between simple modules of the algebras $\mathcal{H}(G_n^\vee,\mathfrak{s}^\vee)$ and $\mathcal{H}(G_{n'}'^\vee,\mathfrak{s}'^\vee)$ attached to Bernstein series of enhanced L-parameters of G_n and $G'_{n'}$, respectively. We studied in [4] the case where $\mathfrak{s} = [T,\mathrm{triv}]_{G_n}$ (with T a maximal torus in G and triv the trivial character of T) and $n' \in \{n, n+1\}$, via the Kazhdan-Lusztig parametrization of the simple modules of $\mathcal{H}(G_n,\mathfrak{s})$ and $\mathcal{H}(G'_{n'},\mathfrak{s}')$ (in this case, we have $\mathfrak{s}' = [T',\mathrm{triv}]_{G'_{n'}}$ where T' is a maximal torus in G').

We thank the anonymous referee for helpful suggestions that allowed to improve the readability of this chapter.

2. Representations of Finite Reductive Groups

Let k be a finite field and let $\overline{k}$ be a fixed algebraic closure of k. Let $\underline{\mathbb{G}}$ be a connected reductive algebraic group defined over k, and let $\mathbb{G} = \underline{\mathbb{G}}(k)$ denote the group of the k-rational points of $\underline{\mathbb{G}}$. The usual scalar product $(\ ,\ )_{\mathbb{G}}$ on the space of class functions on $\mathbb{G}$ is defined by

$$(f_1, f_2)_{\mathbb{G}} := |\mathbb{G}|^{-1} \sum_{g \in \mathbb{G}} f_1(g)\, \overline{f_2(g)}. \tag{2.1}$$

Let $\mathrm{Irr}(\mathbb{G})$ be the set of equivalence classes of irreducible representations of $\mathbb{G}$. For $\pi \in \mathrm{Irr}(\mathbb{G})$, we will also write π for its character. By Corollary 7.7 of [20], for any $\pi \in \mathrm{Irr}(\mathbb{G})$, there exists a k-rational maximal torus $\underline{\mathbb{T}}$ of $\underline{\mathbb{G}}$ and a character θ of $\mathbb{T}$ such that π occurs in the Deligne-Lusztig virtual character $R_{\underline{\mathbb{T}}}^{\underline{\mathbb{G}}}(\theta)$, *i.e.*, $(\pi, R_{\underline{\mathbb{T}}}^{\underline{\mathbb{G}}}(\theta))_{\mathbb{G}} \neq 0$.

Deligne and Lusztig introduced in [20] the following two regularity conditions: a character θ of $\mathbb{T}$ is said to be

- *in general position* if its stabilizer in $(\mathrm{N}_{\underline{\mathbb{G}}}(\underline{\mathbb{T}})/\underline{\mathbb{T}})(k)$ is trivial,
- *non-singular* if it is not orthogonal to any coroot.

If the center of $\underline{\mathbb{G}}$ is connected, then θ is non-singular if and only if it is in general position (see Proposition 5.16 in [20]).

We denote by $\underline{\mathbb{G}}^\vee$ the reductive connected group with root datum dual to that of $\underline{\mathbb{G}}$. The $\mathbb{G}$-conjugacy classes of pairs $(\underline{\mathbb{T}}, \theta)$ as above are in one-to-one correspondence with the $\mathbb{G}^\vee$-conjugacy classes of pairs $(\underline{\mathbb{T}}^\vee, s)$ where $\underline{\mathbb{T}}^\vee$ is a k-rational maximal torus of $\underline{\mathbb{G}}^\vee$ and s is a semisimple element of $\mathbb{G}^\vee$ belonging to $\underline{\mathbb{T}}^\vee$.

The *Lusztig series* $\mathcal{E}(\mathbb{G}, s)$ is defined to be the set

$$\left\{\pi \in \mathrm{Irr}(\mathbb{G})\colon (\pi, R_{\underline{\mathbb{T}}}^{\underline{\mathbb{G}}}(\theta))_{\mathbb{G}} \neq 0, \text{ where } (\underline{\mathbb{T}}, \theta)_{\mathbb{G}} \text{ corresponds to } (\underline{\mathbb{T}}^\vee, s)_{\mathbb{G}^\vee}\right\}.$$

The set $\mathrm{Irr}(\mathbb{G})$ decomposes into a disjoint union

$$\mathrm{Irr}(\mathbb{G}) = \bigsqcup_{(s)} \mathcal{E}(\mathbb{G}, s), \tag{2.2}$$

where (s) runs over the $\mathbb{G}^\vee$-conjugacy classes of semisimple elements of $\mathbb{G}^\vee$. If θ is the trivial character of $\mathbb{T}$, then the representation π is called *unipotent.* By definition, $\mathcal{E}(\mathbb{G}, 1)$ is the subset of unipotent representations in $\mathrm{Irr}(\mathbb{G})$. We denote by $\underline{\mathbb{G}}^\vee(s)$ the centralizer of s in $\underline{\mathbb{G}}^\vee$, and by $\mathbb{G}^\vee(s)$ the group of k-rational points of $\underline{\mathbb{G}}^\vee(s)$.

Remark 2.1: If $\pi \in \mathcal{E}(\mathbb{G}, s)$ and $\underline{\mathbb{G}}^\vee(s)$ is a torus, then $\pi = \pm R_{\underline{\mathbb{T}}}^{\underline{\mathbb{G}}}(\theta)$ with θ in general position (see Lemma 3.6.9 in [13]).

If the center of $\underline{\mathbb{G}}^\vee$ is connected then the group $\underline{\mathbb{G}}^\vee(s)$ is always connected. However, in general it may be disconnected, and we extend the notion of Deligne-Lusztig character as follows. We denote by $\underline{\mathbb{G}}^\vee(s)^\circ$ the identity component of $\underline{\mathbb{G}}^\vee(s)$, and by $\mathbb{G}^\vee(s)^\circ$ the group of k-rational points of $\underline{\mathbb{G}}^\vee(s)^\circ$. For $\underline{\mathbb{T}}^\vee$ a k-rational maximal torus of $\underline{\mathbb{G}}^\vee(s)^\circ$ and $\theta^\vee$ a character of $\mathbb{T}^\vee$, we set

$$R_{\underline{\mathbb{T}}^\vee}^{\underline{\mathbb{G}}^\vee(s)}(\theta^\vee) := \mathrm{Ind}_{\mathbb{G}^\vee(s)^\circ}^{\mathbb{G}^\vee(s)}(R_{\underline{\mathbb{T}}^\vee}^{\underline{\mathbb{G}}^\vee(s)^\circ}(\theta^\vee)). \tag{2.3}$$

We define $\mathcal{E}(\mathbb{G}^\vee(s), 1)$ to be the set of irreducible constituents of $R_{\underline{\mathbb{T}}^\vee}^{\underline{\mathbb{G}}^\vee(s)}(1)$.

By Theorem 4.23 of [39] when the center of $\underline{\mathbb{G}}^\vee$ is connected, and by §12 of [41] in general, there is a bijection

$$\mathcal{E}(\mathbb{G}, s) \xrightarrow{1-1} \mathcal{E}((\mathbb{G}^\vee(s), 1), \quad \pi \mapsto \pi^{\mathrm{u}}. \tag{2.4}$$

The bijection (2.4) satisfies the following properties:

(1) It sends a Deligne-Lusztig character $R_{\underline{\mathbb{T}}}^{\underline{\mathbb{G}}}(\theta)$ in $\mathbb{G}$ (up to a sign) to a Deligne-Lusztig character $R_{\underline{\mathbb{T}}^\vee}^{\underline{\mathbb{G}}^\vee(s)}(1)$, where 1 denotes the trivial character of $\underline{\mathbb{T}}^\vee$ (see §12 of [41]).
(2) It preserves cuspidality in the following sense: if $s \in \mathbb{G}^\vee$, and $\mathcal{E}(\mathbb{G}, s)$ contains a cuspidal representation π_{c}, then
   (a) the largest k-split torus in the center of $\underline{\mathbb{G}}^\vee(s)$ coincides with the largest k-split torus in the center of $\underline{\mathbb{G}}^\vee$ (see (8.4.5) of [39]),
   (b) the unipotent representation $\pi_{\mathrm{c}}^{\mathrm{u}}$ is cuspidal.
(3) The dimension of every representation $\pi \in \mathcal{E}(\mathbb{G}, s)$ is given by

$$\dim(\pi) = \frac{|\mathbb{G}|_{p'}}{|\underline{\mathbb{G}}^\vee(s)(k)|_{p'}} \dim(\pi^{\mathrm{u}}), \tag{2.5}$$

where $|\mathbb{G}|_{p'}$ is the largest prime-to-p factor of the order of $\mathbb{G}$ (see Remark 13.24 of [21]).

2.1. *Centralizers of semisimple elements in classical groups*

Let $\underline{\mathbb{G}}$ be a classical group of rank n over $\overline{k}$, *i.e.*, $\underline{\mathbb{G}}$ is a symplectic group, a special orthogonal group or an orthogonal group. Let $\underline{\mathbb{T}}_n \simeq \overline{k}^\times \times \cdots \times \overline{k}^\times$ be a k-rational maximal torus of $\underline{\mathbb{G}}$. For $s = (\lambda_1, \ldots, \lambda_n) \in \mathbb{T}_n$, we denote by $\nu_\lambda(s)$ the number of the λ_i's which are equal to λ, and by $\langle\lambda\rangle$ the set of all roots in $\overline{k}$ of the irreducible polynomial of λ over k. By §1.B of [7], the centralizer $C_{\mathbb{G}}(s)$ of s in $\mathbb{G}$ decomposes as a product

$$C_{\mathbb{G}}(s) = \prod_{\langle\lambda\rangle\subset\{\lambda_1,\ldots,\lambda_n\}} \mathbb{G}_{(\lambda)}(s), \tag{2.6}$$

where $\mathbb{G}_{(\lambda)}(s)$ is a reductive quasi-simple group of rank equal to $\nu_\lambda(s)\cdot|\langle\lambda\rangle|$. The followings hold:

(i) if $\mathbb{G} = \mathrm{Sp}_{2n}$, then $\mathbb{G}_1(s) \simeq \mathrm{Sp}_{2\nu_1(s)}$ and $\mathbb{G}_{-1}(s) \simeq \mathrm{Sp}_{2\nu_{-1}(s)}$,
(ii) if $\mathbb{G} = \mathrm{SO}_{2n+1}$, then $\mathbb{G}_1(s) \simeq \mathrm{SO}_{2\nu_1(s)+1}$ and $\mathbb{G}_{-1}(s) \simeq \mathrm{O}^{\pm}_{2\nu_{-1}(s)}$,
(iii) if $\mathbb{G} = \mathrm{O}^{\pm}_{2n}$, then $\mathbb{G}_1(s) \simeq \mathrm{O}^{\pm}_{2\nu_1(s)}$ and $\mathbb{G}_{-1}(s) \simeq \mathrm{O}^{\pm}_{2\nu_{-1}(s)}$,
(iv) if $\mathbb{G} = \mathrm{U}_n$ or if $\lambda \neq \pm 1$, then $\mathbb{G}_{(\lambda)}(s)$ is either a general linear group or a unitary group.

We set

$$\mathbb{G}_{\neq}(s) := \prod_{\langle\lambda\rangle\subset\{\lambda_1,\ldots,\lambda_m\},\lambda\neq\pm 1} \mathbb{G}_{(\lambda)}(s). \tag{2.7}$$

By (2.6), we have

$$C_{\mathbb{G}^\vee}(s) = \mathbb{G}_1(s)^\vee \times \mathbb{G}_{-1}(s) \times \mathbb{G}_{\neq}(s), \tag{2.8}$$

and hence a bijection

$$\begin{array}{ccc} \mathcal{E}(C_{\mathbb{G}^\vee}(s), 1) & \simeq & \mathcal{E}(\mathbb{G}_1(s)^\vee, 1) \times \mathcal{E}(\mathbb{G}_{-1}(s), 1) \times \mathcal{E}(\mathbb{G}_{\neq}(s), 1) \\ \pi^{\mathrm{u}} & \mapsto & \pi_1^{\mathrm{u}} \otimes \pi_{-1}^{\mathrm{u}} \otimes \pi_{\neq}^{\mathrm{u}} \end{array}. \tag{2.9}$$

We obtain a one-to-one correspondence

$$\begin{array}{rccc} \mathcal{L}_s\colon & \mathcal{E}(\mathbb{G}, s) & \to & \mathcal{E}(\mathbb{G}_1(s)^\vee, 1) \times \mathcal{E}(\mathbb{G}_{-1}(s), 1) \times \mathcal{E}(\mathbb{G}_{\neq}(s), 1) \\ & \pi & \mapsto & \pi_1^{\mathrm{u}} \otimes \pi_{-1}^{\mathrm{u}} \otimes \pi_{\neq}^{\mathrm{u}} \end{array}. \tag{2.10}$$

2.2. *The Howe correspondence over finite fields*

We suppose that the characteristic of k is odd. Let N be a positive integer, and let W be a vector space over k of dimension $2N$, equipped with a nondegenerate alternated bilinear form $\langle\,,\,\rangle$. A pair of reductive subgroups of $\mathrm{Sp}(W) = \mathrm{Sp}_{2N}(k)$, where each one is the centralizer of the other, is called *reductive dual pair*. We study irreducible dual pairs, because these are the building blocks of all the others. Such pairs are of the following kinds:

- pairs of type I:
  - $(\mathrm{Sp}_{2n}(k), \mathrm{O}_{N'}(k))$ with $nN' = N$;
  - $(\mathrm{U}_n(k), \mathrm{U}_{n'}(k))$ with $nn' = 2N$;
- pairs of type II:
  - $(\mathrm{GL}_n(k), \mathrm{GL}_{n'}(k))$ with $nn' = 2N$.

A reductive dual pair $(\mathbb{G}, \mathbb{G}')$ is said to be in the *stable range* (with $\mathbb{G}'$ smaller) if the defining vector space for $\mathbb{G}$ has a totally isotropic subspace of dimension greater or equal than the dimension the defining vector space for $\mathbb{G}'$, e.g. the pairs $(\mathrm{Sp}_{2n}(k), \mathrm{O}_{2n'}(k))$ such that $n \geq 2n'$.

Howe introduced a correspondence $\Theta_{n,n'} \colon \mathfrak{R}(\mathbb{G}_n) \to \mathfrak{R}(\mathbb{G}_{n'})$ between the categories of complex representations of these subgroups. It is obtained from a particular representation of $\mathrm{Sp}_{2N}(k)$, called the *Weil* (or *oscillator*) *representation.* In order to define this representation, we must introduce the Heisenberg group. This is the group with underlying set $H(W) = W \times k$ and product

$$(w_1, t_1) \cdot (w_2, t_2) = \left(w_1 + w_2, t_1 + t_2 + \frac{1}{2}\langle w_1, w_2 \rangle \right). \tag{2.11}$$

Let ϱ be an irreducible representation of $H(W)$. Its restriction to the center $\mathrm{Z}_{H(W)} \simeq k$ of $H(W)$ equals ψ_ϱ, for a certain character ψ_ϱ of k.

By the Stone-von-Neumann Theorem (see Theorem 2.I.2 in [49]), for any non-trivial character ψ of $\mathrm{Z}_{H(W)}$ there exists (up to equivalence) a unique irreducible representation ϱ of $H(W)$ such that $\psi_\varrho = \psi$. This representation is known as the Heisenberg representation. It depends on ψ, so we denote it by ϱ_ψ.

The natural action of $\mathrm{Sp}(W)$ on $H(W)$ fixes the elements of its center. Hence, for a fixed character ψ of k, the representations ϱ_ψ and $g \cdot \varrho_\psi$ agree on $\mathrm{Z}_{H(W)}$, for any $g \in \mathrm{Sp}(W)$. The Stone-von-Neumann Theorem implies that there is an operator $\omega_\psi(g)$ verifying

$$\varrho_\psi(g \cdot w, t) = \omega_\psi(g) \varrho_\psi(w, t) \omega_\psi(g)^{-1}. \tag{2.12}$$

This defines a projective representation ω_ψ of $\mathrm{Sp}(W)$, which can be lifted to an actual representation of $\mathrm{Sp}(W)$, known as the Weil representation.

The restriction of ω_ψ of $\mathrm{Sp}_{2N}(k)$ to $\mathbb{G}_n \times \mathbb{G}'_{n'}$ is

$$\omega_{\mathbb{G}_n, \mathbb{G}'_{n'}} = \sum_{\substack{\pi \in \mathrm{Irr}(\mathbb{G}) \\ \pi' \in \mathrm{Irr}(\mathbb{G}')}} \mathrm{m}_{\pi,\pi'} \, \pi \otimes \pi', \quad \text{where } \mathrm{m}_{\pi,\pi'} \in \mathbb{Z}_{\geq 0}. \tag{2.13}$$

Define $\Theta_{n'} \colon \mathbb{Z}\,\mathrm{Irr}(\mathbb{G}_n) \to \mathbb{Z}\,\mathrm{Irr}(\mathbb{G}'_{n'})$ by

$$\Theta_{n'}(\pi) := \{\pi' \in \mathrm{Irr}(\mathbb{G}'_{n'}) \, : \, \mathrm{m}_{\pi,\pi'} \neq 0\}, \quad \text{for } \pi \in \mathrm{Irr}(\mathbb{G}_n). \tag{2.14}$$

The occurrence of a irreducible representation π of $\mathbb{G}_n$ in the Howe correspondence for $(\mathbb{G}_n, \mathbb{G}'_{n'_\pi})$ with n'_π minimal (i.e., such that $\Theta_{\mathbb{G}'_{n'}}(\pi) = 0$ for any $n' < n'_\pi$) is referred to as the *first occurrence.*

2.2.1. *The Howe correspondence for unipotent representations*

Between members of a dual pair, the only ones having cuspidal unipotent representations are: $\mathrm{GL}_1(k)$, $\mathrm{Sp}_{2(a^2+a)}(k)$, $\mathrm{U}_{(a^2+a)/2}(k)$ (which have a unique such representation, say τ_a), and $\mathrm{O}_{2a^2}(k)$ (which has two: τ_a^{I} and $\tau_a^{\mathrm{II}} = \tau_a^{\mathrm{I}} \otimes \mathrm{sign}$).

From now on, we will only consider pairs formed by a symplectic group $\mathrm{Sp}_{2n}(k)$ and an orthogonal group $\mathrm{O}_{2n'}^{\epsilon}(k)$, where $\epsilon \in \{\pm\}$, with $\mathrm{O}_{2n'}^{+}(k)$ split and $\mathrm{O}_{2n'}^{-}(k)$ nonsplit. The following results were established in [1]:

(1) if π is a cuspidal irreducible representation of $\mathbb{G}_n = \mathrm{Sp}_{2n}(k)$, then $\Theta_{n'_\pi}(\pi)$ is a singleton $\{\pi'\}$ with π' cuspidal irreducible,
(2) if $\pi \in \mathrm{Irr}(\mathbb{G}_n)$ is unipotent then any $\pi' \in \Theta_{n'}(\pi)$ is unipotent,
(3) the representation τ_a of $\mathrm{Sp}_{2(a^2+a)}(k)$ corresponds to τ_a^{II} if ϵ is the sign of $(-1)^a$ and to τ_{a+1}^{I} otherwise.

Thus, the Howe correspondence between cuspidal unipotent representations is described by the function $\theta\colon \mathbb{N} \to \mathbb{N}$, defined by

$$\theta(a) := \begin{cases} a & \text{if } \epsilon \text{ is the sign of } (-1)^a \\ a+1 & \text{otherwise.} \end{cases} \tag{2.15}$$

By [7], the Howe correspondence for unipotent representations of the dual pair $(\mathrm{Sp}_{2n}(k), \mathrm{O}_{2n'}^{\epsilon}(k))$ induces a correspondence between the parabolically induced representations $\mathrm{i}_{\mathrm{Sp}_{2(a^2+a)}(k)\otimes\mathbb{T}}^{\mathrm{Sp}_{2n}(k)}(\tau_a \otimes 1)$ and $\mathrm{i}_{\mathrm{O}_{2\theta(a)^2}^{\epsilon}(k)\otimes\mathbb{T}'}^{\mathrm{O}_{2n'}^{\epsilon}(k)}(\tau'_{\theta(a)} \otimes 1)$, where $\tau'_{\theta(a)} \in \{\tau_{\theta(a)}^{\mathrm{I}}, \tau_{\theta(a)}^{\mathrm{II}}\}$, and $\mathbb{T}$, $\mathbb{T}'$ are products of $\mathrm{GL}_1(k)$'s.

It induces a correspondence between the endomorphism algebras of these parabolically induced representations, and hence a correspondence Ω_{N_a,N'_a} between irreducible representations of Weyl groups of types B_{N_a} and $\mathrm{B}_{N'_a}$, where $N_a := n - a^2 - a$ and $N'_a := n - \theta(a)^2$.

The following conjectural explicit description of Ω_{N_a,N'_a} for pairs $(\mathrm{Sp}_{2n}(k), \mathrm{O}_{2n'}^{\epsilon}(k))$ was formulated in [7], and confirmed in [55], and in [45], independently.

Let us introduce first some definitions and notation: $\lambda := (\lambda_1 \geq \lambda_2 \geq \cdots \geq \lambda_l)$ where $\lambda_i \in \mathbb{Z}_{\geq 0}$ is called a partition of n if $|\lambda| := \lambda_1+\lambda_2+\cdots+\lambda_l = n$. We write $\lambda \dashv n$. We denote by $\mathcal{P}(n)$ set of partitions of n, and we write $\lambda \cup \mu$ for the partition of $n+m$ with parts $\lambda_1, \ldots, \lambda_l, \mu_1, \ldots, \mu_l$. The usual order on partitions is defined by

$$\lambda \leq \lambda' \quad \text{if and only if} \quad \lambda_1 + \cdots + \lambda_i \leq \lambda'_1 + \cdots + \lambda'_i, \text{ for all } i \in \mathbb{N}.$$

We define another order on partitions as follows: for λ, λ' partitions of possibly different integers, we write

$$\lambda \preceq \lambda' \quad \text{if and only if} \quad \lambda'_{i+1} \le \lambda_i \le \lambda'_i, \text{ for all } i \in \mathbb{N}. \tag{2.16}$$

It says that $\lambda \preceq \lambda'$ if the Young diagram of λ is contained in the one of λ' and that we can go from the first to the second by adding at most one box per column. For instance, the partitions $\lambda = (4,1,1)$ of 6, and $\lambda' = (4,4,1,1)$ of 10 verify $\lambda \preceq \lambda'$. We observe that if two partitions $\lambda = (\lambda_1 \ge \lambda_2 \ge \cdots \ge \lambda_l)$ and $\lambda' = (\lambda_1 \ge \lambda_2 \ge \cdots \ge \lambda_{l'})$ satisfy $\lambda \preceq \lambda'$ and $|\lambda| = |\lambda'|$, then we have $\lambda = \lambda'$. Indeed, by (2.16), we have $l \le l'$, and

$$|\lambda| = \sum_{i=1}^{l} \lambda_i \le \sum_{i=1}^{l} \lambda'_i = |\lambda| - (\lambda'_{l+1} + \cdots + \lambda'_{l'}),$$

thus, $l' = l$ and $\lambda_i = \lambda'_i$ for all $i \in \{1, \ldots, l\}$.

Irreducible characters of a Weyl group of type B_n or C_n are known to be parametrized by bipartitions of n (see for instance [19]). We denote by $\chi_{\xi,\eta}$ the irreducible character which corresponds to the bipartition (ξ, η) of n.

In the rest of this section we consider pairs $(\mathrm{Sp}_{2n}(k), \mathrm{O}^{\epsilon}_{2n'}(k))$ in the stable range (*i.e.*, such that $n \ge 2n'$). We write $N_a(\zeta) := N_a - |\zeta|$ if ζ is a partition such that $|\zeta| \le N_a$. Then we have:

(1) Cases $(\mathrm{Sp}_{2n}, \mathrm{O}^{+}_{2n'})$ with a even and $(\mathrm{Sp}_{2n}, \mathrm{O}^{-}_{2n'})$ with a odd:

$$\Omega_{N_a,N'_a} = \sum_{r=0}^{\min(N_a,N'_a)} \sum_{(\xi,\zeta)\in\mathcal{P}_2(r)} \sum_{\eta,\eta'} \chi_{\xi,\eta} \otimes \chi_{\xi,\eta'},$$

where the third sum is over the partitions $\eta \dashv N_a(\xi)$ and $\eta' \dashv N'_a(\xi)$ such that $\zeta \preceq \eta$ and $\zeta \preceq \eta'$.

(2) Cases $(\mathrm{Sp}_{2n}, \mathrm{O}^{+}_{2n'})$ with a odd and $(\mathrm{Sp}_{2n}, \mathrm{O}^{+}_{2n'})$ with a even:

$$\Omega_{N_a,N'_a} = \sum_{r=0}^{\min(N_a,N'_a)} \sum_{(\xi,\zeta)\in\mathcal{P}_2(r)} \sum_{\xi',\eta'} \chi_{\xi',\eta} \otimes \chi_{\xi,\eta'},$$

where the third sum is over the partitions $\xi' \dashv N_a(\eta)$ and $\eta' \dashv N'_a(\xi)$ such that $\xi \preceq \xi'$ and $\eta \preceq \eta'$.

In general, there exist representations $\pi \in \mathrm{Irr}(\mathbb{G})$ such that $\Theta_{n'}(\pi)$ contains more than one element. Hence, one may wonder if it would be possible to extract a *one-to-one correspondence*. Several approaches to this questions were considered:

- definition of the eta correspondence for dual pairs $(\mathrm{Sp}_{2n}(k), \mathrm{O}_{N'}(k))$ in the stable range in [28],
- construction of a one-to-one correspondence for unipotent representations of pairs of type II and of pairs in the stable range of the form $(\mathrm{O}^{\epsilon}_{2a^2}(k), \mathrm{Sp}_{2(a^2+a+N)}(k))$ or $(\mathrm{Sp}_{2(a^2+a+2)}(k), \mathrm{O}^{\epsilon}_{2a^2+N}(k))$ [6],
- construction of a one-to-one theta correspondence for unipotent representations of irreducible pairs of type I in the stable range in [22],
- extension of both eta and theta correspondences to all irreducible pairs of type I in [57].

Let $(\xi', \eta') \in \mathcal{P}_2(N'_a)$. We set

$$\Theta_{\xi',\eta'} := \left\{\chi_{\xi,\eta} \,:\, \chi_{\xi,\eta} \otimes \chi_{\xi',\eta'} \text{ occurs in } \Omega_{N_a,N'_a}\right\}. \tag{2.17}$$

Recall that a partition is called *symplectic* if each odd part appears with even multiplicity. There is a bijection between symplectic partitions of $2n$ and unipotent conjugacy classes of $\underline{\mathbb{G}}_n = \mathrm{Sp}_{2n}(\overline{k})$. We denote by $\mathcal{O}_\lambda$ the unipotent class associated to the symplectic partition λ. The Springer correspondence maps $\chi_{\xi,\eta}$ to a pair $(\mathcal{O}_\lambda, \rho)$ for some irreducible character ρ of $\mathrm{C}_{\underline{\mathbb{G}}_n}(u)/\,\mathrm{C}_{\underline{\mathbb{G}}_n}(u)^\circ$, where $u \in \mathcal{O}_\lambda$. In this situation we write $\lambda_{\xi,\eta} := \lambda$.

Definition 2.2: We say that $\chi_{\xi_{\max},\eta_{\max}} \in \Theta_{\xi',\eta'}$ is *maximal* if

$$\lambda_{\xi,\eta} \leq \lambda_{\xi_{\max},\eta_{\max}} \quad \text{for all } (\eta, \xi) \in \Theta_{\xi',\eta'}.$$

Remark 2.3: Since the order is not total, it is not clear a priori that a maximal representation exists, and if so, that it is unique.

Definition 2.4: We say that $\chi_{\xi_{\min},\eta_{\min}} \in \Theta_{\xi',\eta'}$ is *minimal* if

$$\lambda_{\xi_{\min},\eta_{\min}} \leq \lambda_{\xi,\eta} \quad \text{for all } (\eta, \xi) \in \Theta_{\xi',\eta'}.$$

A representation is said to be *extremal* if it is either maximal or minimal.

Let $(\xi', \eta') \in \mathcal{P}_2(N'_a)$. By [22],

(1) there exists a unique maximal representation $\chi_{\xi_{\max},\eta_{\max}} \in \Theta_{\xi',\eta'}$, it is given by

$$\xi_{\max} := \xi' \quad \text{and} \quad \eta_{\max} := (N_a - N'_a + \eta'_1 + \eta'_2, \eta'_3, \ldots, \eta'_l).$$

(2) there exists a unique minimal representation $\chi_{\xi_{\min},\eta_{\min}} \in \Theta_{\xi',\eta'}$, it is given by

$$\xi_{\min} := \xi' \quad \text{and} \quad \eta_{\min} := (N_a - N'_a) \cup \eta'.$$

3. Smooth Representations of p-adic Reductive Groups

Let F be a non Archimedean local field and W_F its absolute Weil group. We denote by $I_F, P_F \subset W_F$ the inertia group and the wild inertia group of F, respectively, and by k_F the residual field of F (a finite field with q elements, where q is a power of a prime number p). We denote by $W_F^{\mathrm{t}} := W_F \rtimes I_F$ the tame Weil group of F.

Let $\mathbf{G}$ be a connected reductive algebraic group defined over F. We denote by G the group of the F-rational points of $\mathbf{G}$. Let L be a Levi subgroup of a parabolic subgroup P of G and let $\mathfrak{X}_{\mathrm{nr}}(L)$ denote the group of unramified characters of L. Let σ be an irreducible supercuspidal smooth representation of L and $\mathcal{O}$ the set of equivalence classes of representations L of the form $\sigma \otimes \chi$, with $\chi \in X_{\mathrm{nr}}(L)$. We write $\mathfrak{s} := (L, \mathcal{O})_G = [L, \sigma]_G$ for the G-conjugacy class of the pair $(L, \mathcal{O})$ and $\mathfrak{B}(G)$ for the set of such classes $\mathfrak{s}$. We set $\mathfrak{s}_L := (L, \mathcal{O})_L$.

We denote by $\mathfrak{R}^{\mathfrak{s}}(G)$ the full subcategory of $\mathfrak{R}(G)$ whose objects are the representations (π, V) such that every irreducible G-subquotient of π has its supercuspidal support in $\mathfrak{s}$. The categories $\mathfrak{R}^{\mathfrak{s}}(G)$ are indecomposable and split the full smooth category $\mathfrak{R}(G)$ in a direct product:

$$\mathfrak{R}(G) = \prod_{\mathfrak{s} \in \mathfrak{B}(G)} \mathfrak{R}^{\mathfrak{s}}(G). \tag{3.1}$$

Let $\mathrm{Irr}^{\mathfrak{s}}(G)$ denote the set of irreducible objects of the category $\mathfrak{R}^{\mathfrak{s}}(G)$. As a direct consequence of (3.1), we have

$$\mathrm{Irr}(G) = \prod_{\mathfrak{s} \in \mathfrak{B}(G)} \mathrm{Irr}^{\mathfrak{s}}(G). \tag{3.2}$$

3.1. *Depth-zero supercuspidal representations of p-adic groups*

We denote by $\mathcal{B}(\mathbf{G}, F)$ the Bruhat-Tits building of G and by $\mathcal{B}_{\mathrm{r}}(\mathbf{G}, F) := \mathcal{B}(\mathbf{G}/\mathrm{Z}_{\mathbf{G}}, F)$ the reduced Bruhat-Tits building of G. We have $\mathcal{B}(\mathbf{G}, F) = \mathcal{B}_{\mathrm{r}}(\mathbf{G}, F) \times (X_*(\mathrm{Z}_G) \otimes_{\mathbb{Z}} \mathbb{R})$, where $X_*(\mathrm{Z}_G)$ is the set of F-algebraic cocharacters of Z_G. For each point x in $\mathcal{B}(\mathbf{G}, F)$, we denote by $G_{x,0}$ the parahoric subgroup of G associated to x, by $G_{x,0+}$ the pro-p unipotent radical of $G_{x,0}$, and $\mathbb{G}_{x,0}$ the quotient $G_{x,0}/G_{x,0+}$, (the points of) a connected reductive group over the residue field of F. Let $[x]$ denote the image of x in $\mathcal{B}_{\mathrm{r}}(\mathbf{G}, F)$ and $G_{[x]}$ the stabilizer of $[x]$ under the action G on $\mathcal{B}_{\mathrm{r}}(\mathbf{G}, F)$.

A *depth-zero* representation of G is a representation which admits nonzero invariant vectors under the action of the pro-p unipotent radical $G_{x,0+}$ of a parahoric subgroup $G_{x,0}$ of G.

Let π be any irreducible depth-zero supercuspidal representation of G. There exists a vertex $x \in \mathcal{B}_{\mathrm{r}}(\mathbf{G}, F)$ and an irreducible cuspidal representation π_x of $\mathbb{G}_x$, such that the restriction of π to $G_{x,0}$ contains the inflation $\mathrm{infl}(\pi_x)$ of π_x (see §1–2 of [50] or Proposition 6.6 of [53]). The normalizer $\mathrm{N}_G(G_{x,0})$ of $G_{x,0}$ in G is a totally disconnected group that is compact mod center, which, by the proof of (5.2.8) in [17], coincides with the fixator $G_{[x]}$ of $[x]$ under the action of G on $\mathcal{B}_{\mathrm{r}}(\mathbf{G}, F)$. Let $\widetilde{\pi}_x$ denote an extension of $\mathrm{infl}(\pi_x)$ to $G_{[x]}$. Then π is compactly induced from a representation of $G_{[x]}$:

$$\pi = \text{c-Ind}_{G_{[x]}}^{G}(\widetilde{\pi}_x). \tag{3.3}$$

The representation π is called *unipotent* if π_x is unipotent.

3.2. *Depth-zero supercuspidal representations of p-adic classical groups*

Example 3.1: Let E be an unramified quadratic extension of F, and let V be a vector space over E of dimension n. There are two classes of hermitian forms on V. They are distinguished by the parity of the valuation of the determinant of the form: we denote the form by $\langle\, , \rangle^+$ if the valuation is even, and by $\langle\, , \rangle^-$ otherwise. We denote the corresponding unitary groups by $\mathrm{U}_n^{\pm}(F)$. The group $\mathrm{U}_n^{+}(F)$ is quasi-split. If n is odd, then the groups $\mathrm{U}_n^{+}(F)$ and $\mathrm{U}_n^{-}(F)$ are isomorphic, if n is even, then $\mathrm{U}_n^{-}(F)$ is an inner form of $\mathrm{U}_n^{+}(F)$.

The standard maximal parahoric subgroups of $G = \mathrm{U}_n^{\pm}(F)$ are the groups $G_{x,0}$ such that $\mathbb{G}_{x,0}$ is isomorphic to the product of two unitary groups $\mathrm{U}_{n_1}(k_F)$ and $\mathrm{U}_{n_2}(k_F)$ with $n_1 + n_2 = n$. By §2.2 and (3.3), the supercuspidal unipotent representations of $\mathrm{U}_n(F)$ are the representations $\pi = \text{c-Ind}_{G_{[x]}}^{L}(\widetilde{\pi}_x)$, where $\pi_x = \tau_{a_1} \otimes \tau_{a_2}$ with $a_1^2 + a_1 = 2n_1$ and $a_2^2 + a_2 = 2n_2$. For $i \in \{1, 2\}$, we denote by λ_i the partition of n_i defined by $\lambda_i := (k_i, k_i - 1, k_i - 2, \ldots, 1,)$.

Example 3.2: Let $G = \mathrm{SO}_{2n+1}(F)$. The standard maximal parahoric subgroups $G_{x,0}$ of G have type $\mathrm{D}_{n_1} \times \mathrm{B}_{n_2}$ with $n_1 + n_2 = n$. We have $\mathbb{G}_{x,0} \simeq \mathrm{SO}_{2n_1}(k_F) \times \mathrm{SO}_{2n_2+1}(k_F)$. The corresponding depth-zero supercuspidal representation of G in compactly induced from the inflation to $G_{x,0}$ of an irreducible cuspidal representation $\pi_x^1 \otimes \pi_x^2$ of $\mathbb{G}_{x,0}$. We have $\pi_x^1 \in \mathcal{E}(\mathrm{SO}_{2n_1}, (s_1))$ and $\pi_x^2 \in \mathcal{E}(\mathrm{SO}_{2n_2+1}, (s_2))$. By 2.1, the group W_{s_1} is of one of the following types

(1) $\mathrm{D}_{a_1^2} \times \mathrm{D}_{b_1^2}$ with $a_1^2 + b_1^2 = n_1$ and $a_1, b_1 \in \mathbb{Z}_{\geq 0}$,

(2) $\mathrm{A}^2_{a_1-1} \times \mathrm{A}^2_{\ell_1-1}$ with $a_1 = \frac{c_1+c_1^2}{2}$, $\ell_1 = \frac{d_1+d_1^2}{2}$, $a_1 + \ell_1 = n_1$ and $c_1, d_1 \in \mathbb{Z}_{\geq 0}$,

(3) $\mathrm{D}_{a_1^2} \times \mathrm{A}_{a_1-1}$ with $a_1 = \frac{c_1+c_1^2}{2}$, $a_1^2 + a_1 = n_1$ and $a_1, c_1 \in \mathbb{Z}_{\geq 0}$,

and W_{s_2} is of one of the following types

(1) $\mathrm{D}_{a_2^2} \times \mathrm{C}_{b_2^2+b_2}$ with $a_2^2 + b_2^2 + b_2 = n_2$ and $a_2, b_2 \in \mathbb{Z}_{\geq 0}$,

(2) $\mathrm{A}^2_{a_2-1} \times \mathrm{A}^2_{\ell_3-1}$ with $a_2 = \frac{c_2+c_2^2}{2}$, $\ell_2 = \frac{d_2+d_2^2}{2}$, $a_2 + \ell_1 = n_2$ and $c_2, d_2 \in \mathbb{Z}_{\geq 0}$,

(3) $\mathrm{D}_{a_2^2} \times \mathrm{A}_{a_2-1}$ with $a_2 = \frac{c_2+c_2^2}{2}$, $a_2^2 + a_2 = n_2$ and $a_2, c_2 \in \mathbb{Z}_{\geq 0}$,

(4) $\mathrm{C}_{b_2^2+b_2} \times \mathrm{A}_{a_2-1}$ with $a_2 = \frac{c_2+c_2^2}{2}$, $b_2^2 = b_2 + a_2 = n_2$ and $b_2, c_2 \in \mathbb{Z}_{\geq 0}$.

4. Affine Hecke Algebras

Let $\mathcal{R} := (X, R, Y, R^\vee)$ be a root datum. We denote by W the group generated by the s_α for $\alpha \in R$. It is a finite Weyl group. We write $S := \{s_\alpha : \alpha \in \Delta\}$. Then (W, S) is a finite Coxeter system.

We choose, for every $s \in S$, a complex number q_s, such that

$$q_s = q_{s'} \quad \text{if } s \text{ and } s' \text{ are conjugate in } W. \tag{4.1}$$

Let $\mathbf{q}\colon S \to \mathbb{C}$ be the function $s \mapsto q_s$. We define a new $\mathbb{C}$-algebra $\mathcal{H}(W, \mathbf{q})$ which has a vector space basis $\{T_w : w \in W\}$. Here T_1 is the unit element and the T_W satisfy the following quadratic relations and braid relations

$$(T_s + 1)(T_s - q_s) = 0 \quad \text{and} \quad \underbrace{T_s T_{s'} T_s \cdots}_{m(s,s') \text{ terms}} = \underbrace{T_{s'} T_s T_{s'} \cdots}_{m(s,s') \text{ terms}} \quad \text{for any } s, s' \in S. \tag{4.2}$$

Fix $q \in \mathbb{R}_{>1}$ and let $\lambda, \lambda^*\colon R \to \mathbb{C}$ be functions such that

- if s_α and s_β are conjugate in W, then $\lambda(\alpha) = \lambda(\beta)$ and $\lambda^*(\alpha) = \lambda^*(\beta)$,
- if $\alpha^\vee \notin 2Y$, then $\lambda^*(\alpha) = \lambda(\alpha)$.

We note that $\alpha^\vee \in 2Y$ is only possible for short roots α in a type B component of the root system R.

For $\alpha \in R$ we write

$$q_{s_\alpha} := q^{\lambda(\alpha)} \quad \text{and (if } \alpha^\vee \in m^\vee_{\max}\text{)} \quad q_{s'_\alpha} := q^{\lambda^*(\alpha)}. \tag{4.3}$$

We denote by ℓ the usual length function on W. Let $\mathcal{H}(W, \mathbf{q})$ denote the Iwahori-Hecke algebra of W. It has a basis $\{T_w : w \in W\}$ such that

$$\begin{aligned} T_w T_v &= T_{wv} \text{ if } \ell(w) + \ell(v) = \ell(wv), \\ (T_{s_\alpha} + 1)(T_{s_\alpha} - q_{s_\alpha}) &= \ 0 \ \text{ if } \alpha \in \Delta. \end{aligned}$$

Let $\{\theta_x : x \in X\}$ denote the standard basis of $\mathbb{C}[X]$. Then the affine Hecke algebra $\mathcal{H}(\mathcal{R}, \lambda, \lambda^*, q)$ is the vector space $\mathbb{C}[X] \otimes_{\mathbb{C}} \mathcal{H}(W, \mathbf{q})$ such that $\mathbb{C}[X]$ and $\mathcal{H}(W, \mathbf{q})$ are embedded as subalgebras, and for $\alpha \in \Delta$ and $x \in X$:

$$\theta_x T_{s_\alpha} - T_{s_\alpha} \theta_{s_\alpha(x)} =$$
$$\left((q^{\lambda(\alpha)} - 1) + \theta_{-\alpha}\left(q^{\frac{\lambda(\alpha)+\lambda^*(\alpha)}{2}} - q^{\frac{\lambda(\alpha)-\lambda^*(\alpha)}{2}}\right)\right) \frac{\theta_x - \theta_{s_\alpha(x)}}{\theta_0 - \theta_{-2\alpha}}.$$

When $\alpha^\vee \notin 2Y$, the cross relation simplifies to

$$\theta_x T_{s_\alpha} - T_{s_\alpha} \theta_{s_\alpha(x)} = (q^{\lambda(\alpha)} - 1) \frac{\theta_x - \theta_{s_\alpha(x)}}{\theta_0 - \theta_{-\alpha}}.$$

4.1. *Affine Hecke algebras and Bernstein blocks*

Let $\mathfrak{s} = [L, \sigma]_G \in \mathfrak{B}(G)$. We write $\mathfrak{s}_L := [L, \sigma]_L$ and denote by $W^{\mathfrak{s}}$ the extended finite Weyl group $\mathrm{N}_G(\mathfrak{s}_L)/L$. The set of roots of G with respect to L contains a root system $\Sigma^{\mathfrak{s}}$, namely the set of roots for which the associated Harish-Chandra μ-function has a zero on the orbit $\mathfrak{X}_{\mathrm{nr}}(L) \cdot \sigma$. This induces a semi-direct decomposition

$$W^{\mathfrak{s}} = W(\Sigma^{\mathfrak{s}}) \rtimes \mathfrak{R}^{\mathfrak{s}}, \tag{4.4}$$

where $W(\Sigma^{\mathfrak{s}})$ is the Weyl group of $\Sigma^{\mathfrak{s}}$ and $\mathfrak{R}^{\mathfrak{s}}$ is the stabilizer in $W^{\mathfrak{s}}$ of the set of positive roots. We have

$$L^1 = \bigcap_{\chi \in \mathfrak{X}_{\mathrm{nr}}(L)} \ker(\chi), \tag{4.5}$$

and we suppose that the restriction of σ to L^1 is multiplicity free.

By [59], there exists a 2-cocycle

$$\natural^{\mathfrak{s}} : \mathfrak{R}^{\mathfrak{s}} \times \mathfrak{R}^{\mathfrak{s}} \longrightarrow \mathbb{C}[\mathfrak{X}_{\mathrm{nr}}(L) \cdot \sigma]^\times, \tag{4.6}$$

such that we have a bijection

$$\mathrm{Irr}^{\mathfrak{s}}(G) \longrightarrow \mathrm{Irr}(\mathcal{H}(G, \mathfrak{s})), \tag{4.7}$$

where $\mathcal{H}(G, \mathfrak{s})$ is the twisted extended affine Hecke algebra of the form

$$\mathcal{H}(G, \mathfrak{s}) := \mathcal{H}(\mathcal{R}, \lambda, \lambda^*, q) \rtimes \mathbb{C}[\mathfrak{R}^{\mathfrak{s}}, \natural^{\mathfrak{s}}]. \tag{4.8}$$

4.2. *Affine Hecke algebras and enhanced L-parameters*

We denote by $W'_F := W_F \times \mathrm{SL}_2(\mathbb{C})$ the Weil-Deligne group of F and by $G^\vee$ the complex connected reductive group with root datum dual to that of G. A *Langlands parameter* — or L-parameter — is a morphism $\varphi\colon W'_F \to G^\vee$ such that

- $\varphi|_{\mathrm{SL}_2(\mathbb{C})}$ is morphism of algebraic groups,
- $\varphi(w)$ is a semisimple element of $G^\vee$, for any $w \in W_F$.

Let $\mathrm{Z}_{G^\vee}$ denote the center of $G^\vee$. We denote by $G^\vee_{\mathrm{der}}$ the derived group of $G^\vee$ and by $G^\vee_{\mathrm{sc}}$ the simply connected cover of $G^\vee_{\mathrm{der}}$. Let $\mathbf{Z}$ be a sufficiently large finite central subgroup of $\mathbf{G}$ (for example, the center $\mathrm{Z}_{\mathbf{G}_{\mathrm{der}}}$ of the derived group of $\mathbf{G}$). We set $G^\vee_Z := (G/Z)^\vee$.

For φ a given L-parameter, we define

$$S_\varphi := \mathrm{C}_{G^\vee}(\varphi(W'_F)). \tag{4.9}$$

An *enhanced L-parameter* is a pair (φ, ρ) where φ is an L-parameter for G and $\rho \in \mathrm{Irr}(\mathcal{S}_\varphi)$, with $\mathcal{S}_\varphi := S_\varphi/S^\circ_\varphi$. For φ a given L-parameter, the representation ρ is called an *enhancement* of φ. We define an action of $G^\vee$ on the set of enhanced L-parameters by:

$$g \cdot (\varphi, \rho) := (g\varphi g^{-1}, {}^g\rho), \quad \text{for } g \in G^\vee, \tag{4.10}$$

where ${}^g\rho\colon h \mapsto \rho(g^{-1}hg)$. We denote by Φ_{e} set of $G^\vee$-conjugacy classes of enhanced L-parameters.

Let us consider G as an inner twist of a quasi-split group. Via the Kottwitz isomorphism it is parametrized by a character of $\mathrm{Z}^{W_F}_{G^\vee_{\mathrm{sc}}}$, say ζ_G. We say that $(\varphi, \rho) \in \Phi_{\mathrm{e}}$ is relevant for G if $\mathrm{Z}^{W_F}_{G^\vee_{\mathrm{sc}}}$ acts on ρ as ζ_G. We denote by $\Phi_{\mathrm{e}}(G)$ the subset of Φ_{e} which is relevant for G.

By applying (4.9) to the restriction $\varphi|_{W_F}$ of φ to W_F, we define the (possibly disconnected) complex reductive group

$$\mathcal{G}_\varphi := \mathrm{C}_{G^\vee}(\varphi(W_F \times \{1\})). \tag{4.11}$$

We denote by $\mathcal{G}^\circ_\varphi$ the identity component of $\mathcal{G}_\varphi$, and we set

- $u = u_\varphi := \varphi\left(1, \left(\begin{smallmatrix} 1 & 1 \\ 0 & 1 \end{smallmatrix}\right)\right)$ (a unipotent element);
- $A_{\mathcal{G}_\varphi}(u_\varphi) := \pi_0(\mathrm{C}_{\mathcal{G}_\varphi}(u))$. We have $\mathcal{S}_\varphi \simeq A_{\mathcal{G}_\varphi}(u_\varphi)$.

An enhanced L-parameter $(\varphi, \rho) \in \Phi_{\mathrm{e}}$ is called *cuspidal* if the following properties hold:

- φ is discrete (i.e., $\varphi(W'_F)$ is not contained in any proper Levi subgroup of $G^\vee$),

- (u_φ, ρ) is a *cuspidal pair* in $\mathcal{G}_\varphi$ (see [40], and Remark 4.6, Definition 4.7 and Definition 4.11 in [5]).

We denote by $\Phi_{\mathrm{e,cusp}}(G)$ the set of $G^\vee$-conjugacy of cuspidal enhanced L-parameters for G.

Example 4.1: Let $G = \mathrm{Sp}_{2n}(F)$ and let $\iota_{G^\vee}\colon G^\vee \hookrightarrow \mathrm{GL}_{2n+1}(\mathbb{C})$ be the standard embedding. For every L-parameter φ for G, we define

$$I_\varphi := \{\tau \in \mathrm{Irr}(W_F) \,:\, \tau \text{ occurs in } \iota_{G^\vee} \circ \varphi|_{W_F}\}. \tag{4.12}$$

We have $I_\varphi = I_\varphi^{\mathrm{O}} \sqcup I_\varphi^{\mathrm{S}} \sqcup I_\varphi^{\mathrm{GL}}$, where I_φ^{O} (resp. I_φ^{S}) is the subset of I_φ formed by the representations that are orthogonal (resp. symplectic), and I_φ^{GL} is the maximal subset of I_φ which is formed by representations τ that are not selfdual and satisfy $\tau^\vee \in I_\varphi^{\mathrm{GL}}$ for every $\tau \in I_\varphi^{\mathrm{GL}}$. Hence, we have the following decomposition

$$\iota_{G^\vee} \circ \varphi = \bigoplus_{\tau \in I_\varphi^{\mathrm{O}}} \tau \boxtimes M_\tau \oplus \bigoplus_{\tau \in I_\varphi^{\mathrm{S}}} \tau \boxtimes M_\tau \oplus \bigoplus_{\tau \in I_\varphi^{\mathrm{GL}}} (\tau \oplus \tau^\vee) \boxtimes M_\tau \tag{4.13}$$

where M_τ is a multiplicity space of τ. Let m_τ denote the dimension of M_τ. We have

$$\mathcal{G}_\varphi \simeq \prod_{\tau \in I_\varphi^{\mathrm{O}}} \mathrm{Sp}_{m_\tau}(\mathbb{C}) \times \prod_{\tau \in I_\varphi^{\mathrm{S}}} \mathrm{O}_{m_\tau}(\mathbb{C}) \times \prod_{\tau \in I_\varphi^{\mathrm{GL}}} \mathrm{GL}_{m_\tau}(\mathbb{C}). \tag{4.14}$$

By [52], if the enhanced L-parameter (φ, ρ) is cuspidal, then we have

$$m_\tau = \begin{cases} d_\tau^2 + d_\tau \quad (\text{with } d_\tau \in \mathbb{Z}_{\geq 0}) & \text{if } \tau \in I_\varphi^{\mathrm{O}} \\ d_\tau^2 \quad (\text{with } d_\tau \in \mathbb{Z}_{> 0}) & \text{if } \tau \in I_\varphi^{\mathrm{S}} \\ 1 & \text{if } \tau \in I_\varphi^{\mathrm{GL}}. \end{cases} \tag{4.15}$$

Let $\mathcal{C} \in \mathfrak{U}(\mathcal{G}_\varphi^\circ)$ and let $\mathcal{E}$ irreducible be an $\mathcal{G}_\varphi^\circ$-equivariant local system on $\mathcal{C}$. By [40], the IC-sheaf $\mathcal{F}_\rho := \mathrm{IC}(\mathcal{C}, \mathcal{E}_\rho)$ occurs as a summand of $\mathrm{i}_{\mathcal{L}\subset\mathcal{P}}^{\mathcal{G}^\circ}(\mathrm{IC}(\mathcal{C}_{\mathrm{cusp}}, \mathcal{E}_{\mathrm{cusp}}))$, for some triple $(\mathcal{P}, \mathcal{L}, (\mathcal{C}_{\mathrm{cusp}}, \mathcal{E}_{\mathrm{cusp}}))$, where $\mathcal{P}$ is a parabolic subgroup of $\mathcal{G}^\circ$ with Levi subgroup $\mathcal{L}$ and $(\mathcal{C}_{\mathrm{cusp}}, \mathcal{E}_{\mathrm{cusp}})$ is a cuspidal unipotent pair in $\mathcal{L}$. Moreover, the triple $(\mathcal{P}, \mathcal{L}, \mathcal{C}_{\mathrm{cusp}}, \mathcal{E}_{\mathrm{cusp}})$ is unique up to $\mathcal{G}^\circ$-conjugation.

Let $\rho^\circ \in \mathrm{Irr}(A_{\mathcal{G}^\circ}(u))$. The cuspidal support of (u, ρ°), denoted by $\mathrm{Sc}^{\mathcal{G}^\circ}(u, \rho^\circ)$, is defined to be

$$(\mathcal{L}, (v, \rho^\circ_{\mathrm{cusp}}))_{\mathcal{G}^\circ}, \quad \text{where } v \in \mathcal{C}_{\mathrm{cusp}} \text{ and } \rho^\circ_{\mathrm{cusp}} \leftrightarrow \mathcal{E}_{\mathrm{cusp}}. \tag{4.16}$$

We set $\mathcal{T} := Z^\circ_{\mathcal{L}}$ and $\mathcal{M} := C_{\mathcal{G}}(\mathcal{T})$. The cuspidal support of (u, ρ) is a (well-defined) triple $(\mathcal{M}, v, \rho_{\text{cusp}})_{\mathcal{G}}$, where ρ°_{cusp} occurs in the restriction of ρ_{cusp} to $A_{\mathcal{G}^\circ}(u)$. The cuspidal support of (φ, ρ) is defined to be

$$\mathrm{Sc}(\varphi, \rho) := (C_{{}^LG}(\mathcal{T}), (\varphi_v, \rho_{\text{cusp}})). \tag{4.17}$$

Recall from Section 3.3.1 of [29] that the group of unramified characters of L is naturally isomorphic to $(Z_{L^\vee} \rtimes I_F)^\circ_{W_F}$. We consider this as an object on the Galois side of the local Langlands correspondence and we write

$$\mathfrak{X}_{\text{nr}}({}^LL) := (Z_{L^\vee \rtimes I_F})^\circ_{W_F}. \tag{4.18}$$

Given $(\varphi, \rho) \in \Phi_{\text{e}}(L)$ and $z \in (Z_{L^\vee \rtimes I_F})_{W_F}$, we define $(z \cdot \varphi, \rho) \in \Phi_{\text{e}}(L)$ by

$$z \cdot \varphi := \varphi \text{ on } I_F \times \mathrm{SL}_2(\mathbb{C}) \text{ and } (z \cdot \varphi)(\mathrm{Fr}_F) := \widetilde{z}\varphi(\mathrm{Fr}_F), \tag{4.19}$$

where $\widetilde{z} \in (Z_{L^\vee \rtimes I_F})_{W_F}$ represents z.

An inertial equivalence class for $\Phi_{\text{e}}(G)$ is the $G^\vee$-conjugacy class $\mathfrak{s}^\vee$ of a pair $({}^LL, \mathfrak{s}^\vee_L)$, where L is a Levi subgroup of G, ${}^LL = L^\vee \rtimes W_F$ and $\mathfrak{s}^\vee_L$ is a $\mathfrak{X}_{\text{nr}}({}^LL)$-orbit in $\Phi_{\text{e,cusp}}(L)$.

The *Bernstein series* $\Phi^{\mathfrak{s}^\vee}_{\text{e}}(G)$ associated to $\mathfrak{s}^\vee \in \mathfrak{B}(G^\vee)$ is defined to be the fiber of $\mathfrak{s}^\vee$ under the map Sc defined in (4.17). By [8], the set $\Phi_{\text{e}}(G)$ of $G^\vee$-conjugacy classes of enhanced L-parameters for G is partitioned, analogously to (3.2), as

$$\Phi_{\text{e}}(G) = \coprod_{\mathfrak{s}^\vee \in \mathfrak{B}(G^\vee)} \Phi^{\mathfrak{s}^\vee}(G). \tag{4.20}$$

In [9], we canonically associated an affine Hecke algebra (possibly extended with a finite R-group) $\mathcal{H}(G^\vee, \mathfrak{s}^\vee)$ to every Bernstein series $\Phi^{\mathfrak{s}^\vee}_{\text{e}}(G)$ of enhanced Langlands parameters for G. While we considered only the first case of (4.9), our construction applies equally well to both the other cases. We showed that the simple modules of this algebra are naturally in bijection with the elements of the Bernstein series $\Phi^{\mathfrak{s}^\vee}_{\text{e}}(G)$ and that the set of central characters of the algebra is naturally in bijection with the collection of cuspidal supports of these enhanced Langlands parameters.

We summarise the construction of $\mathcal{H}(G^\vee, \mathfrak{s}^\vee)$. By applying (4.9) to the restriction $\varphi|_{I_F}$ of φ to the inertia group $I_F \subset W_F$, we define the (possibly disconnected) complex reductive group

$$\mathcal{J}_\varphi := C_{G^\vee}(\varphi(I_F \times \{1\})). \tag{4.21}$$

Let $R(\mathcal{J}^\circ, \mathcal{T})$ be the set of $\alpha \in X^*(\mathcal{T}) \backslash \{0\}$ which appear in the adjoint action of $\mathcal{T}$ on the Lie algebra of $\mathcal{J}^\circ_\varphi$. By Proposition 3.9 in [9], $R(\mathcal{J}^\circ, \mathcal{T})$ is a root system. We denote by $W^\circ_{\mathfrak{s}^\vee}$ its Weyl group. Let $W_{\mathfrak{s}^\vee} := \mathrm{N}_{G^\vee}(\mathfrak{s}^\vee)/L^\vee$. We have $W_{\mathfrak{s}^\vee} = W^\circ_{\mathfrak{s}^\vee} \rtimes \mathfrak{R}_{\mathfrak{s}^\vee}$, where

$$\mathfrak{R}_{\mathfrak{s}^\vee} := \left\{w \in W_{\mathfrak{s}^\vee} \,:\, w(R(\mathcal{J}^\circ, \mathcal{T})^+) \subset R(\mathcal{J}^\circ, \mathcal{T})^+\right\}.$$

We define a root datum

$$\mathcal{R}_{\mathfrak{s}^\vee} := (R_{\mathfrak{s}^\vee}, X^*(T_{\mathfrak{s}^\vee}), R^\vee_{\mathfrak{s}^\vee}, X_*(T_{\mathfrak{s}^\vee})),$$

where $T_{\mathfrak{s}^\vee} \simeq \mathfrak{s}^\vee_L$ and

$$R_{\mathfrak{s}^\vee} = \{m_\alpha \alpha \,:\, \alpha \in R(\mathcal{J}^\circ, \mathcal{T})_{\mathrm{red}} \subset X^*(T_{\mathfrak{s}^\vee})\},$$

with $m_\alpha \in \mathbb{Z}_{>0}$. The group $W_{\mathfrak{s}^\vee}$ acts on $\mathcal{R}_{\mathfrak{s}^\vee}$. In [9] we defined $W_{\mathfrak{s}^\vee}$-invariant functions

$$\lambda\colon m_{\mathfrak{s}^\vee} \to \mathbb{Q}_{>0} \quad \text{and} \quad \lambda^*\colon \{m_\alpha\alpha \in m_{\mathfrak{s}^\vee} : m_\alpha\alpha \in 2X_*(T_{\mathfrak{s}^\vee})\} \to \mathbb{Q}. \tag{4.22}$$

We choose semisimple subgroups $\mathcal{J}_i \subset \mathcal{J}^\circ$, normalized by $\mathrm{N}_G(\mathfrak{s})$, such that the derived group $\mathcal{J}^\circ_{\mathrm{der}}$ is the almost direct product of the $\mathcal{J}_i$, and the multiplication map

$$m_{\mathcal{J}^\circ}\colon \mathrm{Z}^\circ_{\mathcal{J}^\circ} \times \mathcal{J}_1 \times \cdots \times \mathcal{J}_d \to \mathcal{J}^\circ \tag{4.23}$$

is a surjective group homomorphism with finite central kernel. It induces a decomposition

$$\mathrm{Lie}(\mathcal{J}) = \mathrm{Z}_{\mathrm{Lie}(\mathcal{J}^\circ)} \oplus \mathrm{Lie}(\mathcal{J}_1) \oplus \cdots \oplus \mathrm{Lie}(\mathcal{J}_d). \tag{4.24}$$

We obtain an orthogonal, $W_{\mathfrak{s}^\vee}$-stable decomposition

$$R(\mathcal{J}^\circ, \mathcal{T}) = m_1 \sqcup \cdots \sqcup m_d, \quad \text{where } m_i := R(\mathcal{J}_i, \mathcal{J}_i \cap \mathcal{T}). \tag{4.25}$$

We let $\vec{r} := (m_1, \ldots, m_d)$ be an array of variables, corresponding to (4.23) in the sense that m_i is relevant for $\mathcal{J}_i$ and m_i only. We have

$$\mathcal{H}(G^\vee, \mathfrak{s}^\vee) = \mathcal{H}(\mathcal{R}, \lambda, \lambda^*, \vec{r}) \rtimes \mathbb{C}[\mathfrak{R}^{\mathfrak{s}^\vee}, \natural^{\mathfrak{s}^\vee}], \tag{4.26}$$

where $\mathcal{H}(\mathcal{R}, \lambda, \lambda^*, \vec{r})$ is a generalized version (with d indeterminates) of the algebras considered in §4 and $\natural^{\mathfrak{s}^\vee} : \mathfrak{R}^{\mathfrak{s}^\vee} \times \mathfrak{R}^{\mathfrak{s}^\vee} \to \mathbb{C}^\times$ is 2-cocycle. The cocycle is trivial when G is an inner twist of $\mathrm{GL}_n(F)$, a pure inner twist of a quasi-split classical group or of the group $\mathrm{GSpin}_n(F)$ (see [9], [10]), and when G is the exceptional group of type G_2 (see [12]).

5. Howe Correspondence

5.1. *A correspondence between Hecke algebras of p-adic groups*

We will consider here only the case of dual pairs formed by a p-adic symplectic group and a p-adic orthogonal group of even dimension. The other cases can be treated in a similar way. Let n and n' be two fixed non-negative integers, and let W_n be a symplectic space of (even) dimension $2n$ over F. The corresponding group of isometries is the symplectic group $G_n := \mathrm{Sp}(W_n)$. Let $V_{n'}$ be a quadratic space of dimension $2n'$ over F (i.e., a space endowed with a non-degenerate symmetric F-bilinear form). We denote by $\eta_{n'}$ the character of $F^\times$ associated to $F(\sqrt{\Delta_{n'}})/F$, where $\Delta_{n'}$ is the discriminant of the form. The orthogonal group $G'_{n'} := \mathrm{O}(V_{n'})$ is the corresponding group of isometries, and $(G_n, G'_{n'})$ is a dual pair in $\mathrm{Sp}(W_{2nn'})$. Fixing a non-trivial additive character ψ of F, we obtain the so-called Weil representation ω of the metaplectic cover $\mathrm{Mp}(W_{2nn'})$ of $\mathrm{Sp}_{W_{2nn'}}$. We can define a splitting $G_n \times G'_{n'} \to \mathrm{Mp}(W_{2nn'})$ (see [36]). By means of this splitting we obtain a Weil representation of $G_n \times G'_{n'}$, which we denote $\omega_{n,n'}$.

For every $\pi \in \mathrm{Irr}(G_n)$, the maximal π-isotypic quotient of $\omega_{n,n'}$ is of the form $\pi \otimes \Theta_{n'}(\pi)$, where $\Theta_{n'}(\pi)$ is a smooth representation of $G'_{n'}$, called the *full theta lift* of π to $W_{n'}$. The representation $\Theta_{n'}(\pi)$, when non-zero, has a unique irreducible quotient, denoted by $\theta_{n'}(\pi)$, which is called the *small theta lift* of π. Similarly, if $\pi' \in \mathrm{Irr}(G'_{n'})$, the maximal π'-isotypic quotient of $\omega_{n,n'}$ is of the form $\Theta_n(\pi') \otimes \pi'$, where $\Theta_n(\pi')$ is a smooth representation of G_n, called the full theta lift of π' to V_n. The representation $\Theta_n(\pi')$, when non-zero, has a unique irreducible quotient, denoted $\theta_n(\pi')$, which is called the small theta lift of π'. These statements were first formulated by Howe in [33], and were proved by Waldspurger in [60] when the residual characteristic of F is odd, and by Gan and Takeda [26] in general.

We have $V_{n'} = V^{\mathrm{an}} \oplus H^{n'}$, where V^{an} is an anisotropic F-vector subspace, and H is the hyperbolic plane. For $m' \geq 0$, let $V_{m'} := V^{\mathrm{an}} \oplus H^{m'}$. The collections $\mathfrak{T} := \{W_m : m \geq 0\}$ and $\mathfrak{T}' := \{V_{m'} : m' \geq 0\}$ are called Witt tower of vector spaces. Let $G = G_n$, with n fixed. One can then consider the tower of theta correspondences associated to the tower of reductive dual pairs $(G, G'_{m'})_{m' \geq 0}$. For an irreducible smooth representation π of G, we thus have the representation $\theta_{m'}(\pi)$. The smallest non-negative integer $n'_{\min}$ such that $\theta_{n'_{\min}}(\pi) \neq 0$ is called the *first occurrence index* of π for the Witt tower $\mathfrak{T}'$, and the representation $\theta^0(\pi) := \theta_{n'_{\min}}(\pi)$ is called the first occurrence of π for this Witt tower.

Similarly, for $G' = G_{n'}$ with n' fixed, one can then consider the tower of theta correspondences associated to the tower of reductive dual pairs $(G_m, G')_{m\geq 0}$. For an irreducible smooth representation π' of G', we have the representation $\theta_{m'}(\pi)$. The smallest non-negative integer $n'_{\min}$ such that $\theta_{n_{\min}}(\pi') \neq 0$ is called the *first occurrence index* of π' for the Witt tower $\mathfrak{T}$, and the representation $\theta^0(\pi') := \theta_{n_{\min}}(\pi)$ is called the first occurrence of π' for this Witt tower.

By Chapter 3 in [49], such $n'_{\min}$ and $n_{\min}$ exist and $n'_{\min} \leq n'$ and $n_{\min} \leq n$. Moreover, we have $\Theta_{m'}(\pi) \neq 0$ for any $m' \geq n'_{\min}$ and $\Theta_m(\pi') \neq 0$ for any $m \geq n_{\min}$. By Theorem 2.5 in [35] (see also Chapter 3 in [49]), if the representation π is supercuspidal then $\Theta_{n'_{\min}}(\pi)$ is irreducible (and thus is equal to $\theta_{n'_{\min}}(\pi)$) and supercuspidal. Similarly, if π' is supercuspidal then $\Theta_{n_{\min}}(\pi')$ is irreducible (and thus is equal to $\theta_{n_{\min}}(\pi')$) and supercuspidal.

A Levi subgroup L_n of a parabolic subgroup P_n of G_n is isomorphic to $G_{n_0} \times \mathrm{GL}_{n_1}(F) \times \cdots \times \mathrm{GL}_{n_r}(F)$, where $n_0 + n_1 + \cdots + n_r = n$, and any $\pi \in \mathrm{Irr}(G_n)$ is isomorphic to a subquotient of the parabolic induced representation $\mathrm{i}_{L_n,P_n}(\sigma)$ where $\sigma \in \mathrm{Irr}_{\mathrm{sc}}(L_n)$. We have $\sigma \simeq \pi_{n'_0} \otimes \tau_1 \otimes \cdots \otimes \tau_r$, where $\pi_{n'_0} \in \mathrm{Irr}_{\mathrm{sc}}(G'_{n'_0})$ and $\tau_i \in \mathrm{Irr}_{\mathrm{sc}}(\mathrm{GL}_{m_i}(F))$ for $1 \leq i \leq r$.

A Levi subgroup $L'_{n'}$ of a parabolic subgroup $P'_{n'}$ of $G'_{n'}$ is isomorphic to $G'_{n'_0} \times \mathrm{GL}_{n'_1}(F) \times \cdots \times \mathrm{GL}_{n'_{r'}}(F)$, where $n'_0 + n'_1 + \cdots + n'_{r'} = n'$ and any $\pi' \in \mathrm{Irr}(G'_{n'})$ is isomorphic to a subquotient of the representation $\mathrm{i}_{L'_{n'},P'_{n'}}(\sigma')$, where $\sigma' \simeq \pi'_{n'_0} \otimes \tau'_1 \otimes \cdots \otimes \tau'_{r'} \in \mathrm{Irr}_{\mathrm{sc}}(L'_{n'})$ such that $\pi'_{n'_0} \in \mathrm{Irr}_{\mathrm{sc}}(G'_{n'_0})$ and $\tau'_j \in \mathrm{Irr}_{\mathrm{sc}}(\mathrm{GL}_{n'_j}(F))$ for $1 \leq j \leq r'$. The G_n-conjugacy class $(L_n, \sigma)_{G_n}$ of the pair (L_n, σ) is uniquely determined by π, is called the supercuspidal support of π, and denoted by $\mathrm{Sc}(\pi)$. Similarly, the $G'_{n'}$-conjugacy class of the pair $(L'_{n'}, \sigma'_{n'})$ is uniquely determined by π', is called the supercuspidal support of π', and denoted by $\mathrm{Sc}(\pi')$. We set

$$\overline{n} := n - n_0 \quad \overline{n}' := n' - n'_0, \quad d := n - n', \quad \text{and} \quad d_0 := n_0 - n'_0. \tag{5.1}$$

By [35], if $\Theta_{n'}(\pi) \neq \{0\}$ and $\pi' := \theta_{n'}(\pi)$, then

(1) if $\overline{n}' \leq \overline{n}$, the supercuspidal support of π' is

$$\left(L'_{n'}, \pi'_{n'_0} \otimes \eta_{n'}^{-1}\tau_1 \otimes \cdots \otimes \eta_{n'}^{-1}\tau_\ell \otimes \nu_F^{d-1} \otimes \nu_F^{d-2} \otimes \cdots \otimes \nu_F^{d_0}\right)_{G'_{n'}},$$

where $\pi'_{n'_0} := \theta^0(\pi_{n_0})$.

(2) if $\overline{n}' \geq \overline{n}$, the supercuspidal support of π is

$$\left(L_n, \pi_{n_0} \otimes \eta_{n'}\tau'_1 \otimes \cdots \otimes \eta_{n'}\tau'_{r'} \otimes \eta_{n'}\nu_F^{-d-1} \otimes \eta_{n'}\nu_F^{-d-2} \otimes \cdots \otimes \eta_{n'}\nu_F^{-d_0}\right)_{G_n},$$

where $\pi_{n_0} := \theta^0(\pi'_{n'_0})$.

Let $L \in \{L_n, L'_{n'}\}$. Recall that a character of L is said to be unramified if it is trivial on all the compact subgroups of L. We denote by $\mathfrak{X}_{\mathrm{nr}}(L)$ the group of unramified characters of L.

Proposition 5.1: *For any $\pi \in \mathrm{Irr}^{\mathfrak{s}}(G_n)$, where $\mathfrak{s} = [L_n, \sigma]_{G_n}$, we have $\theta_{n'}(\pi) \in \mathrm{Irr}^{\theta(\mathfrak{s})}(G_n)$, where*

$$\theta(\mathfrak{s}) := [L'_{n'}, \sigma']_{G'_{n'}}. \tag{5.2}$$

Proof: Let $\pi \in \mathrm{Irr}^{\mathfrak{s}}(G_n)$, where $\mathfrak{s} = [L_n, \sigma]_{G_n}$ and $\pi' := \theta_{n'}(\pi)$. Thus, we have $\mathrm{Sc}(\theta_{n'}(\pi)) = (L'_{n'}, \sigma')_{G'_{n'}}$. We set $\mathfrak{s}' := [L'_{n'}, \sigma']_{G'_{n'}}$. Let $\chi \in \mathfrak{X}_{\mathrm{nr}}(L_n)$, and let $\pi_\chi \in \mathrm{Irr}(G_n)$ such that $\mathrm{Sc}(\pi_\chi) = (L_n, \chi \otimes \sigma)_{G_n}$. The character χ has the following form: $\chi = \mathrm{triv} \otimes \chi_1 \otimes \cdots \otimes \chi_r$, where $\chi_i \in \mathfrak{X}_{\mathrm{nr}}(\mathrm{GL}_{n_i}(F))$ for $1 \le i \le r$. By applying the previous formula (case (1)) to the representation π_χ we see that, if $\overline{n}' \le \overline{n}$, then the supercuspidal support of $\theta_{n'}(\pi_\chi)$ is

$$\begin{aligned}&\left(L'_{n'}, \pi'_{n'_0} \otimes \chi_1 \eta_{n'}^{-1} \tau_1 \otimes \cdots \otimes \chi_r \eta_{n'}^{-1} \tau_r \otimes \nu_F^{n'-n} \otimes \cdots \otimes \nu_F^{\overline{n}'-\overline{n}-1}\right)_{G'_{n'}} \\ &\quad = (L'_{n'}, (\chi \otimes \underbrace{\mathrm{triv} \otimes \cdots \otimes \mathrm{triv}}_{n_0 - n'_0 \text{ times}}) \otimes \sigma')_{G'_{n'}}.\end{aligned}$$

The character $\chi \otimes \mathrm{triv}^{\otimes(n_0 - n'_0)}$ of $L'_{n'}$ being unramified, we have $\theta_{n'}(\pi_\chi) \in \mathrm{Irr}^{\mathfrak{s}'}(G'_{n'})$. Since every representation in $\mathrm{Irr}^{\mathfrak{s}}(G_n)$ is of the form π_χ for some $\chi \in \mathfrak{X}_{\mathrm{nr}}(L_n)$, we get

$$\theta_{n'}(\mathrm{Irr}^{\mathfrak{s}}(G_n)) \subset \mathrm{Irr}^{\mathfrak{s}'}(G'_{n'}).$$

On the other hand, if $\overline{n}' \ge \overline{n}$, by definition of the representation π_χ its supercuspidal support is $(L_n, \chi \otimes \sigma)_{G_n}$, hence, by the formula (case (2)) applied to π, it equals

$$\left(L_n, \pi_{n_0} \otimes \eta_{n'} \chi_1 \tau'_1 \otimes \cdots \otimes \eta_{n'} \chi_{r'} \tau'_{r'} \otimes \eta_{n'} \nu_F^{n'-n} \otimes \cdots \otimes \eta_{n'} \nu_F^{\overline{m}-\overline{n}-1}\right)_{G_n}.$$

Thus, by applying the formula (case (2)) to π_χ, we get

$$\mathrm{Sc}(\theta_{n'}(\pi_\chi)) = (L'_{n'}, \pi'_{n'_0} \otimes \chi_1 \tau'_1 \otimes \cdots \otimes \chi_{r'} \tau'_{r'})_{G'_{n'}} = (L'_{n'}, \chi' \otimes \sigma')_{G'_{n'}},$$

where $\chi' := \mathrm{triv} \otimes \chi_1 \otimes \cdots \otimes \chi_{r'} \in \mathfrak{X}_{\mathrm{nr}}(L'_{n'})$. Hence, $\theta_{n'}(\pi_\chi) \in \mathrm{Irr}^{\mathfrak{s}'}(G'_{n'})$. □

Theorem 5.2: *For every $\mathfrak{s} \in \mathfrak{B}(G)$, the Howe correspondence $\pi \mapsto \theta_{n'}(\pi)$ for the reductive dual pair $(G_n, G'_{n'})$ induces a correspondence $E \mapsto \theta(E)$ between subsets of simple modules of the extended affine Hecke algebras $\mathcal{H}(G_n, \mathfrak{s})$ and $\mathcal{H}(G'_{n'}, \theta(\mathfrak{s}))$.*

Proof: By [31], the algebras $\mathcal{H}(G_n, \mathfrak{s})$ for any $\mathfrak{s} \in \mathfrak{B}(G_n)$ and $\mathcal{H}(G'_{n'}, \mathfrak{s}')$ for any $\mathfrak{s}' \in \mathfrak{B}(G'_{n'})$ are extended affine Hecke algebras (that is, the 2-cocycles are trivial in the case of classical groups). Then the result follows from the combination of Proposition 5.1 with (3.2). □

5.2. *A correspondence between completions of Hecke algebras*

To study representations of G it is often useful to consider various group algebras of G. There is the Hecke algebra $\mathcal{H}(G)$, defined to be the convolution algebra of locally constant, compactly supported functions $f: G \to \mathbb{C}$. The category $\mathfrak{R}(G)$ is equivalent to the category of nondegenerate $\mathcal{H}(G)$-modules. By letting G act on $\mathcal{H}(G)$ by left translation, we obtain from (3.1) a decomposition

$$\mathcal{H}(G) = \bigoplus_{\mathfrak{s} \in \mathfrak{B}(G)} \mathcal{H}(G)^{\mathfrak{s}}, \tag{5.3}$$

with $\mathcal{H}(G)^{\mathfrak{s}} \in \mathfrak{R}^{\mathfrak{s}}(G)$. The spaces $\mathcal{H}(G)^{\mathfrak{s}}$ are two-sided ideals of $\mathcal{H}(G)$.

From the point of view of noncommutative geometry, for the study of tempered representations of G, it is interesting to use the reduced C^*-algebra $C^*_{\mathrm{r}}(G)$, whose spectrum coincides with the tempered dual $\mathrm{Irr}^{\mathrm{t}}(G)$ of G. Analogously, we have the following decomposition of $C^*_{\mathrm{r}}(G)$:

$$C^*_{\mathrm{r}}(G) = \bigoplus_{\mathfrak{s} \in \mathfrak{B}(G)} C^*_{\mathrm{r}}(G)^{\mathfrak{s}}. \tag{5.4}$$

Theorem 5.3: *We suppose that $n' = n$ or $n' = n + 1$. Then the theta correspondence $\pi \mapsto \theta_{n'}(\pi)$ for the reductive dual pair $(G_n, G'_{n'})$ induces a correspondence between subsets of simple modules of $C^*_{\mathrm{r}}(G)^{\mathfrak{s}}$ and $C^*_{\mathrm{r}}(G)^{\theta(\mathfrak{s})}$.*

Proof: By Theorem 1.2 in [23], if $\pi \in \mathrm{Irr}^{\mathrm{t}}(G_n)$, then $\theta_{n'}(\pi) \in \mathrm{Irr}^{\mathrm{t}}(G'_{n'})$. On the other hand, by Theorem 6.2 in [58], when G is G_n or $G'_{n'}$, then $C^*_{\mathrm{r}}(G)^{\mathfrak{s}}$ is Morita equivalent to the reduced C^*-completion of $\mathcal{H}(G, \mathfrak{s})$ for each $\mathfrak{s} \in \mathfrak{B}(G)$. Thus, the result follows from Theorem 5.2. □

Remark 5.4: Mesland and Şengün already observed in [46] that certain cases of the theta correspondence can be described by using objects of C^*-algebraic nature. More precisely, they constructed Hilbert C^*-bimodules over C^*-algebras of groups in equal rank dual pairs and showed that the Rieffel induction functors implemented by these modules coincide with the theta correspondence.

5.3. *The Howe correspondence for depth-zero representations*

In [54], Pan proved that the theta correspondence preserves depth zero representations for type I reductive dual pairs (G, G') over p-adic fields of odd residual characteristic and is compatible with the theta correspondence for finite reductive dual pairs in the following sense. Let $(\pi, \mathcal{V})$ be a depth-zero representation of G. There exists a point x of $\mathcal{B}(\mathbf{G}, F)$ such that π has non-zero invariant vectors under $G_{x,0+}$ and an irreducible cuspidal representation π_x of $\mathbb{G}_{x,0}$ which is contained in $\mathcal{V}^{G_{x,0+}}$. If $(\pi', \mathcal{V}') \in \mathrm{Irr}(G')$ corresponds to $(\pi, \mathcal{V})$ via the theta correspondence, then there exists a point x' of $\mathcal{B}(\mathbf{G}', F)$ such that π' has non-zero invariant vectors under $G'_{x',0+}$ and an irreducible cuspidal representation $\pi_{x'}$ of $\mathbb{G}'_{x',0}$ which is contained in $\mathcal{V}^{G'_{x',0+}}$, and the representation $\pi_{x'}$ corresponds to π_x via the theta correspondence for reductive dual pair $(\mathbb{G}_{x,0}, \mathbb{G}'_{x',0})$ over k_F.

5.4. *The Howe correspondence for positive-depth supercuspidal representations*

Loke and Ma gave in [38] a description of the local theta correspondence between tame supercuspidal representations in terms of the supercuspidal data. Tame supercuspidal representations were constructed in [61]. Let $\mathcal{D}(G_{n'})$ (resp. $\mathcal{D}(G'_{n'})$) be a set of supercuspidal data for G_n (resp. $G'_{n'}$). For a given supercuspidal Yu datum Σ, let $[\pi_\Sigma]$ denote the isomorphism class of the supercuspidal representation attached to Σ. In [30] an equivalence relation $\sim$ on $\mathcal{D}(G_n)$ was defined so that the map from $\overline{\mathcal{D}}(G_n) := \mathcal{D}(G_n)/\sim$ to $\mathrm{Irr}_{\mathrm{sc}}(G_n)$ given by $[\Sigma] \mapsto [\pi_\Sigma]$ is a bijection. We set

$$\overline{\mathcal{D}}_{\mathfrak{T}} := \bigcup_n \overline{\mathcal{D}}(G_n).$$

Using the moment maps and theta correspondences over finite fields, a *theta lift for supercuspidal data* was defined in [38]. It is a map

$$\vartheta_{n',\mathfrak{T}'} : \overline{\mathcal{D}}(G_{n'}) \hookrightarrow \overline{\mathcal{D}}_{\mathfrak{T}'}. \tag{5.5}$$

Suppose that $\Sigma \in \mathcal{D}(G_n)$ and $[\Sigma'] := \vartheta_{n,\mathfrak{T}'}([\Sigma]) \in \overline{\mathcal{D}}_{n'}(G')$ for some n'. By [38], we have $\theta_n(\pi_\Sigma) = \pi'_{\Sigma'}$, and, conversely, if π and π' are irreducible supercuspidal representations of G_n and $G'_{n'}$, respectively, then there exists $\Sigma \in \mathcal{D}(G_n)$ such that $\pi = \pi_\Sigma$ and $\pi' = \pi'_{\Sigma'}$, where $[\Sigma'] := \vartheta_{n',\mathfrak{T}'}([\Sigma])$.

5.5. *A correspondence between Hecke algebras for enhanced L-parameters*

Let $\mathfrak{s} = [L_n, \sigma]_{G_n} \in \mathfrak{B}(G_n)$. When F has characteristic zero, a local Langlands correspondence $\sigma \mapsto (\varphi_\sigma, \rho_\sigma)$ is available via endoscopy, due to Arthur and Mœglin (see [2] and [48]). By [27], this result extends to the case where F has positive characteristic.

We set

$$\mathfrak{s}^\vee := [L_n^\vee, (\varphi_\sigma, \rho_\sigma)]_{G_n^\vee}. \tag{5.6}$$

By [9], for $G \in \{G_n, G'_{n'}\}$, the algebra $\mathcal{H}(G, \mathfrak{s})$ is canonically isomorphic to the algebra $\mathcal{H}(G^\vee, \mathfrak{s}^\vee)$. Thus, by Theorem 5.2, the Howe correspondence $\pi \mapsto \theta_{n'}(\pi)$ for the reductive dual pair $(G_n, G'_{n'})$ induces a correspondence $E \mapsto \theta(E)$ between subsets of simple modules of the extended affine Hecke algebras $\mathcal{H}(G_n^\vee, \mathfrak{s}^\vee)$ and $\mathcal{H}(G'^\vee_{n'}, \theta(\mathfrak{s})^\vee)$, and hence a correspondence

$$(\varphi, \rho) \mapsto (\theta_{n'}(\varphi), \theta_{n'}(\rho)) \tag{5.7}$$

between subsets of the series $\Phi_{\mathrm{e}}^{\mathfrak{s}^\vee}(G_n)$ and $\Phi_{\mathrm{e}}^{\theta(\mathfrak{s})^\vee}(G'_{n'})$.

References

1. J. Adams and A. Moy, "Unipotent representations and reductive dual pairs over finite fields", Trans. Amer. Math. Soc. **340** (1) (1993) 309–321.
2. J. Arthur, *The endoscopic classification of representations: orthogonal and symplectic groups*, Colloquium Publications **61**, Amer. Math. Soc., 2013.
3. H. Atobe and W. T. Gan, "Local theta correspondence of tempered representations and Langlands parameters", Invent. Math. **210** (2017), no. 2, 341–415.
4. A.-M. Aubert, "Description de la correspondance de Howe en terme de classification de Kazhdan-Lusztig", Invent. Math. **103** (1991), 379–415.
5. A.-M. Aubert, "Local Langlands and Springer correspondences", in *Representations of Reductive p-adic Groups*, A.-M. Aubert, M. Mishra, A. Roche, S. Spallone (Eds.), Progress in Mathematics **328**, Birkhäuser, 2019, pp. 1–37.
6. A.-M. Aubert, W. Kraśkiewicz and T. Przebinda, "Howe correspondence and Springer correspondence for dual pairs over a finite field", Proc. Sympos. Pure Math. **92** (2016) 17–44.
7. A.-M. Aubert, J. Michel and R. Rouquier, "Correspondance de Howe pour les groupes réductifs sur les corps finis", Duke Math. J. **83** No. 2 (1996), 353–397.
8. A.-M. Aubert, A. Moussaoui and M. Solleveld, "Generalizations of the Springer correspondence and cuspidal Langlands parameters", Manuscripta Math. **157** (2018), no. 1-2, 121–192.
9. A.-M. Aubert, A. Moussaoui and M. Solleveld, "Affine Hecke algebras for Langlands parameters", arXiv:1701.03593, 2017.

10. A.-M. Aubert, A. Moussaoui and M. Solleveld, "Affine Hecke algebras for classical p-adic groups", arXiv:2211.08196, 2022.
11. A.-M. Aubert and T. Przebinda, "A reverse engineering approach to the Weil representation", Central European J. Math. **12** (2014), 1500–1585.
12. A.-M. Aubert and Y. Xu, "Hecke algebras for p-adic reductive groups and local Langlands correspondences for Bernstein blocks", Adv. Math. **436** (2024), Paper No. 109384, 45 pp.
13. A.-M. Aubert and Y. Xu, "The explicit local Langlands correspondence for G_2", arXiv:2208.12391, 2022.
14. P. Bakić and M. Hanzer, "Theta correspondence for p-adic dual pairs of type I", J. Reine Angew. Math. **776** (2021), 63–117.
15. J. Bernstein, "Le "centre" de Bernstein", in: *Representations of reductive groups over a local field*, Travaux en Cours, 1–32, Ed. P. Deligne, Hermann, Paris, 1984.
16. N. Bourbaki, *Lie Groups and Lie Algebras*, Elements of Math., Chapters 4–6, Springer-Verlag, Berlin, 2002.
17. F. Bruhat and J. Tits, "Groupes réductifs sur un corps local. II. Schémas en groupes. Existence d'une donnée radicielle valuée", Inst. Hautes Études Sci. Publ. Math. (1984), **60**, 197–376.
18. C. Bushnell and P. Kutzko, "Smooth representations of reductive p-adic groups: structure theory via types", Proc. London Math. Soc. **77**:3 (1998), 582–634.
19. R. Carter, *Finite Groups of Lie Type, Conjugacy classes and complex characters*, Wiley Classics Library, 1993.
20. P. Deligne and G. Lusztig, "Representations of reductive groups over finite fields", Ann. of Math. (2) **103**:1 (1976), 103–161.
21. F. Digne and J. Michel, *Representations of finite groups of Lie type*, London Mathematical Society Student Texts **95**, Cambridge University Press, Cambridge, 2020.
22. J. Epequin, "Extremal unipotent representations for the finite Howe correspondence", Journal of Algebra **535** (2019) 480–502.
23. W. T. Gan and S. Takeda, "On the Howe duality conjecture in classical theta correspondence", in *Advances in the theory of automorphic forms and their L-functions*, 105–117, Contemp. Math. **664**, Amer. Math. Soc., Providence, RI, 2016.
24. W. T. Gan and A. Ichino, "Formal degrees and local theta correspondence", Invent. Math. **195** (2014), 509–672.
25. W. T. Gan and G. Savin, "Representations of metaplectic groups I: Epsilon dichotomy and local Langlands correspondence", Compos. Math. **148** (2012), no. 6, 1655–1694.
26. W. T. Gan and S. Takeda, "A proof of the Howe duality conjecture", J. Amer. Math. Soc. **29** (2016), 473–493.
27. R. Ganapathy and S. Varma, "On the local Langlands correspondence for split classical groups over local function fields", J. Inst. Math. Jussieu **16** (2017), 987–107.
28. S. Gurevich and R. Howe, "Small representations of finite classical groups",

in *Representation theory, number theory, and invariant theory*, Progr. Math., Vol. **323**, 2017, pp. 209–234.

29. T. J. Haines, "The stable Bernstein center and test functions for Shimura varieties", pp. 118–186 in: *Automorphic Forms and Galois Representations*, London Math. Soc. Lecture Note Series **415**, Cambrige University Press, 2014.
30. J. Hakim and F. Murnaghan, "Distinguished tame supercuspidal representations", IMRP (2008).
31. V. Heiermann, "Opérateurs d'entrelacement et algèbres de Hecke avec paramètres d'un groupe réductif p-adique: le cas des groupes classiques", Selecta Math. (N.S.) **17**:3 (2011), 713–756.
32. V. Heiermann, "Local Langlands correspondence for classical groups and affine Hecke algebras", Math. Z. (2017) **287**, 1029–1052.
33. R. Howe, "θ-series and invariant theory", in: Automorphic forms, representations and L-functions, American Mathematical Society, 1979, 275–285.
34. D. Kazhdan and G. Lusztig, "Proof of the Deligne-Langlands conjecture for Hecke algebras", Invent. Math. **87**:1 (1987), 153–215.
35. S. Kudla, "On the local theta-correspondence", Invent. Math. **83** (1986), 229–255.
36. S. Kudla, "Splitting metaplectic covers of dual reductive pairs", Isr. J. Math. **87** (1994), 361–401.
37. S. Kudla and S. Rallis, "On first occurrence in the local theta correspondence", in *Automorphic representations, L-functions and applications: progress and prospects*, Ohio State Univ. Math. Res. Inst. Publ., **11**, de Gruyter, Berlin, 2005, 273–308.
38. H. Y. Loke and J.-J. Ma, "Local theta correspondences between supercuspidal representations", Ann. Sci. Éc. Norm. Supér. (4) **51**:4 (2018), 927–991.
39. G. Lusztig, *Characters of reductive groups over a finite field*, Ann. Math. Stud., Princeton, 1984.
40. G. Lusztig, "Intersection cohomology complexes on a reductive group", Invent. Math. **75.2** (1984), 205–272.
41. G. Lusztig, "On the representations of reductive groups with disconnected centre", Astérisque, tome **168** (1988), 157–166.
42. G. Lusztig, "Cuspidal local systems and graded Hecke algebras", Publ. Math. Inst. Hautes Études Sci. **67** (1988), 145–202.
43. G. Lusztig, "Classification of unipotent representations of simple p-adic groups", Int. Math. Res. Not. (1995), 517–589.
44. G. Lusztig, "Classification of unipotent representations of simple p-adic groups, II", Represent. Theory **6** (2002), 243–289.
45. J.-J. Ma, C. Qiu, and J. Zou, "Generic Hecke algebra and theta correspondence over finite fields", Adv. Math. **438** (2024), Paper No. 109444, 44 pp.
46. B. Mesland and M.H. Şengün, "Equal rank local theta correspondence as a strong Morita equivalence", Selecta Math. (N.S.) **30**:4 (2024), Paper No. 72.
47. C. Mœglin, "Multiplicité 1 dans les paquets d'Arthur aux places p-adiques", pp. 333–374 in: *On certain L-functions*, Clay Math. Proc. **13**, Amer. Math. Soc., 2011.

48. C. Mœglin, "Paquets stables des séries discrètes accessibles par endoscopie tordue: leur paramètre de Langlands", Contemp. Math. **614** (2014), 295–336.
49. C. Mœglin, M.-F. Vignéras, and J.-L. Waldspurger, *Correspondances de Howe sur un corps p-adique*, Lecture Notes in Math. **1291**, Springer.
50. L. Morris, "Tamely ramified supercuspidal representations", Ann. Scien. Éc. Norm. Sup. 4e série, **29** (1996), 639–667.
51. L. Morris, "Level zero G-types", Compos. Math. **118** (1999), 135–157.
52. A. Moussaoui, "Centre de Bernstein enrichi pour les groupes classiques", Represent. Theory **21** (2017), 172–246.
53. A. Moy and G. Prasad, "Jacquet functors and unrefined minimal K-types", Comment. Math. Helv. **71** (1996), no. 1, 98–121.
54. S.-Y. Pan, "Local theta correspondence of depth zero representations and theta dichotomy", J. Math. Soc. Japan **54** (2002), 793–845.
55. S.-Y. Pan, "Howe correspondence of unipotent characters for a finite symplectic/even-orthogonal dual pair", Amer. J. Math. **146**:3 (2024), 813–869.
56. S.-Y. Pan, "Lusztig correspondence and Howe correspondence for finite reductive dual pairs", Math. Ann. (2024).
57. S.-Y. Pan, "On theta and eta correspondences for finite symplectic/orthogonal dual pair", arXiv:2006.06241, 2020.
58. M. Solleveld, "On completions of Hecke algebras", in: A.-M. Aubert, M. Mishra, A. Roche, S. Spallone (Eds.), *Representations of Reductive p-adic Groups*, Progress in Mathematics **328**, Birkhäuser, 2019, pp. 207–262.
59. M. Solleveld, "Endomorphism algebras and Hecke algebras for reductive p-adic groups", J. Algebra **606** (2022), 371–470.
60. J.-L. Waldspurger, "Démonstration d'une conjecture de dualité de Howe dans le cas p-adique, $p \neq 2$", in: Festschrift in honor of I. Piatetski-Shapiro on the occasion of his sixtieth birthday, Part I, Weizmann, Jerusalem, 1990, 267–324.
61. J.-K. Yu, "Construction of tame supercuspidal representations", J. Amer. Math. Soc. **14** (2001), no. 3, 579–622.

https://doi.org/10.1142/9789819825448_0002

Progresses on the Local Gan-Gross-Prasad Conjecture

Cheng Chen
School of Mathematics, University of Minnesota,
Minneapolis, MN 55455, USA
cheng.chen@imj-prg.fr

This chapter is a survey that reviews an approach toward the local Gan-Gross-Prasad conjecture. This approach is based on Mœglin and Waldspurger's works [25], [26], [27], [28], [29], [20] and [12]. Compared to the method in these articles, the approach in this chapter modifies the step for the reduction to tempered cases using the multiplicity formula proved in [9] and [10]. This modification provides a picture that treats Bessel and Fourier-Jacobi cases parallelly.

Contents

1. Introduction

The branching problem originally examines the rules of decomposing the restriction of an irreducible representation of a group to a subgroup. The nature of the branching problem varies when one speculates on different types of groups and representations. The branching problem for permutation groups $\mathfrak{S}_{n-1} \subset \mathfrak{S}_n$ and their finite-dimensional representations π_{n-1}, π_n can be described using classification by the Young tableaux $\mathbf{a} = (a_1, \cdots, a_{n-1})$ of π_{n-1}, $\mathbf{b} = (b_1, \cdots, b_n)$ of π_n. (Here $\{a_i\}$,$\{b_j\}$ are descending series of non-negative integers satisfying $\sum_{i=1}^{n-1} a_i = n-1$, $\sum_{j=1}^{n} b_j = n$.) The representation π_{n-1} occurs in the decomposition of $\pi_n|_{\mathfrak{S}_{n-1}}$ if and only if

$$b_1 \geq a_1 \geq b_2 \geq \cdots \geq a_{n-1} \geq b_n.$$

Moreover, $\dim \mathrm{Hom}_{\mathfrak{S}_{n-1}}(\pi_n|_{\mathfrak{S}_{n-1}}, \pi_{n-1}) \leqslant 1$, which is known as *multiplicity-one theorem*. Similar results can be formulated for compact classical groups using the highest weights, e.g., $\mathrm{U}(n-1) \subset \mathrm{U}(n)$, $\mathrm{SO}(n-1) \subset \mathrm{SO}(n)$. The branching problem can be applied to studying symmetry-breaking in Hamiltonian mechanics and quantum mechanics, which speculates how a perturbation affects the fate of the physical systems. Symmetry-breaking in quantum mechanics also needs results on the branching problem for non-compact groups, such as the Lorentz group $\mathrm{SO}(1,3)$. This motivates the formulation of theories on the branching problem of representations of reductive groups over local fields.

Unlike the finite group or compact group situations, the restriction does not fully decompose when studying the branching problem in settings of admissible representations of reductive groups over local fields. Over a local field of characteristic zero, some studies, such as the local theory of automorphic forms, generate more interest in the irreducible quotients of the restriction. B. Gross and D. Prasad formulated a nice conjecture on the branching problem for admissible representations of special orthogonal groups in [15]. They later generalized the conjecture to a more general context, applicable for Bessel models of special orthogonal groups in [16]. In [14], W. T. Gan, Gross, and Prasad extended the conjecture to other classical groups, precisely describing the multiplicities of Bessel and Fourier-Jacobi models. This conjecture is closely related to the study of automorphic forms and L-functions, and is called the *local Gan-Gross-Prasad conjecture*, and we abbreviate it as the *local GGP conjecture*.

For several pairs of groups related to Bessel and Fourier-Jacobi models in automorphic forms, the local GGP conjecture speculated representations in the Vogan L-packet, the union of L-packets associated with a given L-parameter for all pure inner forms of the group. The *multiplicity* of a representation is defined as the dimension of Bessel or Fourier-Jacobi models the representation has. Gan-Gross-Prasad conjectured that, for every generic L-parameter, there is exactly one representation in the Vogan L-packet with multiplicity equal to one. Moreover, based on the parameterization of Vogan L-packets by Vogan's conjecture in [24], Gan-Gross-Prasad also conjectured that this unique representation is characterized by the local root numbers arising from arithmetics.

Recent years have witnessed great progress in this conjecture. Over non-Archimedean local fields of characteristic zero, J.-L. Waldspurger did the prior work and completed the proof for tempered L-parameters of special orthogonal groups using local trace formula and endoscopy in [25], [26], [27] and [28]. For this case, C. Mœglin and Waldspurger proved the conjecture for generic L-parameters in [20]. R. Beuzart-Plessis provided a proof for tempered L-parameters of unitary groups over non-Archimedean local fields in [2] and [3]. Based on this result, Gan and A. Ichino proved the conjecture for the Bessel and Fourier-Jacobi model of unitary groups in [12]. H. Atobe proved the conjecture for the Fourier-Jacobi models of symplectic-metaplectic groups in [1].

Over Archimedean local fields, i.e. $\mathbb{R}$ and $\mathbb{C}$, most results have been proved. H. He proved the conjecture in unitary Bessel cases for discrete series L-parameters in [17]. Beuzart-Plessis proved the uniqueness of the multiplicity-one representation for tempered L-parameters in unitary Bessel cases in [4]. H. Xue completed the proof for unitary Bessel cases in [32] and [33]. Z.-L. Luo proved the uniqueness of the multiplicity-one representation for tempered L-parameters in special orthogonal cases in [19]. Luo and the author completed the proof for tempered special orthogonal cases in [11]. Based on these results, the author completed the proof for generic L-parameters in [8]. Xue proved the Fourier-Jacobi cases for unitary groups in [34], and the author reduced the conjecture for symplectic-metaplectic groups to basic tempered cases in [9]. The remaining open problems are the basic tempered symplectic-metaplectic cases.

Recent progress of the relative Langlands program ([22], [5]) calls for the generalization of the conjecture to other spherical varieties. C. Wan and L. Zhang formulated a parallel conjecture for several strongly tempered spherical varieties in [31] and proved it in some cases. This chapter serves as a retrospective survey on this topic, summarizing some of the previous works on the local Gan-Gross-Prasad conjecture, and presents the results in the following picture that work uniformly in all cases.

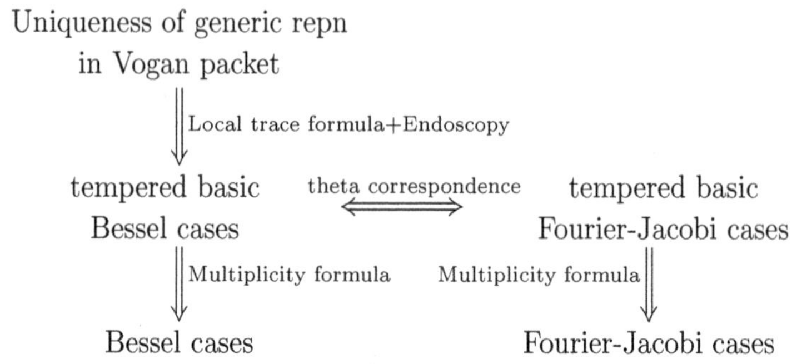

The chapter's structure is the following: Section 2 reviews the statement of the local GGP conjecture. Section 3 reviews Waldspurger's trace formula approach for the tempered Bessel cases and its generalizations. Section 4 reviews the theta correspondence approach to build the bridge between multiplicities of basic tempered Bessel and Fourier-Jacobi models. Section 5 reviews the reduction from generic to tempered cases.

Acknowledgments

This chapter was partially completed while the author was participating in the workshop held at the Institute for Mathematical Sciences, National University of Singapore, in 2022. The author is supported in part by the Research Assistantship from the NSF grant DMS-19018024 and DMS-220089. The author thanks the School of Mathematics of the University of Minnesota for providing the Ordway Fellowship.

2. Statement of the Conjecture

This section aims to give a complete statement of the local GGP conjecture following [14] for both Bessel and Fourier-Jacobi models over local fields of characteristic zero.

2.1. *Models and multiplicity*

Bessel and Fourier-Jacobi models are important objects studied in the local theory of automorphic forms. In classical works, these models are defined over a pair of towers of isometry groups of nondegenerate ϵ-hermitian spaces.

Let F be a local field of characteristic zero and $E = F$ or E is a quadratic field extension of F. Throughout the chapter, we use the bold letters like

$\mathbf{G}$, $\mathbf{H}$ to denote algebraic groups over F and use the Latin letters to denote the groups of their F-points like $G = \mathbf{G}(F)$, $H = \mathbf{H}(F)$.

In the unitary Bessel cases, $\epsilon = 1$, $[E : F] = 2$, and the towers are

$$\mathcal{T}_o = \{\mathbf{U}(V) : \dim_E V \text{ is odd}\}, \quad \mathcal{T}_e = \{\mathbf{U}(W) : \dim_E W \text{ is even}\};$$

in special orthogonal Bessel cases, $\epsilon = 1$, $E = F$, and the towers are

$$\mathcal{T}_o = \{\mathbf{SO}(V) : \dim_E V, \text{ is odd}\}, \quad \mathcal{T}_e = \{\mathbf{SO}(W) : \dim_E W \text{ is even}\}.$$

In skew-unitary Fourier-Jacobi cases, $\epsilon = -1$, $[E : F] = 2$, and the towers are

$$\mathcal{T} = \{\mathbf{U}(V)\}, \quad \widetilde{\mathcal{T}} = \{\mathbf{U}(W)\};$$

in symplectic-metaplectic Fourier-Jacobi cases, $\epsilon = -1$, $E = F$, and the towers are

$$\mathcal{T} = \{\mathbf{Sp}(V)\}, \quad \widetilde{\mathcal{T}} = \{\mathbf{Mp}(W)\}.$$

In each case, two $\mathbf{G}_V$ and $\mathbf{H}_W$ in different towers are called *relevant* if and only if

$$\begin{aligned} &V \subset W, \quad \text{and } V^{\perp} \text{ is split over } E, \textit{or} \\ &W \subset V, \quad \text{and } \quad W^{\perp} \text{ is split over } E. \end{aligned}$$

When $\mathbf{G}_V$, $\mathbf{H}_W$ are relevant, we call (V, W) an *relevant pair.*

We denote by $\Pi_F(\mathbf{G}_V)$ the set of equivalence classes of irreducible and admissible representations of $G_V = \mathbf{G}_V(F)$, and we require it to be Casselman-Wallach ([6], [30]) when F is Archimedean.

Given relevant $\mathbf{G}_V$ and $\mathbf{H}_W$, we define a *multiplicity* $m(\pi_V, \pi_W)$ for $\pi_V \in \Pi_F(\mathbf{G}_V)$ and $\pi_W \in \Pi_F(\mathbf{H}_W)$. Since $\mathbf{G}_V$ and $\mathbf{H}_W$ are relevant, we may assume $W \subset V$, then we have a split S such that $V = W \perp S$, where $\perp$ denotes the orthogonal direct sum of quadratic spaces. Then $\dim_E S$ is odd in Bessel cases and $\dim S$ is even in Fourier-Jacobi cases. When $S \neq 0$, we set $r = \lfloor \frac{\dim_E S}{2} \rfloor$. We choose a basis S over E, and denote it by $\{z_i\}_{i=-r}^{r}$ in Bessel cases and by $\{z_i\}_{1 \leqslant |i| \leqslant r}$ in Fourier-Jacobi cases such that

$$\begin{aligned} &(z_i, z_j) = \delta_i^{-j}, i = 1, \ldots, r, j = -1, \cdots, -r, \text{ and} \\ &(z_i, z_0) = 0, i = \pm 1, \ldots, \pm r, \text{ in Bessel cases,} \end{aligned}$$

where $(\cdot, \cdot)$ is the ϵ-hermitian form on V.

Then we define $\mathbf{N}$ as the unipotent radical of the parabolic group $\mathrm{P}_{V,r}$ stabilizing the full totally isotropic flag

$$X_1 \subset \cdots \subset X_r$$

of S, where $X_i = \mathrm{Span}_E(z_{r-i+1}, \cdots, z_r)$. Fix an additive character ψ_F of F and let ξ be the generic character of $N = \mathbf{N}(F)$ defined by

$$\xi_N(n) = \psi_E\left(\sum_{i=b}^{r-1}(z_i, z_{-(i-1)})\right), \quad n \in N,$$

where

$$b = \begin{cases} 0 & \text{in Bessel cases,} \\ 1 & \text{in Fourier-Jacobi cases,} \end{cases} \quad \text{and} \quad \psi_E(z) = \psi_F(\mathrm{Re}(z)),\ z \in E.$$

Here

$$\mathrm{Re}(z) = \frac{1}{|\mathrm{Gal}(E/F)|} \sum_{\tau \in \mathrm{Gal}(E/F)} z^\tau \text{ for } z \in E.$$

In Bessel cases, we define the multiplicity

$$m(\pi_V, \pi_W) = \dim \mathrm{Hom}_{G_W \ltimes N}(\pi_V|_{G_W \ltimes N} \widehat{\otimes} \pi_W, \xi),$$

where ξ is the character of $G_W \ltimes N$ induced from the character ξ_N of N.

We denote by ω_{V,ψ_F} the Weil representation with respect to ψ_F of

$$\begin{cases} \mathrm{Mp}(V) \ltimes \mathcal{H}(V) & \text{when } V \text{ is a symplectic space,} \\ \mathrm{U}(V) \ltimes \mathcal{H}(V) & \text{when } V \text{ is a skew-hermitian space,} \end{cases}$$

where $\mathcal{H}(W)$ is the Heisenberg group $W \oplus F$. In Fourier-Jacobi cases, since $\pi_W \in \Pi_F(\mathbf{H}_W)$, the completed tensor product $\pi_W \widehat{\otimes} \overline{\omega_{W,\psi_F}}$ is an irreducible representation of $G_W^J = G_W \ltimes \mathcal{H}(W)$. We define the multiplicity

$$m(\pi_V, \pi_W) = \begin{cases} \dim \mathrm{Hom}_{G_W^J \ltimes N}(\pi_V|_{G_W^J \ltimes N} \widehat{\otimes}(\pi_W \widehat{\otimes} \overline{\omega_{W,\psi_F}}), \xi) & \text{when } W \subsetneq V, \\ \dim \mathrm{Hom}_{G_V}(\pi_V \widehat{\otimes}(\pi_W \widehat{\otimes} \overline{\omega_{W,\psi_F}})|_{G_V}, 1) & \text{when } W = V, \end{cases}$$

in Fourier-Jacobi cases, where ξ is the character of $G_W^J \ltimes N$ induced from ξ_N.

In [4], the multiplicity is equivalently defined using the *Gan-Gross-Prasad triple* (G, H, ξ) associated to a relevant pair (V, W), where $G = \mathbf{G}(F)$, $H = \mathbf{H}(F)$ for

$$\mathbf{G} = \begin{cases} \mathbf{G}_V \times \mathbf{H}_W & \text{in Bessel cases,} \\ \mathbf{G}_V \times (\mathbf{G}_W \ltimes \mathcal{H}(W)) & \text{in Fourier-Jacobi cases,} \end{cases}$$

$$\mathbf{H} = \begin{cases} \Delta\mathbf{H}_W \ltimes \mathbf{N} & \text{in Bessel cases,} \\ \Delta\mathbf{G}_W^J \ltimes \mathbf{N} & \text{in Fourier-Jacobi and } W \subsetneq V \text{ cases,} \\ \Delta\mathbf{G}_W & \text{in Fourier-Jacobi and } W = V \text{ cases,} \end{cases}$$

and ξ be the character of H induced from ξ_N. In this setting, the multiplicity can be equivalently defined as

$$m(\pi) = \dim \mathrm{Hom}_H(\pi, \xi),$$

and it is compatible with the previous definition in the sense of

$$m(\pi_V, \pi_W) = \begin{cases} m(\pi_V \boxtimes \pi_W) & \text{in Bessel cases,} \\ m(\pi_V \boxtimes (\pi_W \widehat{\otimes} \omega_{W,\psi_F})) & \text{in Fourier-Jacobi case.} \end{cases}$$

This notation helps treat some arguments of Bessel and Fourier-Jacobi cases uniformly. Moreover, the generalization of the conjecture to spherical varieties is based on this notation.

A GGP triple (G, H, ξ) associated to a relevant pair (V, W) is called *basic* if ξ is the trivial character. In Bessel cases, (G, H, ξ) is basic if $\dim_E V = \dim_E W + 1$. In Fourier-Jacobi cases, (G, H, ξ) is basic if $\dim_E V = \dim_E W$ (known as the *equal-rank cases*) or $\dim_F V = \dim_F W + 2$ (known as the *almost equal-rank cases*).

2.2. *Vogan L-packets*

Vogan L-packet for classical groups

When $\mathbf{G}_V$ is a classical group, R. Langlands classifies the representations in $\Pi_F(\mathbf{G}_V)$ with L-parameters $\varphi_V : \mathrm{WD}_F \to {}^L\mathbf{G}_V$, where WD_F is the Weil-Deligne group. Given an L-parameter φ_V, the local Langlands correspondence gives us an L-packet $\Pi_{F,\varphi_V}(\mathbf{G}_V)$. The Vogan L-packet is defined as

$$\Pi^{\mathrm{Vogan}}_{F,\varphi_V} = \coprod_{\alpha \in H^1(F, \mathbf{G}_V)} \Pi_{F,\varphi_V}(\mathbf{G}_{V_\alpha}).$$

An L-parameter φ_V is called *generic* if $\Pi^{\mathrm{Vogan}}_{F,\varphi_V}$ contains a generic representation. D. Vogan conjectured it over local fields in [24] and confirmed over Archimedean fields that, fixing a Whittaker datum, there exists a bijection

$$\Pi^{\mathrm{Vogan}}_{F,\varphi_V} \longleftrightarrow \text{characters of } \mathcal{S}_{\varphi_V},$$

where $\mathcal{S}_\varphi = \pi_0(\mathrm{Cent}_{\widehat{\mathbf{G}}_V}(\mathrm{Im}(\varphi_V)))$ and $\widehat{\mathbf{G}}_V$ are the dual groups of $\mathbf{G}_V$.

Vogan L-packet for metaplectic groups

Let W be a quadratic space over F satisfying $\dim W = 2n + 1$ and $\mathrm{disc}(W) = 1$. When $\mathbf{G}_V = \mathbf{Mp}(2n)$, the L-parameter $\widetilde{\varphi}_V : \mathrm{WD}_F \to {}^L\mathbf{G}_V =$

$\mathrm{Sp}(2n,\mathbb{C})$ corresponds to an L-parameter $\varphi_W : \mathrm{WD}_F \to {}^L\mathbf{SO}(W) = \mathrm{Sp}(2n,\mathbb{C})$ of $\mathbf{SO}(W)$. The parameter $\widetilde{\varphi}_V$ is called *generic* if φ_W is generic.

The Shimura-Waldspurger's correspondence gives a bijection

$$\theta_{V_{2n}} : \coprod_{\substack{\dim W=2n+1\\ \mathrm{disc}(W)=1}} \Pi_F(\mathbf{SO}(W)) \to \Pi_F^{\mathrm{gen}}(\mathbf{Mp}(2n)),$$

where $\Pi_F^{\mathrm{genu}}(\mathbf{Mp}(2n))$ is the set of equivalence classes of irreducible genuine representations. Using the Shimura-Waldspurger's correspondence, the Vogan L-packet is defined as

$$\Pi_{F,\widetilde{\varphi}_V}^{\mathrm{Vogan}} = \theta_{V_{2n}}(\Pi_{F,\varphi_V}^{\mathrm{Vogan}}),$$

and there is a bijection

$$\Pi_{F,\widetilde{\varphi}_V}^{\mathrm{Vogan}} \longleftrightarrow \text{characters of } \mathcal{S}_{\widetilde{\varphi}_V},$$

when $\widetilde{\varphi}_V$ is generic. Here $\mathcal{S}_{\widetilde{\varphi}_V} = \pi_0(\mathrm{Cent}_{\widehat{\mathbf{G}}_V}(\mathrm{Im}(\widetilde{\varphi}_V)))$.

2.3. *Statement of the conjecture*

Let $\mathbf{G}_V, \mathbf{H}_W$ be a relevant pair of reductive groups and $W \subset V$. For generic L-parameters φ_V, φ_W of $\mathbf{G}_V, \mathbf{H}_W$ respectively, Gan-Gross-Prasad defined the *relevant Vogan packet*

$$\Pi_{F,\varphi_V\times\varphi_W}^{\mathrm{Vogan,rel}} = \coprod_{\alpha\in H^1(F,\mathbf{H}_W)} \Pi_{F,\varphi_V}(\mathbf{G}_{V_\alpha}) \times \Pi_{F,\varphi_W}(\mathbf{H}_{W_\alpha}).$$

Conjecture 1: (Local Gan-Gross-Prasad conjecture) *Given generic L-parameters φ_V of $\mathbf{G}_V$ and φ_W of $\mathbf{H}_W$, the following results hold.*

(1) (Multiplicity-one) *We*

$$\sum_{(\pi_V,\pi_W)\in\Pi_{F,\varphi_V\times\varphi_W}^{\mathrm{Vogan,rel}}} m(\pi_V,\pi_W) = 1.$$

This means there is exactly one pair $(\pi_V,\pi_W) \in \Pi_{F,\varphi_V\times\varphi_W}^{\mathrm{Vogan,rel}}$ such that $m(\pi_V,\pi_W) = 1$, and for other pairs $(\pi_V,\pi_W) \in \Pi_{F,\varphi_V\times\varphi_W}^{\mathrm{Vogan,rel}}$, $m(\pi_V,\pi_W) = 0$.

(2) (Epsilon-dichotomy) *This unique pair (π_V,π_W) can be characterized by the distinguished characters*

$$\chi_{\varphi_V}^W, \chi_{\varphi_W}^V$$

defined by local root numbers as in [14].

The distinguished characters defined in [14] are only dependent on the semisimplifications of the parameters φ_V, φ_W.

Definition 2.1: Let $\mathbf{G}_V$ be a classical group, then ${}^L\mathbf{G}_V$ has a standard representation $\mathrm{std}_{\mathbf{G}_V} : {}^L\mathbf{G}_V \to \mathrm{GL}(M_V)$ on a complex space M_V ([14, p. 22]). Given an L-parameter $\varphi_V : \mathrm{WD}_F \to {}^L G_V$, the *semisimplification* $\varphi_V^{ß}$ of φ_V is the composition $\mathrm{std}_{\mathbf{G}_V} \circ \varphi_V : \mathrm{WD}_F \to \mathbf{GL}(M_V)$.

An example. Let $F = \mathbb{R}$. Consider F-groups

$$\mathbf{G}_{V_1} = \mathbf{SO}(3,0), \mathbf{G}_{V_2} = \mathbf{SO}(1,2)$$

and their F-subgroups $\mathbf{G}_{W_1} = \mathbf{SO}(2,0)$, $\mathbf{G}_{W_2} = \mathbf{SO}(0,2)$.

We study the branching problem for restricting representations of G_{V_1} to G_{W_1} and representations of G_{V_2} to G_{W_2}.

The Langlands dual groups of $\mathbf{G}_{V_1}$ and $\mathbf{G}_{V_2}$ are both equal to $\mathrm{Sp}(2,\mathbb{C})$. For an L-parameter $\varphi_V : \mathrm{WD}_{\mathbb{R}} \to \mathrm{Sp}(2,\mathbb{C})$, the semisimplification $\varphi_V^{ß}$ is a two-dimensional representation of $\mathrm{WD}_{\mathbb{R}}$. The Langlands dual groups of $\mathbf{G}_{W_1}$ and $\mathbf{G}_{W_2}$ are both equal to $\mathrm{SO}(2,\mathbb{C})$. For an L-parameter $\varphi_W : \mathrm{WD}_{\mathbb{R}} \to \mathrm{SO}(2,\mathbb{C})$, the semisimplification $\varphi_V^{ß}$ is also a two-dimensional representation of $\mathrm{WD}_{\mathbb{R}}$.

The Weil-Deligne group $\mathrm{WD}_{\mathbb{R}}$ is equal to the Weil group $\mathcal{W}_{\mathbb{R}} = \mathbb{C}^\times \coprod \mathbb{C}^\times j$. For any non-negative integer l, we denote by $\varphi_l^{ß}$ the two-dimensional $\mathrm{WD}_{\mathbb{R}}$-representation

$$z \mapsto \begin{pmatrix} (\frac{z}{|z|})^l & \\ & (\frac{z}{|z|})^{-l} \end{pmatrix}, \quad j \mapsto \begin{pmatrix} 0 & 1 \\ (-1)^l & 0 \end{pmatrix}.$$

This is also known as the L-parameter of the discrete series representation D_l of $\mathrm{GL}(2,\mathbb{R})$ when $l \geqslant 1$.

When l is even, $\varphi_l^{ß}$ is the semisimplication of two L-parameters of $\mathbf{SO}(2,0)$ and $\mathbf{SO}(0,2)$ corresponding to the representations $\phi_l : z \mapsto z^{l/2}$ and $\phi_{-l} : z \mapsto z^{-l/2}$ where $z \in S^1 \cong \mathrm{SO}(2,0) \cong \mathrm{SO}(0,2)$, respectively.

When l is odd, $\varphi_l^{ß}$ is the semisimplication of an L-parameter of $\mathbf{SO}(3,0)$ and $\mathbf{SO}(1,2)$. The L-packet of $\mathbf{SO}(3,0)$ contains the l-dimensional representation F_l with highest weight $\frac{l-1}{2}$, and the L-packet of $\mathbf{SO}(3,0)$ contains the discrete series representation of $\mathrm{SO}(1,2) = \mathrm{PGL}(2,\mathbb{R})$ induced from the representation D_l of $\mathrm{GL}(2,\mathbb{R})$.

Take an odd integer $n > 0$ and an even integer $l > 0$. By computing the weights for the maximal torus $\mathrm{SO}(2,0)$ of $\mathrm{SO}(3,0)$, ϕ_l occurs in the decomposition of $F_n|_{\mathrm{SO}(2,0)}$ if and only if $l < n$. By the K-types of the discrete series D_n, we obtain that ϕ_l occurs in the decomposition of $F_n|_{\mathbf{SO}(0,2)}$ if and only if $l > n$.

We notice that exactly one of the multiplicities

$$m(F_n, \phi_l), m(D_n, \phi_l)$$

is equal to one. Moreover, the dichotomy is precisely given by the value of the local root number

$$-\varepsilon(1/2, \varphi_l^{\mathrm{B}} \otimes \varphi_n^{\mathrm{B}}, \psi_F) = \begin{cases} 1 & \text{if } l < n, \\ -1 & \text{if } l > n. \end{cases}$$

3. Tempered Bessel Cases

The first breakthrough of the local GGP conjecture was made by Waldspurger on special orthogonal Bessel cases over non-Archimedean local fields using the trace formula and endoscopy. This section briefly introduces Waldpurger's proof and its generalization. The author recommends the reader to read [4] for further analytic details.

3.1. *Trace formula approach*

This section briefly introduces the trace formula method based on [25], [26], [27], and [28].

Harish-Chandra's study on trace characters lays the foundations of the trace formula. Let $\mathbf{G}$ be a reductive group over a local field F with characteristic zero. For any $\pi \in \Pi_F(\mathbf{G})$, Harish-Chandra proved that the trace distribution

$$\Theta_\pi(f) = \mathrm{Tr}\left(\int_G \pi(g)f(g)dg\right)$$

is a locally integral function that is smooth on the regular semisimple locus $G_{\mathrm{reg,B}}$ of $G = \mathbf{G}(F)$. This function is known as the *Harish-Chandra character* and we write it as Θ_π. The growth of the Harish-Chandra character is controlled by the property that $D^G(x)^{1/2}\Theta_\pi$ is locally bounded, where D^G is the Weyl discriminant given by $D^G(x) = |\det(1 - \mathrm{Ad}(x)|_{\mathfrak{g}/\mathfrak{g}_x})|$ for $x \in G_{\mathrm{B}}$. Moreover, for admissible representations $\pi_i (1 \leqslant i \leqslant l)$ that are nonequivalent (infinitesimally nonequivalent in Archimedean cases), their Harish-Chandra characters Θ_{π_i} are linearly independent. The trace formula approach aims to study spectral properties of the representation π via the geometric properties Θ_π.

For a reductive Lie algebra $\mathfrak{g}$ over F, we denote by $\mathcal{N}(\mathfrak{g}(F))$ the set of nilpotent orbits on $\mathfrak{g}(F)$ and denote by $\mathcal{N}_{\text{reg}}(\mathfrak{g}(F))$ the subset of regular elements in it. By Harish-Chandra character expansion, for every $x \in G_{\mathrm{ss}}$, there exist complex numbers $c_{\mathcal{O}_x,\Theta_\pi}$, indexed by $\mathcal{O}_x \in \mathcal{N}(\mathfrak{g}_x(F))$, such that

$$\Theta_\pi(x\exp(Y)) = \sum_{\mathcal{O}_x \in \mathcal{N}(\mathfrak{g}_x(F))} c_{\mathcal{O}_x,\Theta_\pi}\widehat{\mu_{\mathcal{O}_x}}(Y),$$

when F is non-Archimedean, and

$$\Theta_\pi(x\exp(Y)) = \sum_{\mathcal{O}_x \in \mathcal{N}(\mathfrak{g}_x(F))} c_{\mathcal{O}_x,\Theta_\pi}\widehat{\mu_{\mathcal{O}_x}}(Y) + O(|Y|_{\mathfrak{g}_x}),$$

when $F = \mathbb{R}$,

for $Y \in \mathfrak{g}_x(F)$ sufficiently near zero. Here $\widehat{\mu_{\mathcal{O}_x}}(Y)$ is the Fourier transform of the orbital integrals on $\mathcal{O}_x$ and $|\cdot|_{\mathfrak{g}_x}$ is a norm on the Lie algebra $\mathfrak{g}(F)$. The numbers $c_{\mathcal{O}_x,\Theta_\pi}$ are called *germs* in the Harish-Chandra character expansion of Θ_π at x. Smooth functions on the regular semisimple locus satisfying this property are called *quasi-characters*.

In the context of special orthogonal Bessel cases over non-Archimedean local fields, Waldspurger defined a geometric multiplicity for quasi-characters by

$$m_{\text{geom}}(\Theta) = \int_{\Gamma(G,H)} D^G(x)^{1/2} c_\Theta(x)\Delta(x)^{-1/2}dx, \tag{3.1.1}$$

where $\Gamma(G,H)$, equipped with a measure, is the space of semisimple G-conjugacy classes that intersect H satisfying some other properties, $\Delta(x) = D^G(x)D^H(x)^{-2}$ and $c_\Theta(x) = c_{\mathcal{O}_x,\Theta}$ for uniformly chosen $\mathcal{O}_x \in \mathcal{N}(\mathfrak{g}_x(F))$. By studying the local trace formula concerning the triple (G,H,ξ), Waldspurger proved that, for a tempered representation π of G,

$$m(\pi) = m_{\text{geom}}(\Theta_\pi). \tag{3.1.2}$$

Then one can study $m(\pi)$ via Θ_π.

Beuzart-Plessis generalized this result to tempered unitary cases over any local fields in [3] and [4]. In unitary cases, the definition of geometric multiplicity is

$$m_{\text{geom}}(\Theta) = \lim_{s\to 0^+} \int_{\Gamma(G,H)} D^G(x)^{1/2} c_\Theta(x)\Delta(x)^{s-1/2}dx.$$

Therefore, the limit process needs to be taken into consideration. Luo proved this special orthogonal cases over Archimedean local fields in [19].

3.2. *Multiplicity-one in Vogan L-packets*

This section introduces Waldspurger's approach for the multiplicity-one part of the conjecture. Let (G, H, ξ) be a GGP triple of Bessel type. For a tempered L-parameter φ of G, set the quasi-character

$$\Theta_{\Pi_{F,\varphi}(G)} = \Sigma_{\pi \in \Pi_{F,\varphi}(G)} \Theta_\pi.$$

Then, from (3.1.2), the multiplicity-one part of the local GGP conjecture reduces to showing that

$$\sum_{\alpha \in H^1(F,H)} m_{\text{geom}}(\Theta_{\Pi_{F,\varphi}(G_\alpha)}) = 1. \tag{3.2.1}$$

Definition 3.1: Let $\mathbf{G}$ be a classical group over F.

- When F is non-Archimedean, we denote by $r(\mathbf{G})$ the F-rank of $\mathbf{G}$.
- When F is Archimedean, we let $q(\mathbf{G}) = \frac{1}{2}(\dim G - \dim K) - c(G)$, where K is the maximal compact subgroup of G and $c(G)$ is half of the split rank of fundamental Cartan subgroup of G.
- We denote by $\mathbf{G}_{qs}$ a quasi-split pure inner form of $\mathbf{G}$, which is unique up to isomorphism.

The Kottiwitz sign $e(\mathbf{G})$ is defined as

$$e(\mathbf{G}) = \begin{cases} (-1)^{r(\mathbf{G})-r(\mathbf{G}_{qs})}, & \text{when } F \text{ is non-Archimedean,} \\ (-1)^{q(\mathbf{G})-q(\mathbf{G}_{qs})}, & \text{when } F \text{ is Archimedean.} \end{cases}$$

From Shelstad's works, the stable character $\sum_{\pi \in \Pi_{F,\varphi}(G)} \Theta_\pi$ has the following properties:

- $\Theta_{\Pi_{F,\varphi}(G_\alpha)}$ is a stable character, i.e. it has constant values on stable conjugacy classes (conjugacy classes of $\mathbf{G}_\alpha(\bar{F})$).
- There is an identity

$$\Theta_{\Pi_{F,\varphi}(\mathbf{G}_\alpha)} = e(\mathbf{G}_\alpha)\Theta_{\Pi_{F,\varphi}(\mathbf{G}_{qs})}.$$

Subtle computation of the integral expansion of the left-hand side of (3.2.1) implies that it is equal to the number of generic representations in $\Pi_{F,\varphi}(\mathbf{G}_{qs})$ with respective to the given Whittaker datum (see [28], [4, §12.6]), which is equal to one by results in the local Langlands correspondence.

3.3. *Epsilon-dichotomy*

This section introduces Waldspurger's approach towards epsilon-dichotomy and its simplification over $\mathbb{R}$.

To prove the epsilon-dichotomy, Waldspurger expressed the geometric multiplicity in terms of the stable geometric multiplicity of endoscopic groups using the endoscopic character relation. Then, using the local trace formula on twisted reductive groups and twisted endoscopy, Waldspurger computed the stable geometric multiplicity of endoscopic groups using twisted multiplicities, which can be computed directly using the functional equation of Rankin-Selberg integrals.

The *stable geometric multiplicity* is defined as

$$m_{\text{geom}}^{\text{Stab}}(\Theta) = \int_{\Gamma(G,H)} D^G(x)^{1/2} c_\Theta^{\text{Stab}}(x) \Delta(x)^{-1/2} dx,$$

where the stable germ $c_\Theta^{\text{Stab}}(x) = \frac{1}{|\mathcal{N}_{\text{reg}}(\mathfrak{g}_x(F))|} \sum_{\mathcal{O} \in \mathcal{N}_{\text{reg}}(\mathfrak{g}_x(F))} c_{\Theta,\mathcal{O}}(x)$.

For every tempered parameter φ_V of $\mathbf{SO}(V)$, an element $s_V \in \mathcal{S}_{\varphi_V}$ gives a decomposition

$$\varphi_V^{\beta} = \varphi_{V_+}^{\beta} \oplus \varphi_{V_-}^{\beta},$$

where $\varphi_{V_+}^{\beta} = (\varphi_V^{\beta})^{s=+1}$, $\varphi_{V_-}^{\beta} = (\varphi_V^{\beta})^{s=-1}$ are the eigenspaces with respect to eigenvalues $s = 1$ and $s = -1$, respectively and $\varphi_{V_\pm}$ are L-parameters of $\mathbf{SO}(V_\pm)$.

Consider the endoscopic group $\mathbf{SO}(V_+) \times \mathbf{SO}(V_-)$ such that the underlying spaces of $\varphi_{V_\pm}$ are those of the standard representations of ${}^L\mathbf{SO}(V_\pm)$. For every $\alpha \in H^1(F, \mathbf{SO}(V))$, let

$$\Theta_{\Pi_{F,\varphi_V}(\mathbf{SO}(V_\alpha))}^{s_V} = \sum_{\pi_\alpha \in \Pi_{F,\varphi_V}(\mathbf{SO}(V_\alpha))} \chi_{\pi_{V_\alpha}}(s_V) \Theta_{\pi_{V_\alpha}}.$$

The endoscopic character relation shows that $e(G_{V_\alpha})\Theta_{\Pi_{F,\varphi_V}(\mathbf{SO}(V_\alpha))}^{s_V}$ is the endoscopic transfer of the stable character $\Theta_{\Pi_{F,\varphi_{V_-}}(\mathbf{SO}(V_-))}^{1} \times \Theta_{\Pi_{F,\varphi_{V_+}}(\mathbf{SO}(V_+))}^{1}$.

We can introduce similar notions for the quadratic space W. Following the computation of geometric multiplicity using the endoscopic character relation in [27, Prop. 3.3], we have the following relation between geometric multiplicity and stable geometric multiplicity of endoscopic groups:

$$\sum_{\alpha\in H^1(F,H)} m_{\mathrm{geom}}(\Theta^{s_V}_{\Pi_{F,\varphi_V}(\mathbf{SO}(V_\alpha))}, \Theta^{s_W}_{\Pi_{F,\varphi_W}(\mathbf{SO}(W_\alpha))})$$

$$= m^S_{\mathrm{geom}}(\Theta_{\Pi_{F,\varphi_{V_-}}(\mathbf{SO}(V_-))}, \Theta_{\Pi_{F,\varphi_{W_+}}(\mathbf{SO}(W_+))})$$

$$\cdot\, m^S_{\mathrm{geom}}(\Theta_{\Pi_{F,\varphi_{V_+}}(\mathbf{SO}(V_+))}, \Theta_{\Pi_{F,\varphi_{W_-}}(\mathbf{SO}(W_-))}). \quad (3.3.1)$$

For the explicit computation of the stable geometric multiplicity m^S_{geom}, Waldspurger used the local trace formula and endoscopy on twisted spaces when F is non-Archimedean.

Definition 3.2: A *reductive twisted group* over F is a pair $(\mathbf{G}, \widetilde{\mathbf{G}})$ where $\mathbf{G}$ is a connected reductive group over F and $\widetilde{\mathbf{G}}$ is an F-variety which is a $\mathbf{G}$-bi-torsor such that $\widetilde{\mathbf{G}}(F) \neq 0$. $\widetilde{\mathbf{G}}$ is called a $\mathbf{G}$-twisted space.

In the setting of the Bessel cases of the local Gan-Gross-Prasad conjecture for special orthogonal groups, the twisted reductive group considered is

$$(\mathbf{G}, \widetilde{\mathbf{G}}) = (\mathbf{GL}_{N_V}(F) \times \mathbf{GL}_{N_W}(F), \widetilde{\mathbf{GL}}_{N_V}(F) \times \widetilde{\mathbf{GL}}_{N_W}(F)),$$

where N_V, N_W are the dimensions of the standard representations of ${}^L\mathbf{G}_V$, ${}^L\mathbf{G}_W$, respectively, and $\widetilde{\mathbf{GL}}_{N_V}(V)$ is the space of nontrivial bilinear forms $F^{N_V} \times F^{N_V} \to F$.

Let $\mathbf{N}$ be the unipotent group in the Rankin-Selberg integral in [18] for $\mathbf{GL}_{N_V}(F) \times \mathbf{GL}_{N_W}(F)$ and $\xi_N : \mathbf{N}(F) \to F$ be the associated character. We let

$$(\mathbf{H}, \widetilde{\mathbf{H}}) = (\Delta\mathbf{GL}_{N_W}(F) \ltimes \mathbf{N}, \Delta\widetilde{\mathbf{GL}}_{N_W}(F) \ltimes \mathbf{N}).$$

Let $\xi, \widetilde{\xi}$ be characters of $\mathbf{H}(F), \widetilde{\mathbf{H}}(F)$ induced from ξ_N. Then, for $\widetilde{G} = \widetilde{\mathbf{G}}(F)$ and $\widetilde{H} = \widetilde{\mathbf{H}}(F)$, $((G, \widetilde{G}), (H, \widetilde{H}), (\xi, \widetilde{\xi}))$ is the Gan-Gross-Prasad triple for twisted reductive groups.

A smooth representation $(\pi, \widetilde{\pi})$ of $(G, \widetilde{G})$ consists of a smooth representation $\pi : G \to \mathrm{GL}(V_\pi)$ and a mapping $\widetilde{\pi} : \widetilde{G} \to \mathrm{GL}(V_\pi)$ such that

$$\widetilde{\pi}(g\widetilde{\gamma}g') = \pi(g)\widetilde{\pi}(\widetilde{\gamma})\pi(g'), \quad g, g' \in G, \widetilde{\gamma} \in \widetilde{G}.$$

Following from some regularity arguments, the distribution $\Theta_{\widetilde{\pi}}(\widetilde{f}) = \mathrm{tr}(\widetilde{\pi}(\widetilde{f}))$ $(\widetilde{f} \in C_c^\infty(\widetilde{G}))$ is a smooth function on the strongly-regular semisimple locus of $\widetilde{G}(F)$. Waldspurger constructed the geometric $m_{\mathrm{geom}}(\Theta_{\widetilde{\pi}})$ using a similar integral as (3.1.1). Using a twisted local trace formula, for every tempered representation $(\pi, \widetilde{\pi})$ of $(G, \widetilde{G})$, Waldspurger proved that

$$m_{\mathrm{geom}}(\Theta_{\widetilde{\pi}}) = m(\widetilde{\pi}).$$

Here the spectral multiplicity $m(\widetilde{\pi})$ is defined to be the trace of the map

$$l \in \mathrm{Hom}_H(\pi, \xi) \mapsto \widetilde{l} = \widetilde{\xi}(h)^{-1} \cdot l \circ \widetilde{\pi}(\widetilde{h}) \in \mathrm{Hom}_H(\pi, \xi),$$

which is independent of the choice of $\widetilde{h} \in \widetilde{H}$. By [18], $\dim \mathrm{Hom}_H(\pi, \xi) \leqslant 1$, and a non-zero element in $\mathrm{Hom}_H(\pi, \xi)$ is constructed by the Rankin-Selberg integral when $\dim \mathrm{Hom}_H(\pi, \xi) = 1$. From the functional equation of the Rankin-Selberg integral, $m(\widetilde{\pi})$ is equal to the gamma factor, which is equal to the epsilon factor since π is unitary.

Via a similar process as (3.3.1), the twisted geometric multiplicity $m_{\mathrm{geom}}(\widetilde{\pi})$ is related to the stable geometric multiplicity of endoscopic groups via the multiplicity formula obtained from twisted endoscopy.

For generic L-parameters φ_V, φ_W of $\mathbf{SO}(V), \mathbf{SO}(W)$, let $\chi_{\pi_V^0}, \chi_{\pi_W^0}$ be the characters corresponding to the unique pair (π_V^0, π_W^0) in $\Pi^{\mathrm{Vogan}}_{F, \varphi_V \times \varphi_W}$ with multiplicity equal to one. From (3.1.2) and (3.3.1), the function

$$\chi_{\pi_V^0}(s_V)\chi_{\pi_W^0}(s_W) = \sum_{\substack{\alpha \in H^1(F,H) \\ \pi_V \in \Pi_{F,\varphi_V}(\mathbf{SO}(V_\alpha)) \\ \pi_W \in \Pi_{F,\varphi_W}(\mathbf{SO}(W_\alpha))}} \chi_{\pi_{V_\alpha}}(s_V)\chi_{\pi_{W_\alpha}}(s_W) m(\pi_{V_\alpha}, \pi_{W_\alpha})$$

can be expressed as stable geometric multiplicity, which can be computed from twisted spectral multiplicity via twisted endoscopy and twisted trace formula.

For unitary cases over non-Archimedean local fields, Beuzart-Plessis proved the twisted local trace formula and applied it to prove the conjecture in [2]. The twisted local trace formula was not established over Archimedean local fields. In [11], we followed the idea of C. Wan to simplify Waldspurger's proof over $\mathbb{R}$ so that we only use the local trace formula.

When $F = \mathbb{C}$, there is exactly one representation in the Vogan packet, so epsilon-dichotomy is automatic. Let $F = \mathbb{R}$. When $\dim V > 3$, the L-parameter φ_V of $\mathbf{SO}(V)$ can be classified as *parabolic type* or *endoscopic type*.

The L-parameter φ_V is of parabolic type if there is a decomposition

$$\varphi_V^{\mathfrak{B}} = \varphi_V^{\mathbf{GL}} \oplus \varphi_{V_0}^{\mathfrak{B}} \oplus (\varphi_V^{\mathbf{GL}})^\vee, \quad V = V_0 \perp (X_V \oplus X_V^\vee),$$

where $X_V \neq 0$ and the L-parameter $\varphi_V^{\mathbf{GL}} : \mathrm{WD}_{\mathbb{R}} \to {}^L\mathbf{GL}(X_V)$ does not contain any self-dual irreducible components. In this case, let

$\Pi_{F,\varphi_V^{\mathbf{GL}}}(\mathbf{GL}(X_V)) = \{\sigma\}$, and then the map

$$\pi_0 \mapsto \sigma \ltimes \pi_0 = \mathrm{Ind}_{P_{X_V}}^{G_V}(\delta_{P_{X_V}}^{1/2}\sigma\widehat{\otimes}\pi_0) \tag{3.3.2}$$

defines an isomorphism

$$\Pi_{F,\varphi_{V_0}}^{\mathrm{Vogan}} \to \Pi_{F,\varphi_V}^{\mathrm{Vogan}},$$

where $\mathbf{P}_{X_V}$ is the parabolic subgroup of $\mathbf{G}_V$ stabilizing X_V and $P_{X_V} = \mathbf{P}_{X_V}(\mathbb{R})$.

The L-parameter φ_V is of endoscopic type if there exists $s \in \mathcal{S}_{\varphi_V}$ such that the endoscopic group $\mathbf{G}'_V = \mathbf{SO}(V_+) \times \mathbf{SO}(V_-)$ of $\mathbf{G}_V = \mathbf{SO}(V)$ determined by s is smaller than $\mathbf{SO}(V)$, that is,

$$^L\mathbf{G}'_V \subsetneq {}^L\mathbf{G}_V, \quad \mathrm{Cent}_{\widehat{\mathbf{G}}_V}(s)^o = \widehat{\mathbf{G}}'_V. \tag{3.3.3}$$

We used properties of geometric multiplicity in [19] to prove the equality

$$m((\sigma_V \ltimes \pi_V) \boxtimes (\sigma_W \ltimes \pi_W)) = m(\pi_V \boxtimes \pi_W),$$

for tempered representations $\sigma_V, \sigma_W, \pi_V, \pi_W$. This allows us to reduce the conjecture to smaller cases when φ_V is of parabolic type.

We set

$$m_{V,W}^S := \sum_{\alpha\in H^1(\mathbb{R},H)} \sum_{\substack{\pi_V\in\Pi_{F,\varphi_V}(\mathbf{SO}(V_\alpha)) \\ \pi_W\in\Pi_{F,\varphi_W}(\mathbf{SO}(W_\alpha))}} \chi_\pi(1_{\varphi_V}, -1_{\varphi_W}) m(\pi_V, \pi_W).$$

Using (3.1.2) and (3.3.1), we obtain that

$$m_{V,W}^S = m_{\mathrm{geom}}^S(\Theta_{\Pi_{F,\varphi_V}(\mathbf{SO}(V))}, \Theta_{\Pi_{F,\varphi_W}(\mathbf{SO}(W))})$$

and

$$\sum_{\alpha\in H^1(\mathbb{R},H)} \sum_{\substack{\pi_V\in\Pi_{F,\varphi_V}(\mathbf{SO}(V_\alpha)) \\ \pi_W\in\Pi_{F,\varphi_W}(\mathbf{SO}(W_\alpha))}} \chi_{\pi_{V_\alpha}}(s_V)\chi_{\pi_{W_\alpha}}(s_W) m(\pi_{V_\alpha}, \pi_{W_\alpha}) = m_{V_-,W_+}^S m_{V_+,W_-}^S. \tag{3.3.4}$$

The left-hand side is equal to

$$\chi_{\pi_V^0}(s_V)\chi_{\pi_W^0}(s_W), \tag{3.3.5}$$

where $\chi_{\pi_V^0}, \chi_{\pi_W^0}$ are the characters corresponding to the unique pair (π_V^0, π_W^0) in $\Pi_{F,\varphi_V\times\varphi_W}^{\mathrm{Vogan}}$ with multiplicity equal to one. Assuming the local Gan-Gross-Prasad conjecture for small cases, (3.3.4) implies that (3.3.5) is equal to $\chi^{W}_{\varphi_V}(s_V)\chi^{V}_{\varphi_W}(s_W)$ when s satisfies (3.3.3). When φ_V is of endoscopic type, $s_V \in \mathcal{S}_{\varphi_V}$ not satisfying (3.3.3) is a nontrivial subgroup of $\mathcal{S}_{\varphi_V}$, so

$$\chi_{\pi^0_V} \times \chi_{\pi^0_W} = \chi^{W}_{\varphi_V} \times \chi^{V}_{\varphi_W}, \text{ for all } s_V \times s_W \in \mathcal{S}_{\varphi_V} \times \mathcal{S}_{\varphi_W}.$$

This allows us to reduce the conjecture to smaller cases. Parallel arguments can also be established for unitary Bessel cases to obtain a uniform proof of epsilon-dichotomy for tempered Bessel cases over Archimedean local fields.

4. Basic Tempered Fourier-Jacobi Cases

Gan and Ichino proved the basic tempered Fourier-Jacobi cases for unitary groups over non-Archimedean local fields from the basic tempered Bessel cases using the theory of theta correspondence. Atobe proved the corresponding result for symplectic-metaplectic groups, which involved more details. This section takes the symplectic-metaplectic cases as an example and shows how the basic tempered Bessel and Fourier-Jacobi cases are related to each other.

4.1. *Theta correspondence*

This section reviews some facts about the theta correspondence. We refer to [13] for the definitions of *reductive dual pairs* and *theta lifts.* The reductive dual pairs used in the proof are

(1) $(\mathrm{U}(V), \mathrm{U}(W))$, for hermitian spaces V, W;
(2) $(\mathrm{Sp}(V), \mathrm{O}(W))$, for a symplectic space V and an even-dimensional quadratic space W;
(3) $(\mathrm{Mp}(V), \mathrm{O}(W))$, for a symplectic space V and an odd-dimensional quadratic space W.

For reductive dual pairs (G, H) in (1)–(3), the theta lift $\Theta_{\psi_F}(\pi)$ of $\pi \in \Pi_F(G)$ is a unique representation such that

$$\omega_{V\otimes W,\psi_F} / \bigcap_{f\in \mathrm{Hom}_{\widetilde{G}}(\omega_{V\otimes W,\psi_F}|_{\widetilde{G}},\pi)} \ker(f) = \pi \otimes \Theta_{\psi_F}(\pi).$$

Here $\widetilde{G}$ is an appropriate cover of G. Moreover, $\Theta_{\psi_F}(\pi)$ has a unique irreducible quotient $\theta_{\psi_F}(\pi)$.

The diagram

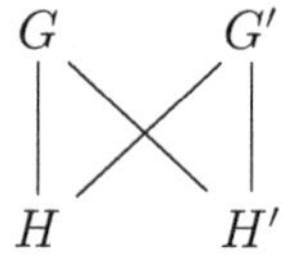

is called a seesaw diagram if and only if $H \subset G$, $H' \subset G'$ and (G, H'), (G', H) are reductive dual pairs. Under consistency conditions of the reductive dual pairs, for every $\pi \in \Pi_F(H)$ and $\sigma \in \Pi_F(H')$, there is a seesaw identity

$$\mathrm{Hom}(\Theta_{\psi_F}(\sigma), \pi) = \mathrm{Hom}(\Theta_{\psi_F}(\pi), \sigma). \tag{4.1.1}$$

We write the diagram for the seesaw identity as

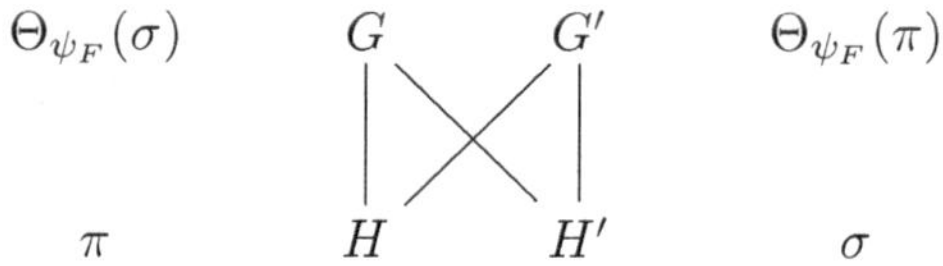

4.2. *Bridges between basic tempered Bessel and Fourier-Jacobi cases*

This section introduces the bridge between basic tempered Bessel and Fourier-Jacobi cases built by the seesaw identities in [12] and [1]. To avoid confusion in notations, this section only reviews the bridge between special orthogonal Bessel cases and symplectic-metaplectic Fourier-Jacobi cases over non-Archimedean cases. A bridge in unitary cases can be established similarly.

Let F be a non-Archimedean local field. For a quadratic space W over F satisfying $\dim W = 2n+1$ and $\mathrm{disc}(W) = 1$. The pair $(\mathrm{SO}(W), \mathrm{Sp}(V_{2n}))$ is a reductive dual pair, where V_{2n} is the $2n$-dimensional symplectic space over F. Let $\sigma \in \Pi_F(\mathbf{SO}(W))$, then $\mathrm{Ind}_{\mathrm{SO}(W)}^{\mathrm{O}(W)}(\sigma)$ decomposes into two representations $\sigma', \sigma' \circ \mathrm{sgn} \in \Pi_F(\mathrm{O}(W))$.

$$\theta_{V_{2n}} : \coprod_{\substack{\dim V = \dim W + 1 \\ \mathrm{disc}(W)=1}} \Pi_F(\mathbf{SO}(W)) \to \Pi_F^{\mathrm{genu}}(\mathbf{Mp}(2n)),$$

is defined as $\theta_{V_{2n}}(\sigma) = \theta_{\psi_F}(\sigma')$.

Consider the seesaw diagram

$$
\begin{array}{cccc}
\Theta_{\psi_F}(\sigma') \boxtimes \omega_{V_{2n},\psi_F^{-1}} & \mathrm{Mp}(2n) \times_{\pm 1} \mathrm{Mp}(2n) & \mathrm{O}(W') & \Theta_{\psi_F}(\pi) \\
 & | \qquad \qquad \times \qquad \qquad | & & \\
\pi & \mathrm{Sp}(2n) & \mathrm{O}(W) \times \mathrm{O}(L) & \sigma' \boxtimes 1
\end{array}
\tag{4.2.1}
$$

where $\dim W = 2n+1$, $\mathrm{disc}(W) = 1$, L is an anisotropic line $W' = W \perp L$. Then, the seesaw identity reads

$$\mathrm{Hom}_{\mathrm{O}(W)}(\Theta_{\psi_F}(\pi), \sigma') = \mathrm{Hom}_{\mathrm{Sp}(2n)}(\Theta_{\psi_F}(\sigma') \widehat{\otimes} \omega_{V_{2n},\psi_F}, \pi). \tag{4.2.2}$$

It is known over non-Archimedean local fields that $\Theta_{\psi_F}(\sigma') = \theta_{\psi_F}(\sigma')$ for tempered σ', then

$$\mathrm{Hom}_{\mathrm{Sp}(2n)}(\Theta_{\psi_F}(\sigma') \widehat{\otimes} \omega_{V_{2n},\psi_F}, \pi) = \mathrm{Hom}_{\mathrm{Sp}(2n)}(\theta_{\psi_F}(\sigma') \widehat{\otimes} \omega_{V_{2n},\psi_F}, \pi).$$

Hence, the dimension of the right-hand side of (4.2.2) is related to the multiplicity

$$m(\theta_{\psi_F}(\sigma'), \pi^\vee)$$

of the basic tempered Fourier-Jacobi case for $(\mathrm{Sp}(V_{2n}), \mathrm{Mp}(V_{2n}))$.

Then, we relate the dimension of the left-hand side of (4.2.2) to the multiplicity of a basic tempered Bessel case. $\theta_{\psi_F}(\pi)$ is the unique quotient of $\Theta_{\psi_F}(\pi)$,

$$\mathrm{Hom}_{\mathrm{O}(W)}(\Theta_{\psi_F}(\pi), \sigma') = \mathrm{Hom}_{\mathrm{O}(W)}(\theta_{\psi_F}(\pi), \sigma').$$

Since $\theta_{\psi_F}(\sigma' \circ \mathrm{sgn}) = 0$, from the seesaw identity for a similar diagram as (4.2.1),

$$\mathrm{Hom}_{\mathrm{O}(W)}(\theta_{\psi_F}(\pi), \sigma' \circ \mathrm{sgn}) = 0.$$

Using the Frobenius reciprocity, we obtain that

$$
\begin{aligned}
&\mathrm{Hom}_{\mathrm{O}(W)}(\theta_{\psi_F}(\pi), \sigma') \\
&= \mathrm{Hom}_{\mathrm{O}(W)}(\theta_{\psi_F}(\pi), \mathrm{Ind}_{\mathrm{SO}(W)}^{\mathrm{O}(W)}(\sigma'|_{\mathrm{SO}(W)})) \\
&= \mathrm{Hom}_{\mathrm{SO}(W)}(\theta_{\psi_F}(\pi)|_{\mathrm{SO}(W)}, \sigma'|_{\mathrm{SO}(W)}).
\end{aligned}
$$

The dimension of this space is equal to the multiplicity

$$m(\theta_{\psi_F}(\pi)|_{\mathrm{SO}(W')}, (\sigma'|_{\mathrm{SO}(W)})^\vee)$$

of the basic tempered Bessel case $(\mathrm{SO}(W'), \mathrm{SO}(W))$ if

$$\theta_{\psi_F}(\pi)|_{\mathrm{SO}(W')}$$

is an irreducible tempered representation. From Prasad's conjecture, the semisimplification of the L-parameter of $\theta_{\psi_F}(\pi)$ has a summand of the trivial representation. This implies the irreducibility of $\theta_{\psi_F}(\pi)|_{\mathrm{SO}(W')}$.

Over Archimedean local fields, the main difficulty for generalizing this approach is that $\Theta_{\psi_F}(\sigma') = \theta_{\psi_F}(\sigma')$ was only established in stable-range situations. An ongoing work of R. Chen, J.-L. Zou and the author aims to refine this method using multiplicity inequalities so that we only use this result in stable-range cases.

5. Generic Cases

The local Gan-Gross-Prasad conjecture can be proved from the basic tempered cases using a structure theorem for representation in generic L-packets and a multiplicity formula for induced representations.

5.1. *Structure theorem*

The following theorem provides a classification of representations in generic L-packets.

Theorem 5.1: *For* $\mathbf{G}_V = \mathbf{U}(V), \mathbf{SO}(V), \mathbf{Sp}(V), \mathbf{Mp}(V)$, *every representation* π_V *in the* L*-packet of a generic parameter can be expressed as*

$$\pi_V = \sigma_V \ltimes \pi_{V_0},$$

where $\pi_{V_0} \in \Pi_{F,\mathrm{temp}}(\mathbf{G}_{V_0})$ *and*

$$\sigma_V = |\det|^{s_{V,1}} \sigma_{V,1} \times \cdots \times |\det|^{s_{V,l_V}} \sigma_{V,l_V} \tag{5.1.1}$$

for $\sigma_{V,i} \in \Pi_{F,\mathrm{temp}}(\mathrm{Res}_{E/F}\mathbf{GL}_{n_i})$ *and* $s_{V,1} \geq \cdots \geq s_{V,l_V} > 0$. *Here* $\Pi_{F,\mathrm{temp}}(\mathbf{G})$ *denote the set of equivalent classes of tempered representations of* $\mathbf{G}(F)$.

When $\mathbf{G}_V$ is not metaplectic, the structure theorem follows from [23, Theorem 6.2] over Archimedean local fields and the standard model conjecture [7] and [21] using the arguments in [20] over non-Archimedean local fields. When $\mathbf{G}_V$ is metaplectic, the structure theorem follows from the corresponding special orthogonal case and properties of theta correspondence.

Definition 5.2: For $\pi_V \in \Pi_F(\mathrm{G}_V)$ in a generic packet, we define the *leading index* $\mathrm{LI}(\pi_V)$ as s_1.

5.2. *Multiplicity formula*

In [20], Mœglin and Waldspurger's proved a multiplicity formula in the special orthogonal Bessel cases over non-Archimedean local fields, showing that for given relevant pair (V, W) and isotropic spaces X_V, X_W, we have

$$m(\sigma_V \ltimes \pi_V, \sigma_W \ltimes \pi_W) = m(\pi_V, \pi_W), \tag{5.2.1}$$

for $\sigma_V \in \Pi_F(\mathbf{GL}(X_V))$, $\sigma_W \in \Pi_F(\mathbf{GL}(X_W))$, $\sigma_V \in \Pi_F(\mathbf{G}_V)$ and $\sigma_W \in \Pi_F(\mathbf{G}_W)$ in generic packets. Here $\ltimes$ denotes the parabolic induction (for non-metaplectic cases, see (3.3.2)).

Variations of the multiplicity formula were proved in [12], [33] and [34] for different cases. In [9], the author gives a uniform approach toward the multiplicity formula over Archimedean local fields.

It is worth mentioning that the local Gan-Gross-Prasad conjecture implies (5.2.1) when $\sigma_V \ltimes \pi_V$ and $\sigma_W \ltimes \pi_W$ are irreducible. Proofs in [20], [12], [9] and [10] suggest that the formula also holds when $\sigma_V \ltimes \pi_V$ and $\sigma_W \ltimes \pi_W$ are reducible. Results in [9] suggested that this formula has a harmonic analysis nature and can be proved without using the local Gan-Gross-Prasad conjecture for tempered L-parameters.

Many proofs of (5.2.1) follow from Mœglin-Waldspurger's approach in [20] for special orthogonal Bessel cases over non-Archimedean local fields. Mœglin-Waldspurger's approach used the following three ingredients and a sophisticated mathematical induction to prove (5.2.1) from the conjecture for tempered parameters. Given σ_V, σ_W, π_V, π_W as in (5.2.1). Let $W^+ = W \perp (X \oplus X^\vee)$.

(1) (Reduction to basic cases) If $\dim W^+ = \dim V + 1$, then

$$m(\pi_V, \pi_W) = m(\sigma_W \ltimes \pi_W, \pi_V) \tag{5.2.2}$$

when $\pi_V^\vee$ does not belong to the Bernstein component associated to $\sigma_W \otimes \pi_{W'}$, where $\pi_{W'}$ is any irreducible representation of $\mathrm{SO}(W')$. Here W' is the quadratic space satisfying $V = W' \perp (X \oplus X^\vee)$.

(2) (Basic forms of the first inequality) If $\dim W^+ = \dim V + 1$, then

$$m(\pi_V, \pi_W) \geqslant m(|\det|^s \sigma_W \ltimes \pi_W, \pi_V),$$

when σ_W is tempered and $s \geqslant \mathrm{LI}(\pi_V)$.

(3) (Basic forms of the second inequality)

$$m(\pi_V, \pi_W) \leqslant m(\sigma_V \ltimes \pi_V, \sigma_W \ltimes \pi_W).$$

Over non-Archimedean local fields, this approach for the multiplicity formula was also successful for unitary Bessel cases [12] and for Fourier-Jacobi cases in [10].

In [9], the author modified these three ingredients to hold over Archimedean local fields, and the proof is independent of the results in the proof of the local Gan-Gross-Prasad conjecture for tempered parameters.

Theorem 5.3: *The multiplicity formula* (5.2.1) *holds in Bessel and Fourier-Jacobi cases over any local fields of characteristic zero.*

5.3. *Reduction to basic tempered cases*

Using the structure theorem and the multiplicity formula, one can prove the local Gan-Gross-Prasad conjecture from the basic tempered cases. The proof for the multiplicity formula over non-Archimedean local fields still depends on results in the proof for tempered cases of the conjecture.

Using the structure theorem, for every π_V, π_W in generic L-packets, we have

$$\pi_V = \sigma_V \ltimes \pi_{V_0}, \quad \pi_W = \sigma_W \ltimes \pi_{W_0}$$

for $\pi_{V_0} \in \Pi_{F,\mathrm{temp}}(\mathbf{G}_{V_0})$, $\pi_{W_0} \in \Pi_{F,\mathrm{temp}}(\mathbf{G}_W)$ and σ_V, σ_W given as in (5.1.1). From (5.2.1), we have

$$m(\pi_V, \pi_W) = m(\pi_{V_0}, \pi_{W_0}).$$

This reduces the generic cases to the tempered cases. To further reduce to basic tempered cases, we apply (5.2.2) and find an appropriate σ'_W such that

$$m(\pi_{V_0}, \pi_{W_0}) = m(\pi_{V_0}, \sigma'_W \ltimes \pi_{W_0}).$$

This reduces the tempered cases to the basic tempered cases.

References

1. H. Atobe, "The local theta correspondence and the local Gan–Gross–Prasad conjecture for the symplectic-metaplectic case", *Math. Annalen*, 371(1):225–295, 2018.
2. R. Beuzart-Plessis, "Expression d'un facteur epsilon de paire par une formule intégrale", *Canad. J. Math.* 66(5):993–1049, 2014.
3. R. Beuzart-Plessis, "La conjecture locale de Gross-Prasad pour les représentations tempérées des groupes unitaires", *Mém. Soc. Math. Fr. (N.S.)*, no. 149, 2016.

4. R. Beuzart-Plessis, *A local trace formula for the Gan-Gross-Prasad conjecture for unitary groups: the Archimedean case*, Astérisque, 2019.
5. D. Ben-Zvi, Y. Sakellaridis, and A. Venkatesh, "Relative Langlands Duality", arXiv preprint arXiv:2409.04677v1, 2024.
6. W. Casselman, "Canonical extensions of Harish-Chandra modules to representations of $\mathfrak{g}$", *Canad. J. Math.*, 41(3):385–438, 1989.
7. W. Casselman and F. Shahidi, "On irreducibility of standard modules for generic representations", *Ann. Sci. École Norm. Sup.*, 31:561–589, 1998.
8. C. Chen, "The local Gross-Prasad conjecture over Archimedean local fields", arXiv preprint arXiv:2102.11404, 2021.
9. C. Chen, "Multiplicity formula for induced representations: Bessel and Fourier-Jacobi models over Archimedean local fields", arXiv preprint arXiv:2308.02912, 2023.
10. C. Chen, D.-H. Jiang, D.-W. Liu, and Z. Lei, "Arithmetic Branching Law and Generic L-packets", *AMS Represent. Theory*, 28:328–365, 2024.
11. C. Chen and Z.-L. Luo, "The local Gross-Prasad conjecture over R: Epsilon dichotomy", arXiv preprint arXiv:2204.01212.
12. W. T. Gan and A. Ichino, "The Gross–Prasad conjecture and local theta correspondence", *Invent. Math.*, 206(3):705–799, 2016.
13. W. T. Gan, S. Kudla, and S. Takeda, *The Local Theta Correspondence*, preliminary version, available at `https://sites.google.com/view/grothendieck-jr/theta-book`.
14. W. T. Gan, B. Gross, and D. Prasad, "Symplectic local root numbers, central critical L-values, and restriction problems in the representation theory of classical groups, Sur les conjectures de Gross et Prasad. I", *Astérisque*, 346:1–109, 2012.
15. B. Gross and D. Prasad, "On the decomposition of a representation of SO_n when restricted to SO_{n-1}", *Canad. J. Math.*, 44(5):974–1002, 1992.
16. B. Gross and D. Prasad, "On irreducible representations of $SO_{2n+1} \times SO_{2m}$", *Canad. J. Math.*, 46(5):930–950, 1994.
17. H.-Y. He, "On the Gan-Gross-Prasad conjecture for $U(p,q)$", *Invent. Math.*, 209:837–884, 2017.
18. H. Jacquet, I. I. Piatetskii-Shapiro and J. A. Shalika, "Rankin-Selberg Convolutions", *Amer. J. Math.* 105(2), 367–464, 1983.
19. Z.-L. Luo, *A local trace formula for the local Gross-Prasad conjecture for special orthogonal groups*, Thesis, The University of Minnesota, 2021.
20. C. Mœglin and J.-L. Waldspurger, "La conjecture locale de Gross-Prasad pour les groupes spéciaux orthogonaux: le cas général", *Astérisque*, 347:167–216, 2012.
21. G. Muic, "A proof of Casselman-Shahidi's conjecture for quasi-split classical groups", *Can. Math. Bull.*, 43:90–99, 2000.
22. Y. Sakellaridis and A. Venkatesh, *Periods and harmonic analysis on spherical varieties*, Astérisque 396, 2017.
23. D. Vogan, "Gelfand-Kirillov dimension for Harish-Chandra modules", *Invent. Math.*, 48(1):75–98, 1978.

24. D. Vogan, "The local Langlands conjecture", *Contemporary Math.*, 145:305–305, 1993.
25. J.-L. Waldspurger, "Une formule intégrale reliée à la conjecture locale de Gross–Prasad", *Compositio Math.*, 146(5):1180–1290, 2010.
26. J.-L. Waldspurger, "Calcul d'une valeur d'un facteur ε par une formule intégrale", *Astérisque*, 347:1–102, 2012.
27. J.-L. Waldspurger, "La conjecture locale de Gross-Prasad pour les représentations tempérées des groupes spéciaux orthogonaux", *Astérisque,* 347:103–165, 2012.
28. J.-L. Waldspurger, "Une formule intégrale reliée à la conjecture locale de Gross-Prasad, 2e partie: extension aux représentations tempérées", *Astérisque,* 346:171–312, 2012.
29. J.-L. Waldspurger, "Une variante d'un résultat de Aizenbud, Gourevitch, Rallis et Schiffmann", *Astérisque*, 346:313–318, 2012.
30. N. Wallach, "Real reductive groups II". *Bull. Amer. Math. Soc.* 30:157–158, 1994.
31. W. Chen and Z. Lei, "Multiplicities for Strongly Tempered Spherical Varieties", arXiv preprint arXiv:2204.07977, 2022.
32. H. Xue, Bessel models for real unitary groups: the tempered case, *Duke Math. J.* 172(5): 995–1031, 2023.
33. H. Xue, Bessel models for unitary groups and Schwartz homology, preprint.
34. H. Xue, Fourier-Jacobi models for real unitary groups, preprint.

https://doi.org/10.1142/9789819825448_0003

Symmetry Breaking for PGL(2) over Non-Archimedean Local Fields

Corina Ciobotaru

Department of Mathematics
Aarhus University
Ny Munkegade 118
8000 Aarhus C, Denmark
cociobotaru@math.au.dk

Jan Frahm

Department of Mathematics
Aarhus University
Ny Munkegade 118
8000 Aarhus C, Denmark
frahm@math.au.dk

For a quadratic extension $\mathbb{E}/\mathbb{F}$ of non-Archimedean local fields we construct explicit holomorphic families of intertwining operators between principal series representations of $\mathrm{PGL}(2,\mathbb{E})$ and $\mathrm{PGL}(2,\mathbb{F})$, also referred to as *symmetry breaking operators*. These families are given in terms of their distribution kernels which can be viewed as distributions on $\mathbb{E}$ depending holomorphically on the principal series parameters. For all such parameters we determine the support of these distributions, and we study their mapping properties. This leads to a classification of all intertwining operators between principal series representations, not necessarily irreducible. As an application, we show that every Steinberg representation of $\mathrm{PGL}(2,\mathbb{E})$ contains a Steinberg representation of $\mathrm{PGL}(2,\mathbb{F})$ as a direct summand of Hilbert spaces.

Contents

Introduction

For a quadratic extension $\mathbb{E}/\mathbb{F}$ of local fields, the pair $(G, G') = (\mathrm{PGL}(2,\mathbb{E}), \mathrm{PGL}(2,\mathbb{F}))$ of reductive groups has the strong Gelfand property, i.e.

$$\dim \mathrm{Hom}_{G'}(\pi|_{G'}, \tau) \leq 1$$

for all irreducible smooth admissible representations π of G and τ of G', additionally assumed to be of moderate growth in the Archimedean case. For non-Archimedean fields, this result is due to D. Prasad [16] where he also determines for which representations the multiplicities are equal to 1 in terms of epsilon factors.

More recently, for the Archimedean case, T. Kobayashi [11] advocated to not only study the abstract multiplicities $\dim \mathrm{Hom}_{G'}(\pi|_{G'}, \tau)$, but to explicitly construct and classify intertwining operators in $\mathrm{Hom}_{G'}(\pi|_{G'}, \tau)$, which he also refers to as *symmetry breaking operators*. Together with B. Speh he studied the pair of real reductive groups $(G, G') = (\mathrm{O}(1, n+1), \mathrm{O}(1, n))$ in great detail, see [12, 13] and also [3, 4, 8] for similar results for other pairs of groups. The key players in their construction are meromorphic families of symmetry breaking operators between principal series representations.

They construct and study these families in terms of their distribution kernels which they identify with distributions on $\mathbb{R}^n$, determine the optimal holomorphic renormalization of these families, their supports and the mapping properties of the corresponding symmetry breaking operators using mostly techniques from analysis. Such results have applications to, for instance, branching laws for unitary representations [21], conformal geometry [10], partial differential equations [6, 14] and automorphic forms [1, 7].

The aim of this chapter is to carry out a similar analysis for the pair of reductive groups $(G, G') = (\mathrm{PGL}(2, \mathbb{E}), \mathrm{PGL}(2, \mathbb{F}))$ where $\mathbb{E}/\mathbb{F}$ is a quadratic extension of non-Archimedean local fields. For this, we first establish a similar correspondence between intertwining operators between principal series of G and G' and distributions/generalized functions on the field $\mathbb{E}$ which have certain invariance properties. Next, we construct explicit holomorphic families of such invariant distributions for all principal series representations of G and G'. We further determine for all principal series parameters the support of these distributions and also study mapping properties of the corresponding symmetry breaking operators. This leads to a full classification of symmetry breaking operators between all principal series of G and G', not necessarily irreducible. Generically, the space of symmetry breaking operators is one-dimensional, as predicted by the results of D. Prasad [16]. However, for some finite set of parameters where at least one of the principal series representations is reducible, the dimension is two, and we find an explicit basis in these cases.

We remark that we do not assume anything about the (residue) characteristic of $\mathbb{E}$ and $\mathbb{F}$. In fact, most of our proofs are quite elementary and computational and also work in (residue) characteristic 2. For some general statements about composition series of principal series and local gamma factors we refer to the book of Bushnell–Henniart [2] which has no restrictions about the (residue) characteristic, see page 3, Section *Notation* in [2].

Let us describe our results in more detail.

Symmetry breaking operators between principal series

Let $\mathbb{E}$ be a non-Archimedean local field, $\mathcal{O}_\mathbb{E}$ its ring of integers and $\mathcal{O}_\mathbb{E}^\times$ the group of units. Write $q_\mathbb{E}$ for the cardinality of the residue field of $\mathbb{E}$, $\nu_\mathbb{E} : \mathbb{E} \to \mathbb{Z} \cup \{\infty\}$ for the valuation of $\mathbb{E}$, and choose a uniformizer $\varpi_\mathbb{E}$ for $\mathcal{O}_\mathbb{E}$. For a multiplicative character $\chi \in \widehat{\mathcal{O}_\mathbb{E}^\times}$ of $\mathcal{O}_\mathbb{E}^\times$ and $s \in \mathbb{C}/\frac{2\pi i}{\ln q_\mathbb{E}}\mathbb{Z}$ we denote by χ_s the multiplicative character of $\mathbb{E}^\times$ given by

$$\chi_s(x) = \chi(\varpi_\mathbb{E}^{-\nu_\mathbb{E}(x)} x)|x|_\mathbb{E}^s \qquad (x \in \mathbb{E}^\times).$$

The character χ_s gives rise to a character of the Borel subgroup B of $G = \mathrm{PGL}(2,\mathbb{E})$ consisting of upper triangular matrices by

$$\chi_s \begin{pmatrix} a & b \\ 0 & 1 \end{pmatrix} = \chi_s(a) \qquad (a \in \mathbb{E}^\times, b \in \mathbb{E}).$$

Denote by $\pi_{\chi,s}$ the corresponding principal series representations induced from χ_s (smooth normalized parabolic induction):

$$\pi_{\chi,s} = \mathrm{Ind}_B^G(\chi_s).$$

For a quadratic extension $\mathbb{E}/\mathbb{F}$ of non-Archimedean local fields we consider the pair of reductive groups $(G, G') = (\mathrm{PGL}(2,\mathbb{E}), \mathrm{PGL}(2,\mathbb{F}))$. We write $\tau_{\eta,t}$ for the principal series representation of G' with parameters $\eta \in \widehat{\mathcal{O}_\mathbb{F}^\times}$ and $t \in \mathbb{C}/\frac{2\pi i}{\ln q_\mathbb{F}}\mathbb{Z}$. The main object of study in this chapter is the space

$$\mathrm{Hom}_{G'}(\pi_{\chi,s}|_{G'}, \tau_{\eta,t})$$

of intertwining operators between principal series of G and G', also referred to as *symmetry breaking operators* in the Archimedean setting.

To construct and classify such operators explicitly, we give an equivalent description in terms of their distribution kernels (see Propositions 1.3 and 1.4). More precisely, a symmetry breaking operator $A \in \mathrm{Hom}_{G'}(\pi_{\chi,s}|_{G'}, \tau_{\eta,t})$ corresponds to a distribution $K \in \mathcal{D}'(\mathbb{E})$ on $\mathbb{E}$ such that (in the distribution sense)

$$Af(\overline{n}_y) = \int_\mathbb{E} K(x) f(\overline{n}_{x+y})\, dx \qquad (f \in \mathrm{Ind}_B^G(\chi_s), y \in \mathbb{F}),$$

where $\mathbb{E} \to \overline{N}$, $x \mapsto \overline{n}_x$ parameterizes the unipotent radical $\overline{N}$ of a Borel subgroup of G opposite to B. This establishes a bijection

$$\mathrm{Hom}_{G'}(\pi_{\chi,s}|_{G'}, \tau_{\eta,t}) \to \mathcal{D}'(\mathbb{E})_{s,t}^{\chi,\eta} \subseteq \mathcal{D}'(\mathbb{E}), \quad A \mapsto K$$

onto a subspace $\mathcal{D}'(\mathbb{E})_{s,t}^{\chi,\eta}$ of $\mathcal{D}'(\mathbb{E})$ consisting of distributions satisfying certain equivariance properties (see Proposition 2.3). We call these distributions *symmetry breaking kernels.*

Construction and classification of symmetry breaking kernels

The main ingredient in the description of $\mathcal{D}'(\mathbb{E})_{s,t}^{\chi,\eta}$ is a family of distributions $K_{s,t}^{\chi,\eta}$ that depends meromorphically on $(s,t) \in \mathbb{C}$ and is contained in

$\mathcal{D}'(\mathbb{E})^{\chi,\eta}_{s,t}$ whenever it is defined. More precisely, we put

$$K^{\chi,\eta}_{s,t}(x) := \chi_{s-\frac{1}{2}}(x^2)\cdot(\chi_{s-\frac{1}{2}}\eta_{t+\frac{1}{2}})\left(\frac{x_2}{x_1^2-x_2^2\alpha^2}\right) \quad (x = x_1+x_2\alpha \in \mathbb{E}-\mathbb{F}).$$

This function is locally integrable on $\mathbb{E}$ whenever $\Re(2s \pm t + \frac{1}{2}) > 0$ (see Lemma 3.3) and therefore defines a family of distributions on $\mathbb{E}$. The main result of Section 3 is its meromorphic extension to all $(s,t) \in \mathbb{C}^2$. To state this result, let

$$\backslash\!\backslash = \{(\chi,s,\eta,t) \in \widehat{\mathcal{O}_{\mathbb{E}}^\times} \times \mathbb{C} \times \widehat{\mathcal{O}_{\mathbb{F}}^\times} \times \mathbb{C} : \chi|_{\mathcal{O}_{\mathbb{F}}^\times}\cdot\eta = 1, 2s+t+\tfrac{1}{2} \in \tfrac{2\pi i}{\ln q_{\mathbb{F}}}\mathbb{Z}\},$$
$$/\!/ = \{(\chi,s,\eta,t) \in \widehat{\mathcal{O}_{\mathbb{E}}^\times} \times \mathbb{C} \times \widehat{\mathcal{O}_{\mathbb{F}}^\times} \times \mathbb{C} : \chi|_{\mathcal{O}_{\mathbb{F}}^\times}\cdot\eta^{-1} = 1, 2s-t+\tfrac{1}{2} \in \tfrac{2\pi i}{\ln q_{\mathbb{F}}}\mathbb{Z}\}.$$

(This notation follows [12], the idea being that in the (s,t)-plane the sets consist of parallel lines with direction indicated by the symbols $\backslash\!\backslash$ and $/\!/$.) Note that we can alternatively write

$$\backslash\!\backslash = \{(\chi,s,\eta,t) \in \widehat{\mathcal{O}_{\mathbb{E}}^\times} \times \mathbb{C} \times \widehat{\mathcal{O}_{\mathbb{F}}^\times} \times \mathbb{C} : \chi_s|_{\mathcal{O}_{\mathbb{F}}^\times}\cdot\eta_t = |\cdot|_{\mathbb{F}}^{-\frac{1}{2}}\},$$
$$/\!/ = \{(\chi,s,\eta,t) \in \widehat{\mathcal{O}_{\mathbb{E}}^\times} \times \mathbb{C} \times \widehat{\mathcal{O}_{\mathbb{F}}^\times} \times \mathbb{C} : \chi_s|_{\mathcal{O}_{\mathbb{F}}^\times}\cdot\eta_t^{-1} = |\cdot|_{\mathbb{F}}^{-\frac{1}{2}}\}.$$

We call a family of distributions meromorphic/holomorphic if its pairing with every test function is meromorphic/holomorphic.

Theorem A: (see Theorem 3.10) *The family of distributions $K^{\chi,\eta}_{s,t} \in \mathcal{D}'(\mathbb{E})$ has a meromorphic extension to $(s,t) \in \mathbb{C}^2$ for all χ and η with simple poles at $(\chi,s,\eta,t) \in \backslash\!\backslash \cup /\!/$. For $(\chi,s,\eta,t) \notin \backslash\!\backslash \cup /\!/$ we have $K^{\chi,\eta}_{s,t} \in \mathcal{D}'(\mathbb{E})^{\chi,\eta}_{s,t}$.*

Since $\backslash\!\backslash$ resp. $/\!/$ is precisely the zero set of the local non-Archimedean L-factor $L(\frac{1}{2},\chi_s|_{\mathbb{F}^\times}\cdot\eta_t)$ resp. $L(\frac{1}{2},\chi_s|_{\mathbb{F}^\times}\cdot\eta_t^{-1})$ (see (3.12) for the definition), the renormalized family

$$\widetilde{K}^{\chi,\eta}_{s,t} := L(\tfrac{1}{2},\chi_s|_{\mathbb{F}^\times}\cdot\eta_t)^{-1}L(\tfrac{1}{2},\chi_s|_{\mathbb{F}^\times}\cdot\eta_t^{-1})^{-1}\cdot K^{\chi,\eta}_{s,t}$$

is holomorphic in $(s,t) \in \mathbb{C}^2$ and satisfies $\widetilde{K}^{\chi,\eta}_{s,t} \in \mathcal{D}'(\mathbb{E})^{\chi,\eta}_{s,t}$ for all parameters (χ,s,η,t). Its values at $(\chi,s,\eta,t) \in \backslash\!\backslash \cup /\!/$ are obtained explicitly (see the residue identities in Theorem 3.11). This allows us to determine the set of parameters (χ,s,η,t) for which $\widetilde{K}^{\chi,\eta}_{s,t} = 0$ and also the support of the distribution $\widetilde{K}^{\chi,\eta}_{s,t} \in \mathcal{D}'(\mathbb{E})$ for all parameters.

Theorem B: (see Theorem 3.12) $\widetilde{K}^{\chi,\eta}_{s,t} = 0$ *if and only if* $(\chi, s, \eta, t) \in L$ *with the discrete set* $L \subseteq //$ *given by* (3.13). *Moreover, we have*

$$\operatorname{supp} \widetilde{K}^{\chi,\eta}_{s,t} = \begin{cases} \emptyset & \textit{for } (\chi, s, \eta, t) \in L, \\ \{0\} & \textit{for } (\chi, s, \eta, t) \in // - L, \\ \mathbb{F} & \textit{for } (\chi, s, \eta, t) \in \backslash\backslash - //, \\ \mathbb{E} & \textit{else.} \end{cases}$$

Theorems A and B are proven by explicitly evaluating the distribution $K^{\chi,\eta}_{s,t}$ on a test function using elementary computations of p-adic integrals (see in particular Section 3.2).

Using the family $\widetilde{K}^{\chi,\eta}_{s,t}$, we determine in Section 4 the space $\mathcal{D}'(\mathbb{E})^{\chi,\eta}_{s,t}$ of symmetry breaking kernels, and hence the space $\operatorname{Hom}_{G'}(\pi_{\chi,s}|_{G'}, \tau_{\eta,t})$ of symmetry breaking operators, for all parameters. For $(\chi, s, \eta, t) \notin L$ the kernel $K^{\chi,\eta}_{s,t}$ is a non-zero element of $\mathcal{D}'(\mathbb{E})^{\chi,\eta}_{s,t}$. For $(\chi, s, \eta, t) \in L$ we can regularize the family $(s', t') \mapsto \widetilde{K}^{\chi,\eta}_{s',t'}$ along two different hyperplanes in $\mathbb{C}^2$ to obtain two linearly independent kernels. Regularizing along $//$, i.e. along $\{(s', t') : 2s' - t' + \frac{1}{2} \in \frac{2\pi i}{\ln q_{\mathbb{F}}}\mathbb{Z}\} \subseteq \mathbb{C}^2$, yields the Dirac delta distribution $\delta \in \mathcal{D}'(\mathbb{E})$ at 0. Regularizing along the hyperplane $\mathbb{C} \times \{t\} \subseteq \mathbb{C}^2$ yields a symmetry breaking kernel

$$\overset{\approx}{K}{}^{\chi,\eta}_{s,t} := (1 - q_{\mathbb{F}}^{-2z})^{-1} \widetilde{K}^{\chi,\eta}_{s+z,t}\Big|_{z=0},$$

whose support equals $\mathbb{E}$. In particular, δ and $\overset{\approx}{K}{}^{\chi,\eta}_{s,t}$ are linearly independent.

Theorem C: (see Corollary 4.4)

$$\operatorname{Hom}_{G'}(\pi_{\chi,s}|_{G'}, \tau_{\eta,t}) \simeq \mathcal{D}'(\mathbb{E})^{\chi,\eta}_{s,t} = \begin{cases} \mathbb{C}\widetilde{K}^{\chi,\eta}_{s,t} & \textit{for } (\chi, s, \eta, t) \notin L, \\ \mathbb{C}\delta \oplus \mathbb{C}\overset{\approx}{K}{}^{\chi,\eta}_{s,t} & \textit{for } (\chi, s, \eta, t) \in L. \end{cases}$$

Note that this is not a contradiction to the multiplicity-one property of the pair (G, G') since for all $(\chi, s, \eta, t) \in L$ at least one of the representations $\pi_{\chi,s}$ and $\tau_{\eta,t}$ is reducible.

Mapping properties of symmetry breaking operators

For $\chi^2 = 1$, the representation $\pi_{\chi,s}$ contains a unique one-dimensional subspace invariant under the maximal compact subgroup $K = \mathrm{PGL}(2, \mathcal{O}_{\mathbb{E}})$. Let $\varphi_{\chi,s}$ be the unique element of this subspace with $\varphi_{\chi,s}(1) = 1$.

If additionally $s \in -\frac{1}{2} + \frac{\pi i}{\ln q_{\mathbb{E}}}\mathbb{Z}$, then the subspace spanned by $\varphi_{\chi,s}$ is also G-invariant. The quotient of $\pi_{\chi,s}$ by this one-dimensional subrepresentation is irreducible and isomorphic to the unique irreducible subrepresentation $\pi^0_{\chi,-s}$ of $\pi_{\chi,-s}$. It is also referred to as *Steinberg representation.* All other representations $\pi_{\chi,s}$ are irreducible (see Theorem 2.1 for details).

We use the notation $\psi_{\eta,t}$ for the corresponding vector in $\tau_{\eta,t}$ if $\eta^2 = 1$ and $\tau^0_{\eta,t}$ for the Steinberg representation which occurs as subrepresentation of $\tau_{\eta,t}$ for $\eta^2 = 1$ and $t \in \frac{1}{2} + \frac{\pi i}{\ln q_{\mathbb{F}}}\mathbb{Z}$.

In Section 5 we compute the action of the symmetry breaking operator $\widetilde{A}^{\chi,\eta}_{s,t}$ with kernel $\widetilde{K}^{\chi,\eta}_{s,t}$ on the vectors $\varphi_{\chi,s}$ and also find the image of this operator in some cases.

Theorem D: (see Corollary 5.4 and Theorem 5.6)

(1) Assume $\chi^2 = 1$*, then* $\widetilde{A}^{\chi,\eta}_{s,t}\varphi_{\chi,s} = 0$ *unless* $\eta = \chi|_{\mathcal{O}_{\mathbb{F}}^\times}$ *in which case we have*

$$\widetilde{A}^{\chi,\eta}_{s,t}\varphi_{\chi,s} = (1 - q_{\mathbb{F}}^{-1})\psi_{\eta,t} \times \begin{cases} (1 - q_{\mathbb{E}}^{-(2s+1)}) & \text{for } \mathbb{E}/\mathbb{F} \text{ unramified,} \\ (1 - q_{\mathbb{E}}^{-(2s+1)})(1 + q_{\mathbb{F}}^{t-\frac{1}{2}}) & \text{for } \mathbb{E}/\mathbb{F} \text{ ramified.} \end{cases}$$

(2) For all parameters (χ, s, η, t) *with* $\chi|_{\mathcal{O}_{\mathbb{F}}^\times} \cdot \eta = \chi|_{\mathcal{O}_{\mathbb{F}}^\times} \cdot \eta^{-1} = 1$*, the image of the operator* $\widetilde{A}^{\chi,\eta}_{s,t} : \pi_{\chi,s}|_{G'} \to \tau_{\eta,t}$ *is given in Theorem 5.6.*

Similar statements for the symmetry breaking operators with distribution kernel δ and $\overset{\approx}{K}{}^{\chi,\eta}_{s,t}$ in the case $(\chi, s, \eta, t) \in L$ can be found in Remark 5.5 as well as Propositions 5.9 and 5.10.

Application: Discrete components in the restriction of unitary representations

Finally, in Section 6 we study the restriction of some unitary representations of G to G'. In particular, we are interested in unitary representations of G' that occur as direct summands in the restriction.

For $\chi^2 = 1$ and $s \in (0, \frac{1}{2})$, the representations $\pi_{\chi,s}$ are irreducible and unitarizable, and their completions (also denoted by $\pi_{\chi,s}$) make up the *complementary series.* At the endpoint $s = \frac{1}{2}$, the Steinberg representation $\pi^0_{\chi,\frac{1}{2}}$, which is a proper subrepresentation of $\pi_{\chi,\frac{1}{2}}$, is also irreducible and unitarizable. $\pi^0_{\chi,\frac{1}{2}}$ is in fact a discrete series representation for G, i.e. it occurs discretely in the Plancherel decomposition of $L^2(G)$.

Theorem E: (see Theorem 6.1) *Assume that* $\chi^2 = 1$ *and* $\eta = \chi_{\mathcal{O}_{\mathbb{F}}^\times}$.

(1) *For* $s \in (\frac{1}{4}, \frac{1}{2})$ *and* $t = 2s - 1$, *the restriction of the irreducible unitary representation* $\pi_{\chi,s}$ *of* G *to* G' *contains* $\tau_{\eta,t}$ *as a direct summand.*

(2) *For* $s = t = \frac{1}{2}$, *the restriction of the irreducible unitary representation* $\pi^0_{\chi,\frac{1}{2}}$ *of* G *to* G' *contains* $\tau^0_{\eta,\frac{1}{2}}$ *as a direct summand.*

Note that the restriction to G' of a unitary principal series, i.e. $\pi_{\chi,s}$ for $s \in i\mathbb{R}$, is tempered and hence cannot contain complementary series. (The explicit decomposition can be obtained using Mackey theory and the Plancherel formula for $\mathrm{PGL}(2,\mathbb{F})$ and should consist of a direct integral of unitary principal series and a direct sum of discrete series.) It is expected that the restriction $\pi_{\chi,s}|_{G'}$ for $s \in [-\frac{1}{4}, \frac{1}{4}]$ has the same decomposition as for $s \in i\mathbb{R}$ (see [15] for an instance of this in the Archimedean case). The previous statement shows that crossing the point $\pm\frac{1}{4}$ in the complementary series adds a complementary series representation of G' to the decomposition. The endpoint representation $\pi^0_{\chi,\frac{1}{2}}$ is again tempered, and the same is true for the corresponding direct summand $\tau^0_{\eta,\frac{1}{2}}$ in its restriction.

For the Archimedean case $\mathbb{E}/\mathbb{F} = \mathbb{C}/\mathbb{R}$, the corresponding statement is due to Speh–Venkataramana [19]. Our proof follows essentially the same idea.

Relation to previous work and outlook

Most of this chapter is inspired by the monograph [12] by Kobayashi–Speh which treats in detail the same problem for spherical principal series representations of the pair of real reductive groups $(G, G') = (\mathrm{O}(1, n+1), \mathrm{O}(1, n))$. For instance, our Theorems A and B are the analog of Theorem 1.5 in [12], Theorem C corresponds to Theorem 1.9 in [12], and Theorem D is the counterpart of Theorems 1.10 and 1.11 in [12]. Note that the case $n = 2$ essentially corresponds to $(\mathrm{PGL}(2,\mathbb{C}), \mathrm{PGL}(2,\mathbb{R}))$.

In their subsequent book [13] they also study the case of unramified principal series representations, but it seems that there is no complete classification of symmetry breaking operators between not necessarily unramified principal series of $(G, G') = (\mathrm{PGL}(2,\mathbb{C}), \mathrm{PGL}(2,\mathbb{R}))$, which would be the Archimedean counterpart of our results. It is reasonable that such a classification could be achieved using the same methods.

We also remark that there are more sophisticated methods to study multiplicities between irreducible representations of p-adic groups. However, the point of this chapter is to present an elementary and rather explicit approach to the problem of constructing and classifying symmetry breaking operators between principal series representations, using similar methods as those that have been applied in the Archimedean situation in the literature (see e.g. [3, 4, 8, 12, 13]).

A possible application of the results of this chapter is the analytic continuation of branching laws for unitary representations of the pair $(G, G') = (\mathrm{PGL}(2, \mathbb{E}), \mathrm{PGL}(2, \mathbb{F}))$. In the Archimedean case, this was done e.g. in [21]. We hope to return to this in a future work.

Acknowledgements

We thank Paul Nelson for his help with the evaluation of the integral in Appendix B. The first named author was partially supported by The Mathematisches Forschungsinstitut Oberwolfach (MFO, Oberwolfach Research Institute for Mathematics). She would like to thank MFO for the perfect working conditions they provide. As well, she was supported by the European Union's Horizon 2020 research and innovation program under the Marie Sklodowska-Curie grant agreement No. 754513, The Aarhus University Research Foundation, and a research grant from the Villum Foundation (Grant No. VIL53023). The second named author was partially supported by a research grant from the Villum Foundation (Grant No. 00025373). We also thank the referee for making kind suggestions and comments which improved the readability of the chapter.

Notation

For a complex vector space V we denote by $V^\vee = \mathrm{Hom}_\mathbb{C}(V, \mathbb{C})$ its dual space. If π is a representation of a group G on V, then $\pi^\vee$ denotes the contragredient representation of G on $V^\vee$. For two sets X and Y we denote by $X - Y$ their set-theoretic difference.

1. Symmetry Breaking Operators and Distribution Kernels

We generalize the characterization in [12], Chapter 3, of symmetry breaking operators between induced representations in terms of their distribution kernels from the case of real reductive groups to the case of reductive groups over arbitrary local fields.

1.1. *Distributional sections*

For a locally compact group G let $d_G^\ell g$ resp. $d_G^r g$ denote a left resp. right Haar measure on G. In the case where G is unimodular, we usually drop the superscript and write $d_G g$ or simply dg.

Now let G be an algebraic group defined over a local field $\mathbb{F}$ and let H be an algebraic subgroup. For a finite-dimensional continuous complex representation (ρ_V, V) of H let

$$\mathcal{V} := G \times_H V := (G \times V)/_{(g,v)\sim(gh,\rho_V(h^{-1})v)}$$

denote the corresponding complex homogeneous vector bundle over G/H. Note that a section of $\mathcal{V}$ can be identified with a function $f : G \to V$ such that $f(gh) = \rho_V(h)^{-1}f(g)$ for all $g \in G$, $h \in H$. We will write $C(G/H, \mathcal{V})$ for the space of continuous sections of $\mathcal{V}$ and $C_c(G/H, \mathcal{V})$ for the subspace of $C(G/H, \mathcal{V})$ of sections with compact support.

Moreover, we write $C^\infty(G/H, \mathcal{V})$ for the space of smooth sections in the case where $\mathbb{F}$ is Archimedean and locally constant sections in the case where $\mathbb{F}$ is non-Archimedean. Finally, put $C_c^\infty(G/H, \mathcal{V}) := C^\infty(G/H, \mathcal{V}) \cap C_c(G/H, \mathcal{V})$. If $\mathbb{F}$ is Archimedean, we endow $C^\infty(G/H, \mathcal{V})$ with the usual Fréchet topology of compact convergence of all derivatives, and $C_c^\infty(G/H, \mathcal{V})$ with the LF topology as an inductive limit of the subspaces of sections with support contained in a fixed compact set.

The group G acts on both $C(G/H, \mathcal{V})$, $C_c(G/H, \mathcal{V})$, $C^\infty(G/H, \mathcal{V})$ and $C_c^\infty(G/H, \mathcal{V})$ by left translation in the following way: for each $g \in G$ and any $f \in C(G/H, \mathcal{V})$ we denote $(g \cdot f)(xH) := f(g^{-1}xH)$, for any $xH \in G/H$.

We want to extend the above G-action on smooth sections to distributional sections. To do that, let

$$\Delta_{G/H} := \frac{(\Delta_G)|_H}{\Delta_H} : H \to \mathbb{R}_+,$$

where Δ_G and Δ_H are the modular functions of G and H. Consider the line bundle

$$\Omega_{G/H} := G \times_H \mathbb{C}_{\Delta_{G/H}} := (G \times \mathbb{C})/_{(g,z)\sim(gh,\Delta_{G/H}(h^{-1})z)} \text{ over } G/H.$$

The following theorem provides a G-invariant integral on $C_c(G/H, \Omega_{G/H})$.

Theorem 1.1: *Let G be a locally compact group and H a closed subgroup. Then the mapping*

$$C_c(G) \to C_c(G/H, \Omega_{G/H}), \quad f \mapsto \tilde{f},$$

where

$$\tilde{f}(x) = \int_H f(xh)\Delta_{G/H}(h)d_H^\ell h \quad (x \in G),$$

is surjective. Moreover, the linear map

$$C_c(G/H, \Omega_{G/H}) \to \mathbb{C}, \quad \tilde{f} \mapsto \int_{G/H} \tilde{f}(x)\, d_{G/H}(xH) := \int_G f(g)\, d^\ell_G g$$

is well-defined, giving a unique (up to scalar multiples) G-invariant integral on $C_c(G/H, \Omega_{G/H})$. In particular, $d_{G/H}$ satisfies the iterated integral formula

$$\int_G f(g)\, d^\ell_G g = \int_{G/H} \left(\int_H f(xh) \Delta_{G/H}(h)\, d^\ell_H h \right) d_{G/H}(xH) \quad (f \in C_c(G)).$$

The proof of Theorem 1.1 is essentially identical to the proof for the case $\Delta_{G/H} \equiv 1$ which can for instance be found in [5], Theorem 2.49.

In view of the integral $d_{G/H}$, we define the *dualizing bundle of* $\mathcal{V}$ to be the homogeneous vector bundle $\mathcal{V}^* := G \times_H V^*$ associated with the representation $(\rho_{V^*}, V^*) := (\rho_V^\vee \otimes \Delta_{G/H}, V^\vee \otimes \mathbb{C})$, where $(\rho_V^\vee, V^\vee)$ denotes the contragredient representation of (ρ_V, V) on $V^\vee := \mathrm{Hom}(V, \mathbb{C})$. In this way, we obtain a canonical non-degenerate G-invariant pairing

$$C^\infty(G/H, \mathcal{V}) \otimes C_c^\infty(G/H, \mathcal{V}^*) \to \mathbb{C},$$
$$f_1 \otimes f_2 \mapsto \langle f_1, f_2 \rangle := \int_{G/H} \langle f_1(g), f_2(g) \rangle\, d_{G/H}(gH),$$

so we can view $C^\infty(G/H, \mathcal{V})$ as a subspace of the dual space $C_c^\infty(G/H, \mathcal{V}^*)^\vee$.

This motivates to define the distributional sections of $\mathcal{V}$ as the dual object to compactly supported smooth sections of $\mathcal{V}^*$

$$\mathcal{D}'(G/H, \mathcal{V}) := C_c^\infty(G/H, \mathcal{V}^*)^\vee.$$

The G-action on $\mathcal{D}'(G/H, \mathcal{V})$ given by

$$\langle g \cdot u, \varphi \rangle = \langle u, g^{-1} \cdot \varphi \rangle \qquad (g \in G, u \in \mathcal{D}'(G/H, \mathcal{V}), \varphi \in C_c^\infty(G/H, \mathcal{V}^*))$$

extends the G-action on $C^\infty(G/H, \mathcal{V})$. Note that $(\mathcal{V}^*)^* = \mathcal{V}$, so $\mathcal{D}'(G/H, \mathcal{V}^*) = C_c^\infty(G/H, \mathcal{V})^\vee$.

1.2. *Distribution kernels of symmetry breaking operators*

Now let G' be another algebraic subgroup of G and H' an algebraic subgroup of G'. For a finite-dimensional continuous complex representation (ρ_W, W) of H' we write $\mathcal{W} = G' \times_{H'} W$ for the corresponding vector bundle over G'/H'. Denote by $\mathrm{Hom}_{G'}(C_c^\infty(G/H, \mathcal{V}), C^\infty(G'/H', \mathcal{W}))$ the space of G'-equivariant linear maps $C_c^\infty(G/H,\mathcal{V}) \to C^\infty(G'/H',\mathcal{W})$ which are additionally continuous in the case where $\mathbb{F}$ is Archimedean.

Definition 1.2: A element $T \in \mathrm{Hom}_{G'}(C_c^\infty(G/H,\mathcal{V}), C^\infty(G'/H',\mathcal{W}))$ is called a *symmetry breaking operator.*

In order to construct and classify symmetry breaking operators, we relate them to their distribution kernels. These turn out to be elements of the space $\mathcal{D}'(G/H,\mathcal{V}^*) \otimes W$ which are invariant under the action of H'. Here, H' acts on $\mathcal{D}'(G/H,\mathcal{V}^*)$ by restricting the action of G. We identify $K \in \mathcal{D}'(G/H,\mathcal{V}^*) \otimes W$ with a linear map $C_c^\infty(G/H,\mathcal{V}) \to W$, so that for $f \in C_c^\infty(G/H,\mathcal{V})$ we have $\langle K, f\rangle \in W$. Then the diagonal action of $h' \in H'$ on $K \in \mathcal{D}'(G/H,\mathcal{V}^*) \otimes W$ can be written as

$$\langle (h',h')\cdot K, f\rangle = \rho_W(h')\langle K, (h')^{-1}\cdot f\rangle \qquad (f \in C_c^\infty(G/H,\mathcal{V})). \tag{1.1}$$

Denote by $(\mathcal{D}'(G/H,\mathcal{V}^*) \otimes W)^{\Delta(H')}$ the subspace of all H'-invariant elements.

The following statement readily generalizes [12], Proposition 3.2:

Proposition 1.3: *The following map is a linear isomorphism:*

$$\mathrm{Hom}_{G'}(C_c^\infty(G/H,\mathcal{V}), C^\infty(G'/H',\mathcal{W})) \to (\mathcal{D}'(G/H,\mathcal{V}^*) \otimes W)^{\Delta(H')},$$
$$T \mapsto K_T := \mathrm{ev}_1 \circ T,$$

where $\mathrm{ev}_1 : C^\infty(G'/H',\mathcal{W}) \to W$, $f \mapsto f(1)$ *is evaluation at the identity* $1 \in G'$.

Proof: With (1.1) it is easy to see that $K_T \in (\mathcal{D}'(G/H,\mathcal{V}^*) \otimes W)^{\Delta(H')}$ whenever $T \in \mathrm{Hom}_{G'}(C_c^\infty(G/H,\mathcal{V}), C^\infty(G'/H',\mathcal{W}))$, so the map under consideration is defined. Now, for $T \in \mathrm{Hom}_{G'}(C_c^\infty(G/H,\mathcal{V}), C^\infty(G'/H',\mathcal{W}))$ and $f \in C_c^\infty(G/H,\mathcal{V})$ note that for $g' \in G'$:

$$(Tf)(g') = ((g')^{-1}\cdot(Tf))(1) = (T((g')^{-1}\cdot f))(1) = \langle K_T, (g')^{-1}\cdot f\rangle. \tag{1.2}$$

This implies that the map $T \mapsto K_T$ is injective. To show surjectivity, we define for a given $K_T \in (\mathcal{D}'(G/H,\mathcal{V}^*) \otimes W)^{\Delta(H')}$ and $f \in C_c^\infty(G/H,\mathcal{V})$ a function $Tf : G' \to W$ by the right-hand side of (1.2), i.e. $(Tf)(g') := \langle K_T, (g')^{-1}\cdot f\rangle$. Since f is smooth/locally constant with compact support, the map $g' \mapsto (g')^{-1}\cdot f$ is smooth/locally constant and hence Tf is also smooth/locally constant. Moreover, for all $g' \in G'$, $h' \in H'$:

$$\begin{aligned}(Tf)(g'h') &= \langle K_T, (g'h')^{-1} \cdot f\rangle = \langle K_T, (h')^{-1}(g')^{-1} \cdot f\rangle \\ &= \rho_W(h')^{-1}\langle (h',h') \cdot K_T, (g')^{-1} \cdot f\rangle = \rho_W(h')^{-1}\langle K_T, (g')^{-1} \cdot f\rangle \\ &= \rho_W(h')^{-1} Tf(g'),\end{aligned}$$

so $Tf \in C^\infty(G'/H', \mathcal{W})$. It remains to show that $f \mapsto Tf$ is G'-intertwining, and in the Archimedean case additionally continuous. The intertwining property is shown as follows:

$$(T(g' \cdot f))(x) = \langle K_T, x^{-1}g' \cdot f\rangle = \langle K_T, ((g')^{-1}x)^{-1} \cdot f\rangle = ((g')^{-1} \cdot Tf)(x)$$

for all $f \in C_c^\infty(G/H, \mathcal{V})$ and $g', x \in G'$. Finally, continuity of T is a consequence of the continuity of K_T as a linear map $C_c^\infty(G/H, \mathcal{V}) \to W$ if $\mathbb{F}$ is Archimedean. □

It follows from the previous proof that the operator $T \in \operatorname{Hom}_{G'}(C_c^\infty(G/H, \mathcal{V}), C^\infty(G'/H', \mathcal{W}))$ is given in terms of its distribution kernel $K_T \in (\mathcal{D}'(G/H, \mathcal{V}^*) \otimes W)^{\Delta(H')}$ by

$$\begin{aligned}Tf(g') = \langle K_T, (g')^{-1} \cdot f\rangle &= \int_{G/H} K_T(g)\big(f(g'g)\big)\, d(gH) \\ &= \int_{G/H} K_T((g')^{-1}g)\big(f(g)\big)\, d(gH),\end{aligned}$$

where the last two expressions have to be understood in the sense of generalized functions.

1.3. *Symmetry breaking operators between principal series*

Now assume that $G/H = G/P$ is a flag variety of a reductive algebraic group G over the local field $\mathbb{F}$, i.e. $P \subseteq G$ is a parabolic subgroup defined over $\mathbb{F}$. We assume additionally that $G' \subseteq G$ is a reductive subgroup and that $G'/H' = G'/P'$ is a flag variety for G'. In particular, both G/P and G'/P' are compact.

Fix an opposite parabolic subgroup $\overline{P}$ of P and denote by $\overline{N}$ its unipotent radical. Note that if we fix a Levi subgroup of P, then there is a unique opposite parabolic subgroup $\overline{P}$ with the same Levi. Note that $\overline{N} \cap P = \{e\}$ and $\overline{N}P$ is open and dense in G.

While a distribution on G/P is in general not uniquely determined by its restriction to the open dense subset $\overline{N}P/P \subseteq G/P$, we show that this is indeed the case for the kernel K_T given by Proposition 1.3 of a symmetry breaking operator $T \in \operatorname{Hom}_{G'}(C^\infty(G/P,\mathcal{V}), C^\infty(G'/P',\mathcal{W}))$ under the following condition:

$$P'\overline{N}P = G. \tag{1.3}$$

To make this precise, consider for open subsets $\Omega' \subseteq \Omega \subseteq G/P$ the restriction map

$$r^{\Omega}_{\Omega'} : \mathcal{D}'(\Omega,\mathcal{V}^*) \to \mathcal{D}'(\Omega',\mathcal{V}^*)$$

which is dual to the extension map

$$i^{\Omega}_{\Omega'} : C_c^\infty(\Omega',\mathcal{V}) \to C_c^\infty(\Omega,\mathcal{V}), \quad (i^{\Omega}_{\Omega'}\varphi)(x) = \begin{cases} \varphi(x) & \text{for } x \in \Omega', \\ 0 & \text{else.} \end{cases} \tag{1.4}$$

Of particular importance will be the restriction to subsets of the form $\Omega = UP/P$ with $U \subseteq \overline{N}$ open. On $UP/P \simeq U$ the bundle $\mathcal{V}$ is trivial, so we obtain an isomorphism

$$\begin{aligned} &\gamma_U : C_c^\infty(U) \otimes V \to C_c^\infty(UP/P,\mathcal{V}), \\ &(\gamma_U\varphi)(x) := \rho_V(p^{-1})(\varphi(\overline{n})) \qquad (x = \overline{n}p \in UP), \end{aligned} \tag{1.5}$$

where we view $C_c^\infty(U) \otimes V$ as the space of compactly supported V-valued smooth functions on U. The corresponding dual map into the space of V^*-valued distributions on U is the isomorphism given by

$$\gamma_U^* : \mathcal{D}'(UP/P,\mathcal{V}^*) \to \mathcal{D}'(U) \otimes V^*, \quad \langle \gamma_U^* v, \varphi\rangle := \langle v, \gamma_U\varphi\rangle, \tag{1.6}$$

where $v \in \mathcal{D}'(UP/P,\mathcal{V}^*)$ and $\varphi \in C_c^\infty(U) \otimes V$, and

$$\mathcal{D}'(U) = C_c^\infty(U)^\vee$$

denotes the space of distributions on U.

Now, for every $g \in G$, the action of g^{-1} on $C^\infty(G/P,\mathcal{V})$ maps $C_c^\infty(g\overline{N}P/P,\mathcal{V})$ to $C_c^\infty(\overline{N}P/P,\mathcal{V})$, so the dual map $g\cdot : \mathcal{D}'(\overline{N}P/P,\mathcal{V}^*) \to \mathcal{D}'(g\overline{N}P/P,\mathcal{V}^*)$ makes the following diagram commute:

$$\begin{array}{ccc} \mathcal{D}'(G/P,\mathcal{V}^*) & \xrightarrow{g\cdot} & \mathcal{D}'(G/P,\mathcal{V}^*) \\ \Big\downarrow r^{G/P}_{\overline{N}P/P} & & \Big\downarrow r^{G/P}_{g\overline{N}P/P} \\ \mathcal{D}'(\overline{N}P/P,\mathcal{V}^*) & \xrightarrow{g\cdot} & \mathcal{D}'(g\overline{N}P/P,\mathcal{V}^*). \end{array} \tag{1.7}$$

Write $\overline{N}_g = \overline{N} \cap g\overline{N}P$ and denote by $\pi(g)$ the following composition:

$$\mathcal{D}'(\overline{N}) \otimes V^* \xrightarrow{(\gamma^*_{\overline{N}})^{-1}} \mathcal{D}'(\overline{N}P/P, \mathcal{V}^*) \xrightarrow{g\cdot} \mathcal{D}'(g\overline{N}P/P, \mathcal{V}^*)$$
$$\xrightarrow{r^{g\overline{N}P/P}_{\overline{N}_gP/P}} \mathcal{D}'(\overline{N}_gP/P, \mathcal{V}^*) \xrightarrow{\gamma^*_{\overline{N}_g}} \mathcal{D}'(\overline{N}_g) \otimes V^*.$$

Proposition 1.4: *Under the assumption* (1.3), *the map*

$$\gamma^*_{\overline{N}} \circ r^{G/P}_{\overline{N}P/P} : (\mathcal{D}'(G/P, \mathcal{V}^*) \otimes W)^{\Delta(P')} \to \mathcal{D}'(\overline{N}) \otimes V^* \otimes W$$

induces a bijection onto the subspace

$$(\mathcal{D}'(\overline{N}) \otimes V^* \otimes W)^{\Delta(P')} := \{K \in \mathcal{D}'(\overline{N}) \otimes V^* \otimes W \mid$$
$$\big[\pi(p) \otimes \rho_W(p)\big]K = K|_{\overline{N}_p} \textit{ for all } p \in P'\}. \quad (1.8)$$

Proof: Let us prove that given $A \in (\mathcal{D}'(G/P, \mathcal{V}^*) \otimes W)^{\Delta(P')}$ the distribution

$$K = \gamma^*_{\overline{N}} \circ r^{G/P}_{\overline{N}P/P}(A)$$

is indeed contained in $(\mathcal{D}'(\overline{N}) \otimes V^* \otimes W)^{\Delta(P')}$. In fact, by the definition of $\pi(p)$ we get for every $p \in P'$:

$$\pi(p)K = \gamma^*_{\overline{N}_p} \circ r^{p'\overline{N}P/P}_{\overline{N}_pP/P}((p,1) \cdot r^{G/P}_{\overline{N}P/P}(A)).$$

By the commutativity of the diagram (1.7) this can be written as

$$\gamma^*_{\overline{N}_p} \circ r^{p'\overline{N}P/P}_{\overline{N}_pP/P} \circ r^{G/P}_{p\overline{N}P/P}((p,1) \cdot A),$$

which, by the functoriality of restriction, equals

$$\gamma^*_{\overline{N}_p} \circ r^{G/P}_{\overline{N}_pP/P}((p,1) \cdot A).$$

Now, $A \in (\mathcal{D}'(G/P, \mathcal{V}^*) \otimes W)^{\Delta(P')}$ implies $(p,p) \cdot A = A$, thus $(p,1) \cdot A = \rho_W(p)^{-1}A$, so the above expression equals

$$\rho_W(p)^{-1}\big[\gamma^*_{\overline{N}_p} \circ r^{G/P}_{\overline{N}_pP/P}(A)\big].$$

Finally, γ^*_U is functorial in U, so this can be written as

$$\rho_W(p)^{-1}\big[\gamma^*_{\overline{N}} \circ r^{G/P}_{\overline{N}P/P}(A)\big]|_{\overline{N}_p} = \rho_W(p)^{-1}K|_{\overline{N}_p}$$

and hence $K \in (\mathcal{D}'(\overline{N}) \otimes V^* \otimes W)^{\Delta(P')}$.

To see that the above map is surjective we use the sheaf property of distributions. For a fixed $K \in (\mathcal{D}'(\overline{N}) \otimes V^* \otimes W)^{\Delta(P')}$ and every $p \in P'$ define

$$K_p := \rho_W(p)\big[(p,1)\cdot(\gamma^*_{\overline{N}})^{-1}(K)\big] \in \mathcal{D}'(p\overline{N}P/P, \mathcal{V}^*) \otimes W.$$

We claim that K_p and $K_{p'}$ agree on $p\overline{N}P/P \cap p'\overline{N}P/P$ for all $p, p' \in P'$. In fact, acting on both K_p and $K_{p'}$ by the inverse of p', we may and do assume that $p' = e$. To show that $K_p = K_e$ on $p\overline{N}P/P \cap \overline{N}P/P = \overline{N}_pP/P$ we note that by the definition of $\pi(p)$ we have

$$K_p|_{\overline{N}_pP/P} = \rho_W(p)\big[(p,1)\cdot(\gamma^*_{\overline{N}})^{-1}(K)\big]|_{\overline{N}_pP/P} = (\gamma^*_{\overline{N}_p})^{-1}\big[\pi(p)\otimes\rho_W(p)\big](K).$$

By the assumption on K we have $\big[\pi(p) \otimes \rho_W(p)\big](K) = K|_{\overline{N}_p}$, so the expression above equals

$$(\gamma^*_{\overline{N}_p})^{-1}(K|_{\overline{N}_p}) = K_e|_{\overline{N}_pP/P}$$

by the functoriality of γ^*_U in U. This shows that $\{K_p\}_{p\in P'}$ is a family of distributions that pairwise agree on intersections. By the assumption (1.3), the domains $p\overline{N}P/P$ ($p \in P$) cover G/P, so by the sheaf property of distributions there exists a (unique) $A \in \mathcal{D}'(G/P, \mathcal{V}^*)\otimes W$ such that $A|_{p\overline{N}P/P} = K_p$. That A is in fact contained in $(\mathcal{D}'(G/P, \mathcal{V}^*)\otimes W)^{\Delta(P')}$ follows by restricting it to the cover $p\overline{N}P/P$, $p \in P'$, and using the equivariance property of K.

Finally, we show that the above map is injective. Assume $A \in (\mathcal{D}'(G/P, \mathcal{V}^*) \otimes W)^{\Delta(P')}$ such that $\gamma^*_{\overline{N}} \circ r^{G/P}_{\overline{N}P/P}(A) = 0$. Since $\gamma^*_{\overline{N}}$ is an isomorphism, this is equivalent to $r^{G/P}_{\overline{N}P/P}(A) = 0$. But $(p,p)\cdot A = A$ for every $p \in P'$, so the diagram (1.7) implies

$$r^{G/P}_{p\overline{N}P/P}(A) = r^{G/P}_{p\overline{N}P/P}((p,p)\cdot A) = (p,p)\cdot r^{G/P}_{\overline{N}P/P}(A) = 0.$$

Since the open sets $p\overline{N}P/P$ ($p \in P'$) cover G/P, this shows $A = 0$. □

We interpret $\mathcal{D}'(\overline{N})$ as a space of generalized functions on $\overline{N}$ by identifying a locally integrable function $f : \overline{N} \to \mathbb{C}$ with a distribution $f \in \mathcal{D}'(\overline{N})$ by

$$\langle f, \varphi\rangle = \int_{\overline{N}} f(\overline{n})\varphi(\overline{n})\, d\overline{n} \qquad (\varphi \in C^\infty_c(\overline{N})),$$

where $d\overline{n}$ denotes a fixed Haar measure on $\overline{N}$. It is natural to use the Haar measure on $\overline{N}$, because after suitable normalization of $d\overline{n}$ we have

$$\int_{G/P} h(xP)\, d_{G/P}(xP) = \int_{\overline{N}} h(\overline{n})\, d\overline{n} \qquad (h \in C(G/P, \Omega_{G/P})). \tag{1.9}$$

This follows from Theorem 1.1 and the integral formula

$$\int_G h(g)\,dg = \int_{\overline{N}} \int_P h(\overline{n}p)\,d_P^r p\,d_{\overline{N}}^\ell \overline{n} = \int_{\overline{N}} \int_P h(\overline{n}p)\Delta_{G/P}(p)\,d_P^\ell p\,d_{\overline{N}}^\ell \overline{n}.$$

Using the language of generalized functions on $\overline{N}$, the condition in (1.8) can be rewritten as follows:

Lemma 1.5: *Fix $g \in G$ and for $\overline{n} \in \overline{N}_g$ write $g^{-1}\overline{n} = \overline{n}(g^{-1}\overline{n})p(g^{-1}\overline{n}) \in \overline{N}P$. Then the map $\pi(g) : \mathcal{D}'(\overline{N}) \otimes V^* \to \mathcal{D}'(\overline{N}_g) \otimes V^*$ is given by*

$$(\pi(g)u)(\overline{n}) = \rho_{V^*}(p(g^{-1}\overline{n}))^{-1}u(\overline{n}(g^{-1}\overline{n})) \qquad (\overline{n} \in \overline{N}_g) \tag{1.10}$$

in the sense of generalized functions.

Proof: Unwrapping the definition of $\pi(g)$, we have for any $\phi \in C_c^\infty(\overline{N}_g) \otimes V$:

$$\langle \pi(g)u, \phi \rangle = \left\langle u, \gamma_{\overline{N}}^{-1} \circ g^{-1} \circ \iota_{\overline{N}_g P/P}^{g\overline{N}P/P} \circ \gamma_{\overline{N}_g} \phi \right\rangle.$$

If now u is given by a locally integrable V-valued function on $\overline{N}$, this equals

$$\int_{\overline{N}} u(\overline{n}) \left(\gamma_{\overline{N}}^{-1} \circ g^{-1} \circ \iota_{\overline{N}_g P/P}^{g\overline{N}P/P} \circ \gamma_{\overline{N}_g} \phi \right) (\overline{n})\,d\overline{n}.$$

By (1.9) the integral over $\overline{N}$ can also be written as an integral over the open subset $\overline{N}P/P$ of G/P:

$$\int_{\overline{N}P/P} (\gamma_{\overline{N}} u)(xP) \left(g^{-1} \circ \iota_{\overline{N}_g P/P}^{g\overline{N}P/P} \circ \gamma_{\overline{N}_g} \phi \right) (xP)\,d_{G/P}(xP),$$

where we abuse notation and write $\gamma_{\overline{N}}$ also for its extension to $L^1_{\mathrm{loc}}(\overline{N}) \otimes V$, the locally integrable V-valued functions on $\overline{N}$, by the same formula. Since the integral over G/P is G-invariant, this equals

$$\int_{g\overline{N}P/P} (\gamma_{\overline{N}} u)(g^{-1}xP) \left(\iota_{\overline{N}_g P/P}^{g\overline{N}P/P} \circ \gamma_{\overline{N}_g} \phi \right) (xP)\,d_{G/P}(xP).$$

Note that it suffices to integrate over $\overline{N}_g P/P$ since this set contains the support of $\gamma_{\overline{N}_g}\phi$. Finally, rewriting the integral over $\overline{N}_g P/P$ as an integral over $\overline{N}_g$ using (1.9) we obtain

$$\int_{\overline{N}_g} (\gamma_{\overline{N}} u)(g^{-1}\overline{n})\phi(\overline{n})\,d\overline{n}.$$

Since

$$(\gamma_{\overline{N}} u)(g^{-1}\overline{n}) = \rho_{V^*}(p(g^{-1}\overline{n}))^{-1}u(\overline{n}(g^{-1}\overline{n}))$$

by definition, the proof is complete. □

2. Symmetry Breaking Operators for Principal Series of PGL(2)

We apply the theory developed in the previous section to characterize symmetry breaking operators between principal series of PGL(2) in terms of their distribution kernels.

2.1. *Principal series of* PGL(2)

Let $\mathbb{E}$ be a non-Archimedean local field with absolute value $|\cdot|_{\mathbb{E}}$. We denote by $\mathcal{O}_{\mathbb{E}} = \{u \in \mathbb{E} : |u|_{\mathbb{E}} \leq 1\}$ the ring of integers, by $\mathcal{O}_{\mathbb{E}}^{\times} = \{x \in \mathbb{E} : |x|_{\mathbb{E}} = 1\}$ the group of units, and we choose a uniformizer $\varpi_{\mathbb{E}}$ for $\mathcal{O}_{\mathbb{E}}$. Write $q_{\mathbb{E}}$ for the cardinality of the residue field $\mathcal{O}_{\mathbb{E}}/\varpi_{\mathbb{E}}\mathcal{O}_{\mathbb{E}}$. In what follows, we normalize Haar measure dx on $\mathbb{E}$ so that $\mathcal{O}_{\mathbb{E}}$ has measure 1. Further, we normalize the absolute value $|\cdot|_{\mathbb{E}}$ so that $d(ax) = |a|_{\mathbb{E}}\, dx$ for all $a \in \mathbb{E}^{\times}$.

We consider the group $G = \mathrm{PGL}(2, \mathbb{E})$ and abuse notation by writing an element of G as a 2×2 matrix with entries in $\mathbb{E}$ which represents the corresponding equivalence class in $\mathrm{PGL}(2, \mathbb{E})$. Let B be the standard Borel subgroup of G consisting of upper triangular matrices.

For a fixed character χ of $\mathcal{O}_{\mathbb{E}}^{\times}$ (i.e. $\chi \in \widehat{\mathcal{O}_{\mathbb{E}}^{\times}}$) and $s \in \mathbb{C}$ we denote by χ_s the character of $\mathbb{E}^{\times}$ given by

$$\chi_s(x) := \chi(\varpi_{\mathbb{E}}^{-\nu_{\mathbb{E}}(x)} x)|x|_{\mathbb{E}}^{s} \qquad (x \in \mathbb{E}^{\times}).$$

Clearly, every character of $\mathbb{E}^{\times}$ is of this form since $\mathbb{E}^{\times} = \mathcal{O}_{\mathbb{E}}^{\times} \cdot \{\varpi_{\mathbb{E}}^{k} : k \in \mathbb{Z}\}$. The character χ_s defines a character of B, also denoted by χ_s, by

$$\chi_s \begin{pmatrix} a & b \\ 0 & 1 \end{pmatrix} = \chi_s(a).$$

Note that χ_s only depends on $s + \frac{2\pi i}{\ln q_{\mathbb{E}}}\mathbb{Z}$.

Let $(\pi_{\chi,s}, \mathrm{Ind}_B^G(\chi_s))$ denote the smooth representation of G induced from the character χ_s of B, i.e. $\pi_{\chi,s}$ acts by left translation on the space

$$\mathrm{Ind}_B^G(\chi_s) := \{f \in C^{\infty}(G) : f(gb) = \chi_{s+\frac{1}{2}}(b)^{-1} f(g),\ \forall g \in G, b \in B\}.$$

The representations $\pi_{\chi,s}$ are called the *principal series of* G. If $\chi = 1$, they are called *unramified*, otherwise they are called *ramified.* Note that $\pi_{\chi,s}$ can also be realized on the space $C^{\infty}(G/B, \mathcal{V})$, where $\mathcal{V}$ is the homogeneous vector bundle induced from the character $\chi_{s+\frac{1}{2}} = \chi_s \otimes \Delta_{G/B}^{1/2}$ of B. Since for $a \in \mathbb{E}^{\times}$ and $b \in \mathbb{E}$

$$\Delta_B \begin{pmatrix} a & b \\ 0 & 1 \end{pmatrix} = |a|_{\mathbb{E}}^{-1} \qquad \text{and} \qquad \Delta_{G/B} \begin{pmatrix} a & b \\ 0 & 1 \end{pmatrix} = \left(\frac{\Delta_G|_B}{\Delta_B}\right) \begin{pmatrix} a & b \\ 0 & 1 \end{pmatrix} = |a|_{\mathbb{E}},$$

the dualizing bundle $\mathcal{V}^*$ is induced from

$$(\chi_{s+\frac{1}{2}})^\vee \otimes \Delta_{G/B} = \overline{\chi}_{-s-\frac{1}{2}} \cdot \Delta_{G/B} = \overline{\chi}_{-s+\frac{1}{2}},$$

where $\chi^\vee = \overline{\chi}$ is the complex conjugate character of χ and $\overline{\chi}_{-s+\frac{1}{2}} = \overline{\chi}_{-s} \otimes \Delta_{G/B}^{1/2}$.

From [2], §9.6 (to be applied with $\chi_2 = \chi_1^{-1}$) we have the following result:

Theorem 2.1: *The representation $\pi_{\chi,s}$ is irreducible unless $\chi^2 = 1$ and $s \in \pm\frac{1}{2} + \frac{i\pi}{\ln q_{\mathbb{E}}}\mathbb{Z}$. In this case, it has a unique irreducible subrepresentation and the quotient of $\pi_{\chi,s}$ by this subrepresentation is irreducible. More precisely:*

(1) If $\chi^2 = 1$ and $s \in -\frac{1}{2} + \frac{i\pi}{\ln q_{\mathbb{E}}}\mathbb{Z}$, the representation $\pi_{\chi,s}$ has a unique one-dimensional subrepresentation given by the character $\chi_{s+\frac{1}{2}} \circ \det : G \to \mathbb{C}^\times$.

(2) If $\chi^2 = 1$ and $s \in \frac{1}{2} + \frac{i\pi}{\ln q_{\mathbb{E}}}\mathbb{Z}$, the representation $\pi_{\chi,s}$ has a unique one-dimensional quotient given by the character $\chi_{s-\frac{1}{2}} \circ \det : G \to \mathbb{C}^\times$.

The proper infinite-dimensional quotients/subrepresentations of $\pi_{\chi,s}$ are called *Steinberg representations.* They can most easily be described using the standard intertwining operator. Recall that for sufficiently large $\Re(s)$ the standard intertwining operator $T_{\chi,s} : \pi_{\chi,s} \to \pi_{\overline{\chi},-s}$ is defined by the absolutely convergent integral

$$T_{\chi,s}f(g) = \int_{\overline{N}} f(gw\overline{n})\, d\overline{n} \qquad (f \in \operatorname{Ind}_B^G(\chi_s)),$$

where

$$w = \begin{pmatrix} 0 & 1 \\ 1 & 0 \end{pmatrix} \in G \tag{2.1}$$

denotes a representative of the longest Weyl group element. This family of operators can be extended meromorphically to all $s \in \mathbb{C}$. Since for $x, y \in \mathbb{F}$, $y \neq 0$, we can decompose

$$\overline{n}_x w \overline{n}_y = \overline{n}_{x+1/y} \begin{pmatrix} y & 1 \\ 0 & y^{-1} \end{pmatrix},$$

it follows that

$$T_{\chi,s}f(\overline{n}_x) = \int_{\mathbb{E}} \chi_{s+\frac{1}{2}}(y^2)^{-1} f(\overline{n}_{x+\frac{1}{y}})\, dy = \int_{\mathbb{E}} \chi_{s-\frac{1}{2}}(y^2) f(\overline{n}_{x-y})\, dy, \tag{2.2}$$

i.e. on $\overline{N} \simeq \mathbb{E}$ the operator $T_{\chi,s}$ is given by convolution with the function $\chi_{s-\frac{1}{2}}(y^2)$. In particular, for $\chi^2 = 1$ and $s \in \frac{1}{2} + \frac{i\pi}{\ln q_{\mathbb{E}}}\mathbb{Z}$ we obtain an intertwining operator

$$T_{\chi,s} : \pi_{\chi,s} \to \pi_{\chi,-s}, \quad T_{\chi,s} f(\overline{n}_x) = \int_{\mathbb{E}} f(\overline{n}_y)\, dy.$$

The image of this operator are the functions which are constant on $\overline{N}$, i.e. the one-dimensional irreducible subrepresentation of $\pi_{\chi,-s}$, and its kernel is the Steinberg representation.

2.2. *The branching problem*

Let $\mathbb{E}/\mathbb{F}$ be a quadratic field extension, i.e. $\mathbb{E} = \mathbb{F}(\alpha)$ with $\alpha \in \mathbb{E} - \mathbb{F}$, $\alpha^2 \in \mathbb{F}^\times - \mathbb{F}^{\times 2}$. After possibly multiplying α by a power of $\varpi_{\mathbb{F}}$, we may assume that either $\alpha^2 \in \mathcal{O}_{\mathbb{F}}^\times$ (in the case where $\mathbb{E}/\mathbb{F}$ is unramified, i.e. $q_{\mathbb{E}} = q_{\mathbb{F}}^2$) or $\alpha^2 \in \varpi_{\mathbb{F}}\mathcal{O}_{\mathbb{F}}^\times$ (in the case where $\mathbb{E}/\mathbb{F}$ is ramified, i.e. $q_{\mathbb{E}} = q_{\mathbb{F}}$). Then the absolute values $|\cdot|_{\mathbb{E}}$ and $|\cdot|_{\mathbb{F}}$ are related by

$$|x_1 + x_2\alpha|_{\mathbb{E}} = |x_1^2 - x_2^2\alpha^2|_{\mathbb{F}} \qquad (x_1, x_2 \in \mathbb{F}).$$

We consider the pair of groups $(G, G') = (\mathrm{PGL}(2,\mathbb{E}), \mathrm{PGL}(2,\mathbb{F}))$ and study the restriction of representations of G to G'. With the same notation as for $\mathbb{E}$, we consider for a character η of $\mathcal{O}_{\mathbb{F}}^\times$ and $t \in \mathbb{C}$ the principal series $(\tau_{\eta,t}, \mathrm{Ind}_{B'}^{G'}(\eta_t))$. An intertwining operator $A \in \mathrm{Hom}_{G'}(\pi_{\chi,s}|_{G'}, \tau_{\eta,t})$ is called *symmetry breaking operator.*

2.3. *Symmetry breaking kernels for* PGL(2)

We now use the results from Section 1 to identify symmetry breaking operators for $(G, G') = (\mathrm{PGL}(2,\mathbb{E}), \mathrm{PGL}(2,\mathbb{F}))$ with their distribution kernels. Putting $(\rho_V, V) := (\chi_{s+\frac{1}{2}}, \mathbb{C})$ and $(\rho_W, W) := (\eta_{t+\frac{1}{2}}, \mathbb{C})$, we have

$$\begin{aligned}\mathrm{Hom}_{G'}(\pi_{\chi,s}|_{G'}, \tau_{\eta,t}) &:= \mathrm{Hom}_{G'}(C^\infty(G/B, \mathcal{V}), C^\infty(G'/B', \mathcal{W}))\\ &= \mathrm{Hom}_{G'}(C_c^\infty(G/B, \mathcal{V}), C^\infty(G'/B', \mathcal{W})),\end{aligned}$$

so Proposition 1.3 applies. Moreover, the assumption (1.3) is satisfied in this situation:

Lemma 2.2: *For $(G, G') = (\mathrm{PGL}(2,\mathbb{E}), \mathrm{PGL}(2,\mathbb{F}))$ with $B \subseteq G$ the Borel subgroup of upper triangular matrices, $\overline{N}$ the subgroup of lower triangular unipotent matrices and $B' = B \cap G'$, we have $B'\overline{N}B = G$.*

Proof: Recall that one can realize G/B as the projective line $P^1(\mathbb{E})$. The action of G on $P^1(\mathbb{E})$ is given by

$$\begin{pmatrix} a & b \\ c & d \end{pmatrix} \cdot \left[\begin{smallmatrix} x \\ y \end{smallmatrix}\right] = \left[\begin{smallmatrix} ax+by \\ cx+dy \end{smallmatrix}\right], \qquad \left[\begin{smallmatrix} x \\ y \end{smallmatrix}\right] \in P^1(\mathbb{E}), \begin{pmatrix} a & b \\ c & d \end{pmatrix} \in G,$$

and B is the stabilizer of the point $\left[\begin{smallmatrix} 1 \\ 0 \end{smallmatrix}\right] \in P^1(\mathbb{E})$. Hence, the set $\overline{N}B/B$ consists of all $\left[\begin{smallmatrix} 1 \\ x \end{smallmatrix}\right] \in P^1(\mathbb{E})$, $x \in \mathbb{E}$. Consequently, the complement of $\overline{N}B/B$ in $P^1(\mathbb{E})$ only consists of the single point $\left[\begin{smallmatrix} 0 \\ 1 \end{smallmatrix}\right]$, and B' clearly contains matrices that do not stabilize this point. □

It follows that Proposition 1.4 also applies, so we obtain

$$\mathrm{Hom}_{G'}(\pi_{\chi,s}|_{G'}, \tau_{\eta,t}) \simeq (\mathcal{D}'(\overline{N}) \otimes V^* \otimes W)^{\Delta(B')}.$$

Since both V and W are one-dimensional, the right-hand side is a space of scalar-valued distributions on $\overline{N}$. To make the condition in (1.8) more precise, we use Lemma 1.5. Note that $\rho_{V^*} = \overline{\chi}_{-s+\frac{1}{2}}$, so the condition for $K \in \mathcal{D}'(\overline{N})$ to be contained in $(\mathcal{D}'(\overline{N}) \otimes V^* \otimes W)^{\Delta(B')}$ becomes

$$K(\overline{n}(b^{-1}\overline{n})) = \overline{\chi}_{-s+\frac{1}{2}}(p(b^{-1}\overline{n}))\eta_{t+\frac{1}{2}}(b)^{-1}K(\overline{n}) \qquad \text{for all } b \in B', \overline{n} \in \overline{N}_b. \tag{2.3}$$

Using the isomorphism

$$\mathbb{E} \to \overline{N}, \quad x \mapsto \overline{n}_x := \begin{pmatrix} 1 & 0 \\ x & 1 \end{pmatrix},$$

we can identify $(\mathcal{D}'(\overline{N}) \otimes V^* \otimes W)^{\Delta(B')}$ with a space of distributions on $\mathbb{E}$ which we denote $\mathcal{D}'(\mathbb{E})^{\chi,\eta}_{s,t}$. Then we can compute both $\overline{n}(b^{-1}\overline{n}_x)$ and $p(b^{-1}\overline{n}_x)$ explicitly for $b \in B'$ and $x \in \mathbb{E}$ such that $b^{-1}\overline{n}_x \in \overline{N}B$ and in this way make (2.3) more explicit. We first consider $b = \mathrm{diag}(a, 1)$ with $a \in \mathbb{F}^\times$, then

$$b^{-1}\overline{n}_x = \begin{pmatrix} a & 0 \\ 0 & 1 \end{pmatrix}^{-1} \overline{n}_x = \overline{n}_{ax} \begin{pmatrix} a^{-1} & 0 \\ 0 & 1 \end{pmatrix} \qquad \text{for all } x \in \mathbb{E},$$

so $\overline{N}_b = \overline{N}$ and

$$\overline{n}(b^{-1}\overline{n}_x) = \overline{n}_{ax} \qquad \text{and} \qquad p(b^{-1}\overline{n}_x) = \begin{pmatrix} a^{-1} & 0 \\ 0 & 1 \end{pmatrix},$$

and (2.3) becomes

$$K(ax) = \overline{\chi}_{-s+\frac{1}{2}}(a^{-1})\eta_{t+\frac{1}{2}}(a)^{-1}K(x) \qquad (x \in \mathbb{E}, a \in \mathbb{F}^\times).$$

Next, we consider $b = \begin{pmatrix} 1 & y \\ 0 & 1 \end{pmatrix} \in N'$ with $y \in \mathbb{F}$ and decompose

$$b^{-1}\overline{n}_x = \begin{pmatrix} 1 & y \\ 0 & 1 \end{pmatrix}^{-1} \overline{n}_x = \overline{n}_{(1-xy)^{-1}x} \begin{pmatrix} 1-xy & -y \\ 0 & (1-xy)^{-1} \end{pmatrix} \qquad \text{for } x \neq y^{-1},$$

so $\overline{N}_b = \overline{N} - \{\overline{n}_{y^{-1}}\}$ and

$$\overline{n}(b^{-1}\overline{n}_x) = \overline{n}_{(1-xy)^{-1}x} \qquad \text{and} \qquad p(b^{-1}\overline{n}_x) = \begin{pmatrix} 1-xy & -y \\ 0 & (1-xy)^{-1} \end{pmatrix},$$

and (2.3) becomes

$$K(\tfrac{x}{1-xy}) = \overline{\chi}_{-s+\frac{1}{2}}((1-xy)^2)K(x) \qquad (y \in \mathbb{F}, x \in \mathbb{E} - \{y^{-1}\}).$$

Since $\overline{\chi}_{-s+\frac{1}{2}} = (\chi_{s-\frac{1}{2}})^{-1}$, we obtain the following characterization of distribution kernels of symmetry breaking operators between $\pi_{\chi,s}$ and $\tau_{\eta,t}$:

Proposition 2.3: *$\mathcal{D}'(\mathbb{E})^{\chi,\eta}_{s,t}$ consists of the distributions $u \in \mathcal{D}'(\mathbb{E})$ satisfying the following two conditions (in the sense of generalized functions):*

(a) For $a \in \mathbb{F}^\times$: $u(ax) = \chi_{s-\frac{1}{2}}(a)\eta_{t+\frac{1}{2}}(a)^{-1}u(x)$ for all $x \in \mathbb{E}$.
(b) For $b \in \mathbb{F}^\times$: $u(\frac{x}{1-bx}) = \chi_{s-\frac{1}{2}}((1-bx)^2)^{-1}u(x)$ for all $x \in \mathbb{E} - \{b^{-1}\}$.

Remark 2.4: The two conditions in Proposition 2.3 are meant in the sense of generalized functions. We rephrase them in the language of distributions by applying both sides to a test function $\varphi \in C_c^\infty(\mathbb{E})$.

(a) Since $d(ax) = |a|_{\mathbb{E}}\,dx$, for all $u \in L^1_{\text{loc}}(\mathbb{E})$ and $\varphi \in C_c^\infty(\mathbb{E})$ we have

$$\begin{aligned}\langle u(a\cdot), \varphi\rangle &= \int_{\mathbb{E}} u(ax)\varphi(x)\,dx = |a|_{\mathbb{E}}^{-1}\int_{\mathbb{E}} u(x)\varphi(a^{-1}x)\,dx \\ &= |a|_{\mathbb{E}}^{-1}\langle u, \varphi(a^{-1}\cdot)\rangle.\end{aligned}$$

Therefore, condition (a) can be rewritten as

$$\langle u, \varphi(a^{-1}\cdot)\rangle = |a|_{\mathbb{E}}\chi_{s-\frac{1}{2}}(a)\eta_{t+\frac{1}{2}}(a)^{-1}\langle u, \varphi\rangle = (\chi_{s+\frac{1}{2}}\eta^{-1}_{t+\frac{1}{2}})(a)\langle u, \varphi\rangle. \tag{2.4}$$

(b) For the second condition, note that, since $\frac{dx}{|x|_{\mathbb{E}}}$ is a Haar measure on the multiplicative group $\mathbb{E}^\times$ and $\mathbb{E} - \mathbb{E}^\times = \{0\}$ is of measure zero, we have the following integral formula:

$$\int_{\mathbb{E}} \psi\left(\frac{x}{1-bx}\right) |1-bx|_{\mathbb{E}}^{-2}\, dx = \int_{\mathbb{E}} \psi\left(\frac{1}{x^{-1}-b}\right) |x^{-1}-b|_{\mathbb{E}}^{-2} |x|_{\mathbb{E}}^{-2}\, dx$$
$$\overset{(y=x^{-1})}{=} \int_{\mathbb{E}} \psi\left(\frac{1}{y-b}\right) |y-b|_{\mathbb{E}}^{-2}\, dy \overset{(z=y-b)}{=} \int_{\mathbb{E}} \psi(z^{-1}) |z|_{\mathbb{E}}^{-2}\, dz$$
$$\overset{(w=z^{-1})}{=} \int_{\mathbb{E}} \psi(w)\, dw$$

for $\psi \in L^1(\mathbb{E})$, so

$$\langle u(\tfrac{\cdot}{1-b\cdot}), \varphi\rangle = \langle u, |1+b\cdot|_{\mathbb{E}}^{-2} \varphi(\tfrac{\cdot}{1+b\cdot})\rangle.$$

With this formula, condition (b) applied to a test function $\varphi \in C_c^\infty(\mathbb{E} - \{b^{-1}\})$ reads

$$\langle u, |1+b\cdot|_{\mathbb{E}}^{2} \varphi(\tfrac{\cdot}{1+b\cdot})\rangle = \langle u, \chi_{s-\frac{1}{2}}((1-b\cdot)^2)^{-1}\varphi\rangle. \tag{2.5}$$

We conclude that the construction and classification of symmetry breaking operators $A \in \operatorname{Hom}_{G'}(\pi_{\chi,s}|_{G'}, \tau_{\eta,t})$ can be carried out in terms of their distribution kernels $K \in \mathcal{D}'(\mathbb{E})_{s,t}^{\chi,\eta}$.

3. A Holomorphic Family of Symmetry Breaking Kernels

We find a family of symmetry breaking kernels $K_{s,t}^{\chi,\eta} \in \mathcal{D}'(\mathbb{E})_{s,t}^{\chi,\eta}$ that depends meromorphically on $s,t \in \mathbb{C}$ and normalize it to a holomorphic family $\widetilde{K}_{s,t}^{\chi,\eta}$.

3.1. *Classification of symmetry breaking kernels on* $\mathbb{E}^\times$

We first classify the possible restrictions of a symmetry breaking kernel $u \in \mathcal{D}'(\mathbb{E})_{s,t}^{\chi,\eta}$ to $\mathbb{E}^\times = \mathbb{E} - \{0\}$. For this consider the representative w of the longest Weyl group element defined in (2.1). Since $w\overline{N} = Nw$, the conditions on a distribution on G/B to be N'-invariant are easier to study on the open dense subset $w\overline{N}B/B = NwB/B$ than on $\overline{N}B/B$. We therefore first apply the action of w to a possible symmetry breaking kernel in $(\mathcal{D}'(G/B, \mathcal{V}^*) \otimes W)^{\Delta(B')}$.

In the coordinates $\overline{n}_x \in \overline{N}$, $x \in \mathbb{E}$, this amounts to decomposing $w\overline{n}_x$ into $\overline{N}B$, if possible. For $x \in \mathbb{E}^\times$ we find

$$\begin{aligned} w\overline{n}_x &= \begin{pmatrix} 0 & 1 \\ 1 & 0 \end{pmatrix} \begin{pmatrix} 1 & 0 \\ x & 1 \end{pmatrix} = \begin{pmatrix} x & 1 \\ 1 & 0 \end{pmatrix} = \begin{pmatrix} 1 & 0 \\ 1/x & 1 \end{pmatrix} \begin{pmatrix} x & 0 \\ 0 & -x^{-1} \end{pmatrix} \begin{pmatrix} 1 & 1/x \\ 0 & 1 \end{pmatrix} \\ &\equiv \begin{pmatrix} 1 & 0 \\ 1/x & 1 \end{pmatrix} \begin{pmatrix} -x^2 & 0 \\ 0 & 1 \end{pmatrix} \begin{pmatrix} 1 & 1/x \\ 0 & 1 \end{pmatrix}, \end{aligned} \tag{3.1}$$

where the last identity is modulo multiples of the identity matrix. Hence the action of w on $u \in \mathcal{D}'(G/B,\mathcal{V}^*)$ is given by

$$\pi_{\overline{\chi},-s}(w)u(\overline{n}_x) = \overline{\chi}_{-s+\frac{1}{2}}(-x^2)^{-1}u(\overline{n}_{1/x}) = \chi_{s-\frac{1}{2}}(-1)\cdot\chi_{s-\frac{1}{2}}(x^2)u(\overline{n}_{1/x})$$

for $x \in \mathbb{E}^\times$. This motivates the definition of the following involutive linear isomorphism:

$$\kappa : \mathcal{D}'(\mathbb{E}^\times) \to \mathcal{D}'(\mathbb{E}^\times), \quad (\kappa u)(x) := \chi_{s-\frac{1}{2}}(x^2)u(\tfrac{1}{x}).$$

Lemma 3.1: *If $u \in \mathcal{D}'(\mathbb{E})^{\chi,\eta}_{s,t}$, then $v = \kappa(u|_{\mathbb{E}^\times})$ satisfies*

(a) For $a \in \mathbb{F}^\times$: $v(ax) = (\chi_{s-\frac{1}{2}}\eta_{t+\frac{1}{2}})(a)v(x)$ for all $x \in \mathbb{E}^\times$.
(b) For $y \in \mathbb{F}$: $v(x+y) = v(x)$ for all $x \in \mathbb{E}^\times - \{-y\}$.

Proof: If $u \in \mathcal{D}'(\mathbb{E})^{\chi,\eta}_{s,t}$, then for $a \in \mathbb{F}^\times$:

$$\begin{aligned}(\kappa u)(ax) &= \chi_{s-\frac{1}{2}}((ax)^2)u(\tfrac{1}{ax})\\ &= \chi_{s-\frac{1}{2}}(a^2)\chi_{s-\frac{1}{2}}(x^2)\chi_{s-\frac{1}{2}}(\tfrac{1}{a})\eta_{t+\frac{1}{2}}(\tfrac{1}{a})^{-1}u(\tfrac{1}{x})\\ &= \chi_{s-\frac{1}{2}}(a)\eta_{t+\frac{1}{2}}(a)(\kappa u)(x),\end{aligned}$$

and for $y \in \mathbb{F}$:

$$\begin{aligned}(\kappa u)(x+y) &= \chi_{s-\frac{1}{2}}((x+y)^2)u(\tfrac{1}{x+y}) = \chi_{s-\frac{1}{2}}((x+y)^2)u(\tfrac{1/x}{1+y/x})\\ &= \chi_{s-\frac{1}{2}}((x+y)^2)\chi_{s-\frac{1}{2}}((1+\tfrac{y}{x})^2)^{-1}u(\tfrac{1}{x})\\ &= \chi_{s-\frac{1}{2}}(x^2)u(\tfrac{1}{x}) = (\kappa u)(x).\end{aligned}$$

□

It follows from Lemma 3.1(b) that if $u \in \mathcal{D}'(\mathbb{E})^{\chi,\eta}_{s,t}$, then $v := \kappa(u|_{\mathbb{E}^\times})$ is of the form $v(x) = w(x_2)$ for some $w \in \mathcal{D}'(\mathbb{F})$, where $x = x_1 + x_2\alpha \in \mathbb{E}^\times$. Moreover, Lemma 3.1(a) implies that w is homogeneous of degree $\chi_{s-\frac{1}{2}}|_{\mathbb{F}^\times}\cdot\eta_{t+\frac{1}{2}}$. Homogeneous distribution are classified in [9], §2.3, and we recall the result in Appendix A. From Theorem A.1 we conclude that unless $\chi|_{\mathcal{O}_\mathbb{F}^\times}\cdot\eta = 1$ and $2s+t+\frac{1}{2} \in \frac{2\pi i}{\ln q_\mathbb{F}}\mathbb{Z}$, i.e. $(\chi,s,\eta,t) \in \setminus\!\!\setminus$, we have

$$v(x_1+\alpha x_2) = \text{const}\times(\chi_{s-\frac{1}{2}}\eta_{t+\frac{1}{2}})(x_2),$$

so only the x_2-coordinate of $x = x_1 + \alpha x_2$ contributes to v. Hence $u|_{\mathbb{E}^\times} = \kappa(v)$ is of the form

$$u(x) = \text{const}\times\chi_{s-\frac{1}{2}}(x^2)(\chi_{s-\frac{1}{2}}\eta_{t+\frac{1}{2}})((\tfrac{1}{x})_2),$$

where $\frac{1}{x} = (\frac{1}{x})_1 + (\frac{1}{x})_2\alpha$ with $(\frac{1}{x})_1, (\frac{1}{x})_2 \in \mathbb{F}$. Note that

$$\frac{1}{x} = \frac{1}{x_1+x_2\alpha} = \frac{x_1 - x_2\alpha}{x_1^2 - x_2^2\alpha^2},$$

so $(\frac{1}{x})_2 = -(x_1^2 - x_2^2\alpha^2)^{-1}x_2$ and hence

$$u(x) = \text{const} \times \chi_{s-\frac{1}{2}}(x^2)(\chi_{s-\frac{1}{2}}\eta_{t+\frac{1}{2}})(\tfrac{x_2}{x_1^2 - x_2^2\alpha^2}).$$

(Note that we omit the minus sign in $(\frac{1}{x})_2$ since it only contributes the constant factor $(\chi_{s-\frac{1}{2}}\eta_{t+\frac{1}{2}})(-1)$.) In the case where $(\chi, s, \eta, t) \in \backslash\backslash$ we have by the same arguments that

$$v(x_1 + \alpha x_2) = \text{const} \times \delta(x_2)$$

and hence $u|_{\mathbb{E}^\times} = \kappa(v)$ is of the form

$$\begin{aligned} u(x) &= \text{const} \times \chi_{s-\frac{1}{2}}(x^2)\delta((\tfrac{1}{x})_2) = \text{const} \times \chi_{s-\frac{1}{2}}(x^2)\delta(\tfrac{x_2}{x_1^2 - x_2^2\alpha^2}) \\ &= \text{const} \times |x_1^2 - x_2^2\alpha^2|_{\mathbb{F}}\chi_{s-\frac{1}{2}}(x^2)\delta(x_2) = \text{const} \times \chi_s(x^2)\delta(x_2) \\ &= \text{const} \times \chi_s(x_1^2)\delta(x_2). \end{aligned}$$

If we denote by $\mathcal{D}'(\mathbb{E}^\times)^{\chi,\eta}_{s,t}$ the space of all distributions $u \in \mathcal{D}'(\mathbb{E}^\times)$ satisfying the conditions of Proposition 2.3 wherever it makes sense, this shows:

Proposition 3.2: $\dim \mathcal{D}'(\mathbb{E}^\times)^{\chi,\eta}_{s,t} = 1$ *for all* (χ, s, η, t). *More precisely,* $\mathcal{D}'(\mathbb{E}^\times)^{\chi,\eta}_{s,t}$ *is spanned by the distribution*

$$u(x) = \begin{cases} \eta_{t+\frac{1}{2}}(x_1^2)^{-1}\delta(x_2) & \textit{if } (\chi, s, \eta, t) \in \backslash\backslash, \\ \chi_{s-\frac{1}{2}}(x^2)(\chi_{s-\frac{1}{2}}\eta_{t+\frac{1}{2}})(\tfrac{x_2}{x_1^2 - x_2^2\alpha^2}) & \textit{else.} \end{cases}$$

3.2. *The meromorphic family*

We define a function $K^{\chi,\eta}_{s,t}$ on $\mathbb{E} - \mathbb{F}$ by

$$K^{\chi,\eta}_{s,t}(x) = \chi_{s-\frac{1}{2}}(x^2)(\chi_{s-\frac{1}{2}}\eta_{t+\frac{1}{2}})(\tfrac{x_2}{x_1^2 - x_2^2\alpha^2}), \qquad \text{where } x = x_1 + x_2\alpha \in \mathbb{E}^\times.$$

Note that since $x^2 = (x_1^2 - x_2^2\alpha^2)\frac{x_1 + x_2\alpha}{x_1 - x_2\alpha}$, this can also be written as

$$\begin{aligned} K^{\chi,\eta}_{s,t}(x) &= \chi\left(\frac{x_1 + x_2\alpha}{x_1 - x_2\alpha}\right)\chi_{s-\frac{1}{2}}(x_2)\eta_{t+\frac{1}{2}}(\tfrac{x_2}{x_1^2 - x_2^2\alpha^2}) \\ &= \chi\left(\frac{x_1 + x_2\alpha}{x_1 - x_2\alpha}\right)\chi_0(x_2)\eta_0(\tfrac{x_2}{x_1^2 - x_2^2\alpha^2})|x|_{\mathbb{E}}^{-t-\frac{1}{2}}|x_2|_{\mathbb{F}}^{2s+t-\frac{1}{2}}. \end{aligned}$$

Note that $K^{\chi,\eta}_{s,t}$ only depends on s modulo $\frac{2\pi i}{\ln q_{\mathbb{E}}}\mathbb{Z}$ and on t modulo $\frac{2\pi i}{\ln q_{\mathbb{F}}}\mathbb{Z}$.

Since

$$|K^{\chi,\eta}_{s,t}(x)| = |x|_{\mathbb{E}}^{-\Re(t+\frac{1}{2})}|x_2|_{\mathbb{F}}^{\Re(2s+t-\frac{1}{2})},$$

it is clear that $K_{s,t}^{\chi;\eta}$ is locally integrable on $\mathbb{E}$ for $-\Re(t+\frac{1}{2}), \Re(2s+t-\frac{1}{2}) \geq 0$. In fact, it is locally integrable for a slightly larger range of parameters:

Lemma 3.3: *$K_{s,t}^{\chi;\eta}$ is locally integrable on $\mathbb{E}$ if and only if $\Re(2s\pm t+\frac{1}{2}) > 0$.*

Proof: This will follow from the computations in Sections 3.2.2 and 3.2.3. □

In particular, $K_{s,t}^{\chi;\eta}$ is a distribution in $\mathcal{D}'(\mathbb{E})$ for $\Re(2s \pm t + \frac{1}{2}) > 0$. To meromorphically extend $K_{s,t}^{\chi;\eta} \in \mathcal{D}'(\mathbb{E})$ in $s, t \in \mathbb{C}$ and determine its residues, we let $\varphi \in C_c^\infty(\mathbb{E})$ and consider $\langle K_{s,t}^{\chi;\eta}, \varphi\rangle$. Since φ is locally constant with compact support, there is a neighborhood $V = \varpi_{\mathbb{F}}^n \mathcal{O}_{\mathbb{F}}$ of 0 in $\mathbb{F}$ such that φ is constant on each coset $x + (V \times V)$. Here, we identify $\mathbb{E} \simeq \mathbb{F} \times \mathbb{F}$ by $x_1 + x_2\alpha \mapsto (x_1, x_2)$. We write

$$\langle K_{s,t}^{\chi;\eta}, \varphi\rangle = \int_{\mathbb{F}\times(\mathbb{F}-V)} K_{s,t}^{\chi;\eta}(x)\varphi(x)\,dx + \int_{(\mathbb{F}-V)\times V} K_{s,t}^{\chi;\eta}(x)\varphi(x)\,dx + \int_{V\times V} K_{s,t}^{\chi;\eta}(x)\varphi(x)\,dx \quad (3.2)$$

and analyze the three terms separately.

3.2.1. *The first term*

The integral over $\mathbb{F} \times (\mathbb{F} - V)$ turns out to be holomorphic in $s, t \in \mathbb{C}$.

Proposition 3.4: *The map*

$$(s,t) \mapsto \int_{\mathbb{F}\times(\mathbb{F}-V)} K_{s,t}^{\chi;\eta}(x)\varphi(x)\,dx$$

defines a holomorphic function in $(s,t) \in \mathbb{C}^2$.

Proof: For $(x_1, x_2) \in \mathbb{F} \times (\mathbb{F} - V)$ we have

$$|x|_{\mathbb{E}} = |x_1^2 - x_2^2\alpha^2|_{\mathbb{F}} > |\varpi_{\mathbb{F}}^{2n}\alpha^2|_{\mathbb{F}} > 0 \qquad \text{and} \qquad |x_2|_{\mathbb{F}} > |\varpi_{\mathbb{F}}^n|_{\mathbb{F}} > 0,$$

so the integral converges absolutely for all $s, t \in \mathbb{C}$. □

3.2.2. *The second term*

In this subsection we prove the following statement for the integral over $(\mathbb{F} - V) \times V$ in (3.2):

Proposition 3.5: *The map*

$$(s,t) \mapsto \int_{(\mathbb{F}-V)\times V} K_{s,t}^{\chi,\eta}(x)\varphi(x)\,dx$$

extends to a meromorphic function in $(s,t) \in \mathbb{C}^2$.

- *If* $\chi|_{\mathcal{O}_{\mathbb{F}}^\times} \cdot \eta \neq 1$ *it is holomorphic in* $(s,t) \in \mathbb{C}^2$.
- *If* $\chi|_{\mathcal{O}_{\mathbb{F}}^\times} \cdot \eta = 1$ *it is holomorphic in* $\{(s,t) \in \mathbb{C}^2 : 2s+t+\frac{1}{2} \notin \frac{2\pi i}{\ln q_{\mathbb{F}}}\mathbb{Z}\}$ *with residue at* $2s+t+\frac{1}{2} \in \frac{2\pi i}{\ln q_{\mathbb{F}}}\mathbb{Z}$ *given by*

$$\lim_{2s+t+\frac{1}{2}\to 0} (1-q_{\mathbb{F}}^{-(2s+t+\frac{1}{2})}) \int_{(\mathbb{F}-V)\times V} K_{s,t}^{\chi,\eta}(x)\varphi(x)\,dx$$
$$= (1-q_{\mathbb{F}}^{-1}) \int_{\mathbb{F}-V} \eta_{t+\frac{1}{2}}(x_1^2)^{-1}\varphi(x_1)\,dx_1.$$

We remark that the last expression can be viewed as the distribution $(1-q_{\mathbb{F}}^{-1})\eta_{t+\frac{1}{2}}(x_1^2)^{-1}\delta(x_2)$ applied to φ.

First note that $x \in (\mathbb{F}-V)\times V$ implies

$$|x|_{\mathbb{E}} = |x_1^2 - x_2^2\alpha^2|_{\mathbb{F}} = |x_1^2|_{\mathbb{F}} = |x_1|_{\mathbb{F}}^2,$$

so the integral under consideration becomes

$$\int_{(\mathbb{F}-V)\times V} \chi\left(\frac{x_1+x_2\alpha}{x_1-x_2\alpha}\right)\chi_0(x_2)\eta_0\left(\frac{x_2}{x_1^2-x_2^2\alpha^2}\right)|x_1|_{\mathbb{F}}^{-2t-1}|x_2|_{\mathbb{F}}^{2s+t-\frac{1}{2}}\varphi(x)\,dx.$$

Since φ has compact support and is constant on all cosets $x+(V\times V)$, the restriction of φ to $(\mathbb{F}-V)\times V$ is a finite linear combination of characteristic functions of translates of $V\times V$: $\varphi|_{(\mathbb{F}-V)\times V} = \sum_\ell c_\ell \chi_{(y_\ell+V)\times V}$ with $c_\ell = \varphi(y_\ell) \in \mathbb{C}$ and $y_\ell \in \mathbb{F}-V$. The contribution of the ℓ'th term to the integral is $\varphi(y_\ell)$ times

$$\int_{y_\ell+\varpi_{\mathbb{F}}^n\mathcal{O}_{\mathbb{F}}}\int_{\varpi_{\mathbb{F}}^n\mathcal{O}_{\mathbb{F}}} \chi\left(\frac{x_1+\alpha x_2}{x_1-\alpha x_2}\right)\chi_0(x_2)\eta_0(x_2)\eta_0(x_1^2-x_2^2\alpha^2)^{-1} \times |x_1|_{\mathbb{F}}^{-2t-1}|x_2|_{\mathbb{F}}^{2s+t-\frac{1}{2}}\,dx_2\,dx_1$$

$$= \int_{y_\ell+\varpi_{\mathbb{F}}^n\mathcal{O}_{\mathbb{F}}} |x_1|_{\mathbb{F}}^{-2t-1}\left(\sum_{j=n}^{\infty}\int_{\varpi_{\mathbb{F}}^j\mathcal{O}_{\mathbb{F}}^\times} \chi\left(\frac{x_1+\alpha x_2}{x_1-\alpha x_2}\right)(\chi_0\eta_0)(x_2) \times \eta_0(x_1^2-x_2^2\alpha^2)^{-1}|x_2|_{\mathbb{F}}^{2s+t-\frac{1}{2}}\,dx_2\right)dx_1 \tag{3.3}$$

$$= \int_{y_\ell+\varpi_{\mathbb{F}}^n\mathcal{O}_{\mathbb{F}}} |x_1|_{\mathbb{F}}^{-2t-1}\left(\sum_{j=n}^{\infty} q_{\mathbb{F}}^{-j(2s+t-\frac{1}{2})}\int_{\varpi_{\mathbb{F}}^j\mathcal{O}_{\mathbb{F}}^\times} \chi\left(\frac{x_1+\alpha x_2}{x_1-\alpha x_2}\right)(\chi\eta)(\varpi_{\mathbb{F}}^{-j}x_2) \times \eta_0(x_1^2-x_2^2\alpha^2)^{-1}\,dx_2\right)dx_1.$$

The inner integral is computed in the following lemma:

Lemma 3.6: *For every $n\in\mathbb{Z}$ there is $N\in\mathbb{Z}$ such that:*

(1) For any $x_1\in\mathbb{F}-\varpi_{\mathbb{F}}^n\mathcal{O}_{\mathbb{F}}$ and any $x_2\in\varpi_{\mathbb{F}}^N\mathcal{O}_{\mathbb{F}}$ we have:

$$\chi\left(\frac{x_1+\alpha x_2}{x_1-\alpha x_2}\right)=1 \qquad \text{and} \qquad \eta_0(x_1^2-\alpha^2x_2^2)=\eta_0(x_1^2).$$

(2) For any $x_1\in\mathbb{F}-\varpi_{\mathbb{F}}^n\mathcal{O}_{\mathbb{F}}$ and any $j\geq N$ we have

$$\int_{\varpi_{\mathbb{F}}^j\mathcal{O}_{\mathbb{F}}^\times} \chi\left(\frac{x_1+\alpha x_2}{x_1-\alpha x_2}\right)(\chi\eta)(\varpi_{\mathbb{F}}^{-j}x_2)\eta_0(x_1^2-x_2^2\alpha^2)^{-1}\,dx_2 = \begin{cases} q_{\mathbb{F}}^{-j}(1-q_{\mathbb{F}}^{-1})\eta_0(x_1^2)^{-1} & \text{if } \chi|_{\mathcal{O}_{\mathbb{F}}^\times}\cdot\eta=1,\\ 0 & \text{if } \chi|_{\mathcal{O}_{\mathbb{F}}^\times}\cdot\eta\neq 1.\end{cases}$$

Proof:

(1) We first consider the term

$$\chi\left(\frac{x_1+\alpha x_2}{x_1-\alpha x_2}\right)=\chi\left(1+\frac{2\alpha x_2}{x_1-\alpha x_2}\right).$$

Since χ is locally constant, it is constant in a neighborhood of 1. But

$$\left|\frac{2\alpha x_2}{x_1-\alpha x_2}\right|_{\mathbb{E}} = |2\alpha|_{\mathbb{E}} \cdot |x_2|_{\mathbb{F}}^2 \cdot |x|_{\mathbb{E}}^{-1} = |2\alpha|_{\mathbb{E}} \cdot |x_2|_{\mathbb{F}}^2 \cdot |x_1|_{\mathbb{F}}^{-2},$$

and if $x_1 \in \mathbb{F} - \varpi_{\mathbb{F}}^n \mathcal{O}_{\mathbb{F}}$ we have $|x_1|_{\mathbb{F}} > |\varpi_{\mathbb{F}}^n|_{\mathbb{F}} > 0$. It follows that for any given neighborhood of 1 we can find $N > 0$ such that $1+\frac{2\alpha x_2}{x_1-\alpha x_2}$ is contained in this neighborhood for all $x_1 \in \mathbb{F}-\varpi_{\mathbb{F}}^n\mathcal{O}_{\mathbb{F}}$ and $x_2 \in \varpi_{\mathbb{F}}^N \mathcal{O}_{\mathbb{F}}$.
Now consider the term $\eta_0(x_1^2 - x_2^2\alpha^2)$. Since η_0 is locally constant, we can for every fixed $y \neq 0$ find $N \in \mathbb{Z}$ such that $\eta_0(x_1^2-x_2^2\alpha^2) = \eta_0(y^2) = \eta_0(x_1^2)$ for all $x_1 \in y + \varpi_{\mathbb{F}}^N \mathcal{O}_{\mathbb{F}}$ and $x_2 \in \varpi_{\mathbb{F}}^N \mathcal{O}_{\mathbb{F}}$. Covering the compact set $\varpi_{\mathbb{F}}^n \mathcal{O}_{\mathbb{F}}^\times$ by finitely many of the sets $y + \varpi_{\mathbb{F}}^N \mathcal{O}_{\mathbb{F}}$ we obtain $N \in \mathbb{Z}$ with the property that $\eta_0(x_1^2 - x_2^2\alpha^2) = \eta_0(x_1^2)$ for all $x_1 \in \varpi_{\mathbb{F}}^n \mathcal{O}_{\mathbb{F}}^\times$ and $x_2 \in \varpi_{\mathbb{F}}^N \mathcal{O}_{\mathbb{F}}$. Finally, using the fact that η_0 is multiplicative, a rescaling argument shows that this in fact holds for all $x_1 \in \varpi_{\mathbb{F}}^j \mathcal{O}_{\mathbb{F}}^\times$ with $j \leq n$, hence for all $x_1 \in \mathbb{F} - \varpi_{\mathbb{F}}^n \mathcal{O}_{\mathbb{F}}$.

(2) By (1) and (2) we find for $x_1 \in \mathbb{F} - \varpi_{\mathbb{F}}^n \mathcal{O}_{\mathbb{F}}$ and $j \geq N$:

$$\begin{aligned}\int_{\varpi_{\mathbb{F}}^j\mathcal{O}_{\mathbb{F}}^\times} \chi\left(\frac{x_1+\alpha x_2}{x_1-\alpha x_2}\right)(\chi\eta)(\varpi_{\mathbb{F}}^{-j}x_2)\eta_0(x_1^2-x_2^2\alpha^2)^{-1}\,dx_2 \\ = \eta_0(x_1^2)^{-1}\int_{\varpi_{\mathbb{F}}^j\mathcal{O}_{\mathbb{F}}^\times}(\chi\eta)(\varpi_{\mathbb{F}}^{-j}x_2)\,dx_2 \\ = |\varpi_{\mathbb{F}}^j|_{\mathbb{F}}\cdot\eta_0(x_1^2)^{-1}\int_{\mathcal{O}_{\mathbb{F}}^\times}(\chi\eta)(x_2)\,dx_2.\end{aligned}$$

This is the integral of the character $\chi|_{\mathcal{O}_{\mathbb{F}}^\times} \cdot \eta$ of the compact abelian group $\mathcal{O}_{\mathbb{F}}^\times$, hence it vanishes if $\chi|_{\mathcal{O}_{\mathbb{F}}^\times} \cdot \eta$ is non-trivial and it equals the volume of $\mathcal{O}_{\mathbb{F}}^\times$ if $\chi|_{\mathcal{O}_{\mathbb{F}}^\times} \cdot \eta$ is the trivial character (see [17], Lemma 4.1). The volume of $\mathcal{O}_{\mathbb{F}}^\times$ is $(1 - q_{\mathbb{F}}^{-1})$, so the claim follows. □

Proof: (Proof of Proposition 3.5) Assume first that $\chi|_{\mathcal{O}_{\mathbb{F}}^\times} \cdot \eta \neq 1$, then (3.3) becomes

$$\int_{y_\ell+\varpi_{\mathbb{F}}^n\mathcal{O}_{\mathbb{F}}} |x_1|_{\mathbb{F}}^{-2t-1}\left(\sum_{j=n}^{N-1} q_{\mathbb{F}}^{-j(2s+t-\frac{1}{2})} \int_{\varpi_{\mathbb{F}}^j\mathcal{O}_{\mathbb{F}}^\times} \chi\left(\frac{x_1+\alpha x_2}{x_1-\alpha x_2}\right)(\chi\eta)(\varpi_{\mathbb{F}}^{-j}x_2) \times \eta_0(x_1^2-x_2^2\alpha^2)^{-1}\,dx_2\right)dx_1. \quad (3.4)$$

This expression is obviously holomorphic in $s,t\in\mathbb{C}$. If $\chi|_{\mathcal{O}_{\mathbb{F}}^\times}\cdot\eta=1$, then in addition to (3.4) we get the term

$$\int_{y_\ell+\varpi_{\mathbb{F}}^n\mathcal{O}_{\mathbb{F}}} |x_1|_{\mathbb{F}}^{-2t-1}\left(\sum_{j=N}^{\infty} q_{\mathbb{F}}^{-j(2s+t-\frac{1}{2})}q_{\mathbb{F}}^{-j}(1-q_{\mathbb{F}}^{-1})\eta_0(x_1^2)^{-1}\right)dx_1$$

$$= \left((1-q_{\mathbb{F}}^{-1})\sum_{j=N}^{\infty} q_{\mathbb{F}}^{-j(2s+t+\frac{1}{2})}\right)\int_{y_\ell+\varpi_{\mathbb{F}}^n\mathcal{O}_{\mathbb{F}}} |x_1|_{\mathbb{F}}^{-2t-1}\eta_0(x_1^2)^{-1}\,dx_1 \quad (3.5)$$

$$= (1-q_{\mathbb{F}}^{-1})\frac{q_{\mathbb{F}}^{-N(2s+t+\frac{1}{2})}}{1-q_{\mathbb{F}}^{-(2s+t+\frac{1}{2})}}\int_{y_\ell+\varpi_{\mathbb{F}}^n\mathcal{O}_{\mathbb{F}}} |x_1|_{\mathbb{F}}^{-2t-1}\eta_0(x_1^2)^{-1}\,dx_1.$$

Putting together (3.3), (3.4) and (3.5) shows that in this case

$$\lim_{2s+t+\frac{1}{2}\to 0}(1-q_{\mathbb{F}}^{-(2s+t+\frac{1}{2})})\int_{(\mathbb{F}-V)\times V} K_{s,t}^{\chi,\eta}(x)\varphi(x)\,dx$$

$$= (1-q_{\mathbb{F}}^{-1})\sum_\ell \varphi(y_\ell)\int_{y_\ell+\varpi_{\mathbb{F}}^n} |x_1|_{\mathbb{F}}^{-2t-1}\eta_0(x_1^2)^{-1}\,dx_1$$

$$= (1-q_{\mathbb{F}}^{-1})\int_{\mathbb{F}-\varpi_{\mathbb{F}}^n\mathcal{O}_{\mathbb{F}}} \eta_{t+\frac{1}{2}}(x_1^2)^{-1}\varphi(x_1)\,dx_1. \qquad \square$$

3.2.3. *The third term*

The integral over $V\times V$ in (3.2) is the most involved term. To state formulas for its residues, we use the constant $c_{\mathbb{F}}>0$ and the gamma factors introduced in Appendix B.

Proposition 3.7: *The map*

$$(s,t)\mapsto\int_{V\times V} K_{s,t}^{\chi,\eta}(x)\varphi(x)\,dx$$

extends to a meromorphic function in $(s,t)\in\mathbb{C}^2$.

- *If* $\chi|_{\mathcal{O}_\mathbb{F}^\times} \cdot \eta^{-1} \neq 1$ *it is holomorphic in* $(s,t) \in \mathbb{C}^2$.
- *If* $\chi|_{\mathcal{O}_\mathbb{F}^\times} \cdot \eta^{-1} = 1$ *and* $\chi|_{\mathcal{O}_\mathbb{F}^\times} \cdot \eta \neq 1$ *it is holomorphic in* $\{(s,t) \in \mathbb{C}^2 : 2s - t + \frac{1}{2} \notin \frac{2\pi i}{\ln q_\mathbb{F}}\mathbb{Z}\}$ *with residue at* $2s - t + \frac{1}{2} \in \frac{2\pi i}{\ln q_\mathbb{F}}\mathbb{Z}$ *given by*

$$\lim_{2s-t+\frac{1}{2}\to 0} (1 - q_\mathbb{F}^{-(2s-t+\frac{1}{2})}) \int_{V\times V} K_{s,t}^{\chi,\eta}(x)\varphi(x)\,dx$$
$$= c_\mathbb{F}(1-q_\mathbb{F}^{-1})\chi_0(-\alpha)^{-2}|\alpha^2|_\mathbb{F}^{-t+\frac{1}{2}}\Gamma((\chi^2|_{\mathcal{O}_\mathbb{F}^\times})_{2t})\Gamma((\chi^{-2})_{-t+\frac{1}{2}})\cdot\varphi(0). \tag{3.6}$$

- *If* $\chi|_{\mathcal{O}_\mathbb{F}^\times} \cdot \eta^{-1} = 1$ *and* $\chi|_{\mathcal{O}_\mathbb{F}^\times} \cdot \eta = 1$ *it is holomorphic in* $\{(s,t) \in \mathbb{C}^2 : 2s \pm t + \frac{1}{2} \notin \frac{2\pi i}{\ln q_\mathbb{F}}\mathbb{Z}\}$ *with residue at* $2s - t + \frac{1}{2} \in \frac{2\pi i}{\ln q_\mathbb{F}}\mathbb{Z}$ *given by* (3.6), *and with residue at* $2s + t + \frac{1}{2} \in \frac{2\pi i}{\ln q_\mathbb{F}}\mathbb{Z}$ *given by*

$$\lim_{2s+t+\frac{1}{2}\to 0} (1 - q_\mathbb{F}^{-(2s+t+\frac{1}{2})}) \int_{V\times V} K_{s,t}^{\chi,\eta}(x)\varphi(x)\,dx = \frac{(1-q_\mathbb{F}^{-1})^2 q_\mathbb{F}^{2tn}}{1-q_\mathbb{F}^{2t}}. \tag{3.7}$$

Since φ is constant on $V \times V$, the third term in (3.2) equals $\varphi(0)$ times

$$\int_{V\times V} \chi\left(\frac{x_1 + x_2\alpha}{x_1 - x_2\alpha}\right)\chi_0(x_2)\eta_0\left(\frac{x_2}{x_1^2 - x_2^2\alpha^2}\right)|x|_\mathbb{E}^{-t-\frac{1}{2}}|x_2|_\mathbb{F}^{2s+t-\frac{1}{2}}\,dx.$$

Decomposing $V = \varpi_\mathbb{F}^n\mathcal{O}_\mathbb{F} = \{0\} \sqcup \bigsqcup_{i=n}^\infty \varpi_\mathbb{F}^i\mathcal{O}_\mathbb{F}^\times$ and neglecting the set $\{0\}$ which is of measure zero, we rewrite this integral as

$$\sum_{i=n}^\infty\sum_{j=n}^\infty |\varpi_\mathbb{F}^{2j} - \varpi_\mathbb{F}^{2i}\alpha^2|_\mathbb{F}^{-t-\frac{1}{2}}|\varpi_\mathbb{F}^i|_\mathbb{F}^{2s+t-\frac{1}{2}}$$
$$\times \int_{\varpi_\mathbb{F}^i\mathcal{O}_\mathbb{F}^\times}(\chi_0\eta_0)(x_2)\int_{\varpi_\mathbb{F}^j\mathcal{O}_\mathbb{F}^\times}\chi\left(\frac{x_1 + x_2\alpha}{x_1 - x_2\alpha}\right)\eta_0(x_1^2 - x_2^2\alpha^2)^{-1}\,dx_1\,dx_2. \tag{3.8}$$

The double integral simplifies as follows:

Lemma 3.8: *For* $i, j \in \mathbb{Z}$ *we have*

$$\int_{\varpi_\mathbb{F}^i\mathcal{O}_\mathbb{F}^\times}(\chi_0\eta_0)(x_2)\int_{\varpi_\mathbb{F}^j\mathcal{O}_\mathbb{F}^\times}\chi\left(\frac{x_1 + \alpha x_2}{x_1 - \alpha x_2}\right)\eta_0(x_1^2 - x_2^2\alpha^2)^{-1}\,dx_1\,dx_2$$
$$= \begin{cases} q_\mathbb{F}^{-2i}(1-q_\mathbb{F}^{-1})\int_{\varpi_\mathbb{F}^{j-i}\mathcal{O}_\mathbb{F}^\times}\chi_0(x_1-\alpha)^{-2}\,dx_1 & \textit{for } \chi|_{\mathcal{O}_\mathbb{F}^\times}\cdot\eta^{-1} = 1, \\ 0 & \textit{for } \chi|_{\mathcal{O}_\mathbb{F}^\times}\cdot\eta^{-1} \neq 1. \end{cases}$$

Proof: Writing

$$\frac{x_1 + x_2\alpha}{x_1 - x_2\alpha} = \frac{\frac{x_1}{x_2} + \alpha}{\frac{x_1}{x_2} - \alpha} \qquad \text{and} \qquad x_1^2 - x_2^2\alpha^2 = x_2^2\Big(\Big(\frac{x_1}{x_2}\Big)^2 - \alpha^2\Big)$$

and substituting $a = \frac{x_1}{x_2}$ with $dx_1 = |x_2|_{\mathbb{F}}\, da$ yields

$$\begin{aligned}
&\int\limits_{\varpi_F^i \mathcal{O}_{\mathbb{F}}^\times} (\chi_0\eta_0)(x_2) \int\limits_{\varpi_F^j \mathcal{O}_{\mathbb{F}}^\times} \chi\left(\frac{x_1 + \alpha x_2}{x_1 - \alpha x_2}\right) \eta_0(x_1^2 - x_2^2\alpha^2)^{-1}\, dx_1\, dx_2 \\
&= q_{\mathbb{F}}^{-i} \int\limits_{\varpi_F^i \mathcal{O}_{\mathbb{F}}^\times} (\chi_0\eta_0^{-1})(x_2) \int\limits_{\varpi_F^{j-i} \mathcal{O}_{\mathbb{F}}^\times} \chi\left(\frac{x_1 + \alpha}{x_1 - \alpha}\right) \eta_0(x_1^2 - \alpha^2)^{-1}\, dx_1\, dx_2 \\
&= q_{\mathbb{F}}^{-i} \int\limits_{\varpi_F^{j-i} \mathcal{O}_{\mathbb{F}}^\times} \chi\left(\frac{x_1 + \alpha}{x_1 - \alpha}\right) \eta_0(x_1^2 - \alpha^2)^{-1} \int\limits_{\varpi_F^i \mathcal{O}_{\mathbb{F}}^\times} (\chi_0\eta_0^{-1})(x_2)\, dx_2\, dx_1. \quad (3.9)
\end{aligned}$$

The inner integral can be evaluated as in the proof of Lemma 3.6:

$$\begin{aligned}
\int\limits_{\varpi_{\mathbb{F}}^i \mathcal{O}_{\mathbb{F}}^\times} (\chi_0\eta_0^{-1})(x_2)\, dx_2 &= q_{\mathbb{F}}^{-i} \int_{\mathcal{O}_{\mathbb{F}}^\times} (\chi\eta^{-1})(x_2)\, dx_2 \\
&= \begin{cases} q_{\mathbb{F}}^{-i}(1 - q_{\mathbb{F}}^{-1}) & \text{for } \chi|_{\mathcal{O}_{\mathbb{F}}^\times} \cdot \eta^{-1} = 1, \\ 0 & \text{for } \chi|_{\mathcal{O}_{\mathbb{F}}^\times} \cdot \eta^{-1} \neq 1. \end{cases}
\end{aligned}$$

Assume that $\chi|_{\mathcal{O}_{\mathbb{F}}^\times} \cdot \eta^{-1} = 1$, i.e. $\eta = \chi|_{\mathcal{O}_{\mathbb{F}}^\times}$, then (3.9) becomes

$$\begin{aligned}
q_{\mathbb{F}}^{-2i}(1 - q_{\mathbb{F}}^{-1}) &\int_{\varpi_{\mathbb{F}}^{j-i} \mathcal{O}_{\mathbb{F}}^\times} \chi\left(\frac{x_1 + \alpha}{x_1 - \alpha}\right) \eta_0(x_1^2 - \alpha^2)^{-1}\, dx_1 \\
&= q_{\mathbb{F}}^{-2i}(1 - q_{\mathbb{F}}^{-1}) \int_{\varpi_{\mathbb{F}}^{j-i} \mathcal{O}_{\mathbb{F}}^\times} \chi_0(x_1 - \alpha)^{-2}\, dx_1. \qquad \square
\end{aligned}$$

This shows Proposition 3.7 for $\chi|_{\mathcal{O}_{\mathbb{F}}^\times} \cdot \eta^{-1} \neq 1$ since in this case the integral vanishes. Now assume that $\chi|_{\mathcal{O}_{\mathbb{F}}^\times} \cdot \eta^{-1} = 1$, then the previous lemma allows to rewrite (3.8) as

$$
\begin{aligned}
&(1-q_{\mathbb{F}}^{-1})\sum_{i=n}^{\infty}\sum_{j=n}^{\infty}|\varpi_{\mathbb{F}}^{2j}-\varpi_{\mathbb{F}}^{2i}\alpha^2|_{\mathbb{F}}^{-t-\frac{1}{2}}|\varpi_{\mathbb{F}}^{i}|_{\mathbb{F}}^{2s+t+\frac{3}{2}}\int_{\varpi_{\mathbb{F}}^{j-i}\mathcal{O}_{\mathbb{F}}^{\times}}\chi_0(x_1-\alpha)^{-2}\,dx_1\\
&=(1-q_{\mathbb{F}}^{-1})\sum_{i=n}^{\infty}|\varpi_{\mathbb{F}}^{i}|_{\mathbb{F}}^{2s-t+\frac{1}{2}}\sum_{j=n}^{\infty}\int_{\varpi_{\mathbb{F}}^{j-i}\mathcal{O}_{\mathbb{F}}^{\times}}\chi_0(x_1-\alpha)^{-2}|x_1^2-\alpha^2|_{\mathbb{F}}^{-t-\frac{1}{2}}\,dx_1\\
&=(1-q_{\mathbb{F}}^{-1})q_{\mathbb{F}}^{-n(2s-t+\frac{1}{2})}\sum_{i=0}^{\infty}q_{\mathbb{F}}^{-i(2s-t+\frac{1}{2})}\sum_{m=-\infty}^{i}\int_{\varpi_{\mathbb{F}}^{-m}\mathcal{O}_{\mathbb{F}}^{\times}}\chi_0(x_1-\alpha)^{-2}\\
&\qquad\times|x_1^2-\alpha^2|_{\mathbb{F}}^{-t-\frac{1}{2}}\,dx_1. \qquad (3.10)
\end{aligned}
$$

Lemma 3.9: *There exists $M>0$ such that for $k\geq M$:*

$$
\int_{\varpi_{\mathbb{F}}^{-k}\mathcal{O}_{\mathbb{F}}^{\times}}\chi_0(x_1-\alpha)^{-2}\,dx_1=\begin{cases}(1-q_{\mathbb{F}}^{-1})q_{\mathbb{F}}^{k} & \text{for } \chi^2|_{\mathcal{O}_{\mathbb{F}}^{\times}}=1,\\ 0 & \text{for } \chi^2|_{\mathcal{O}_{\mathbb{F}}^{\times}}\neq 1.\end{cases}
$$

Proof: Since χ_0 is locally constant on $\mathcal{O}_{\mathbb{E}}^{\times}$ (see e.g. [9], §1.4), there exists $M>0$ such that $\chi_0(1+y)=\chi_0(1)=1$ for all $y\in\varpi_{\mathbb{E}}^{M}\mathcal{O}_{\mathbb{E}}$. After possibly replacing M by $2M$, we have $\chi_0(1+y)=1$ for all $y\in\varpi_{\mathbb{F}}^{M}\mathcal{O}_{\mathbb{E}}$. Recall that we assume $\alpha\in\mathcal{O}_{\mathbb{E}}$. Then $k\geq M$ implies $\varpi_{\mathbb{F}}^{k}\alpha\in\varpi_{\mathbb{F}}^{M}\mathcal{O}_{\mathbb{E}}$, so for all $x_1\in\varpi_{\mathbb{F}}^{-k}\mathcal{O}_{\mathbb{F}}^{\times}$ we obtain

$$
\chi_0(x_1-\alpha)=\chi_0(x_1)\chi_0(1-x_1^{-1}\alpha)=\chi_0(x_1).
$$

This allows us to compute the integral for $k\geq M$:

$$
\int_{\varpi_{\mathbb{F}}^{-k}\mathcal{O}_{\mathbb{F}}^{\times}}\chi_0(x_1-\alpha)^{-2}\,dx_1=\int_{\varpi_{\mathbb{F}}^{-k}\mathcal{O}_{\mathbb{F}}^{\times}}\chi_0(x_1)^{-2}\,dx_1=q_{\mathbb{F}}^{k}\int_{\mathcal{O}_{\mathbb{F}}^{\times}}\chi(u)^{-2}\,du,
$$

which is the integral of the character $\chi^{-2}|_{\mathcal{O}_{\mathbb{F}}^{\times}}$ over the compact group $\mathcal{O}_{\mathbb{F}}^{\times}$ (see [17], Lemma 4.1). □

Note that under the assumption $\chi|_{\mathcal{O}_{\mathbb{F}}^{\times}}\cdot\eta^{-1}=1$, the condition $\chi^2|_{\mathcal{O}_{\mathbb{F}}^{\times}}=1$ is equivalent to $\chi|_{\mathcal{O}_{\mathbb{F}}^{\times}}\cdot\eta=1$. In view of the previous lemma, we split the summation in (3.10) into

$$\begin{aligned}\sum_{i=0}^{\infty}\sum_{m=-\infty}^{i} &= \left(\sum_{i=0}^{M-1}+\sum_{i=M}^{\infty}\right)\sum_{m=-\infty}^{i}\\ &= \sum_{i=0}^{M-1}\left(\sum_{m=-\infty}^{M-1}-\sum_{m=i+1}^{M-1}\right)+\sum_{i=M}^{\infty}\left(\sum_{m=-\infty}^{M-1}+\sum_{m=M}^{i}\right)\\ &= \sum_{i=0}^{\infty}\sum_{m=-\infty}^{M-1}-\sum_{i=0}^{M-2}\sum_{m=i+1}^{M-1}+\sum_{i=M}^{\infty}\sum_{m=M}^{i}.\end{aligned}$$

(Note that in the second double sum, the term for $i = M-1$ vanishes since in this case $\sum_{m=i+1}^{M-1} = \sum_{m=M}^{M-1} = 0$.) Then (3.10) becomes

$$\begin{aligned}&(1-q_{\mathbb{F}}^{-1})q_{\mathbb{F}}^{-n(2s-t+\frac{1}{2})}\\ &\times\left(\sum_{i=0}^{\infty} q_{\mathbb{F}}^{-i(2s-t+\frac{1}{2})}\int_{\varpi_{\mathbb{F}}^{-(M-1)}\mathcal{O}_{\mathbb{F}}}\chi_0(x_1-\alpha)^{-2}|x_1^2-\alpha^2|_{\mathbb{F}}^{-t-\frac{1}{2}}\,dx_1\right.\\ &\qquad-\sum_{i=0}^{M-2} q_{\mathbb{F}}^{-i(2s-t+\frac{1}{2})}\sum_{m=i+1}^{M-1}\int_{\varpi_{\mathbb{F}}^{-m}\mathcal{O}_{\mathbb{F}}^{\times}}\chi_0(x_1-\alpha)^{-2}|x_1^2-\alpha^2|_{\mathbb{F}}^{-t-\frac{1}{2}}\,dx_1\\ &\qquad\left.+\sum_{i=M}^{\infty} q_{\mathbb{F}}^{-i(2s-t+\frac{1}{2})}\sum_{m=M}^{i}\int_{\varpi_{\mathbb{F}}^{-m}\mathcal{O}_{\mathbb{F}}^{\times}}\chi_0(x_1-\alpha)^{-2}|x_1^2-\alpha^2|_{\mathbb{F}}^{-t-\frac{1}{2}}\,dx_1\right).\end{aligned}\tag{3.11}$$

The second term in (3.11) is a finite sum and hence holomorphic in $s, t \in \mathbb{C}$. The first term becomes

$$\frac{(1-q_{\mathbb{F}}^{-1})q_{\mathbb{F}}^{-n(2s-t+\frac{1}{2})}}{1-q_{\mathbb{F}}^{-(2s-t+\frac{1}{2})}}\int_{\varpi_{\mathbb{F}}^{-(M-1)}\mathcal{O}_{\mathbb{F}}}\chi_0(x_1-\alpha)^{-2}|x_1^2-\alpha^2|_{\mathbb{F}}^{-t-\frac{1}{2}}\,dx_1$$

which is holomorphic in $\{(s,t)\in\mathbb{C}^2 : 2s-t+\frac{1}{2}\notin\frac{2\pi i}{\ln q_{\mathbb{F}}}\mathbb{Z}\}$ with the obvious residue at $1-q_{\mathbb{F}}^{-(2s-t+\frac{1}{2})}=0$. Finally, the third term vanishes for $\chi|_{\mathcal{O}_{\mathbb{F}}^{\times}}\cdot\eta\neq 1$, and for $\chi|_{\mathcal{O}_{\mathbb{F}}^{\times}}\cdot\eta = 1$ it becomes

$$\begin{aligned}&(1-q_{\mathbb{F}}^{-1})^2 q_{\mathbb{F}}^{-n(2s-t+\frac{1}{2})}\sum_{i=M}^{\infty} q_{\mathbb{F}}^{-i(2s-t+\frac{1}{2})}\sum_{m=M}^{i} q_{\mathbb{F}}^{-2tm}\\ &= (1-q_{\mathbb{F}}^{-1})^2 q_{\mathbb{F}}^{-n(2s-t+\frac{1}{2})} q_{\mathbb{F}}^{-M(2s+t+\frac{1}{2})}\frac{1}{(1-q_{\mathbb{F}}^{-(2s-t+\frac{1}{2})})(1-q_{\mathbb{F}}^{-(2s+t+\frac{1}{2})})}.\end{aligned}$$

This shows the holomorphicity statements in Proposition 3.7. It remains to prove the claimed residue formulas.

Following the previous discussion, multiplying (3.11) with $(1 - q_{\mathbb{F}}^{-(2s-t+\frac{1}{2})})$ and setting $2s - t + \frac{1}{2} = 0$ yields

$$(1 - q_{\mathbb{F}}^{-1}) \int_{\varpi_{\mathbb{F}}^{-(M-1)}\mathcal{O}_{\mathbb{F}}} \chi_0(x_1 - \alpha)^{-2} |x_1^2 - \alpha^2|_{\mathbb{F}}^{-t-\frac{1}{2}} \, dx_1$$

if $\chi|_{\mathcal{O}_{\mathbb{F}}^\times} \cdot \eta \neq 1$, and in the case $\chi|_{\mathcal{O}_{\mathbb{F}}^\times} \cdot \eta = 1$ additionally

$$(1 - q_{\mathbb{F}}^{-1}) \frac{(1 - q_{\mathbb{F}}^{-1}) q_{\mathbb{F}}^{-M(2s+t+\frac{1}{2})}}{1 - q_{\mathbb{F}}^{-(2s+t+\frac{1}{2})}}.$$

In view of the choice of M in Lemma 3.9, the fact that $2s + t + \frac{1}{2} = 2t$, the previous computations, and the change of variables $x_1 = -y\alpha^2$, this expression equals (in each of the cases $\chi|_{\mathcal{O}_{\mathbb{F}}^\times} \cdot \eta \neq 1$ or $\chi|_{\mathcal{O}_{\mathbb{F}}^\times} \cdot \eta = 1$)

$$\begin{aligned}(1 - q_{\mathbb{F}}^{-1}) &\int_{\mathbb{F}} \chi_0(x_1 - \alpha)^{-2} |x_1^2 - \alpha^2|_{\mathbb{F}}^{-t-\frac{1}{2}} \, dx_1 \\ &= (1 - q_{\mathbb{F}}^{-1}) \chi_0(-\alpha)^{-2} |\alpha^2|_{\mathbb{F}}^{-t+\frac{1}{2}} \int_{\mathbb{F}} \chi_0(1 + y\alpha)^{-2} |1 + y\alpha|_{\mathbb{E}}^{-t-\frac{1}{2}} \, dy.\end{aligned}$$

Evaluating the remaining integral with (B.4) shows (3.6).

On the other hand, multiplying (3.11) with $(1 - q_{\mathbb{F}}^{-(2s+t+\frac{1}{2})})$ and setting $2s + t + \frac{1}{2} = 0$ yields

$$\frac{(1 - q_{\mathbb{F}}^{-1})^2 q_{\mathbb{F}}^{-n(2s-t+\frac{1}{2})}}{1 - q_{\mathbb{F}}^{-(2s-t+\frac{1}{2})}}.$$

Since in this case $2s - t + \frac{1}{2} = -2t$, this shows (3.7).

3.2.4. *Holomorphic renormalization*

For a character η_t of $\mathbb{F}^\times$, $\eta \in \widehat{\mathcal{O}_{\mathbb{F}}^\times}$, $t \in \mathbb{C}$, recall the local non-Archimedean L-factor

$$L(s, \eta_t) = \begin{cases} (1 - q_{\mathbb{F}}^{-s-t})^{-1} & \text{if } \eta = 1, \\ 1 & \text{if } \eta \neq 1, \end{cases} \qquad (s \in \mathbb{C}). \tag{3.12}$$

We define the following renormalization of $K_{s,t}^{\chi,\eta}$:

$$\widetilde{K}_{s,t}^{\chi,\eta} := L(\tfrac{1}{2}, \chi_s|_{\mathbb{F}^\times} \cdot \eta_t)^{-1} L(\tfrac{1}{2}, \chi_s|_{\mathbb{F}^\times} \cdot \eta_t^{-1})^{-1} \cdot K_{s,t}^{\chi,\eta},$$

or more explicitly:

$$\widetilde{K}^{\chi,\eta}_{s,t} = K^{\chi,\eta}_{s,t} \times \begin{cases} (1-q_{\mathbb{F}}^{-(2s+t+\frac{1}{2})}) \\ \times(1-q_{\mathbb{F}}^{-(2s-t+\frac{1}{2})}) & \text{if } \chi|_{\mathcal{O}_{\mathbb{F}}^\times}\cdot\eta=1 \text{ and } \chi|_{\mathcal{O}_{\mathbb{F}}^\times}\cdot\eta^{-1}=1, \\ (1-q_{\mathbb{F}}^{-(2s+t+\frac{1}{2})}) & \text{if } \chi|_{\mathcal{O}_{\mathbb{F}}^\times}\cdot\eta=1 \text{ and } \chi|_{\mathcal{O}_{\mathbb{F}}^\times}\cdot\eta^{-1}\neq 1, \\ (1-q_{\mathbb{F}}^{-(2s-t+\frac{1}{2})}) & \text{if } \chi|_{\mathcal{O}_{\mathbb{F}}^\times}\cdot\eta\neq 1 \text{ and } \chi|_{\mathcal{O}_{\mathbb{F}}^\times}\cdot\eta^{-1}=1, \\ 1 & \text{if } \chi|_{\mathcal{O}_{\mathbb{F}}^\times}\cdot\eta\neq 1 \text{ and } \chi|_{\mathcal{O}_{\mathbb{F}}^\times}\cdot\eta^{-1}\neq 1. \end{cases}$$

With this renormalization we obtain the following result:

Theorem 3.10: *For all $\chi \in \widehat{\mathcal{O}_{\mathbb{E}}^\times}$ and $\eta \in \widehat{\mathcal{O}_{\mathbb{F}}^\times}$ the map $(s,t) \mapsto \widetilde{K}^{\chi,\eta}_{s,t}$ extends to a holomorphic function from $\mathbb{C}^2$ to $\mathcal{D}'(\mathbb{E})$, i.e. for every $\varphi \in C_c^\infty(\mathbb{E})$ the function $\mathbb{C}^2 \to \mathbb{C}$, $(s,t) \mapsto \langle \widetilde{K}^{\chi,\eta}_{s,t}, \varphi\rangle$ is holomorphic. Moreover, $\widetilde{K}^{\chi,\eta}_{s,t} \in \mathcal{D}'(\mathbb{E})^{\chi,\eta}_{s,t}$ for all $s,t \in \mathbb{C}$.*

Proof: Propositions 3.4, 3.5 and 3.7 applied to (3.2) show that $\langle \widetilde{K}^{\chi,\eta}_{s,t}, \varphi\rangle$ is holomorphic in $(s,t) \in \mathbb{C}^2$ for every $\varphi \in C_c^\infty(\mathbb{E})$. It remains to show that $\widetilde{K}^{\chi,\eta}_{s,t} \in \mathcal{D}'(\mathbb{E})^{\chi,\eta}_{s,t}$ for all $s,t \in \mathbb{C}$. First note that by the considerations in Section 3.1, the restriction of $K^{\chi,\eta}_{s,t}$ to $\mathbb{E}^\times$ satisfies the conditions of Proposition 2.3. For $(s,t) \in \mathbb{C}^2$ with $\Re(2s \pm t + \frac{1}{2}) > 0$ the distribution $K^{\chi,\eta}_{s,t}$ is locally integrable by Lemma 3.3, and since $\mathbb{F}$ is of measure zero in $\mathbb{E}$, the conditions of Proposition 2.3 also hold on $\mathbb{E}$ in the distribution sense. Consequently, the renormalized kernel $\widetilde{K}^{\chi,\eta}_{s,t}$ satisfies the conditions of Proposition 2.3 whenever $\Re(2s \pm t + \frac{1}{2}) > 0$, but since it is holomorphic in $(s,t) \in \mathbb{C}^2$ and the conditions of Proposition 2.3 depend holomorphically on (s,t), it satisfies the conditions of Proposition 2.3 for all $(s,t) \in \mathbb{C}^2$, i.e. $\widetilde{K}^{\chi,\eta}_{s,t} \in \mathcal{D}'(\mathbb{E})^{\chi,\eta}_{s,t}$. □

Using the residue calculations from Propositions 3.5 and 3.7 we now find the values of $\widetilde{K}^{\chi,\eta}_{s,t}$ at $(\chi,s,\eta,t) \in /\!/ \cup \backslash\!\backslash$. These special values turn out to be scalar multiples of the Dirac distribution $\delta \in \mathcal{D}'(\mathbb{E})$ at $x=0$ and a holomorphic family of distributions that is supported on $\mathbb{F} \subseteq \mathbb{E}$ which we now define. For $\xi \in \widehat{\mathcal{O}_{\mathbb{F}}^\times}$ recall from Appendix A the nowhere vanishing holomorphic family of distributions $\widetilde{\xi}_\alpha \in \mathcal{D}'(\mathbb{F})$, $\alpha \in \mathbb{C}$. For $\eta \in \widehat{\mathcal{O}_{\mathbb{F}}^\times}$ we define a family of distributions L^η_t on $\mathbb{E}$ by

$$L^\eta_t(x_1 + x_2\alpha) = \widetilde{(\eta^{-2})}_{-2t-1}(x_1)\cdot\delta(x_2) \qquad (t \in \mathbb{C}).$$

Note that the restriction of L^η_t to $\mathbb{E}^\times$ is (up to a scalar) the distribution that occurs in Proposition 3.2 in the case $(\chi,s,\eta,t) \in \backslash\!\backslash - /\!/$. By Theorem A.1,

L_t^η depends holomorphically on $t \in \mathbb{C}$ and

$$L_t^\eta(x_1 + x_2\alpha) = \begin{cases} \eta_{t+\frac{1}{2}}(x_1^2)^{-1} \cdot \delta(x_2) & \text{if } \eta^2 \neq 1, \\ (1 - q_{\mathbb{F}}^{2t})\eta_{t+\frac{1}{2}}(x_1^2)^{-1} \cdot \delta(x_2) & \text{if } \eta^2 = 1 \text{ and } 2t \notin \frac{2\pi i}{\ln q_{\mathbb{F}}}\mathbb{Z}, \\ (1 - q_{\mathbb{F}}^{-1})\delta(x_1) \cdot \delta(x_2) & \text{if } \eta^2 = 1 \text{ and } 2t \in \frac{2\pi i}{\ln q_{\mathbb{F}}}\mathbb{Z}. \end{cases}$$

Theorem 3.11: (Residue Identities)

(1) For $(\chi, s, \eta, t) \in \backslash\backslash$ *we have*

$$\widetilde{K}_{s,t}^{\chi,\eta} = (1 - q_{\mathbb{F}}^{-1}) \cdot L_t^\eta.$$

(2) For $(\chi, s, \eta, t) \in //$ *we have*

$$\widetilde{K}_{s,t}^{\chi,\eta} = c_{\mathbb{F}}(1 - q_{\mathbb{F}}^{-1})\chi_0(-\alpha)^{-2}|\alpha^2|_{\mathbb{F}}^{-t+\frac{1}{2}} \\ \times \Gamma((\chi^2|_{\mathcal{O}_{\mathbb{F}}^\times})_{2t})\Gamma((\chi^{-2})_{-t+\frac{1}{2}})L(\tfrac{1}{2}, \chi_s|_{\mathbb{F}^\times} \cdot \eta_t)^{-1} \cdot \delta,$$

where the constant $c_{\mathbb{F}} > 0$ *and the gamma factors are as in Appendix B.*

We remark that for $(\chi, s, \eta, t) \in \backslash\backslash \cap //$ the expressions in Theorem 3.11(1) and (2) agree.

Proof: These results follow by putting together Propositions 3.4, 3.5 and 3.7. It only remains to match the formulas in Propositions 3.5 and 3.7 in the case $(\chi, s, \eta, t) \in \backslash\backslash$. More precisely, for $(\chi, s, \eta, t) \in \backslash\backslash$ we have to show that

$$(1-q_{\mathbb{F}}^{-1})\int_V \eta_{t+\frac{1}{2}}(x_1^2)^{-1}\varphi(x_1)\,dx_1 = \varphi(0)\times\begin{cases} \frac{(1-q_{\mathbb{F}}^{-1})^2 q_{\mathbb{F}}^{2tn}}{1-q_{\mathbb{F}}^{2t}} & \text{if } \chi|_{\mathcal{O}_{\mathbb{F}}^\times} \cdot \eta^{-1} = 1, \\ 0 & \text{if } \chi|_{\mathcal{O}_{\mathbb{F}}^\times} \cdot \eta^{-1} \neq 1. \end{cases}$$

Since $\varphi(x_1) = \varphi(0)$ for $x_1 \in V$, this amounts to computing the integral

$$\int_V \eta_{t+\frac{1}{2}}(x_1^2)^{-1}\,dx_1$$

with $V = \varpi_{\mathbb{F}}^n\mathcal{O}_{\mathbb{F}}$ which is similar to the computations in the previous subsections. □

Now we can determine the zeros of the function $(s,t) \mapsto \widetilde{K}_{s,t}^{\chi,\eta}$ and the support of $\widetilde{K}_{s,t}^{\chi,\eta}$ for all parameters, where $\widetilde{K}_{s,t}^{\chi,\eta}$ is seen as a distribution in $\mathcal{D}'(\mathbb{E})_{s,t}^{\chi,\eta}$. To simplify the statement, we put

$$L = \{(\chi, s, \eta, t) : (s,t) \in L^{\chi,\eta}\},$$

where $L^{\chi,\eta}$ is a subset of $\mathbb{C}^2/(\frac{2\pi i}{\ln q_{\mathbb{E}}}\mathbb{Z}\times\frac{2\pi i}{\ln q_{\mathbb{F}}}\mathbb{Z})$. (Recall that $\widetilde{K}^{\chi,\eta}_{s,t}$ only depends on (s,t) modulo $\frac{2\pi i}{\ln q_{\mathbb{E}}}\mathbb{Z}\times\frac{2\pi i}{\ln q_{\mathbb{F}}}\mathbb{Z}$.) If $\chi|_{\mathcal{O}_{\mathbb{F}}^\times}\cdot\eta\neq 1$ or $\chi|_{\mathcal{O}_{\mathbb{F}}^\times}\cdot\eta^{-1}\neq 1$ we define

$$L^{\chi,\eta}=\emptyset$$

and for $\chi|_{\mathcal{O}_{\mathbb{F}}^\times}\cdot\eta=\chi|_{\mathcal{O}_{\mathbb{F}}^\times}\cdot\eta^{-1}=1$ we put

$$L^{\chi,\eta}=\begin{cases}\{(0,\frac{1}{2}),(\frac{\pi i}{\ln q_{\mathbb{E}}},\frac{1}{2}+\frac{\pi i}{\ln q_{\mathbb{F}}})\} & \text{if } \chi^2\neq 1 \text{ and}\\ & \mathbb{E}/\mathbb{F} \text{ is unramified,}\\ \{(0,\frac{1}{2}),(\frac{\pi i}{\ln q_{\mathbb{E}}},\frac{1}{2}), & \text{if } \chi^2\neq 1 \text{ and}\\ \quad(\frac{\pi i}{2\ln q_{\mathbb{E}}},\frac{1}{2}+\frac{\pi i}{\ln q_{\mathbb{F}}}),(\frac{3\pi i}{2\ln q_{\mathbb{E}}},\frac{1}{2}+\frac{\pi i}{\ln q_{\mathbb{F}}})\} & \quad\mathbb{E}/\mathbb{F} \text{ is ramified,}\\ \{(-\frac{1}{2},-\frac{1}{2}),(-\frac{1}{2}+\frac{\pi i}{\ln q_{\mathbb{E}}},-\frac{1}{2}+\frac{\pi i}{\ln q_{\mathbb{F}}})\} & \text{if } \chi^2=1 \text{ and}\\ & \mathbb{E}/\mathbb{F} \text{ is unramified,}\\ \{(-\frac{1}{2},-\frac{1}{2}),(-\frac{1}{2}+\frac{\pi i}{\ln q_{\mathbb{E}}},-\frac{1}{2}), & \text{if } \chi^2=1 \text{ and}\\ \quad(\frac{\pi i}{2\ln q_{\mathbb{E}}},\frac{1}{2}+\frac{\pi i}{\ln q_{\mathbb{F}}}),(\frac{3\pi i}{2\ln q_{\mathbb{E}}},\frac{1}{2}+\frac{\pi i}{\ln q_{\mathbb{F}}})\} & \quad\mathbb{E}/\mathbb{F} \text{ is ramified.}\end{cases} \tag{3.13}$$

Note that $L\subseteq /\!/$ and $L\cap \backslash\!\backslash=\emptyset$.

Theorem 3.12: $\widetilde{K}^{\chi,\eta}_{s,t}=0$ *if and only if* $(\chi,s,\eta,t)\in L$. *Moreover, we have*

$$\operatorname{supp}\widetilde{K}^{\chi,\eta}_{s,t}=\begin{cases}\emptyset & \text{for } (\chi,s,\eta,t)\in L,\\ \{0\} & \text{for } (\chi,s,\eta,t)\in /\!/-L,\\ \mathbb{F} & \text{for } (\chi,s,\eta,t)\in \backslash\!\backslash-/\!/,\\ \mathbb{E} & \text{else.}\end{cases}$$

Proof: First note that since $K^{\chi,\eta}_{s,t}|_{\mathbb{E}-\mathbb{F}}$ is a locally integrable and nowhere vanishing function, we have $\operatorname{supp}\widetilde{K}^{\chi,\eta}_{s,t}=\mathbb{E}$ unless we have renormalized by $(1-q_{\mathbb{F}}^{-(2s+t+\frac{1}{2})})$ or $(1-q_{\mathbb{F}}^{-(2s-t+\frac{1}{2})})$ and one of these factors vanishes, i.e. $(\chi,s,\eta,t)\in /\!/\cup\backslash\!\backslash$. In these two cases, Theorem 3.11 expresses $\widetilde{K}^{\chi,\eta}_{s,t}$ as a constant multiple of the Dirac distribution δ or the nowhere vanishing family L^η_t whose support is either $\mathbb{F}$ or $\{0\}$. It remains to check whether the constant vanishes. In Theorem 3.11(1) the constant $(1-q_{\mathbb{F}}^{-1})$ is obviously non-zero, so $\widetilde{K}^{\chi,\eta}_{s,t}\neq 0$ whenever $(\chi,s,\eta,t)\in\backslash\!\backslash$. For $(\chi,s,\eta,t)\in /\!/-\backslash\!\backslash$ Theorem 3.11(2) shows that $\widetilde{K}^{\chi,\eta}_{s,t}=0$ if and only if

$$\Gamma((\chi^2|_{\mathcal{O}_{\mathbb{F}}^\times})_{2t})\Gamma((\chi^{-2})_{-t+\frac{1}{2}})\\ \times\begin{cases}1 & \text{if } \chi|_{\mathcal{O}_{\mathbb{F}}^\times}\cdot\eta\neq 1,\\ (1-q_{\mathbb{F}}^{-(2s+t+\frac{1}{2})}) & \text{if } \chi|_{\mathcal{O}_{\mathbb{F}}^\times}\cdot\eta=1 \text{ and } 2s+t+\frac{1}{2}\notin\frac{2\pi i}{\ln q_{\mathbb{F}}}\mathbb{Z},\end{cases}$$

vanishes. To understand poles and zeros of the gamma factors, we distinguish three cases and apply Lemma B.1:

(1) $\chi^2|_{\mathcal{O}_\mathbb{F}^\times} \neq 1$. In this case, both gamma factors are holomorphic in t and nowhere vanishing.
(2) $\chi^2|_{\mathcal{O}_\mathbb{F}^\times} = 1$ and $\chi^2 \neq 1$. Here, the second gamma factor is holomorphic and nowhere vanishing, and the first gamma factor has simple poles at $t \in \frac{\pi i}{\ln q_\mathbb{F}}\mathbb{Z}$ and simple zeros at $t \in \frac{1}{2} + \frac{\pi i}{\ln q_\mathbb{F}}\mathbb{Z}$. Since $2s - t + \frac{1}{2} \in \frac{2\pi i}{\ln q_\mathbb{F}}\mathbb{Z}$ and $2s + t + \frac{1}{2} \notin \frac{2\pi i}{\ln q_\mathbb{F}}\mathbb{Z}$, the poles at $t \in \frac{\pi i}{\ln q_\mathbb{F}}\mathbb{Z}$ do not appear.
(3) $\chi^2 = 1$. In addition to the zeros of the first gamma factor, the second gamma factor has simple poles at $t \in \frac{1}{2} + \frac{2\pi i}{\ln q_\mathbb{E}}\mathbb{Z}$ and simple zeros at $t \in -\frac{1}{2} + \frac{2\pi i}{\ln q_\mathbb{E}}\mathbb{Z}$. Note that for $q_\mathbb{E} = q_\mathbb{F}^2$ (i.e. $\mathbb{E}/\mathbb{F}$ is unramified) the poles cancel the zeros of the first gamma factor, while for $q_\mathbb{E} = q_\mathbb{F}$ (i.e. $\mathbb{E}/\mathbb{F}$ is ramified) they only cancel half of them.

Since $\chi^2|_{\mathcal{O}_\mathbb{F}^\times} = 1$ is equivalent to $\chi|_{\mathcal{O}_\mathbb{F}^\times} \cdot \eta = 1$, this gives the following description of the set $L^{\chi,\eta}$ of all $(s,t) \in \mathbb{C}^2$ such that $\widetilde{K}_{s,t}^{\chi,\eta} = 0$: $L^{\chi,\eta} = \emptyset$ if $\chi|_{\mathcal{O}_\mathbb{F}^\times} \cdot \eta \neq 1$ or $\chi_{\mathcal{O}_\mathbb{F}^\times} \cdot \eta^{-1} \neq 1$, and in the case $\chi|_{\mathcal{O}_\mathbb{F}^\times} \cdot \eta = \chi_{\mathcal{O}_\mathbb{F}^\times} \cdot \eta^{-1} = 1$ we have

$$L^{\chi,\eta} = \begin{cases} \{(s,t) : 2s - t + \frac{1}{2} \in \frac{2\pi i}{\ln q_\mathbb{F}}\mathbb{Z}, t \in \frac{1}{2} + \frac{\pi i}{\ln q_\mathbb{F}}\mathbb{Z}\} & \text{if } \chi^2 \neq 1, \\ \{(s,t) : 2s - t + \frac{1}{2} \in \frac{2\pi i}{\ln q_\mathbb{F}}\mathbb{Z}, t \in -\frac{1}{2} + \frac{\pi i}{\ln q_\mathbb{F}}\mathbb{Z}\} & \text{if } \chi^2 = 1 \text{ and } q_\mathbb{E} = q_\mathbb{F}^2, \\ \{(s,t) : 2s - t + \frac{1}{2} \in \frac{2\pi i}{\ln q_\mathbb{F}}\mathbb{Z}, t \in \frac{1}{2} + \frac{\pi i}{\ln q_\mathbb{F}}(2\mathbb{Z}+1)\} & \text{if } \chi^2 = 1 \text{ and} \\ \cup \{(s,t) : 2s - t + \frac{1}{2} \in \frac{2\pi i}{\ln q_\mathbb{F}}\mathbb{Z}, t \in -\frac{1}{2} + \frac{2\pi i}{\ln q_\mathbb{F}}\mathbb{Z}\} & \quad q_\mathbb{E} = q_\mathbb{F}. \end{cases}$$

This is the same set as defined above. □

4. Classification of Symmetry Breaking Kernels for PGL(2)

In this section we determine the space $\mathcal{D}'(\mathbb{E})_{s,t}^{\chi,\eta}$ of symmetry breaking kernels explicitly for all (χ, s, η, t), following the methods of [12], Chapter 11.

Recall from Section 3.1 the space $\mathcal{D}'(\mathbb{E}^\times)_{s,t}^{\chi,\eta}$ of all distributions on $\mathbb{E}^\times$ satisfying the same conditions as the distributions in $\mathcal{D}'(\mathbb{E})_{s,t}^{\chi,\eta}$, but only on $\mathbb{E}^\times$. By Proposition 3.2 the space $\mathcal{D}'(\mathbb{E}^\times)_{s,t}^{\chi,\eta}$ is one-dimensional for all (χ, s, η, t). To understand the difference between $\mathcal{D}'(\mathbb{E}^\times)_{s,t}^{\chi,\eta}$ and $\mathcal{D}'(\mathbb{E})_{s,t}^{\chi,\eta}$ we consider the restriction map $\mathcal{D}'(\mathbb{E})_{s,t}^{\chi,\eta} \to \mathcal{D}'(\mathbb{E}^\times)_{s,t}^{\chi,\eta}$. Its kernel is equal to the space $\mathcal{D}'_{\{0\}}(\mathbb{E})_{s,t}^{\chi,\eta}$ of distributions in $\mathcal{D}'(\mathbb{E})_{s,t}^{\chi,\eta}$ whose support is contained in $\{0\}$. This yields an exact sequence

$$0 \to \mathcal{D}'_{\{0\}}(\mathbb{E})_{s,t}^{\chi,\eta} \to \mathcal{D}'(\mathbb{E})_{s,t}^{\chi,\eta} \to \mathcal{D}'(\mathbb{E}^\times)_{s,t}^{\chi,\eta} \tag{4.1}$$

and hence an upper bound for the dimension of $\mathcal{D}'(\mathbb{E})^{\chi,\eta}_{s,t}$:

$$\dim \mathcal{D}'(\mathbb{E})^{\chi,\eta}_{s,t} \leq \dim \mathcal{D}'_{\{0\}}(\mathbb{E})^{\chi,\eta}_{s,t} + \dim \mathcal{D}'(\mathbb{E}^\times)^{\chi,\eta}_{s,t} = \dim \mathcal{D}'_{\{0\}}(\mathbb{E})^{\chi,\eta}_{s,t} + 1. \tag{4.2}$$

The following lemma makes the upper bound more explicit by determining $\mathcal{D}'_{\{0\}}(\mathbb{E})^{\chi,\eta}_{s,t}$, and it also provides a lower bound based on the existence of the non-trivial holomorphic family $\widetilde{K}^{\chi,\eta}_{s,t}$:

Lemma 4.1:

(1) $\mathcal{D}'_{\{0\}}(\mathbb{E})^{\chi,\eta}_{s,t} = \begin{cases} \mathbb{C}\delta & \textit{if } (\chi,s,\eta,t) \in /\!/, \\ \{0\} & \textit{else.} \end{cases}$

(2) $\dim \mathcal{D}'(\mathbb{E})^{\chi,\eta}_{s,t} \geq 1$ *for all* $s,t \in \mathbb{C}$, $\chi \in \widehat{\mathcal{O}^\times_\mathbb{E}}$, $\eta \in \widehat{\mathcal{O}^\times_\mathbb{F}}$.

Proof: Since the space of distributions with support in $\{0\}$ is spanned by δ, it suffices for (1) to show that δ satisfies the conditions in Proposition 2.3 if and only if $(\chi,s,\eta,t) \in /\!/$. For this we use the reformulation in Remark 2.4. First note that δ always satisfies (2.5) since for any $\varphi \in C^\infty_c(\mathbb{E} - \{b^{-1}\})$:

$$\langle \delta, |1+b\cdot|^2_\mathbb{E}\varphi(\tfrac{(\cdot)}{1+b(\cdot)})\rangle = \varphi(0) = \langle \delta, \chi_{s-\frac{1}{2}}((1-b(\cdot))^2)^{-1}\varphi\rangle.$$

Moreover, (2.5) is satisfied by δ if and only if $\chi_{s+\frac{1}{2}}|_{\mathcal{O}^\times_\mathbb{F}} \cdot \eta^{-1}_{t+\frac{1}{2}} = 1$, i.e. $\chi|_{\mathcal{O}^\times_\mathbb{F}} \cdot \eta^{-1} = 1$ and $2(s+\frac{1}{2}) - (t+\frac{1}{2}) \in \frac{2\pi i}{\ln q_\mathbb{F}}\mathbb{Z}$. This is equivalent to $(\chi,s,\eta,t) \in /\!/$.

(2) is clear for all parameters such that $\widetilde{K}^{\chi,\eta}_{s,t} \neq 0$. But if $\widetilde{K}^{\chi,\eta}_{s,t} = 0$ for some $(s,t) = (s_0,t_0)$, we can consider the restriction of the holomorphic function $(s,t) \mapsto \widetilde{K}^{\chi,\eta}_{s,t}$ to a line through (s_0,t_0), say $s \mapsto \widetilde{K}^{\chi,\eta}_{s,t_0}$. This function vanishes at $s = s_0$, but is not identically zero by Theorem 3.12. It follows that for some $k \geq 0$, the function $s \mapsto (s-s_0)^{-k}\widetilde{K}^{\chi,\eta}_{s,t_0}$ is holomorphic and non-zero at $s = s_0$. Its value at $s = s_0$ is contained in $\mathcal{D}'(\mathbb{E})^{\chi,\eta}_{s_0,t_0}$ since the conditions in Proposition 2.3 depend continuously on $s,t \in \mathbb{C}$. □

Together with Lemma 4.1, the inequality (4.2) implies that $\dim \mathcal{D}'(\mathbb{E})^{\chi,\eta}_{s,t} = 1$ for $(\chi,s,\eta,t) \notin /\!/$. By Theorem 3.12 we have $\widetilde{K}^{\chi,\eta}_{s,t} \neq 0$ in this case, so $\mathcal{D}'(\mathbb{E})^{\chi,\eta}_{s,t} = \mathbb{C}\widetilde{K}^{\chi,\eta}_{s,t}$ for $(\chi,s,\eta,t) \notin /\!/$. For $(\chi,s,\eta,t) \in /\!/$ we study the restriction map $\mathcal{D}'(\mathbb{E})^{\chi,\eta}_{s,t} \to \mathcal{D}'(\mathbb{E}^\times)^{\chi,\eta}_{s,t}$.

Lemma 4.2: *The restriction map* $\mathcal{D}'(\mathbb{E})^{\chi,\eta}_{s,t} \to \mathcal{D}'(\mathbb{E}^\times)^{\chi,\eta}_{s,t}$ *is identically zero if* $(\chi,s,\eta,t) \in /\!/ - L$.

Proof: The proof follows the ideas of [12], Chapter 11.4.

For $(\chi, s, \eta, t) \in //$ the normalizing factor $L(\frac{1}{2}, \chi_s|_{\mathbb{F}^\times} \cdot \eta_t^{-1})$ of $\widetilde{K}_{s,t}^{\chi,\eta}$ has a pole of order one. From Theorem 3.10 it follows that $\widetilde{K}_{s,t}^{\chi,\eta}$ is holomorphic in s and t, so the meromorphic function

$$z \mapsto u_z = L(\tfrac{1}{2}, \chi_{s+z}|_{\mathbb{F}^\times} \cdot \eta_{t-2z}^{-1}) \cdot \widetilde{K}_{s+z,t-2z}^{\chi,\eta} = L(\tfrac{1}{2}, \chi_{s+z}|_{\mathbb{F}^\times} \cdot \eta_{t-2z})^{-1} \cdot K_{s+z,t-2z}^{\chi,\eta}$$

has at most a pole of order one at $z = 0$ and hence it can be expanded as

$$u_z = u_{-1} z^{-1} + u_0 + u_1 z + \cdots$$

with $u_j \in \mathcal{D}'(\mathbb{E})$. Since $(\chi, s, \eta, t) \notin L$, Theorem 3.12 implies $\operatorname{supp} u_{-1} = \{0\}$. Hence, $u_z|_{\mathbb{E}^\times}$ is holomorphic at $z = 0$ and $u_0|_{\mathbb{E}^\times} = (u_z|_{\mathbb{E}^\times})|_{z=0}$ is the following distribution:

$$\begin{aligned} u_0|_{\mathbb{E}^\times}(x_1 + x_2\alpha) &= L(\tfrac{1}{2}, \chi_s|_{\mathbb{F}^\times} \cdot \eta_t)^{-1} \cdot K_{s,t}^{\chi,\eta}|_{\mathbb{E}^\times}(x_1 + x_2\alpha) \\ &= \chi_{s-\frac{1}{2}}(x^2)(\chi_{s-\frac{1}{2}}\eta_{t+\frac{1}{2}})(x_1^2 - x_2^2\alpha^2)^{-1} \\ &\qquad \times L(\tfrac{1}{2}, \chi_s|_{\mathbb{F}^\times} \cdot \eta_t)^{-1}(\chi_{s-\frac{1}{2}}\eta_{t+\frac{1}{2}})(x_2). \end{aligned}$$

Note that the first factor $\chi_{s-\frac{1}{2}}(x^2)(\chi_{s-\frac{1}{2}}\eta_{t+\frac{1}{2}})(x_1^2 - x_2^2\alpha^2)^{-1}$ is smooth and nowhere vanishing on $\mathbb{E}^\times$ and the second factor is non-zero by Theorem A.1. It follows that $u_0|_{\mathbb{E}^\times}$ is a non-zero element of $\mathcal{D}'(\mathbb{E}^\times)_{s,t}^{\chi,\eta}$.

For $a \in \mathbb{F}^\times$ and $z \in \mathbb{C}$ we define the operator D_z^a on $\mathcal{D}'(\mathbb{E})$ by

$$D_z^a u(x) = u(ax) - |a|_{\mathbb{F}}^{4z-2} u(x).$$

Then Proposition 2.3 and Theorem 3.10 imply $D_z^a u_z = 0$ for all $a \in \mathbb{F}^\times$ and all $z \in \mathbb{C}$ for which u_z is defined. Since $|a|_{\mathbb{F}}^z = 1 + \ln |a|_{\mathbb{F}} \cdot z + \mathcal{O}(z^2)$ we can write

$$D_z^a = D_0^a + z D_1^a + \mathcal{O}(z^2)$$

with

$$D_0^a u(x) = u(ax) - |a|_{\mathbb{F}}^{-2} u(x) \qquad \text{and} \qquad D_1^a u(x) = -4(\ln |a|_{\mathbb{F}})|a|_{\mathbb{F}}^{-2} u(x).$$

Note that D_1^a is a multiple of the identity, so it commutes with D_0^a. Comparing coefficients of z^{-1} and z^0 in $D_z^a u_z = 0$ yields

$$D_0^a u_{-1} = 0 \ \text{ and } \ D_0^a u_0 + D_1^a u_{-1} = 0. \tag{4.3}$$

If now $v \in \mathcal{D}'(\mathbb{E})_{s,t}^{\chi,\eta}$, then by Proposition 3.2 and the observation above $v|_{\mathbb{E}^\times} = c \cdot u_0|_{\mathbb{E}^\times}$ for some $c \in \mathbb{C}$. This implies $(v - cu_0)|_{\mathbb{E}^\times} = 0$, so $\operatorname{supp}(v - cu_0) \subseteq \{0\}$ and hence $v - cu_0$ is a multiple of δ. But $\delta(ax) = |a|_{\mathbb{F}}^{-2}\delta(x)$ for all $a \in \mathbb{F}^\times$, whence

$$D_0^a(v - cu_0) = 0.$$

On the other hand $v \in \mathcal{D}'(\mathbb{E})^{\chi,\eta}_{s,t}$ implies $D^a_0 v = 0$ by Proposition 2.3, so by (4.3):

$$0 = c \cdot D^a_0 u_0 = -c \cdot D^a_1 u_{-1} = 4c \cdot (\ln |a|_{\mathbb{F}})|a|_{\mathbb{F}}^{-2} u_{-1} \qquad \text{for all } a \in \mathbb{F}^\times.$$

But since $u_{-1} \neq 0$ this gives $c = 0$, so $v|_{\mathbb{E}^\times} = 0$. □

By the previous statement combined with the exact sequence (4.1) and Lemma 4.1, we also find $\dim \mathcal{D}'(\mathbb{E})^{\chi,\eta}_{s,t} = 1$ for $(\chi, s, \eta, t) \in // - L$, and hence, by Theorem 3.12, $\mathcal{D}'(\mathbb{E})^{\chi,\eta}_{s,t} = \mathbb{C}\widetilde{K}^{\chi,\eta}_{s,t}$.

It remains to study the case $(\chi, s, \eta, t) \in L$, and we show that in this case we actually have $\dim \mathcal{D}'(\mathbb{E})^{\chi,\eta}_{s,t} = 2$. One symmetry breaking kernel in $\mathcal{D}'(\mathbb{E})^{\chi,\eta}_{s,t}$ is δ (see Lemma 4.1). We obtain a linearly independent kernel by regularizing $(s,t) \mapsto K^{\chi,\eta}_{s,t}$ along a hyperplane in $\mathbb{C}^2$.

Proposition 4.3: *Let $(\chi, s, \eta, t) \in L$. Then the function*

$$\mathbb{C} \to \mathcal{D}'(\mathbb{E}), \quad z \mapsto (1 - q_{\mathbb{F}}^{-2z})^{-1} \widetilde{K}^{\chi,\eta}_{s+z,t} = (1 - q_{\mathbb{F}}^{-(2s+2z+t+\frac{1}{2})}) K^{\chi,\eta}_{s+z,t}$$

is weakly holomorphic at $z = 0$ and its value at $z = 0$ defines a distribution $\approx{K}^{\chi,\eta}_{s,t}$ in $\mathcal{D}'(\mathbb{E})^{\chi,\eta}_{s,t}$ with support equal to $\mathbb{E}$.

Proof: By Theorem 3.10 the function

$$f : \mathbb{C} \to \mathcal{D}'(\mathbb{E})^{\chi,\eta}_{s,t}, \quad f(z) = \widetilde{K}^{\chi,\eta}_{s+z,t}$$

is weakly holomorphic, and by Theorem 3.12 it satisfies $f(0) = 0$. Therefore,

$$g(z) = (1 - q_{\mathbb{F}}^{-2z})^{-1} f(z)$$

is holomorphic at $z = 0$. Since $f(z) \in \mathcal{D}'(\mathbb{E})^{\chi,\eta}_{s+z,t}$ for all $z \in \mathbb{C}$ and the conditions in Proposition 2.3 depend continuously on s, it follows that $g(z)$ is contained in $\mathcal{D}'(\mathbb{E})^{\chi,\eta}_{s+z,t}$ for all $z \in \mathbb{C}$, in particular $g(0) \in \mathcal{D}'(\mathbb{E})^{\chi,\eta}_{s,t}$. To show that the distribution $g(0)$ has support equal to $\mathbb{E}$, first note that $(\chi, s, \eta, t) \in // - \backslash\backslash$, so $1 - q_{\mathbb{F}}^{-(2s-t+\frac{1}{2})} = 0$ while $1 - q_{\mathbb{F}}^{-(2s+t+\frac{1}{2})} \neq 0$. It follows that

$$g(z) = (1 - q_{\mathbb{F}}^{-(2s+2z+t+\frac{1}{2})}) K^{\chi,\eta}_{s+z,t}$$

with $1 - q_{\mathbb{F}}^{-(2s+2z+t+\frac{1}{2})} \neq 0$ for $z = 0$. Therefore, the restriction of $g(0)$ to $\mathbb{E} - \mathbb{F}$ is a non-trivial scalar multiple of the nowhere vanishing function $K^{\chi,\eta}_{s,t}|_{\mathbb{E}-\mathbb{F}}$. In particular, $\mathbb{E} - \mathbb{F} \subseteq \operatorname{supp} g(0)$ and hence $\operatorname{supp} g(0) = \mathbb{E}$. □

Clearly δ and $\approx{K}^{\chi,\eta}_{s,t}$ are linearly independent since they have different supports. This completes the proof of the following classification result:

Corollary 4.4:

$$\mathcal{D}'(\mathbb{E})^{\chi,\eta}_{s,t} = \begin{cases} \mathbb{C}\widetilde{K}^{\chi,\eta}_{s,t} & \textit{for } (\chi,s,\eta,t)\notin L, \\ \mathbb{C}\delta\oplus\mathbb{C}\overset{\approx}{K}{}^{\chi,\eta}_{s,t} & \textit{for } (\chi,s,\eta,t)\in L. \end{cases}$$

5. Mapping Properties of Symmetry Breaking Operators

In this section we study some mapping properties of symmetry breaking operators between principal series of PGL(2). For this, we denote by $\widetilde{A}^{\chi,\eta}_{s,t}:\pi_{\chi,s}\to\tau_{\eta,t}$ the intertwining operator with distribution kernel $\widetilde{K}^{\chi,\eta}_{s,t}$, i.e.

$$\widetilde{A}^{\chi,\eta}_{s,t}f(\overline{n}_y) = \int_{\mathbb{E}}\widetilde{K}^{\chi,\eta}_{s,t}(x)f(\overline{n}_{x+y})\,dx \qquad (f\in\operatorname{Ind}_B^G(\chi_s),y\in\mathbb{F}), \tag{5.1}$$

where the integral is to be understood in the sense of distributions. Note that $\widetilde{A}^{\chi,\eta}_{s,t}=0$ for $(\chi,s,\eta,t)\in L$. In the same way, we denote by $\overset{\approx}{A}{}^{\chi,\eta}_{s,t}$ the symmetry breaking operator with kernel $\overset{\approx}{K}{}^{\chi,\eta}_{s,t}$ for $(\chi,s,\eta,t)\in L$, and by $C^{\chi,\eta}_{s,t}$ the operator with kernel $\delta(x)$ in the case $(\chi,s,\eta,t)\in /\!/$, i.e. the restriction map

$$C^{\chi,\eta}_{s,t}f(\overline{n}_y) = f(\overline{n}_y) \qquad (f\in\operatorname{Ind}_B^G(\chi_s),y\in\mathbb{F}). \tag{5.2}$$

In the first part we determine the action of symmetry breaking operators on vectors which transform by a character under the action of a maximal compact subgroup, and in the second part we find the images of all symmetry breaking operators in the case where $\chi|_{\mathcal{O}_\mathbb{F}^\times}\cdot\eta=\chi|_{\mathcal{O}_\mathbb{F}^\times}\cdot\eta^{-1}=1$.

5.1. *Evaluation on one-dimensional types*

We first give a self-contained proof of the classification of one-dimensional subrepresentations of the restriction of $\pi_{\chi,s}$ to the maximal compact subgroup $K=\mathrm{PGL}(2,\mathcal{O}_\mathbb{E})$. For the case of p-adic fields, $p\neq 2$, see e.g. [18], Lemma 2.2, Theorem 3.3, §3.2. The proof below works in any (residue) characteristic.

Lemma 5.1: *The restriction of the representation $\pi_{\chi,s}$ of G to the maximal compact subgroup K contains a one-dimensional subrepresentation of K if and only if $\chi^2=1$, i.e. χ is real-valued. In this case the one-dimensional subrepresentation is unique and given by the character $\chi\circ\det:K\to\mathbb{C}^\times$. (Note that since $\chi^2=1$, the composition $\chi\circ\det$ is well-defined on* $\mathrm{PGL}(2,\mathcal{O}_\mathbb{E})$.*) It contains a unique vector $\varphi_{\chi,s}$ with $\varphi_{\chi,s}(1)=1$ which is given by*

$$\varphi_{\chi,s}(kb)=\chi_{s+\frac{1}{2}}(b)^{-1}\chi(\det k)^{-1} \qquad (k\in K,b\in B).$$

Proof: Since $KB = G$, the restriction of $\pi_{\chi,s}$ to K is isomorphic to $\mathrm{Ind}_{K\cap B}^{K}(\chi)$, where we abuse notation and view χ as the character of $K \cap B$ given by

$$\chi\begin{pmatrix} a & b \\ 0 & 1 \end{pmatrix} = \chi(a) \qquad (a \in \mathcal{O}_{\mathbb{E}}^{\times}, b \in \mathcal{O}_{\mathbb{E}}).$$

By Frobenius reciprocity, the multiplicity of a one-dimensional subrepresentation given by the character ρ of K in $\mathrm{Ind}_{K\cap B}^{K}(\chi)$ is equal to the multiplicity of χ in the restriction $\rho|_{K\cap B}$. This can at most be one. It therefore suffices to show that χ can be extended to a character ρ of K if and only if $\chi^2 = 1$, and that in this case the character is given by $\rho = \chi \circ \det$. That $\rho = \chi \circ \det$ extends χ from $K \cap B$ to K in the case $\chi^2 = 1$ is clear, so it remains to prove the converse statement. Assume that ρ is an extension of χ to K. Using the element $w \in K$ defined in (2.1), any matrix in K can be decomposed as follows:

$$\begin{pmatrix} a & b \\ c & d \end{pmatrix} = \begin{cases} w \begin{pmatrix} 1 & c/a \\ 0 & 1 \end{pmatrix} w \begin{pmatrix} a & b \\ 0 & (ad-bc)/a \end{pmatrix} & \text{if } a \in \mathcal{O}_{\mathbb{E}}^{\times}, \\ \begin{pmatrix} 1 & a/c \\ 0 & 1 \end{pmatrix} w \begin{pmatrix} c & d \\ 0 & (bc-ad)/c \end{pmatrix} & \text{else.} \end{cases}$$

(Note that if $a \notin \mathcal{O}_{\mathbb{E}}^{\times}$ then $c \in \mathcal{O}_{\mathbb{E}}^{\times}$, because the determinant $ad - bc$ is in $\mathcal{O}_{\mathbb{E}}^{\times}$.) It follows that

$$\rho\begin{pmatrix} a & b \\ c & d \end{pmatrix} = \begin{cases} \rho(w)^2 \chi(\frac{a^2}{ad-bc}) & \text{if } a \in \mathcal{O}_{\mathbb{E}}^{\times}, \\ \rho(w)\chi(\frac{c^2}{bc-ad}) & \text{else.} \end{cases}$$

Such a matrix with $a \notin \mathcal{O}_{\mathbb{E}}^{\times}$ is contained in K for all $b, c \in \mathcal{O}_{\mathbb{E}}^{\times}$, because in this case $ad - bc \in \mathcal{O}_{\mathbb{E}}$. Therefore, applying ρ to the identity

$$\begin{pmatrix} 0 & b \\ c & d \end{pmatrix} \begin{pmatrix} 0 & b' \\ c' & d' \end{pmatrix} = \begin{pmatrix} bc' & bd' \\ dc' & cb' + dd' \end{pmatrix}$$

yields

$$\rho(w)^2 \chi\left(\frac{cc'}{bb'}\right) = \rho(w)^2 \chi\left(\frac{(bc')^2}{bb'cc'}\right) \qquad \text{for all } b, b', c, c' \in \mathcal{O}_{\mathbb{E}}^{\times}.$$

This shows that $\chi(c^2) = \chi(b^2)$ for all $b, c \in \mathcal{O}_{\mathbb{E}}^{\times}$ and hence $\chi^2 = 1$. Finally, applying ρ to the identity

$$\begin{pmatrix} 1 & b \\ 0 & 1 \end{pmatrix} \begin{pmatrix} 0 & b' \\ c' & d' \end{pmatrix} = \begin{pmatrix} bc' & b' + bd' \\ c' & d' \end{pmatrix}$$

with $b \in \mathcal{O}_{\mathbb{E}}^{\times}$ gives

$$\rho(w)\chi\left(\frac{c'}{b'}\right) = \chi(-b'c')^{-1},$$

so $\rho(w) = \chi(-1)$. This shows that ρ is uniquely determined and in both cases $a \in \mathcal{O}_{\mathbb{E}}^{\times}$ and $a \notin \mathcal{O}_{\mathbb{E}}^{\times}$ given by $\chi \circ \det$. □

We now find an explicit expression for $\varphi_{\chi,s}$ on $\overline{N}$.

Lemma 5.2: *For $x \in \mathbb{E}$ we have*

$$\varphi_{\chi,s}(\overline{n}_x) = \max(1, |x|_{\mathbb{E}})^{-(2s+1)}.$$

Proof: For $|x|_{\mathbb{E}} \leq 1$, the matrix $\overline{n}_x$ is contained in K, so we can choose $b = 1$ in the decomposition into KB to find

$$\varphi_{\chi,s}(\overline{n}_x) = 1.$$

For $|x|_{\mathbb{E}} > 1$ we write

$$\overline{n}_x = \begin{pmatrix} 1 & 0 \\ x & 1 \end{pmatrix} = \begin{pmatrix} \varpi_{\mathbb{E}}^{-v(x)} & -\varpi_{\mathbb{E}}^{v(x)}/x \\ \varpi_{\mathbb{E}}^{-v(x)}x & 0 \end{pmatrix} \begin{pmatrix} \varpi_{\mathbb{E}}^{v(x)} & \varpi_{\mathbb{E}}^{v(x)}/x \\ 0 & \varpi_{\mathbb{E}}^{-v(x)} \end{pmatrix} \in KB,$$

so

$$\varphi_{\chi,s}(\overline{n}_x) = |\varpi_{\mathbb{E}}^{2v(x)}|_{\mathbb{E}}^{-(s+\frac{1}{2})} = |x|_{\mathbb{E}}^{-(2s+1)}.$$ □

Since $\widetilde{A}_{s,t}^{\chi,\eta}$ is G'-intertwining and the vector $\varphi_{\chi,s}$ transforms by the character $\chi \circ \det$ under the action of K, its image $\widetilde{A}_{s,t}^{\chi,\eta}\varphi_{\chi,s}$ transforms by the character $\chi|_{\mathcal{O}_{\mathbb{F}}^{\times}} \circ \det$ under the action of $K' = \mathrm{PGL}(2, \mathcal{O}_{\mathbb{F}})$. But $\tau_{\eta,t}|_{K'}$ contains a one-dimensional subrepresentation given by the character $\chi|_{\mathcal{O}_{\mathbb{F}}^{\times}} \circ \det$ if and only if $\eta = \chi|_{\mathcal{O}_{\mathbb{F}}^{\times}}$, so we find

$$\widetilde{A}_{s,t}^{\chi,\eta}\varphi_{\chi,s} = 0 \qquad \text{if } \chi|_{\mathcal{O}_{\mathbb{F}}^{\times}} \neq \eta.$$

Note that under the assumption $\chi^2 = 1$, we have the following equivalences:

$$\chi|_{\mathcal{O}_{\mathbb{F}}^{\times}} = \eta \quad \Leftrightarrow \quad \chi|_{\mathcal{O}_{\mathbb{F}}^{\times}} \cdot \eta = 1 \quad \Leftrightarrow \quad \chi|_{\mathcal{O}_{\mathbb{F}}^{\times}} \cdot \eta^{-1} = 1.$$

In what follows, assume that these conditions are satisfied, then $\widetilde{A}_{s,t}^{\chi,\eta}\varphi_{\chi,s}$ is a constant multiple of $\psi_{\eta,t}$, the unique vector in $\tau_{\eta,t}$ which transforms by the character $\eta \circ \det$ of K' and is normalized by $\psi_{\eta,t}(1) = 1$. To determine the constant, we evaluate (5.1) at $y = 0$ and use $\psi_{\eta,t}(1) = 1$, so

$$\widetilde{A}_{s,t}^{\chi,\eta}\varphi_{\chi,s} = \widetilde{A}_{s,t}^{\chi,\eta}\varphi_{\chi,s}(1) \cdot \psi_{\eta,t},$$

where

$$\widetilde{A}^{\chi,\eta}_{s,t}\varphi_{\chi,s}(1) = (1-q_{\mathbb{F}}^{-(2s-t+\frac{1}{2})})(1-q_{\mathbb{F}}^{-(2s+t+\frac{1}{2})}) \times \int_{\mathbb{E}} |x|_{\mathbb{E}}^{-t-\frac{1}{2}} |x_2|_{\mathbb{F}}^{2s+t-\frac{1}{2}} \max(1,|x|_{\mathbb{E}})^{-(2s+1)}\,dx.$$

The integral is computed in the following lemma:

Lemma 5.3: *For* $|\alpha^2|_{\mathbb{F}} = 1$*:*

$$\int_{\mathbb{E}} |x|_{\mathbb{E}}^{-t-\frac{1}{2}} |x_2|_{\mathbb{F}}^{2s+t-\frac{1}{2}} \max(1,|x|_{\mathbb{E}})^{-(2s+1)}\,dx = \frac{(1-q_{\mathbb{F}}^{-1})(1-q_{\mathbb{F}}^{-4s-2})}{(1-q_{\mathbb{F}}^{-(2s-t+\frac{1}{2})})(1-q_{\mathbb{F}}^{-(2s+t+\frac{1}{2})})},$$

and for $|\alpha^2|_{\mathbb{F}} = q_{\mathbb{F}}^{-1}$*:*

$$\int_{\mathbb{E}} |x|_{\mathbb{E}}^{-t-\frac{1}{2}} |x_2|_{\mathbb{F}}^{2s+t-\frac{1}{2}} \max(1,|x|_{\mathbb{E}})^{-(2s+1)}\,dx = \frac{(1-q_{\mathbb{F}}^{-1})(1+q_{\mathbb{F}}^{t-\frac{1}{2}})(1-q_{\mathbb{F}}^{-(2s+1)})}{(1-q_{\mathbb{F}}^{-(2s-t+\frac{1}{2})})(1-q_{\mathbb{F}}^{-(2s+t+\frac{1}{2})})}.$$

Proof: We write the integral over $x \in \mathbb{E}$ as a series of integrals over $\varpi_{\mathbb{F}}^m\mathcal{O}_{\mathbb{F}}^\times \times \varpi_{\mathbb{F}}^n\mathcal{O}_{\mathbb{F}}^\times$. Using that $\mathrm{vol}(\varpi_{\mathbb{F}}^m\mathcal{O}_{\mathbb{F}}^\times) = (1-q_{\mathbb{F}}^{-1})q_{\mathbb{F}}^{-m}$ we find

$$= (1-q_{\mathbb{F}}^{-1})^2 \sum_{m\in\mathbb{Z}}\sum_{n\in\mathbb{Z}} q_{\mathbb{F}}^{-m-n} |(\varpi_{\mathbb{F}}^m)^2 - (\varpi_{\mathbb{F}}^n)^2\alpha^2|_{\mathbb{F}}^{-t-\frac{1}{2}} |\varpi_{\mathbb{F}}^n|_{\mathbb{F}}^{2s+t-\frac{1}{2}} \times \max(1,|(\varpi_{\mathbb{F}}^m)^2 - (\varpi_{\mathbb{F}}^n)^2\alpha^2|_{\mathbb{F}})^{-(2s+1)}.$$

Recall that $|(\varpi_{\mathbb{F}}^m)^2 - (\varpi_{\mathbb{F}}^n)^2\alpha^2|_{\mathbb{F}}$ equals $|(\varpi_{\mathbb{F}}^m)^2|_{\mathbb{F}}$ if $n \geq m$ and $|(\varpi_{\mathbb{F}}^n)^2\alpha^2|_{\mathbb{F}}$ if $m > n$, which is true for both cases $|\alpha^2|_{\mathbb{F}} = 1$ or $|\alpha^2|_{\mathbb{F}} = q_{\mathbb{F}}^{-1}$. Then by writing $\sum\limits_{m\in\mathbb{Z}}\sum\limits_{n\in\mathbb{Z}} = \sum\limits_{m\in\mathbb{Z}}(\sum\limits_{m\leq n} + \sum\limits_{m>n})$, we have

$$= (1-q_{\mathbb{F}}^{-1})^2 \sum_{m\in\mathbb{Z}}\sum_{m\leq n} q_{\mathbb{F}}^{-m-n} |\varpi_{\mathbb{F}}^{2m}|_{\mathbb{F}}^{-t-\frac{1}{2}} |\varpi_{\mathbb{F}}^n|_{\mathbb{F}}^{2s+t-\frac{1}{2}} \max(1,|\varpi_{\mathbb{F}}^{2m}|_{\mathbb{F}})^{-(2s+1)}$$

$$+(1-q_{\mathbb{F}}^{-1})^2 \sum_{m\in\mathbb{Z}}\sum_{m>n} q_{\mathbb{F}}^{-m-n} |\varpi_{\mathbb{F}}^{2n}\alpha^2|_{\mathbb{F}}^{-t-\frac{1}{2}} |\varpi_{\mathbb{F}}^n|_{\mathbb{F}}^{2s+t-\frac{1}{2}} \max(1,|\varpi_{\mathbb{F}}^{2n}\alpha^2|_{\mathbb{F}})^{-(2s+1)}$$

$$= (1-q_{\mathbb{F}}^{-1})^2 \sum_{m\in\mathbb{Z}}\sum_{m\leq n} q_{\mathbb{F}}^{2mt} q_{\mathbb{F}}^{-n(2s+t+\frac{1}{2})} \max(1,q_{\mathbb{F}}^{-2m})^{-(2s+1)}$$

$$+ (1-q_{\mathbb{F}}^{-1})^2 |\alpha^2|_{\mathbb{F}}^{-t-\frac{1}{2}} \sum_{m\in\mathbb{Z}}\sum_{m>n} q_{\mathbb{F}}^{-m} q_{\mathbb{F}}^{-n(2s-t-\frac{1}{2})} \max(1,q_{\mathbb{F}}^{-2n}|\alpha^2|_{\mathbb{F}})^{-(2s+1)}.$$

In the first sum we write $n = m + k$ and in the second sum we write $m = n + k + 1$:

$$(1-q_{\mathbb{F}}^{-1})^2 \sum_{m\in\mathbb{Z}} q_{\mathbb{F}}^{-m(2s-t+\frac{1}{2})} \max(1, q_{\mathbb{F}}^{-2m})^{-(2s+1)} \sum_{k=0}^{\infty} q_{\mathbb{F}}^{-k(2s+t+\frac{1}{2})}$$

$$+q_{\mathbb{F}}^{-1}(1-q_{\mathbb{F}}^{-1})^2|\alpha^2|_{\mathbb{F}}^{-t-\frac{1}{2}} \sum_{n\in\mathbb{Z}} q_{\mathbb{F}}^{-n(2s-t+\frac{1}{2})} \max(1, q_{\mathbb{F}}^{-2n}|\alpha^2|_{\mathbb{F}})^{-(2s+1)} \sum_{k=0}^{\infty} q_{\mathbb{F}}^{-k}.$$

Finally, the sums over k are geometric series, and splitting the sums according to $\max(1, q_{\mathbb{F}}^{-2n}|\alpha^2|_{\mathbb{F}})$ and $\max(1, q_{\mathbb{F}}^{-2m})$ we get:

$$\frac{1-q_{\mathbb{F}}^{-1}}{1-q_{\mathbb{F}}^{-(2s-t+\frac{1}{2})}}\left(\frac{1-q_{\mathbb{F}}^{-1}}{1-q_{\mathbb{F}}^{-(2s+t+\frac{1}{2})}} + |\alpha^2|_{\mathbb{F}}^{-t-\frac{1}{2}} q_{\mathbb{F}}^{-1}\right)$$

$$+ q_{\mathbb{F}}^{-(2s+t+\frac{3}{2})} \frac{1-q_{\mathbb{F}}^{-1}}{1-q_{\mathbb{F}}^{-(2s+t+\frac{3}{2})}}\left(\frac{1-q_{\mathbb{F}}^{-1}}{1-q_{\mathbb{F}}^{-(2s+t+\frac{1}{2})}} + |\alpha^2|_{\mathbb{F}}^{-2s-t-\frac{3}{2}} q_{\mathbb{F}}^{-1}\right).$$

Considering the cases $|\alpha^2|_{\mathbb{F}} = 1$ and $|\alpha^2|_{\mathbb{F}} = q_{\mathbb{F}}^{-1}$ separately and simplifying the resulting expression shows the claimed formulas. □

Corollary 5.4: *Assume* $\chi^2 = 1$, *then* $\widetilde{A}_{s,t}^{\chi,\eta}\varphi_{\chi,s} = 0$ *unless* $\eta = \chi|_{\mathcal{O}_{\mathbb{F}}^\times}$ *in which case we have*

$$\widetilde{A}_{s,t}^{\chi,\eta}\varphi_{\chi,s} = (1-q_{\mathbb{F}}^{-1})\psi_{\eta,t} \times \begin{cases} (1-q_{\mathbb{E}}^{-(2s+1)}) & \text{for } \mathbb{E}/\mathbb{F} \text{ unramified,} \\ (1-q_{\mathbb{E}}^{-(2s+1)})(1+q_{\mathbb{F}}^{t-\frac{1}{2}}) & \text{for } \mathbb{E}/\mathbb{F} \text{ ramified.} \end{cases}$$

Note that if $\mathbb{E}/\mathbb{F}$ is unramified we have $\mathcal{O}_{\mathbb{F}}^\times \subseteq \mathcal{O}_{\mathbb{E}}^{\times 2}$, so the condition $\chi^2 = 1$ implies $\chi|_{\mathcal{O}_{\mathbb{F}}^\times} = 1$.

Remark 5.5: Clearly, $C_{s,t}^{\chi,\eta}\varphi_{\chi,s} = \psi_{\eta,t}$ for $(\chi, s, \eta, t) \in //$ and $\chi^2 = 1$. In the case $(\chi, s, \eta, t) \in L$ one can also determine the action of the operator $\approx{A}_{s,t}^{\chi,\eta}$ by taking residues in Corollary 5.4.

5.2. *Images of symmetry breaking operators*

In this section we determine the image of $\widetilde{A}_{s,t}^{\chi,\eta} \in \operatorname{Hom}_{G'}(\pi_{\chi,s}|_{G'}, \tau_{\eta,t})$ for all $s, t \in \mathbb{C}$ in the case where $\chi|_{\mathcal{O}_{\mathbb{F}}^\times} \cdot \eta = \chi|_{\mathcal{O}_{\mathbb{F}}^\times} \cdot \eta^{-1} = 1$. Recall from Section 3.2.4 that in this case

$$\widetilde{K}_{s,t}^{\chi,\eta}(x) = (1-q_{\mathbb{F}}^{2s+t+\frac{1}{2}})(1-q_{\mathbb{F}}^{2s-t+\frac{1}{2}}) \cdot \chi_0(x^2)|x|_{\mathbb{E}}^{-t-\frac{1}{2}}|x_2|_{\mathbb{F}}^{2s+t-\frac{1}{2}}.$$

We recall from Theorem 2.1 that the representation $\tau_{\eta,t}$ is irreducible unless $t \in \pm\frac{1}{2} + \frac{i\pi}{\ln q_{\mathbb{F}}}\mathbb{Z}$. For $t \in -\frac{1}{2} + \frac{i\pi}{\ln q_{\mathbb{F}}}\mathbb{Z}$ it has a one-dimensional subrepresentation spanned by the function $\psi_{\eta,t}$ defined in the previous section. Note that $\psi_{\eta,t}(\overline{n}_y) = 1$ for all $y \in \mathbb{F}$. As explained in Section 2.1, the irreducible subrepresentation $\tau^0_{\eta,t}$ of $\tau_{\eta,t}$ for $t \in \frac{1}{2} + \frac{i\pi}{\ln q_{\mathbb{F}}}\mathbb{Z}$ is the kernel of the intertwining operator

$$T_{\eta,t} : \tau_{\eta,t} \to \tau_{\eta,-t}, \quad T_{\eta,t}f(\overline{n}_y) = \int_{\mathbb{F}} f(\overline{n}_z)\,dz,$$

which maps onto the functions which are constant on $\overline{N}'$, i.e. onto $\mathbb{C}\psi_{\eta,-t}$.

Theorem 5.6: *Assume that $\chi|_{\mathcal{O}_{\mathbb{F}}^\times} \cdot \eta = \chi|_{\mathcal{O}_{\mathbb{F}}^\times} \cdot \eta^{-1} = 1$, and let $s,t \in \mathbb{C}$. The image of $\widetilde{A}^{\chi,\eta}_{s,t}$ is given as follows:*

(1) For $\chi^2 \neq 1$ and $\mathbb{E}/\mathbb{F}$ unramified:

$$\operatorname{Im}(\widetilde{A}^{\chi,\eta}_{s,t}) = \begin{cases} \{0\} & \text{if } (s,t) \in \{(0,\frac{1}{2}), (\frac{\pi i}{\ln q_{\mathbb{E}}}, \frac{1}{2} + \frac{\pi i}{\ln q_{\mathbb{F}}})\} = L^{\chi,\eta}, \\ \mathbb{C}\psi_{\eta,t} & \text{if } (s,t) \in \{(0,-\frac{1}{2}), (\frac{\pi i}{\ln q_{\mathbb{E}}}, -\frac{1}{2} + \frac{\pi i}{\ln q_{\mathbb{F}}})\}, \\ \tau^0_{\eta,t} & \text{if } (s,t) \in \left[\mathbb{C} \times (\frac{1}{2} + \frac{\pi i}{\ln q_{\mathbb{F}}}\mathbb{Z})\right] - L^{\chi,\eta}, \\ \tau_{\eta,t} & \text{else.} \end{cases}$$

(2) For $\chi^2 \neq 1$ and $\mathbb{E}/\mathbb{F}$ ramified:

$$\operatorname{Im}(\widetilde{A}^{\chi,\eta}_{s,t}) = \begin{cases} \{0\} & \text{if } (s,t) \in \{(0,\frac{1}{2}), (\frac{\pi i}{\ln q_{\mathbb{E}}}, \frac{1}{2}), \\ & (\frac{\pi i}{2\ln q_{\mathbb{E}}}, \frac{1}{2} + \frac{\pi i}{\ln q_{\mathbb{F}}}), (\frac{3\pi i}{2\ln q_{\mathbb{E}}}, \frac{1}{2} + \frac{\pi i}{\ln q_{\mathbb{F}}})\} = L^{\chi,\eta}, \\ \mathbb{C}\psi_{\eta,t} & \text{if } (s,t) \in \{(0,-\frac{1}{2}), (\frac{\pi i}{\ln q_{\mathbb{E}}}, -\frac{1}{2}), \\ & (\frac{\pi i}{2\ln q_{\mathbb{E}}}, -\frac{1}{2} + \frac{\pi i}{\ln q_{\mathbb{F}}}), (\frac{3\pi i}{2\ln q_{\mathbb{E}}}, -\frac{1}{2} + \frac{\pi i}{\ln q_{\mathbb{F}}})\}, \\ \tau^0_{\eta,t} & \text{if } (s,t) \in \left[\mathbb{C} \times (\frac{1}{2} + \frac{\pi i}{\ln q_{\mathbb{F}}}\mathbb{Z})\right] - L^{\chi,\eta}, \\ \tau_{\eta,t} & \text{else.} \end{cases}$$

(3) For $\chi^2 = 1$ and $\mathbb{E}/\mathbb{F}$ unramified:

$$\operatorname{Im}(\widetilde{A}^{\chi,\eta}_{s,t}) = \begin{cases} \{0\} & \text{if } (s,t) \in \{(-\frac{1}{2},-\frac{1}{2}), \\ & \qquad (-\frac{1}{2} + \frac{\pi i}{\ln q_{\mathbb{E}}}, -\frac{1}{2} + \frac{\pi i}{\ln q_{\mathbb{F}}})\} = L^{\chi,\eta}, \\ \mathbb{C}\psi_{\eta,t} & \text{if } (s,t) \in \left[\mathbb{C} \times (-\frac{1}{2} + \frac{\pi i}{\ln q_{\mathbb{F}}}\mathbb{Z})\right] - L^{\chi,\eta}, \\ \tau^0_{\eta,t} & \text{if } (s,t) \in \{(-\frac{1}{2},\frac{1}{2}), (-\frac{1}{2} + \frac{\pi i}{\ln q_{\mathbb{E}}}, \frac{1}{2} + \frac{\pi i}{\ln q_{\mathbb{F}}})\}, \\ \tau_{\eta,t} & \text{else.} \end{cases}$$

(4) For $\chi^2 = 1$ *and* $\mathbb{E}/\mathbb{F}$ *ramified:*

$$\mathrm{Im}(\widetilde{A}^{\chi,\eta}_{s,t}) = \begin{cases} \{0\} & \text{if } (s,t) \in \{(-\frac{1}{2}, -\frac{1}{2}), (-\frac{1}{2} + \frac{\pi i}{\ln q_{\mathbb{E}}}, -\frac{1}{2}), \\ & (\frac{\pi i}{2\ln q_{\mathbb{E}}}, \frac{1}{2} + \frac{\pi i}{\ln q_{\mathbb{F}}}), (\frac{3\pi i}{2\ln q_{\mathbb{E}}}, \frac{1}{2} + \frac{\pi i}{\ln q_{\mathbb{F}}})\} = L^{\chi,\eta}, \\ \mathbb{C}\psi_{\eta,t} & \text{if } (s,t) \in \Big[(\mathbb{C} - (-\frac{1}{2} + \frac{\pi i}{\ln q_{\mathbb{E}}}\mathbb{Z})) \times \{-\frac{1}{2}\}\Big] \\ & \cup \{(\frac{\pi i}{2\ln q_{\mathbb{E}}}, -\frac{1}{2} + \frac{\pi i}{\ln q_{\mathbb{F}}}), (\frac{3\pi i}{2\ln q_{\mathbb{E}}}, -\frac{1}{2} + \frac{\pi i}{\ln q_{\mathbb{F}}})\}, \\ \tau^0_{\eta,t} & \text{if } (s,t) \in \Big[(\mathbb{C} \times \{\frac{1}{2} + \frac{\pi i}{\ln q_{\mathbb{F}}}\}) - \\ & \{(\frac{\pi i}{2\ln q_{\mathbb{E}}}, \frac{1}{2} + \frac{\pi i}{\ln q_{\mathbb{F}}}), (\frac{3\pi i}{2\ln q_{\mathbb{E}}}, \frac{1}{2} + \frac{\pi i}{\ln q_{\mathbb{F}}})\}\Big] \\ & \cup \{(-\frac{1}{2}, \frac{1}{2}), (-\frac{1}{2} + \frac{\pi i}{\ln q_{\mathbb{E}}}, \frac{1}{2})\}, \\ \tau_{\eta,t} & \text{else.} \end{cases}$$

The proof follows from Theorem 3.12 together with the following two lemmas treating the cases $t \in \frac{1}{2} + \frac{i\pi}{\ln q_{\mathbb{F}}}\mathbb{Z}$ and $t \in -\frac{1}{2} + \frac{i\pi}{\ln q_{\mathbb{F}}}\mathbb{Z}$ where $\tau_{\eta,t}$ is reducible.

We start with the case $t \in \frac{1}{2} + \frac{i\pi}{\ln q_{\mathbb{F}}}\mathbb{Z}$. The image of $\widetilde{A}^{\chi,\eta}_{s,t}$ is contained in the subrepresentation $\tau^0_{\eta,t}$ if and only if $T_{\eta,t} \circ \widetilde{A}^{\chi,\eta}_{s,t} = 0$. This composition can be expressed in terms of $\widetilde{A}^{\chi,\eta}_{s,-t}$:

Lemma 5.7: *Assume that* $\chi|_{\mathcal{O}^\times_{\mathbb{F}}} \cdot \eta = \chi|_{\mathcal{O}^\times_{\mathbb{F}}} \cdot \eta^{-1} = 1$ *and* $t \in \frac{1}{2} + \frac{\pi i}{\ln q_{\mathbb{F}}}\mathbb{Z}$. *Then*

$$T_{\eta,t} \circ \widetilde{A}^{\chi,\eta}_{s,t} = c_{\mathbb{F}}\Gamma(|\cdot|^{2t}_{\mathbb{F}})\Gamma((\chi^2)_{-t+\frac{1}{2}}) \cdot \widetilde{A}^{\chi,\eta}_{s,-t}$$

as an identity of holomorphic functions in $s \in \mathbb{C}$. *In particular,* $\mathrm{Im}(\widetilde{A}^{\chi,\eta}_{s,t}) \subseteq \tau^0_{\eta,t}$ *if and only if*

$$\begin{cases} \text{always} & \text{if } \chi^2 \neq 1, \\ (s,t) \in \{(-\frac{1}{2}, \frac{1}{2}), (-\frac{1}{2} + \frac{i\pi}{\ln q_{\mathbb{E}}}, \frac{1}{2} + \frac{i\pi}{\ln q_{\mathbb{F}}})\} & \text{if } \chi^2 = 1 \text{ and } \mathbb{E}/\mathbb{F} \text{ is unramified}, \\ t = \frac{1}{2} + \frac{\pi i}{\ln q_{\mathbb{F}}} \text{ or } (s,t) \in \{(-\frac{1}{2}, \frac{1}{2}), (-\frac{1}{2} + \frac{\pi i}{\ln q_{\mathbb{E}}}, \frac{1}{2})\} & \text{if } \chi^2 = 1 \text{ and } \mathbb{E}/\mathbb{F} \text{ is ramified}. \end{cases}$$

Proof: The composition of $\widetilde{A}^{\chi,\eta}_{s,t}$ with $T_{\eta,t}$ is given by

$$\begin{aligned} T_{\eta,t} \circ \widetilde{A}^{\chi,\eta}_{s,t} f(\overline{n}_z) &= \int_{\mathbb{F}} \int_{\mathbb{E}} \widetilde{K}^{\chi,\eta}_{s,t}(x) f(\overline{n}_{x+y})\, dx\, dy \\ &= \int_{\mathbb{E}} \left(\int_{\mathbb{F}} \widetilde{K}^{\chi,\eta}_{s,t}(x+y)\, dy \right) f(\overline{n}_x)\, dx. \end{aligned}$$

On the other hand, $T_{\eta,t} \circ \widetilde{A}^{\chi,\eta}_{s,t} \in \operatorname{Hom}_{G'}(\pi_{\chi,s}|_{G'}, \tau_{\eta,-t})$, so by Corollary 4.4 $T_{\eta,t} \circ \widetilde{A}^{\chi,\eta}_{s,t} = c(\chi, s, \eta, t)\widetilde{A}^{\chi,\eta}_{s,-t}$ for some constant $c(\chi, s, \eta, t)$. (This argument works for generic $s \in \mathbb{C}$ such that $(s,t) \notin L^{\chi,\eta}$, but since $c(\chi, s, \eta, t)$ is meromorphic in s, the formula $T_{\eta,t} \circ \widetilde{A}^{\chi,\eta}_{s,t} = c(\chi, s, \eta, t)\widetilde{A}^{\chi,\eta}_{s,-t}$ is true for all s.) In terms of integral kernels, this means

$$\int_{\mathbb{F}} \widetilde{K}^{\chi,\eta}_{s,t}(x+y)\, dy = c(\chi, s, \eta, t)\widetilde{K}^{\chi,\eta}_{s,-t}(x).$$

Putting $x = \alpha$ this reads (dropping the normalization factors):

$$\int_{\mathbb{F}} \chi_0(y+\alpha)^2 |y+\alpha|_{\mathbb{E}}^{-t-\frac{1}{2}}\, dy = c(\chi, s, \eta, t) \cdot \chi_0(\alpha)^2 |\alpha|_{\mathbb{E}}^{t-\frac{1}{2}}.$$

The remaining integral can be evaluated with Proposition B.4. This shows the claimed formula. For the last statement, we combine Theorem 3.12 and Lemma B.1. □

Now let us consider the case $t \in -\frac{1}{2} + \frac{i\pi}{\ln q_{\mathbb{F}}}\mathbb{Z}$. The image of $\widetilde{A}^{\chi,\eta}_{s,t}$ is contained in the one-dimensional subrepresentation spanned by the function $\psi_{\eta,t}$ if and only if the kernel $\widetilde{K}^{\chi,\eta}_{s,t}(x)$ is independent of x_1, where $x = x_1 + x_2\alpha$.

Lemma 5.8: *Assume that* $\chi|_{\mathcal{O}_{\mathbb{F}}^\times} \cdot \eta = \chi|_{\mathcal{O}_{\mathbb{F}}^\times} \cdot \eta^{-1} = 1$ *and* $t \in -\frac{1}{2} + \frac{\pi i}{\ln q_{\mathbb{F}}}\mathbb{Z}$. *Then the kernel* $\widetilde{K}^{\chi,\eta}_{s,t}(x)$ *is independent of* x_1 *if and only if* $(s,t) \in L^{\chi,\eta}$ *(in which case* $\widetilde{K}^{\chi,\eta}_{s,t} = 0$*), or* $\chi^2 = 1$ *and* $t \in -\frac{1}{2} + \frac{2\pi i}{\ln q_{\mathbb{E}}}\mathbb{Z}$, *or*

$$(s,t) \in \begin{cases} \{(0, -\frac{1}{2}), (\frac{\pi i}{\ln q_{\mathbb{E}}}, -\frac{1}{2} + \frac{\pi i}{\ln q_{\mathbb{F}}})\} & \text{if } \mathbb{E}/\mathbb{F} \text{ is unramified,} \\ \{(0, -\frac{1}{2}), (\frac{\pi i}{\ln q_{\mathbb{E}}}, -\frac{1}{2}), (\frac{\pi i}{2\ln q_{\mathbb{E}}}, -\frac{1}{2} + \frac{\pi i}{\ln q_{\mathbb{F}}}), & \text{if } \mathbb{E}/\mathbb{F} \text{ is ramified.} \\ \qquad (\frac{3\pi i}{2\ln q_{\mathbb{E}}}, -\frac{1}{2} + \frac{\pi i}{\ln q_{\mathbb{F}}})\} & \end{cases}$$

Proof: Let us first assume that $2s \pm t + \frac{1}{2} \notin \frac{2\pi i}{\ln q_{\mathbb{F}}}\mathbb{Z}$, then $\operatorname{supp} \widetilde{K}^{\chi,\eta}_{s,t} = \mathbb{E}$ by Theorem 3.12. On $\mathbb{E} - \mathbb{F}$ the kernel is (up to the non-zero normalization factor) given by

$$K^{\chi,\eta}_{s,t}(x_1 + x_2\alpha) = \chi_0(x^2)|x|_{\mathbb{E}}^{-t-\frac{1}{2}} |x_2|_{\mathbb{F}}^{2s+t-\frac{1}{2}}$$

which is independent of x_1 if and only if it is equal to the same expression for $x_1 = 0$:

$$(\chi^2)_{-t-\frac{1}{2}}(x) = \chi_0(x^2)|x|_{\mathbb{E}}^{-t-\frac{1}{2}} = \chi_0(x_2^2\alpha^2)|x_2\alpha|_{\mathbb{E}}^{-t-\frac{1}{2}}$$

for all $x = x_1 + x_2\alpha$, $x_2 \neq 0$. But since $\chi^2|_{\mathcal{O}_{\mathbb{F}}^\times} = 1$ and

$$|x_2\alpha|_{\mathbb{E}}^{-t-\frac{1}{2}} = |x_2^2\alpha^2|_{\mathbb{F}}^{-t-\frac{1}{2}} = |x_2|_{\mathbb{F}}^{-2t-1}|\alpha^2|_{\mathbb{F}}^{-t-\frac{1}{2}},$$

the right-hand side equals $\chi_0(\alpha^2)|\alpha^2|_{\mathbb{F}}^{-t-\frac{1}{2}} \in \{\pm 1\}$ for all $x_2 \in \mathbb{F}^\times$. This implies that the above condition is equivalent to

$$(\chi^2)_{-t-\frac{1}{2}}(x) = \chi_0(\alpha^2)|\alpha^2|_{\mathbb{F}}^{-t-\frac{1}{2}}$$

for all $x \in \mathbb{E} - \mathbb{F}$, and therefore, by density, also for all $x \in \mathbb{E}^\times$. This implies that the character $(\chi^2)_{-t-\frac{1}{2}}$ of $\mathbb{E}^\times$ is constant, in particular the right-hand side has to be 1. We conclude that $\chi^2 = 1$ and $-t - \frac{1}{2} \in \frac{2\pi i}{\ln q_{\mathbb{E}}}\mathbb{Z}$.

Now, if $2s + t + \frac{1}{2} \in \frac{2\pi i}{\ln q_{\mathbb{F}}}\mathbb{Z}$, then $\widetilde{K}_{s,t}^{\chi,\eta}$ is a non-zero multiple of L_t^η by Theorem 3.11. Since $t \in -\frac{1}{2} + \frac{\pi i}{\ln q_{\mathbb{F}}}\mathbb{Z}$ we observe that $L_t^\eta(x) = (1 - q_{\mathbb{F}}^{-1})\delta(x_2)$ is independent of x_1.

Finally, if $2s - t + \frac{1}{2} \in \frac{2\pi i}{\ln q_{\mathbb{F}}}\mathbb{Z}$, then $\widetilde{K}_{s,t}^{\chi,\eta}(x)$ is a multiple of $\delta(x)$ which is not independent of x_1. It follows that, in this case, $\widetilde{K}_{s,t}^{\chi,\eta}$ is independent of x_1 if and only if it is zero, i.e. $(s,t) \in L^{\chi,\eta}$. □

For $(\chi, s, \eta, t) \in L$, we also determine the image of the two operators $C_{s,t}^{\chi,\eta}$ and $\widetilde{\widetilde{A}}_{s,t}^{\chi,\eta}$. We start with $C_{s,t}^{\chi,\eta}$ which is for all $(\chi, s, \eta, t) \in //$ given by restriction from $\overline{N}$ to $\overline{N}'$ (see (5.2)).

Proposition 5.9: *For all* $(\chi, s, \eta, t) \in //$ *we have*

$$\operatorname{Im}(C_{s,t}^{\chi,\eta}) = \tau_{\eta,t}.$$

Proof: The operator $C_{s,t}^{\chi,\eta}$ is clearly non-zero since there exist compactly supported locally constant functions on $\mathbb{E}$ whose restriction to $\mathbb{F}$ is non-zero. Therefore, its image is $\tau_{\eta,t}$ in all cases where this representation is irreducible. If $\tau_{\eta,t}$ is reducible, it has a unique proper subrepresentation which is either given by $\mathbb{C}\psi_{\eta,t}$ with $\psi_{\eta,t}(\overline{n}_y) = 1$ for all $y \in \mathbb{F}$ or by the kernel $\tau_{\eta,t}^0$ of the operator $T_{\eta,t} : \tau_{\eta,t} \to \tau_{\eta,-t}$ given by integration over $\mathbb{F}$. Since there exist compactly supported locally constant functions on $\mathbb{E}$ whose restriction to $\mathbb{F}$ is non-constant, the image of $C_{s,t}^{\chi,\eta}$ can never be contained in $\mathbb{C}\psi_{\eta,t}$. Similarly, there exist compactly supported locally constant functions on $\mathbb{E}$ whose integral over $\mathbb{F}$ is non-zero, so the image of $C_{s,t}^{\chi,\eta}$ can neither be $\tau_{\eta,t}^0$. □

Now let us study the images of the operators $\widetilde{\widetilde{A}}_{s,t}^{\chi,\eta}$ for $(\chi, s, \eta, t) \in L$. Note that $(\chi, s, \eta, t) \in L$ implies $\chi|_{\mathcal{O}_{\mathbb{F}}^\times} \cdot \eta = \chi|_{\mathcal{O}_{\mathbb{F}}^\times} \cdot \eta^{-1} = 1$ and either $t \in \frac{1}{2} + \frac{\pi i}{\ln q_{\mathbb{F}}}\mathbb{Z}$ or $t \in -\frac{1}{2} + \frac{\pi i}{\ln q_{\mathbb{F}}}\mathbb{Z}$.

Proposition 5.10: *Let* $(\chi, s, \eta, t) \in L$.

(1) *For* $t \in \frac{1}{2} + \frac{\pi i}{\ln q_{\mathbb{F}}}\mathbb{Z}$ *we always have* $\operatorname{Im}(\overset{\approx}{A}{}^{\chi,\eta}_{s,t}) = \tau^0_{\eta,t}$.

(2) *For* $t \in -\frac{1}{2} + \frac{\pi i}{\ln q_{\mathbb{F}}}\mathbb{Z}$ *we always have* $\operatorname{Im}(\overset{\approx}{A}{}^{\chi,\eta}_{s,t}) = \mathbb{C}\psi_{\eta,t}$.

Proof:

(1) From Lemma 5.7 it follows that

$$\begin{aligned} T_{\eta,t} \circ \overset{\approx}{A}{}^{\chi,\eta}_{s,t} &= (1 - q_{\mathbb{F}}^{-2z})^{-1} T_{\eta,t} \circ \widetilde{A}^{\chi,\eta}_{s+z,t}\Big|_{z=0} \\ &= (1 - q_{\mathbb{F}}^{-2z})^{-1} c_{\mathbb{F}} \Gamma(|\cdot|^{2t}_{\mathbb{F}}) \Gamma((\chi^2)_{-t+\frac{1}{2}}) \cdot \widetilde{A}^{\chi,\eta}_{s+z,-t}\Big|_{z=0}. \end{aligned}$$

It can be checked case-by-case using Lemma B.1 that $\Gamma(|\cdot|^{2t}_{\mathbb{F}})\Gamma((\chi^2)_{-t+\frac{1}{2}}) = 0$ in all cases, so the final expression is equal to zero for all z, hence $T_{\eta,t} \circ \overset{\approx}{A}{}^{\chi,\eta}_{s,t} = 0$.

(2) For $t \in -\frac{1}{2} + \frac{\pi i}{\ln q_{\mathbb{F}}}\mathbb{Z}$ and $(s,t) \in L^{\chi,\eta}$ we always have $\chi^2 = 1$, $2s - t + \frac{1}{2} = 0$ and $t \in -\frac{1}{2} + \frac{2\pi i}{\ln q_{\mathbb{E}}}\mathbb{Z}$ by Theorem 3.12. Hence, the kernel $K^{\chi,\eta}_{s,t}$ has the form $K^{\chi,\eta}_{s+z,t}(x) = |x_2|^{2z-2}$. This distribution is clearly independent of x_1 for all $z \in \mathbb{C}$, so the same is true for its value at $z = 0$. This shows the claim. □

6. Application: Discrete Components in the Restriction of Unitary Representations

We recall the unitary structure on the complementary series and the Steinberg representations of $G = \mathrm{PGL}(2, \mathbb{E})$. The construction makes use of the invariant non-degenerate bilinear form

$$\operatorname{Ind}_B^G(\chi_s) \times \operatorname{Ind}_B^G(\overline{\chi}_{-s}) \to \mathbb{C}, \quad (f_1, f_2) \mapsto \int_{G/B} (f_1 \cdot f_2)(x)\, d_{G/B}(xB),$$

where $d_{G/B}(xB)$ denotes the invariant integral on $C(G/B, \Omega_{G/B})$ (see Theorem 1.1). Using (1.9), this form can also be written as an integral over $\overline{N}$:

$$\int_{G/B} (f_1 \cdot f_2)(x)\, d_{G/B}(xB) = \int_{\overline{N}} f_1(\overline{n}) f_2(\overline{n})\, d\overline{n}.$$

This motivates the realization of $\pi_{\chi,s}$ on a space of locally constant functions on $\overline{N} \simeq \mathbb{E}$ by restriction from G/B to the open dense subset $\overline{N}B/B \simeq \mathbb{E}$: for $f \in \operatorname{Ind}_B^G(\chi_s)$ let $f_{\mathbb{E}}(x) = f(\overline{n}_x)$ $(x \in \mathbb{E})$ and put

$$I(\chi, s)^\infty = \{f_{\mathbb{E}} : f \in \operatorname{Ind}_B^G(\chi_s)\}.$$

Then the invariant bilinear form from above translates to

$$I(\chi,s)^\infty \times I(\overline{\chi},-s)^\infty \to \mathbb{C}, \quad (f_1,f_2) \mapsto \int_{\mathbb{E}} f_1(x) f_2(x)\,dx.$$

Since this form is non-degenerate, we can identify $I(\chi,s)^\infty$ with a subspace of the dual space

$$I(\chi,s)^{-\infty} := (I(\overline{\chi},-s)^\infty)^\vee.$$

Note that since $C_c^\infty(\mathbb{E}) \subseteq I(\chi,s)^\infty$, we can view $I(\chi,s)^{-\infty}$ as a subspace of $\mathcal{D}'(\mathbb{E})$.

Recall from Section 2.1 the standard intertwining operators $T_{\chi,s} : \pi_{\chi,s} \to \pi_{\overline{\chi},-s}$. Using (2.2), we obtain their counterparts on $I(\chi,s)^\infty$:

$$T_{\chi,s} : I(\chi,s)^\infty \to I(\overline{\chi},-s)^\infty, \quad T_{\chi,s} f(x') = \int_{\mathbb{E}} \chi_{s-\frac{1}{2}}(x-x')^2 f(x)\,dx.$$

Combining the invariant bilinear form from above with the standard intertwining operator $\widetilde{T}_{\chi,s}$, we obtain an invariant sesquilinear form on $I(\chi,s)^\infty$ whenever $\chi^2 = 1$ and $s \in \mathbb{R}$:

$$\langle f_1, f_2 \rangle_{\chi,s} = \int_{\mathbb{E}} T_{\chi,s} f_1(x') \overline{f_2(x')}\,dx' = \int_{\mathbb{E}}\int_{\mathbb{E}} |x-x'|_{\mathbb{E}}^{2s-1} f_1(x) \overline{f_2(x')}\,dx\,dx'$$

for $f_1, f_2 \in I(\chi,s)^\infty$. To determine for which $s \in \mathbb{R}$ this form is positive definite, we apply the Fourier transform on $\mathbb{E}$ (see Appendix B for its definition and basic properties).

Since the intertwining operator $T_{\chi,s}$ is given by convolution with the character $|\cdot|_{\mathbb{E}}^{2s-1}$, and since the Fourier transform of this character equals $\Gamma(|\cdot|_{\mathbb{E}}^{2s})|\cdot|_{\mathbb{E}}^{-2s}$, we can rewrite the invariant sesquilinear form using the Plancherel formula for the Fourier transform:

$$\langle f_1, f_2 \rangle_{\chi,s} = c_{\mathbb{E}} \Gamma(|\cdot|_{\mathbb{E}}^{2s}) \int_{\mathbb{E}} \widehat{f_1}(x) \overline{\widehat{f_2}(x)} |x|_{\mathbb{E}}^{-2s}\,dx.$$

For $s \in (0,\frac{1}{2})$, the gamma factor is regular (see Lemma B.1) and the function $|x|_{\mathbb{E}}^{-2s}$ is locally integrable and positive, so the form is positive definite. The completion of $I(\chi,s)^\infty$ with respect to this inner product is the Hilbert space

$$I(\chi,s) = \{f \in \mathcal{D}'(\mathbb{E}) : \widehat{f} \in L^2(\mathbb{E}, |x|_{\mathbb{E}}^{-2s}\,dx)\}.$$

The representations $\pi_{\chi,s}$ of $G = \mathrm{PGL}(2,\mathbb{E})$ on $I(\chi,s)$ are unitary and irreducible and they form the complementary series of G. Note that

$$I(\chi,s)^\infty \subseteq I(\chi,s) \subseteq I(\chi,s)^{-\infty}.$$

For $s=\frac{1}{2}$, the kernel $I(\chi,\frac{1}{2})_0^\infty=\ker T_{\chi,\frac{1}{2}}$ of the intertwining operator $T_{\chi,\frac{1}{2}}$ is a proper subrepresentation of codimension one, called the Steinberg representation. Note that

$$I(\chi,\tfrac{1}{2})_0^\infty=\left\{f\in I(\chi,\tfrac{1}{2})^\infty:\int_{\mathbb{E}}f(x)\,dx=0\right\}.$$

The invariant form $\langle\cdot,\cdot\rangle_{\chi,\frac{1}{2}}$ vanishes on $I(\chi,\frac{1}{2})_0^\infty$, but we can renormalize it to obtain an inner product

$$\langle f_1,f_2\rangle'_{\chi,\frac{1}{2}}:=\Gamma(|\cdot|_{\mathbb{E}}^{2s})^{-1}\langle f_1,f_2\rangle_{\chi,s}\big|_{s=\frac{1}{2}}\qquad(f_1,f_2\in I(\chi,\tfrac{1}{2})_0^\infty).$$

Denote by $I(\chi,\frac{1}{2})_0$ the completion of $I(\chi,\frac{1}{2})_0^\infty$. By the previous computation we have

$$\langle f_1,f_2\rangle'_{\chi,\frac{1}{2}}=c_{\mathbb{E}}\int_{\mathbb{E}}\widehat{f_1}(x)\overline{\widehat{f_2}(x)}|x|_{\mathbb{E}}^{-1}\,dx.$$

Note that the integral exists since $f\in I(\chi,\frac{1}{2})_0^\infty$ implies $\widehat{f}(0)=0$, so $\widehat{f}$ vanishes in a neighborhood of 0. It follows that

$$I(\chi,\tfrac{1}{2})_0=\{f\in\mathcal{D}'(\mathbb{E}):\widehat{f}\in L^2(\mathbb{E},|x|_{\mathbb{E}}^{-1}\,dx)\}.$$

Denote by $J(\eta,t)^{\pm\infty}$ and $J(\eta,t)$ the corresponding representation spaces for G'. For $\eta=\chi|_{\mathcal{O}_{\mathbb{F}}^\times}$ and $t=2s+\frac{1}{2}$ the restriction operator

$$C_{s,t}^{\chi,\eta}:I(\chi,s)^\infty\to J(\eta,t)^\infty,\quad C_{s,t}^{\chi,\eta}f=f|_{\mathbb{F}},$$

is G'-intertwining. We consider its transpose

$$(C_{s,t}^{\chi,\eta})^\vee:J(\overline{\eta},-t)^{-\infty}\to I(\overline{\chi},-s)^{-\infty},\quad(C_{s,t}^{\chi,\eta})^\vee f(x_1+x_2\alpha)=f(x_1)\delta(x_2).$$

Theorem 6.1: *Assume that $\chi^2=1$ and put $\eta=\chi|_{\mathcal{O}_{\mathbb{F}}^\times}$.*

(1) For $s\in(\frac{1}{4},\frac{1}{2})$ and $t=2s-\frac{1}{2}$, the intertwining operator

$$(C_{-s,-t}^{\chi,\eta})^\vee:J(\eta,t)^{-\infty}\to I(\chi,s)^{-\infty}$$

restricts to an isometric embedding (up to a scalar) $J(\eta,t)\hookrightarrow I(\chi,s)$. In particular, the restriction of the irreducible unitarizable representation $\pi_{\chi,s}$ of G to G' contains $\tau_{\eta,t}$ as a direct summand.

(2) For $s=t=\frac{1}{2}$, the intertwining operator

$$(C_{-s,-t}^{\chi,\eta})^\vee:J(\eta,t)^{-\infty}\to I(\chi,s)^{-\infty}$$

restricts to an isometric embedding (up to a scalar) $J(\eta,t)_0\hookrightarrow I(\chi,s)_0$. In particular, the restriction of the irreducible unitarizable representation $\pi_{\chi,s}^0$ of G to G' contains $\tau_{\eta,t}^0$ as a direct summand.

Proof: We first note that

$$(\widehat{C^{\chi,\eta}_{-s,-t}})^\vee f(x_1+x_2\alpha)=\widehat{f}(x_1) \qquad \text{for all } f\in J(\eta,t)^{-\infty}, x_1,x_2\in\mathbb{F}.$$

It follows that

$$\begin{aligned}\|(C^{\chi,\eta}_{-s,-t})^\vee f\|^2_{\chi,s} &= c_{\mathbb{E}}\Gamma(|\cdot|^{2s}_{\mathbb{E}})\int_{\mathbb{E}}|\widehat{f}(x_1)|^2|x|_{\mathbb{E}}^{-2s}\,dx\\ &= c_{\mathbb{E}}\Gamma(|\cdot|^{2s}_{\mathbb{E}})\int_{\mathbb{F}}|\widehat{f}(x_1)|^2\left(\int_{\mathbb{F}}|x_1+x_2\alpha|_{\mathbb{E}}^{-2s}\,dx_2\right)dx_1.\end{aligned}$$

Substituting $x_2=x_1y$, the inner integral equals

$$\begin{aligned}\int_{\mathbb{F}}|x_1+x_2\alpha|_{\mathbb{E}}^{-2s}\,dx_2 = |x_1|_{\mathbb{F}}^{1-4s}\int_{\mathbb{F}}|1+y\alpha|_{\mathbb{E}}^{-2s}\,dy&\\ = c_{\mathbb{F}}\Gamma(|\cdot|^{4s-1}_{\mathbb{F}})\Gamma(|\cdot|^{1-2s}_{\mathbb{E}})\cdot|x_1|_{\mathbb{F}}^{1-4s}&\end{aligned}$$

by Proposition B.4. Note that the integral converges if and only if $s>\frac{1}{4}$. Since $1-4s=-2t$, it follows that

$$\|(C^{\chi^\vee,\eta^\vee}_{-s,-t})^\vee f\|^2_{\chi,s}=c_{\mathbb{E}}c_{\mathbb{F}}\Gamma(|\cdot|^{4s-1}_{\mathbb{F}})\Gamma(|\cdot|^{1-2s}_{\mathbb{E}})\Gamma(|\cdot|^{2s}_{\mathbb{E}})\int_{\mathbb{F}}|\widehat{f}(x_1)|^2|x_1|_{\mathbb{F}}^{-2t}\,dx_1$$

which is a non-zero multiple of $\|f\|^2_{\eta,t}$ for $s\in(\frac{1}{4},\frac{1}{2})$. This shows (1). To prove (2), we renormalize by $\Gamma(|\cdot|^{2s}_{\mathbb{E}})$ at $s=\frac{1}{2}$ and find

$$\langle(C^{\chi^\vee,\eta^\vee}_{-s,-t})^\vee f,(C^{\chi^\vee,\eta^\vee}_{-s,-t})^\vee f\rangle^0_{\chi,s}=c_{\mathbb{E}}\Gamma(|\cdot|^{4s-1}_{\mathbb{F}})\Gamma(|\cdot|^{1-2s}_{\mathbb{E}})\langle f,f\rangle^0_{\eta,t}.$$

By Lemma B.1, $\Gamma(|\cdot|^{4s-1}_{\mathbb{F}})$ has a single zero at $s=\frac{1}{2}$ while $\Gamma(|\cdot|^{1-2s}_{\mathbb{E}})$ has a single pole, so the product is regular and non-zero at $s=\frac{1}{2}$. This shows (2).

□

Appendices

A. Homogeneous Distributions

Let $\mathbb{F}$ be a non-Archimedean local field and $q_{\mathbb{F}}$ the cardinality of its residue field. We denote by $C_c^\infty(\mathbb{F})$ the space of compactly supported locally constant functions on $\mathbb{F}$ with values in $\mathbb{C}$. A distribution on $\mathbb{F}$ is by definition a functional $u: C_c^\infty(\mathbb{F})\to\mathbb{C}$. The space of all distributions on $\mathbb{F}$ is denoted by $\mathcal{D}'(\mathbb{F})$. Every locally integrable function u on $\mathbb{F}$ can be identified with the distribution given by

$$\langle u,f\rangle=\int_{\mathbb{F}}u(x)f(x)\,dx \qquad (f\in C_c^\infty(\mathbb{F})).$$

For a multiplicative character ξ of $\mathbb{F}^\times$ we say that a distribution $u \in \mathcal{D}'(\mathbb{F})$ is *homogeneous of degree* ξ if for any test function $f \in C_c^\infty(\mathbb{F})$ and $a \in \mathbb{F}^\times$ we have (see [9], §2.3):

$$\langle u, f(a^{-1}\cdot)\rangle = \xi(a)|a|_{\mathbb{F}}\langle u, f\rangle.$$

This means, u satisfies the following equality in the sense of generalized functions:

$$u(ax) = \xi(a)u(x) \qquad (a \in \mathbb{F}^\times, x \in \mathbb{F}).$$

Every multiplicative character ξ of $\mathbb{F}^\times$ can uniquely be written as

$$\xi(a) =: \chi_s(a) = |a|_{\mathbb{F}}^s \chi(\varpi_{\mathbb{F}}^{-\nu_{\mathbb{F}}(a)} a)$$

for some $s \in \mathbb{C}/\frac{2\pi i}{\ln q_{\mathbb{F}}}\mathbb{Z}$ and $\chi \in \widehat{\mathcal{O}_{\mathbb{F}}^\times}$. One can prove that χ_s is locally integrable on $\mathbb{F}$ if and only if $\Re(s) > -1$ and in this case it defines a homogeneous distribution of degree χ_s (see [9], §2.3). To extend this family of distributions analytically to all $s \in \mathbb{C}$, we renormalize χ_s by

$$\widetilde{\chi}_s(a) := L(1, \chi_s)^{-1}\chi_s(a),$$

where

$$L(t, \chi_s) = \begin{cases} 1 & \text{for } \chi \neq 1, \\ (1 - q_{\mathbb{F}}^{-s-t})^{-1} & \text{for } \chi = 1. \end{cases}$$

Then we have the following classification of homogeneous distributions in terms of $\widetilde{\chi}_s$ (see e.g. [9], §2.3):

Theorem A.1: *For every multiplicative character ξ of $\mathbb{F}^\times$ the space of homogeneous distribution of degree ξ is one-dimensional. More precisely, for $\xi = \chi_s$ it is spanned by $\widetilde{\chi}_s$. For $s \notin -1 + \frac{2\pi i}{\ln q_{\mathbb{F}}}\mathbb{Z}$ we have* $\operatorname{supp}\widetilde{\chi}_s = \mathbb{F}$ *and for $s \in -1 + \frac{2\pi i}{\ln q_{\mathbb{F}}}\mathbb{Z}$ we have $\widetilde{\chi}_s = (1 - q_{\mathbb{F}}^{-1}) \cdot \delta$, where δ denotes the Dirac distribution at $x = 0$.*

B. Evaluation of an Integral

For a quadratic extension $\mathbb{E}/\mathbb{F}$ of non-Archimedean local fields, $\mathbb{E} = \mathbb{F}(\alpha)$ with $\alpha^2 \in \mathbb{F}$, and a multiplicative character χ of $\mathbb{E}^\times$ we evaluate the integrals

$$\int_{\mathbb{F}} \chi(1 + y\alpha)\, dy \qquad \text{and} \qquad \int_{\mathbb{F}} \chi(y + \alpha)\, dy,$$

where dy denotes a Haar measure on the additive group $\mathbb{F}$.

The computation involves the Fourier transform on both $\mathbb{F}$ and $\mathbb{E}$. For this, we use the Haar measure $dx = dx_1\, dx_2$ ($x = x_1 + x_2\alpha$, $x_1, x_2 \in \mathbb{F}$) on

$\mathbb{E}$ which is the product of two copies of the same fixed Haar measure dy on $\mathbb{F}$. Fix a non-trivial additive character ψ of $\mathbb{F}$ and extend it to $\mathbb{E}$ by

$$\psi(x_1 + x_2\alpha) := \psi(x_1) \qquad (x_1, x_2 \in \mathbb{F}).$$

The character ψ can be used to define the Fourier transform $\mathcal{F}f = \widehat{f}$ of functions $f \in C_c^\infty(\mathbb{E})$ and $f \in C_c^\infty(\mathbb{F})$:

$$\widehat{f}(x) := \int_{\mathbb{E}} \psi(xy) f(y)\, dy \qquad (f \in C_c^\infty(\mathbb{E})) \qquad \text{and}$$

$$\widehat{f}(x_1) := \int_{\mathbb{F}} \psi(x_1 y_1) f(y_1)\, dy_1 \qquad (f \in C_c^\infty(\mathbb{F})).$$

The Fourier transform $\mathcal{F}$ gives bijections $C_c^\infty(\mathbb{E}) \to C_c^\infty(\mathbb{E})$ and $C_c^\infty(\mathbb{F}) \to C_c^\infty(\mathbb{F})$, respectively (see [9], §2.4). Its inverse is given by the inversion formula: there exist constants $c_{\mathbb{E}}, c_{\mathbb{F}} > 0$ such that

$$f(x) = c_{\mathbb{E}} \int_{\mathbb{E}} \psi(-xy) \widehat{f}(y)\, dy \qquad (f \in C_c^\infty(\mathbb{E})) \qquad \text{and}$$

$$f(x_1) = c_{\mathbb{F}} \int_{\mathbb{F}} \psi(-x_1 y_1) \widehat{f}(y_1)\, dy_1 \qquad (f \in C_c^\infty(\mathbb{F})).$$

With these definitions, the Fourier transform satisfies the convolution formula

$$\widehat{f * g} = \widehat{f} \cdot \widehat{g}, \qquad \text{where } (f * g)(x) = \int_{\mathbb{E}} f(y) g(x-y)\, dy$$

and the Plancherel formula

$$\int_{\mathbb{E}} f(x) \overline{g(x)}\, dx = c_{\mathbb{E}} \int_{\mathbb{E}} \widehat{f}(y) \overline{\widehat{g}(y)}\, dy,$$

and similar for $\mathbb{F}$.

Using the symplectic form

$$\omega(x_1 + x_2\alpha, y_1 + y_2\alpha) := x_1 y_2 - x_2 y_1 \qquad (x_1, x_2, y_1, y_2 \in \mathbb{F})$$

on the two-dimensional vector space $\mathbb{E}$ over $\mathbb{F}$, we further define the *symplectic Fourier transform* $\mathcal{F}_{\mathrm{symp}} f$ of $f \in C_c^\infty(\mathbb{E})$ by

$$\mathcal{F}_{\mathrm{symp}} f(x) := \int_{\mathbb{E}} \psi(\omega(x, y)) f(y)\, dy.$$

The symplectic Fourier transform appears naturally when studying the convolution of functions on the Heisenberg group in terms of the Fourier transform in the central variable (see e.g. [20], §XII.3.3). Since $\mathcal{F}_{\mathrm{symp}} f(x_1 + x_2\alpha) = \mathcal{F}f(-x_2 + x_1\alpha^{-1})$, the symplectic Fourier transform also maps

$C_c^\infty(\mathbb{E})$ bijectively onto itself. Note that $\omega(ax,y)=\omega(x,\overline{a}y)$ for all $a,x,y\in\mathbb{E}$, where $\overline{a_1+a_2\alpha}=a_1-a_2\alpha$.

All Fourier transforms can be extended uniquely to distributions by duality. More precisely, for $u,f\in C_c^\infty(\mathbb{E})$ we have

$$\int_{\mathbb{E}}\widehat{u}(x)f(x)\,dx=\int_{\mathbb{E}}u(y)\widehat{f}(y)\,dy,$$

so we define for $u\in\mathcal{D}'(\mathbb{E})$:

$$\langle\widehat{u},f\rangle:=\langle u,\widehat{f}\rangle\qquad(f\in C_c^\infty(\mathbb{E})).$$

Similar formulas extend the Fourier transform on $\mathbb{F}$ and the symplectic Fourier transform on $\mathbb{E}$ to distributions. Then for $x\in\mathbb{F}$ we have

$$\int_{\mathbb{F}}\psi(xy)\,dy=c_{\mathbb{F}}^{-1}\cdot\delta(x)$$

in the sense of distributions.

By §2.5 in [9] and Theorem A.1, the Fourier transform of a multiplicative character $\chi(x)|x|_{\mathbb{E}}^{-1}$ of $\mathbb{E}^\times$, which defines a homogeneous distribution of degree $\chi(x)|x|_{\mathbb{E}}^{-1}$ if $\chi\neq 1$, is a homogeneous distribution of degree $\chi(x)^{-1}$ to within a factor, called the gamma factor $\Gamma(\chi)$. The gamma factor $\Gamma(\chi)$ takes the following integral form:

$$\Gamma(\chi)=\lim_{n\to\infty}\int_{q_{\mathbb{E}}^{-n}\le|x|_{\mathbb{E}}\le q_{\mathbb{E}}^{n}}\psi(x)\chi(x)|x|_{\mathbb{E}}^{-1}\,dx,$$

where $q_{\mathbb{E}}$ denotes the order of the residue field of $\mathcal{O}_{\mathbb{E}}$. Then

$$\widehat{\chi}(x)=\Gamma(\chi|\cdot|_{\mathbb{E}})\cdot|x|_{\mathbb{E}}^{-1}\chi(x)^{-1}$$

as distributions. From §2.5 in [9] we further recall (see also [2], §23):

Lemma B.1: *For a multiplicative character χ of $\mathcal{O}_{\mathbb{E}}^\times$ and $s\in\mathbb{C}$ let χ_s be the multiplicative character of $\mathbb{E}^\times$ given by $\chi_s(x):=\chi(\varpi_{\mathbb{E}}^{-\nu_{\mathbb{E}}(x)}x)|x|_{\mathbb{E}}^{s}$ $(x\in\mathbb{E}^\times)$.*

- *(i) If $\chi\neq 1$, then $\Gamma(\chi_s)$ is holomorphic in $s\in\mathbb{C}$ and nowhere vanishing.*
- *(ii) If $\chi=1$, then $\Gamma(\chi_s)$ is meromorphic in $s\in\mathbb{C}$ with simple poles at $s\in\frac{2\pi i}{\ln q_{\mathbb{E}}}\mathbb{Z}$ and simple zeros at $s\in 1+\frac{2\pi i}{\ln q_{\mathbb{E}}}\mathbb{Z}$ and no further singularities or zeros.*

The relevance of the symplectic Fourier transform is its relation to the integrals we would like to evaluate:

Lemma B.2: *For $\varphi \in C_c^\infty(\mathbb{E})$, the following identity holds:*

$$\int_{\mathbb{F}} \varphi(1+y\alpha)\,dy = c_{\mathbb{F}} \int_{\mathbb{F}} \psi(r)\,\mathcal{F}_{\mathrm{symp}}\,\varphi(r\alpha)\,dr.$$

Proof: Indeed, we can compute as follows (in the distribution sense):

$$\begin{aligned}\int_{\mathbb{F}} \psi(r)\,\mathcal{F}_{\mathrm{symp}}\,\varphi(r\alpha)\,dr &= \int_{\mathbb{F}}\int_{\mathbb{E}} \psi(r)\psi(-ry_1)\varphi(y)\,dy\,dr \\ &= \int_{\mathbb{E}} \varphi(y) \int_{\mathbb{F}} \psi(r(1-y_1))\,dr\,dy \\ &= c_{\mathbb{F}}^{-1} \int_{\mathbb{E}} \varphi(y)\delta(1-y_1)\,dy \\ &= c_{\mathbb{F}}^{-1} \int_{\mathbb{F}} \varphi(1+y_2\alpha)\,dy_2. \qquad \square\end{aligned}$$

To be able to use the formula from Lemma B.2, we need to compute $\mathcal{F}_{\mathrm{symp}}\,\chi$ for a multiplicative character χ of $\mathbb{E}^\times$.

Lemma B.3: *The symplectic Fourier transform of a multiplicative character χ of $\mathbb{E}^\times$ is given by*

$$\mathcal{F}_{\mathrm{symp}}\,\chi(x) = |\alpha|_{\mathbb{E}}\chi(\alpha)\Gamma(\chi|\cdot|_{\mathbb{E}}) \cdot |\overline{x}|_{\mathbb{E}}^{-1}\chi(\overline{x})^{-1}.$$

Proof: For $x \in \mathbb{E}$ and $a \in \mathbb{E}^\times$ we have

$$\begin{aligned}\mathcal{F}_{\mathrm{symp}}\,\chi(ax) &= \int_{\mathbb{E}} \psi(\omega(ax,y))\chi(y)\,dy = \int_{\mathbb{E}} \psi(\omega(x,\overline{a}y))\chi(y)\,dy \\ &= |\overline{a}|_{\mathbb{E}}^{-1} \int_{\mathbb{E}} \psi(\omega(x,y))\chi(\overline{a}^{-1}y)\,dy = |\overline{a}|_{\mathbb{E}}^{-1}\chi(\overline{a})^{-1}\,\mathcal{F}_{\mathrm{symp}}\,\chi(x).\end{aligned}$$

Thus, $\mathcal{F}_{\mathrm{symp}}\,\chi$ is again a homogeneous distribution of degree $|\overline{x}|_{\mathbb{E}}^{-1}\chi(\overline{x})^{-1}$. It follows that

$$\mathcal{F}_{\mathrm{symp}}\,\chi(x) = c_\chi |\overline{x}|_{\mathbb{E}}^{-1}\chi(\overline{x})^{-1}$$

for some constant $c_\chi \in \mathbb{C}$. To determine c_χ, put $x = -\alpha$, then

$$\begin{aligned}c_\chi |\alpha|_{\mathbb{E}}^{-1}\chi(\alpha)^{-1} = \mathcal{F}_{\mathrm{symp}}\,\chi(-\alpha) &= \int_{\mathbb{E}} \psi(y_1)\chi(y)\,dy \\ &= \int_{\mathbb{E}} \psi(y)\chi(y)\,dy = \Gamma(\chi|\cdot|_{\mathbb{E}}). \qquad \square\end{aligned}$$

Proposition B.4: *For a multiplicative character χ of $\mathbb{E}^\times$ we have*

$$\int_{\mathbb{F}} \chi(1+y\alpha)\,dy = c_{\mathbb{F}}\chi(-1)\Gamma(\chi^{-1}|_{\mathbb{F}^\times}|\cdot|_{\mathbb{F}}^{-1})\Gamma(\chi|\cdot|_{\mathbb{E}}),$$

$$\int_{\mathbb{F}} \chi(y+\alpha)\,dy = c_{\mathbb{F}}\chi(-\alpha)|\alpha^2|_{\mathbb{F}}\Gamma(\chi^{-1}|_{\mathbb{F}^\times}|\cdot|_{\mathbb{F}}^{-1})\Gamma(\chi|\cdot|_{\mathbb{E}}).$$

Proof: First note that by the coordinate change $y = z\alpha^2$ we have

$$\int_{\mathbb{F}} \chi(y+\alpha)\,dy = \chi(\alpha)|\alpha^2|_{\mathbb{F}} \int_{\mathbb{F}} \chi(1+z\alpha)\,dz,$$

so it suffices to compute the first integral. By the previous Lemmata B.2 and B.3:

$$\begin{aligned}\int_{\mathbb{F}} \chi(1+y\alpha)\,dy &= c_{\mathbb{F}} \int_{\mathbb{F}} \psi(r)\,\mathcal{F}_{\mathrm{symp}}\,\chi(r\alpha)\,dr \\ &= c_{\mathbb{F}}|\alpha|_{\mathbb{E}}\chi(\alpha)\Gamma(\chi|\cdot|_{\mathbb{E}}) \int_{\mathbb{F}} \psi(r)|-r\alpha|_{\mathbb{E}}^{-1}\chi(-r\alpha)^{-1}\,dr \\ &= c_{\mathbb{F}}\chi(-1)\Gamma(\chi|\cdot|_{\mathbb{E}}) \int_{\mathbb{F}} \psi(r)|r|_{\mathbb{F}}^{-2}\chi(r)^{-1}\,dr.\end{aligned}$$

The remaining integral equals $\Gamma(\chi^{-1}|_{\mathbb{F}^\times}|\cdot|_{\mathbb{F}}^{-1})$. □

References

1. Joseph Bernstein and Andre Reznikov, *Subconvexity bounds for triple L-functions and representation theory*, Ann. of Math. (2) **172** (2010), no. 3, 1679–1718.
2. Colin J. Bushnell and Guy Henniart, *The local Langlands conjecture for* GL(2), Textbooks in Mathematics, vol. 335, Springer-Verlag, Berlin, 2006.
3. Jean-Louis Clerc, *Singular conformally invariant trilinear forms, I: The multiplicity one theorem*, Transform. Groups **21** (2016), no. 3, 619–652.
4. ———, *Singular conformally invariant trilinear forms, II: The higher multiplicity case*, Transform. Groups **22** (2017), no. 3, 651–706.
5. Gerald B. Folland, *A course in abstract harmonic analysis*, second ed., Textbooks in Mathematics, CRC Press, Boca Raton, FL, 2016.
6. Jan Frahm, Bent Ørsted, and Genkai Zhang, *An extension problem related to the fractional Branson–Gover operators*, J. Funct. Anal. **278** (2020), no. 5, 108395, 33.
7. Jan Frahm and Feng Su, *Upper bounds for geodesic periods over rank one locally symmetric spaces*, Forum Math. **30** (2018), no. 5, 1065–1077.
8. Jan Frahm and Clemens Weiske, *Symmetry breaking operators for real reductive groups of rank one*, J. Funct. Anal. **279** (2020), no. 5, 108568, 70.
9. Izrail' M. Gel'fand and Mark I. Graev, *Representations of the group of second-order matrices with elements in a locally compact field and special functions on locally compact fields*, Uspehi Mat. Nauk **18** (1963), no. 4 (112), 29–99.

10. Andreas Juhl and Bent Ørsted, *Shift operators, residue families and degenerate Laplacians*, Pacific J. Math. **308** (2020), no. 1, 103–160.
11. Toshiyuki Kobayashi, *A program for branching problems in the representation theory of real reductive groups*, Representations of reductive groups, Progr. Math., vol. 312, Birkhäuser/Springer, Cham, 2015, pp. 277–322.
12. Toshiyuki Kobayashi and Birgit Speh, *Symmetry breaking for representations of rank one orthogonal groups*, Mem. Amer. Math. Soc. **238** (2015), no. 1126.
13. ———, *Symmetry breaking for representations of rank one orthogonal groups II*, Lecture Notes in Mathematics, vol. 2234, Springer, Singapore, 2018.
14. Jan Möllers, Bent Ørsted, and Genkai Zhang, *On boundary value problems for some conformally invariant differential operators*, Comm. Partial Differential Equations **41** (2016), no. 4, 609–643.
15. Jan Möllers and Yoshiki Oshima, *Restriction of most degenerate representations of* $\mathrm{O}(1, N)$ *with respect to symmetric pairs*, J. Math. Sci. Univ. Tokyo **22** (2015), no. 1, 279–338.
16. Dipendra Prasad, *Invariant forms for representations of* GL_2 *over a local field*, Amer. J. Math. **114** (1992), no. 6, 1317–1363.
17. Paul J. Sally, Jr., *An introduction to p-adic fields, harmonic analysis and the representation theory of* SL_2, Lett. Math. Phys. **46** (1998), no. 1, 1–47.
18. Allan J. Silberger, PGL_2 *over the p-adics: its representations, spherical functions, and Fourier analysis*, Lecture Notes in Mathematics, Vol. 166, Springer-Verlag, Berlin-New York, 1970.
19. Birgit Speh and T. N. Venkataramana, *Discrete components of some complementary series*, Forum Math. **23** (2011), no. 6, 1159–1187.
20. Elias M. Stein, *Harmonic analysis: real-variable methods, orthogonality, and oscillatory integrals*, Princeton Mathematical Series, vol. 43, Princeton University Press, Princeton, NJ, 1993, With the assistance of Timothy S. Murphy, Monographs in Harmonic Analysis, III. MR 1232192
21. Clemens Weiske, *Branching of unitary* $\mathrm{O}(1, n+1)$*-representations with nontrivial* $(\mathfrak{g}, K)$*-cohomology*, Ann. Inst. Fourier (Grenoble) **74** (2024), no. 6, 2331–2377.

https://doi.org/10.1142/9789819825448_0004

Invariant Vectors for Depth Zero Supercuspidals of $GSp(4)$

Jonathan Cohen
Department of Mathematics
University of North Texas
225 S. Avenue B, Denton, TX 76201, USA
jonathan.cohen@unt.edu

Let F be a non-Archimedean local field of characteristic zero and (π, V) a depth zero, irreducible, supercuspidal representation of $\mathrm{GSp}(4, F)$ with trivial central character. We calculate the dimensions of the spaces of Klingen-invariant and Siegel-invariant vectors in V of level $\mathfrak{p}^n$ for all $n \geq 0$, illustrating the ideas for residual characteristic 3.

Contents

1. Introduction

This chapter is concerned with a dimension counting problem in p-adic representation theory for the group $\mathrm{GSp}(4)$. Part of the motivation comes from the following problem in the classical theory of Siegel modular forms. If $\Gamma \subset \mathrm{Sp}(4, \mathbb{Q})$ is a congruence subgroup, then the dimension of the space of cusp forms of level Γ (and integer weight k, say) is not known in general.

For example, if

$$\Gamma = \begin{bmatrix} \mathbb{Z} & 4\mathbb{Z} & \mathbb{Z} & \mathbb{Z} \\ \mathbb{Z} & \mathbb{Z} & \mathbb{Z} & \mathbb{Z} \\ \mathbb{Z} & 4\mathbb{Z} & \mathbb{Z} & \mathbb{Z} \\ 4\mathbb{Z} & 4\mathbb{Z} & 4\mathbb{Z} & \mathbb{Z} \end{bmatrix} \cap \mathrm{Sp}(4, \mathbb{Z})$$

is the "Klingen congruence subgroup of level 4" then the associated dimensions were only recently computed in [6]. The method by which this was achieved required, as one of its several inputs, the dimensions of spaces of fixed vectors in all irreducible smooth representations of $\mathrm{GSp}(4, \mathbb{Q}_2)$ for the subgroups

$$\mathrm{Kl}(4) = \begin{bmatrix} \mathbb{Z}_2 & \mathbb{Z}_2 & \mathbb{Z}_2 & \mathbb{Z}_2 \\ 4\mathbb{Z}_2 & \mathbb{Z}_2 & \mathbb{Z}_2 & \mathbb{Z}_2 \\ 4\mathbb{Z}_2 & \mathbb{Z}_2 & \mathbb{Z}_2 & \mathbb{Z}_2 \\ 4\mathbb{Z}_2 & 4\mathbb{Z}_2 & 4\mathbb{Z}_2 & \mathbb{Z} \end{bmatrix} \cap \mathrm{GSp}(4, \mathbb{Z}_2).$$

These dimensions, and more, had been computed in [8]. We remark that the different "shape" of the subgroups in question is an artifact of different conventions for alternating forms in the classical and representation-theoretic contexts.

If one hopes to use the approach taken in [6] for other congruence subgroups, then a necessary component is the determination of dimensions of spaces of fixed vectors in all irreducible smooth representations of $\mathrm{GSp}(4, \mathbb{Q}_p)$ for appropriate local subgroups. In this chapter we are concerned with the two families of subgroups

$$\mathrm{Kl}(p^n) = \begin{bmatrix} \mathbb{Z}_p & \mathbb{Z}_p & \mathbb{Z}_p & \mathbb{Z}_p \\ p^n\mathbb{Z}_p & \mathbb{Z}_p & \mathbb{Z}_p & \mathbb{Z}_p \\ p^n\mathbb{Z}_p & \mathbb{Z}_p & \mathbb{Z}_p & \mathbb{Z}_p \\ p^n\mathbb{Z}_p & p^n\mathbb{Z}_p & p^n\mathbb{Z}_p & \mathbb{Z}_p \end{bmatrix} \cap \mathrm{GSp}(4, \mathbb{Z}_p)$$

$$\mathrm{Si}(p^n) = \begin{bmatrix} \mathbb{Z}_p & \mathbb{Z}_p & \mathbb{Z}_p & \mathbb{Z}_p \\ \mathbb{Z}_p & \mathbb{Z}_p & \mathbb{Z}_p & \mathbb{Z}_p \\ p^n\mathbb{Z}_p & p^n\mathbb{Z}_p & \mathbb{Z}_p & \mathbb{Z}_p \\ p^n\mathbb{Z}_p & p^n\mathbb{Z}_p & \mathbb{Z}_p & \mathbb{Z}_p \end{bmatrix} \cap \mathrm{GSp}(4, \mathbb{Z}_p).$$

At this time, computing the associated dimensions for all representations appears overly ambitious, so we restrict ourselves to the very special case of depth zero supercuspidals (constructed in the next section). We will illustrate these ideas for the case $p = 3$, as this simplifies some matters without obscuring most of the general phenomena.

We give a brief outline of the contents of this chapter. After setting up notation and definitions in Section 2, we show in Section 3 that one class of depth zero supercuspidal representations (all of which are nongeneric) will not have any fixed vectors for $\mathrm{Kl}(\mathfrak{p}^n)$; see Theorem 3.1. We then obtain our main results, the dimension formulas for the remaining representations; see Theorems 3.9 and 3.14. This is the technical heart of the chapter, and requires the determination and enumeration of a certain set of double cosets, together with a collection of character table computations for $\mathrm{GSp}(4, \mathbb{F}_q)$. In Section 4 we give the dimension formulas for $\mathrm{Si}(\mathfrak{p}^n)$; see Theorems 4.5 and 4.9.

2. Setup

Let F be a finite extension of $\mathbb{Q}_p$, with ring of integers $\mathfrak{o}$, maximal ideal $\mathfrak{p}$, uniformizer ϖ, and residue field $\mathbb{F}_q$. Let $\lambda_0 : \mathbb{F}_q^\times \to \mathbb{C}^\times$ denote the nontrivial quadratic character when q is odd. Let $J = \begin{bmatrix} & & & 1 \\ & & 1 & \\ & -1 & & \\ -1 & & & \end{bmatrix}$ and let $G = \mathrm{GSp}(4, F) = \{g \in \mathrm{GL}(4, F) : {}^t g J g = \mu(g) J\}$, $Z = Z(G)$, $T \subset G$ the set of diagonal matrices, and $G^0 = \{g \in G : \mu(g) \in \mathfrak{o}^\times\}$. We will need the two maximal compact subgroups $K = \mathrm{GSp}(4, \mathfrak{o})$ and

$$\mathrm{K}(\mathfrak{p}) = \begin{bmatrix} \mathfrak{o} & \mathfrak{o} & \mathfrak{o} & \mathfrak{p}^{-1} \\ \mathfrak{p} & \mathfrak{o} & \mathfrak{o} & \mathfrak{o} \\ \mathfrak{p} & \mathfrak{o} & \mathfrak{o} & \mathfrak{o} \\ \mathfrak{p} & \mathfrak{p} & \mathfrak{p} & \mathfrak{o} \end{bmatrix} \cap G^0$$

as well as the standard Iwahori subgroup

$$I = \begin{bmatrix} \mathfrak{o}^\times & \mathfrak{o} & \mathfrak{o} & \mathfrak{o} \\ \mathfrak{p} & \mathfrak{o}^\times & \mathfrak{o} & \mathfrak{o} \\ \mathfrak{p} & \mathfrak{p} & \mathfrak{o}^\times & \mathfrak{o} \\ \mathfrak{p} & \mathfrak{p} & \mathfrak{p} & \mathfrak{o}^\times \end{bmatrix} \cap G$$

and the sequences of Klingen congruence subgroups

$$\mathrm{Kl}(\mathfrak{p}^n) = \begin{bmatrix} \mathfrak{o} & \mathfrak{o} & \mathfrak{o} & \mathfrak{o} \\ \mathfrak{p}^n & \mathfrak{o} & \mathfrak{o} & \mathfrak{o} \\ \mathfrak{p}^n & \mathfrak{o} & \mathfrak{o} & \mathfrak{o} \\ \mathfrak{p}^n & \mathfrak{p}^n & \mathfrak{p}^n & \mathfrak{o} \end{bmatrix} \cap G^0$$

and Siegel congruence subgroups

$$\mathrm{Si}(\mathfrak{p}^n) = \begin{bmatrix} \mathfrak{o} & \mathfrak{o} & \mathfrak{o} & \mathfrak{o} \\ \mathfrak{o} & \mathfrak{o} & \mathfrak{o} & \mathfrak{o} \\ \mathfrak{p}^n & \mathfrak{p}^n & \mathfrak{o} & \mathfrak{o} \\ \mathfrak{p}^n & \mathfrak{p}^n & \mathfrak{o} & \mathfrak{o} \end{bmatrix} \cap G^0$$

where n is a positive integer. Let K^+ and $\mathrm{K}(\mathfrak{p})^+$ be the prounipotent radicals of K and $\mathrm{K}(\mathfrak{p})^+$, respectively. Define the matrices

$$s_1 = \begin{bmatrix} & 1 & & \\ 1 & & & \\ & & & 1 \\ & & 1 & \end{bmatrix}, \qquad s_2 = \begin{bmatrix} 1 & & & \\ & & 1 & \\ & -1 & & \\ & & & 1 \end{bmatrix}, \qquad t_{i,j} = \begin{bmatrix} \varpi^{2i+j} & & & \\ & \varpi^{i+j} & & \\ & & \varpi^{i} & \\ & & & 1 \end{bmatrix},$$

$$S(x,y,z) = \begin{bmatrix} 1 & & & \\ x & 1 & & \\ y & & 1 & \\ z & y & -x & 1 \end{bmatrix}, \qquad N(x,y,z) = \begin{bmatrix} 1 & & & \\ & 1 & & \\ x & y & 1 & \\ z & x & & 1 \end{bmatrix}.$$

The images of s_1 and s_2 generate the Weyl group $W := N_G(T)/T$. Let

$$\mathrm{GL}_{2,2}(\mathbb{F}_q) := \left\{(g,h) \in \mathrm{GL}(2,\mathbb{F}_q)^2 : \det(g) = \det(h)\right\}.$$

Given ϖ, there is an isomorphism $f : \mathrm{K}(\mathfrak{p})/\mathrm{K}(\mathfrak{p})^+ \to \mathrm{GL}_{2,2}(\mathbb{F}_q)$ given by

$$\begin{bmatrix} a & * & * & \varpi^{-1}b \\ * & a' & b' & * \\ * & c' & d' & * \\ \varpi c & * & * & d \end{bmatrix} \mapsto \left(\begin{bmatrix} a & b \\ c & d \end{bmatrix}, \begin{bmatrix} a' & b' \\ c' & d' \end{bmatrix}\right).$$

For a representation $\rho_1 \boxtimes \rho_2$ of $\mathrm{GL}(2,\mathbb{F}_q)^2$, we denote by $[\rho_1 \boxtimes \rho_2]$ its restriction to $\mathrm{GL}_{2,2}(\mathbb{F}_q)$. For an irreducible representation ρ of a group, we denote by ω_ρ its central character.

If H is a subgroup of G and $\rho : H \to \mathrm{GL}(W)$ is a representation of H, then $\text{c-Ind}_H^G(\rho)$ is the space of functions $f : G \to W$ such that $f(hg) = \rho(h)f(g)$ for all $h \in H$, $g \in G$, with f being smooth and having compact support modulo H. This space is a smooth representation of G under right translation.

If π is a depth zero supercuspidal representation of G then π arises in one of two ways. Either $\pi = \text{c-Ind}_{ZK}^G(\sigma)$ where $\sigma|_K$ is an inflation of an irreducible cuspidal representation of K/K^+, or $\pi = \text{c-Ind}_{N_G(\mathrm{K}(\mathfrak{p}))}^G(\tau)$ where τ is an irreducible representation of $N_G(\mathrm{K}(\mathfrak{p}))$ and $\tau|_{\mathrm{K}(\mathfrak{p})}$ is an inflation of a cuspidal representation of $\mathrm{K}(\mathfrak{p})/\mathrm{K}(\mathfrak{p})^+$. We will compute $\dim \pi^{\mathrm{Kl}(\mathfrak{p}^n)}$ for all such π. Since π is supercuspidal, it has no fixed vectors for an Iwahori subgroup, and in particular $\pi^{\mathrm{Kl}(\mathfrak{p})} = 0$. We therefore need to consider only fixed vectors for $\mathrm{Kl}(\mathfrak{p}^n)$ with $n \geq 2$. Similarly for $\mathrm{Si}(\mathfrak{p}^n)$.

In the case where $q = 3$, which will be our focus, there are 12 irreducible depth zero supercuspidal representations of $\mathrm{GSp}(4, F)$ having trivial central character. Of these, 4 arise from K and the other 8 from $\mathrm{K}(\mathfrak{p})$. In the notation of [7], the 2 generic cuspidal characters of $\mathrm{GSp}(4, \mathbb{F}_3)$ are $X_4(\Theta)$ and $\lambda_0 X_4(\Theta)$, where Θ is any character of order 10 of the subgroup of order 20 in $\mathbb{F}_{81}^\times$; the 2 nongeneric cuspidal characters of $\mathrm{GSp}(4, \mathbb{F}_3)$ are θ_2 and $\lambda_0\theta_2$. One constructs the 8 relevant representations coming from $\mathrm{K}(\mathfrak{p})$ as follows: there are 3 cuspidal representations of $\mathrm{GL}(2, \mathbb{F}_3)$, which we denote by ρ_1, $\rho_{-1} := \lambda_0\rho_1$, and ρ_0, where ρ_0 is the unique such representation with reducible restriction to $\mathrm{SL}(2, \mathbb{F}_3)$. We note that $\omega_{\rho_1}(-1) = -1$ and $\omega_{\rho_0}(-1) = 1$. The representation $[\rho_0 \boxtimes \rho_0]$ is a sum of two irreducible representations $\sigma_\pm$ of $\mathrm{GL}_{2,2}(\mathbb{F}_q)$, while $[\rho_1 \boxtimes \rho_1]$ and $[\rho_{-1} \boxtimes \rho_1]$ are irreducible. When inflated to $\mathrm{K}(\mathfrak{p})$, each of these 4 representations has two extensions to $N_G(\mathrm{K}(\mathfrak{p}))$, and the corresponding 8 compact inductions are all distinct.

3. Klingen Vectors

We will first consider those depth zero supercuspidals coming from $\mathrm{K}(\mathfrak{p})$. It turns out that these do not admit any Klingen vectors of any level.

Theorem 3.1: *Let* $\pi = c\text{-}Ind_{N_G(\mathrm{K}(\mathfrak{p}))}^G(\tau)$ *as above. Then* $\pi^{\mathrm{Kl}(\mathfrak{p}^n)} = 0$ *for all* $n \geq 0$.

Proof: For $g \in G$ and $n \geq 1$ define

$$R_g := \left(g\mathrm{Kl}(\mathfrak{p}^n)g^{-1} \cap \mathrm{K}(\mathfrak{p})\right)\mathrm{K}(\mathfrak{p})^+ = \left(g\mathrm{Kl}(\mathfrak{p}^n)g^{-1} \cap N_G(\mathrm{K}(\mathfrak{p}))\right)\mathrm{K}(\mathfrak{p})^+.$$

If $f \in \pi^{\mathrm{Kl}(\mathfrak{p}^n)}$ then $f(g) \in \tau^{R_g}$. It suffices to show $\tau^{R_g} = 0$ for g ranging over a set of representatives for $N_G(\mathrm{K}(\mathfrak{p}))\backslash G/\mathrm{Kl}(\mathfrak{p}^n)$. We have $G = IN_G(T)I$ by a well-known decomposition (see [2]). Since $N_G(\mathrm{K}(\mathfrak{p}))$ contains I along with a set of representatives for $N_G(T)/T$, and using the Iwahori factorization of I, we may take $g = t_{i,j}S(x,y,z)$ where $i, j \in \mathbb{Z}$, $x, y, z \in \mathfrak{p}$. Using $s_2 \in \mathrm{Kl}(\mathfrak{p}^n) \cap \mathrm{K}(\mathfrak{p})$ we may assume that $i + j \geq i$, so $j \geq 0$.

If $x = 0$ then let $d \in \mathfrak{o}$ and compute

$$g\begin{bmatrix}1&&&\\&1&&\\&\varpi^j d&1&\\&&&1\end{bmatrix}g^{-1} = \begin{bmatrix}1&&&\\&1&&\\&d&1&\\&&&1\end{bmatrix}.$$

Thus $R_g \supset \begin{bmatrix} 1 & & & \\ & 1 & & \\ & \mathfrak{o} & 1 & \\ & & & 1 \end{bmatrix}$. The image of $\begin{bmatrix} 1 & & & \\ & 1 & & \\ & \mathfrak{o} & 1 & \\ & & & 1 \end{bmatrix}$ in $\mathrm{K}(\mathfrak{p})/\mathrm{K}(\mathfrak{p})^+$ is the unipotent radical of a parabolic subgroup. So $\tau^{R_g} = 0$ since $\tau|_{\mathrm{K}(\mathfrak{p})}$ is cuspidal.

If $x \neq 0$ then we may assume $1 \leq \mathrm{val}(x) \leq i$, since $g = S(\varpi^{-i}x, 0, 0)t_{i,j}S(0, y, z + xy)$, and $S(\varpi^{-i}x, 0, 0) \in \mathrm{K}(\mathfrak{p})$ if $\mathrm{val}(x) > i$. Then let $d \in \mathfrak{p}^{2i+j+1-2\mathrm{val}(x)} \subset \mathfrak{p}^{j+1}$, so $dx \in \mathfrak{p}^{i+j+1}$, and compute

$$g \begin{bmatrix} 1 & & & \\ & 1 & & \\ & d & 1 & \\ & & & 1 \end{bmatrix} g^{-1} = \begin{bmatrix} 1 & & & \\ & 1 & & \\ -\varpi^{-i-j}dx & \varpi^{-j}d & 1 & \\ -\varpi^{-2i-j}dx^2 & -\varpi^{-i-j}dx & & 1 \end{bmatrix}.$$

Thus $R_g \supset \begin{bmatrix} 1 & & & \\ & 1 & & \\ & & 1 & \\ \mathfrak{p} & & & 1 \end{bmatrix}$. The image of $\begin{bmatrix} 1 & & & \\ & 1 & & \\ & & 1 & \\ \mathfrak{p} & & & 1 \end{bmatrix}$ in $\mathrm{K}(\mathfrak{p})/\mathrm{K}(\mathfrak{p})^+$ is the unipotent radical of a parabolic subgroup. So $\tau^{R_g} = 0$ since $\tau|_{\mathrm{K}(\mathfrak{p})}$ is cuspidal. □

Now suppose $\pi = \text{c-Ind}_{ZK}^{G}(\sigma)$. For $g \in G$ and fixed $n \geq 1$ define the objects

$$\begin{aligned} [g] &= ZKg\mathrm{Kl}(\mathfrak{p}^n) \\ R_g &= (g\mathrm{Kl}(\mathfrak{p}^n)g^{-1} \cap K)K^+/K^+ \\ \mathrm{Supp}(\pi) &= \{[g] : \sigma^{R_g} \neq 0\} = \{[g] : \exists f \in \pi^{\mathrm{Kl}(\mathfrak{p}^n)},\ f(g) \neq 0\}. \end{aligned}$$

Then

$$\dim \pi^{\mathrm{Kl}(\mathfrak{p}^n)} = \sum_{[g] \in \mathrm{Supp}(\pi)} \dim \sigma^{R_g}.$$

We therefore require a determination of $\mathrm{Supp}(\pi)$ and of $\dim \sigma^{R_g}$ for $[g] \in \mathrm{Supp}(\pi)$.

3.1. *Determination of* Supp(π)

In this section we obtain a parametrization of $\mathrm{Supp}(\pi)$; in a subsequent section we will compute the values of $\dim \sigma^{R_g}$ via the character table of $K/K^+ = \mathrm{GSp}(4, \mathbb{F}_q)$. Arguing as in the nongeneric case, since $K \supset I$ and $K \supset \langle s_1, s_2 \rangle$, we may take double coset representatives for $ZK\backslash G/\mathrm{Kl}(\mathfrak{p}^n)$ to have the form $t_{i,j}S(x, y, z)$ where $x, y, z \in \mathfrak{p}$. Using $s_2 \in \mathrm{Kl}(\mathfrak{p}^n) \cap K$ we may assume that $i + j \geq i$, so $j \geq 0$. Define the two unipotent radicals

$U_S, U_K \subset \mathrm{GSp}(4, \mathbb{F}_q)$ by

$$U_S = \left\{ \begin{bmatrix} 1 & & & \\ & 1 & & \\ x & y & 1 & \\ z & x & & 1 \end{bmatrix} : x, y, z \in \mathbb{F}_q \right\}, \quad U_K = \left\{ \begin{bmatrix} 1 & & & \\ x & 1 & & \\ y & & 1 & \\ z & y & -x & 1 \end{bmatrix} : x, y, z \in \mathbb{F}_q \right\}.$$

Since σ is cuspidal, $\sigma^{R_g} = 0$ if R_g contains a conjugate of U_S or U_K. Most of the following lemmas are proved by constructing the appropriate matrix identity, as in the previous section; we omit most proofs as they are tedious.

Lemma 3.2: *Consider $[t_{i,j}S(x, y, z)]$ with $j \geq 0$.*

a) If $x \neq 0$ we may assume $1 \leq \mathrm{val}(x) \leq i - 1$. If $y \neq 0$ we may assume $1 \leq \mathrm{val}(y) \leq i+j-1$. If $z \neq 0$ we may assume $1 \leq \mathrm{val}(z) \leq \min(2i+j, n)-1$.

b) If $xy \neq 0$ we may assume $1 \leq \mathrm{val}(y/x) \leq j - 1$.

c) If $xz \neq 0$, we may assume $1 \leq \mathrm{val}(z/x) \leq i + j - 1$.

d) If $yz \neq 0$ we may assume $\min(1, 1+i) \leq \mathrm{val}(z/y) \leq \max(i, \mathrm{val}(y))-1$.

Lemma 3.3: *Let $j \geq 0$ and $[t_{i,j}S(x, y, z)] \in \mathrm{Supp}(\pi)$. Then $2 - n \leq i \leq n - 1$ and $2i + j \geq 1$.*

Proposition 3.4: *Suppose $j \geq 0$ and $[g] := [t_{i,j}S(x, y, z)] \in \mathrm{Supp}(\pi)$.*

a) We have $j \geq 1$.

b) If $i \leq 0$ then $[g] = [t_{i,j}]$.

Proof: a) Suppose $j = 0$. We may assume $xy = 0$ and $1 \leq i \leq n - 1$. Conjugating with s_2 if necessary, we may assume $x = 0$ and compute

$$g \begin{bmatrix} 1 & & & \\ & d_1 & d_2 & \\ & d_3 & d_4 & \\ & & & \Delta \end{bmatrix} g^{-1} = \begin{bmatrix} 1 & & & \\ -\varpi^{-i} y d_2 & d_1 & d_2 & \\ \varpi^{-i} y(1 - d_4) & d_3 & d_4 & \\ \varpi^{-2i}(z(1 - \Delta) - y^2 d_2) & \varpi^{-i} y(d_1 - \Delta) & \varpi^{-i} y d_2 & \Delta \end{bmatrix},$$

where $\Delta = d_1 d_4 - d_2 d_3$. If $1 \leq \mathrm{val}(y) \leq i-1$ then taking $d_4 = d_1^{-1}(1+d_2 d_3)$, $d_3 \in \mathfrak{o}$, $d_2 \in \mathfrak{p}^{2i-\mathrm{val}(y^2)} \subset \mathfrak{p}^{i-\mathrm{val}(y)+1} \subset \mathfrak{p}$, and $d_1 \in 1 + \mathfrak{p}^{i-\mathrm{val}(y)}$, shows $R_g \supset U_S$. So we may assume $y = 0$. If $1 \leq \mathrm{val}(z) < 2i$ then taking d_i so that $\Delta \in 1 + \mathfrak{p}^{2i-\mathrm{val}(z)}$ shows $R_g \supset \begin{bmatrix} 1 & & \\ & \mathrm{SL}(2,\mathbb{F}_3) & \\ & & 1 \end{bmatrix}$. The character table for $\mathrm{GSp}(4, \mathbb{F}_q)$ shows that $X_4(\Theta)$ and θ_2 have no fixed vectors for this subgroup.

b) We may assume $x = 0$. Note $j \geq i + j \geq 2i + j \geq 1$. Then

$$g \begin{bmatrix} d_1 d_4 & d_1 b & & \\ & d_1 & & \\ & d_3 & d_4 & -b \\ & & & 1 \end{bmatrix} g^{-1} = \begin{bmatrix} d_1 d_4 & \varpi^i d_1 b & & \\ & d_1 & & \\ \varpi^{-i-j}(d_4 y(d_1 - 1) + zb) & \varpi^{-j} d_3 & d_4 & -\varpi^i b \\ \varpi^{-2i-j} z(d_1 d_4 - 1) & m & & 1 \end{bmatrix}$$

where $m = \varpi^{-i-j}(y(d_1 - 1) + zd_1 b)$. Let $d_3 \in \mathfrak{p}^j$. If $y \neq 0$ we may assume $\mathrm{val}(y) < \mathrm{val}(z) - i$ and take $b \in \mathfrak{p}^{-i}$, $d_4 = d_1^{-1} \in 1 + bz/y + \mathfrak{p}^{i+j-\mathrm{val}(y)}$, so R_g contains a conjugate of U_K. If $y = 0$ and $z \neq 0$, we may assume $\mathrm{val}(z) < 2i + j \leq i + j$ and take $b \in \mathfrak{p}^{i+j-\mathrm{val}(z)}$, $d_3 \in \mathfrak{p}^j$, $d_4 = 1$ and $d_1 \in 1 + \mathfrak{p}^{2i+j-\mathrm{val}(z)}$, so $R_g \supset U_S$. □

Proposition 3.5: *Suppose $j \geq 0$ and $[t_{i,j}] \in \mathrm{Supp}(\pi)$.*

a) We have $i + j \leq n - 1$.

b) If $i \leq 0$ then $R_{t_{i,j}} = \begin{bmatrix} * & * & 0 & 0 \\ 0 & * & 0 & 0 \\ 0 & * & * & * \\ 0 & 0 & 0 & * \end{bmatrix}$.

c) If $i \geq 1$, $j \geq 1$, and $2i + j \leq n - 1$ then $R_{t_{i,j}} = \begin{bmatrix} * & 0 & 0 & 0 \\ 0 & * & 0 & 0 \\ 0 & * & * & 0 \\ 0 & 0 & 0 & * \end{bmatrix}$.

d) If $i \geq 1$, $j \geq 1$, and $2i + j \geq n$ then $R_{t_{i,j}} = \begin{bmatrix} * & 0 & 0 & 0 \\ 0 & * & 0 & 0 \\ 0 & * & * & 0 \\ * & 0 & 0 & * \end{bmatrix}$.

Define the matrices

$$\begin{aligned} X_k &= S(\varpi^k, 0, 0) \\ Y_k &= S(0, \varpi^k, 0) \\ Z_k &= S(0, 0, \varpi^k). \end{aligned}$$

It is easy to see using integral diagonal matrices that $[t_{i,j} S(x, 0, 0)] = [t_{i,j} X_{\mathrm{val}(x)}]$, $[t_{i,j} S(0, y, 0)] = [t_{i,j} Y_{\mathrm{val}(y)}]$, and $[t_{i,j} S(0, 0, z)] = [t_{i,j} Z_{\mathrm{val}(z)}]$.

Proposition 3.6: *Let $1 \leq j$, $1 \leq i \leq n - 1$, and $1 \leq k \leq n - 1$.*

a) Suppose $k \leq i - 1$. Then $[t_{i,j} X_k] \in \mathrm{Supp}(\pi)$ if and only if $i + j + k \leq n - 1$, in which case

$$R_{t_{i,j} X_k} = \left\{ \begin{bmatrix} a & & & \\ * & a & & \\ & & d & \\ * & & * & d \end{bmatrix} : a, d \in \mathbb{F}_q^\times \right\}.$$

b) Suppose $k \leq i+j-1$. *Then* $[t_{i,j}Y_k] \in \mathrm{Supp}(\pi)$ *if and only if* $2i+j \leq \min(n, 2k) - 1$, *in which case*

$$R_{t_{i,j}Y_k} = \left\{ \begin{bmatrix} a & & & \\ & d & & \\ * & * & a & \\ & * & & d \end{bmatrix} : a, d \in \mathbb{F}_q^\times \right\}.$$

c) Suppose $k \leq 2i+j-1$. *Then* $[t_{i,j}Z_k] \in \mathrm{Supp}(\pi)$ *if and only if* $i+j+1 \leq k$, *in which case*

$$R_{t_{i,j}Z_k} = \left\{ \begin{bmatrix} a & & & \\ & ad & & \\ & * & a/d & \\ * & & & a \end{bmatrix} : a, d \in \mathbb{F}_q^\times \right\}.$$

Now we will rule out some remaining double cosets from lying in $\mathrm{Supp}(\pi)$. For the inequalities appearing here, see Lemma 3.2.

Lemma 3.7: *Suppose* $1 \leq j$, $1 \leq \mathrm{val}(x) \leq i-1$, *and* $1 \leq \mathrm{val}(y) \leq i+j-1$.

a) If $1 \leq \mathrm{val}(y/x) \leq j-1$ *then* $[t_{i,j}S(x,y,0)] \notin \mathrm{Supp}(\pi)$.

b) If $1 \leq \mathrm{val}(z/x) \leq i+j-1$ *then* $[t_{i,j}S(x,0,z)] \notin \mathrm{Supp}(\pi)$.

c) If $1 \leq \mathrm{val}(z) \leq 2i+j-1$ *then* $[t_{i,j}S(0,y,z)] \notin \mathrm{Supp}(\pi)$.

There is one remaining family of double cosets to consider.

Lemma 3.8: *Suppose* $1 \leq \mathrm{val}(x) < i$, $1 \leq \mathrm{val}(y/x) < j$, *and* $1 \leq \mathrm{val}(z/y)$. *Let* $g = t_{i,j}S(x,y,z)$.

a) We have $[g] \in \mathrm{Supp}(\pi)$ *if and only if the following conditions hold:*

i) $i+j < i + \mathrm{val}(y + z/x) = j + \mathrm{val}(xz/y) < n$.

ii) $j < 2\mathrm{val}(y/x)$.

b) If conditions i) and ii) hold then R_g *is conjugate (by a diagonal matrix) to*

$$\left\{ \begin{bmatrix} a & & & \\ v & a & & \\ b & & a & \\ c & b & -v & a \end{bmatrix} \begin{bmatrix} 1 & & & \\ & 1 & & \\ & v & 1 & \\ & & & 1 \end{bmatrix} : a \in \mathbb{F}_q^\times, b, c, v \in \mathbb{F}_q \right\}.$$

We now make the following observation.

Theorem 3.9: *Suppose* $q = 3$, $\sigma = \theta_2$ *or* $\lambda_0\theta_2$, *and* $\pi = c\text{-}Ind_{ZK}^{G}(\sigma)$. *Then* $\pi^{\mathrm{Kl}(\mathfrak{p}^n)} = 0$ *for all* $n \geq 0$.

Proof: For all the subgroups R_g associated to potential $g \in \mathrm{Supp}(\pi)$, we have seen that R_g contains a conjugate of the subgroup $\left\{\begin{bmatrix} 1 & & x & \\ & 1 & & \\ & & 1 & \\ & & & 1 \end{bmatrix} : x \in \mathbb{F}_q\right\}$. So $\sigma^{R_g} = 0$ by the character table in [7]. □

Lemma 3.10: *If* $q = 3$ *and* $\sigma = X_4(\Theta)$ *or* $\lambda_0 X_4(\Theta)$, *then* $\dim \sigma^{R_g}$ *is given in Table 1 below.*

Proof: This is a simple exercise using the character table found in [7]. □

3.2. *Counting double cosets*

We have determined the necessary and sufficient conditions on $[t_{i,j}S(x, y, z)]$ to lie in $\mathrm{Supp}(\pi)$. It remains to consider when two such double cosets are equal, and then to enumerate $\mathrm{Supp}(\pi)$.

The first column of Table 1 below lists the families of double coset representatives (the last to be defined shortly). The conjugacy class of R_g in $\mathrm{GSp}(4, \mathbb{F}_q)$ is an invariant of $[g]$. From the R_g column in Table 1, one sees that the only potential equalities of double cosets occur within a row or else between those of the form $[t_{i,j}X_k]$ and $[t_{i',j'}Y_{k'}]$. Suppose now that $j \geq 1$. If $i \leq 0$ then it is not hard to verify that $[t_{i,j}]$ determines i and j directly. So assume $i \geq 1$. By the Cartan decomposition and the fact that $\mathrm{Kl}(\mathfrak{p}^n) \subset K$, the double coset $[t_{i,j}S(x, y, z)]$ determines i and j. It is easy to check by a direct matrix computation that $[t_{i,j}X_k]$, $[t_{i,j}Y_k]$, and $[t_{i,j}Z_k]$ each determines k uniquely as well. It is also not hard to show directly that $[t_{i,j}X_k] \neq [t_{i,j}Y_{k'}]$.

This just leaves the last family of double cosets $[g] = [t_{i,j}S(x, y, z)]$ considered in Lemma 3.8. It is easy to show by direct computation that $[g]$ determines $\mathrm{val}(x)$, $\mathrm{val}(y)$, and $\mathrm{val}(z)$. Using integral diagonal matrices, we may assume that x and y are powers of ϖ. So it remains only to consider what conditions on a unit $u \in \mathfrak{o}^\times$ are necessary and sufficient to allow $[t_{i,j}S(x, y, uz)] = [t_{i,j}S(x, y, z)]$.

Lemma 3.11: *Let* $[g] := [t_{i,j}S(x, y, uz)]$ *and* $[g_1] := [t_{i,j}S(x, y, z)]$ *lie in* $\mathrm{Supp}(\pi)$, *where* $[g_1]$ *satisfies the conditions of Lemma 3.8 and* $u \in \mathfrak{o}^\times$. *Then* $[g] = [g_1]$ *if and only if* $u \in 1 + \mathfrak{p}^{j-\mathrm{val}(y/x)}$.

Proof: If $\text{val}(u-1) \geq j - \text{val}(y/x)$ then $[g_1] = [g]$ since

$$g_1 \begin{bmatrix} 1 & & & \\ & 1 & & \\ & \frac{y}{x}(1-u) & u & \\ & & & u \end{bmatrix} g^{-1} = \begin{bmatrix} 1 & & & \\ & 1 & & \\ & \varpi^{-j}\frac{y}{x}(1-u) & u & \\ & & & u \end{bmatrix} \in K.$$

If $\text{val}(u-1) < j - \text{val}(y/x)$ then one can show by a tedious matrix computation that $[g_1] \neq [g]$; we omit the details. □

For $\vec{k} = (i, k_x, k_y, k_z, u) \in \mathbb{Z}_+^4 \times \mathfrak{o}^\times$ let $j = i + \text{val}(\varpi^{k_y} + u\varpi^{k_z - k_x}) - (k_x + k_z - k_y)$; we assume that $\varpi^{k_y} + u\varpi^{k_z-k_x} \neq 0$. Then define

$$S(\vec{k}) = t_{i,j} S(\varpi^{k_x}, \varpi^{k_y}, u\varpi^{k_z}) \in G.$$

We have determined the conditions on $\vec{k}$ so that $[S(\vec{k})] \in \text{Supp}(\pi)$, as well as necessary and sufficient conditions for two such double cosets to be distinct. It remains to compute the number of such double cosets. First, we do so for the others.

Proposition 3.12: *The numbers of distinct double cosets of each type, except those of the form $[S(\vec{k})]$, are given in the last column of Table 1.*

Proof: The numbers of double cosets $[t_{i,j}]$ for the first four rows of Table 1 are easy to compute.

Consider the double cosets $[t_{i,j}X_k]$. The conditions on (i, j, k) can be written as $k+1 \leq i \leq n-1-j-k$, $1 \leq j \leq n-2-2k$, $1 \leq k \leq \lfloor \frac{n-3}{2} \rfloor$. The reader can check that

$$\sum_{k=1}^{\lfloor \frac{n-3}{2} \rfloor} \sum_{j=1}^{n-2-2k} (n-1-j-2k) = \left\lfloor \frac{(n-1)(n-3)(2n-7)}{24} \right\rfloor.$$

It is not hard to check that the map $(i, j, k) \mapsto (i, j, i+j+k)$ induces a bijection between the sets of $[t_{i,j}X_k]$ and $[t_{i,j}Z_k]$ lying in $\text{Supp}(\pi)$. Equality with the number of $[t_{i,j}X_k]$ follows.

For $[t_{i,j}Y_k]$, the conditions $1 \leq j$, $1 \leq i$, $k \leq i+j-1$, and $2i+j \leq \min(n, 2k) - 1$ force $i + \lceil \frac{j+1}{2} \rceil \leq k$, $1 \leq i \leq \lfloor \frac{n-j-1}{2} \rfloor$, and $3 \leq j \leq n-3$. The reader can check that

$$\sum_{j=3}^{n-3} \sum_{i=1}^{\lfloor \frac{n-j-1}{2} \rfloor} \sum_{k=i+\lceil \frac{j+1}{2} \rceil}^{i+j-1} (1) = \frac{n-3}{6} \left\lfloor \frac{(n-2)(n-4)}{4} \right\rfloor.$$

This completes the proof. □

We finally count the last family of double cosets, those of the form $[g] = [S(\vec{k})]$. For ease of notation, let $x = \varpi^{k_x}$, $y = \varpi^{k_y}$ and $z = u\varpi^{k_z}$. For a fixed (i, k_x, k_y, k_z), the distinct $[S(\vec{k})]$ correspond to orbits of $1+\mathfrak{p}^{j-\mathrm{val}(y/x)}$ under multiplication on the sets

$$\begin{cases} \mathfrak{o}^\times & \text{if } \mathrm{val}(z) \neq \mathrm{val}(xy) \\ \mathfrak{o}^\times \setminus (-1+\mathfrak{p}) & \text{if } \mathrm{val}(z) = \mathrm{val}(xy) \text{ and } \mathrm{val}(1+u) = 0 \\ (-1+\mathfrak{p}^{\mathrm{val}(1+u)}) \setminus (-1+\mathfrak{p}^{\mathrm{val}(1+u)+1}) & \text{if } \mathrm{val}(z) = \mathrm{val}(xy) \text{ and } \mathrm{val}(1+u) > 0 \end{cases}.$$

This follows from Lemma 3.11. So the number of choices of u yielding distinct $[S(\vec{k})]$ are

$$\begin{cases} 2 * 3^{j-\mathrm{val}(y/x)-1} & \text{if } \mathrm{val}(z) \neq \mathrm{val}(xy) \\ 3^{j-\mathrm{val}(y/x)-1} & \text{if } \mathrm{val}(z) = \mathrm{val}(xy) \text{ and } \mathrm{val}(1+u) = 0 \\ 2 * 3^{j-\mathrm{val}(y/x)-\mathrm{val}(1+u)-1} & \text{if } \mathrm{val}(z) = \mathrm{val}(xy) \text{ and } \mathrm{val}(1+u) > 0 \end{cases}.$$

Note $j - \mathrm{val}(y/x) - \mathrm{val}(1+u) = i - \mathrm{val}(x) > 0$ when $\mathrm{val}(z) = \mathrm{val}(xy)$. The following lemma is a tedious counting exercise; we omit the details of the proof.

Lemma 3.13: *a) The number of double cosets* $[S(\vec{k})] \in \mathrm{Supp}(\pi)$ *with* $\mathrm{val}(z) < \mathrm{val}(xy)$ *is*

$$3^{\lfloor \frac{n-5}{4} \rfloor} a\left(n - 4\left\lfloor \frac{n-5}{4} \right\rfloor\right) - a(n)$$

where

$$a(n) = \left\lfloor \frac{(n-1)(n-3)(2n-7)}{24} \right\rfloor + \frac{n^2+3n+15}{2}.$$

b) The number of double cosets $[S(\vec{k})] \in \mathrm{Supp}(\pi)$ *with* $\mathrm{val}(z) > \mathrm{val}(xy)$ *is the same as the number with* $\mathrm{val}(z) < \mathrm{val}(xy)$.

c) The number of double cosets $[S(\vec{k})] \in \mathrm{Supp}(\pi)$ *with* $\mathrm{val}(z) = \mathrm{val}(xy)$ *and* $\mathrm{val}(1+u) = 0$ *is*

$$3^{\lfloor \frac{n-4}{4} \rfloor} b\left(n - 4\left\lfloor \frac{n-4}{4} \right\rfloor\right) - b(n)$$

where

$$b(n) = \frac{1}{2}\left\lfloor \frac{n^2}{4} \right\rfloor + \frac{3}{2}.$$

d) The number of double cosets $[S(\vec{k})] \in \mathrm{Supp}(\pi)$ *with* $\mathrm{val}(1+u) > 0$ *is*

$$3^{\lfloor \frac{n-6}{4} \rfloor} c\left(n - 4\left\lfloor \frac{n-6}{4} \right\rfloor\right) - c(n)$$

where

$$c(n) = \frac{n-3}{6}\left\lfloor \frac{(n-2)(n-4)}{4} \right\rfloor + \frac{1}{2}\left\lfloor \frac{n^2+2n+14}{2} \right\rfloor.$$

Assembling the above results yields the following theorem.

Theorem 3.14: *Let* $q = 3$, $n \geq 1$, $\sigma = X_4(\Theta)$ *or* $\lambda_0 X_4(\Theta)$, *and* $\pi = \text{c-}Ind_{ZK}^{G}(\sigma)$. *Then*

$$\dim \pi^{\mathrm{Kl}(\mathfrak{p}^n)} = -(n+6)^2 + 3^{\lfloor \frac{n-2}{4} \rfloor} \begin{cases} 112 & \text{if } n \in 4\mathbb{Z} \\ 147 & \text{if } n \in 1+4\mathbb{Z} \\ 65 & \text{if } n \in 2+4\mathbb{Z} \\ 85 & \text{if } n \in 3+4\mathbb{Z} \end{cases}.$$

Table 1. Klingen fixed vector information.

g	i	j	R_g	$\dim \sigma^{R_g}$	$\lvert\{g\}\rvert$
$t_{i,j}$	$[2-n, 0]$	$[1-2i, n-i-1]$	$\begin{bmatrix} * & * & & \\ & * & & \\ & * & * & * \\ & & & * \end{bmatrix}$	1	$\frac{n(n-1)}{2}$
$t_{i,j}$	$[1, n-2]$	$[\max(1, n-2i), n-i-1]$	$\begin{bmatrix} * & & & \\ & * & & \\ & * & * & \\ * & & & * \end{bmatrix}$	1	$\left\lfloor \frac{(n-1)^2}{4} \right\rfloor$
$t_{i,j}$	$[1, \lfloor \frac{n-2}{2} \rfloor]$	$[1, n-2i-1]$	$\begin{bmatrix} * & & & \\ & * & & \\ & * & * & \\ & & & * \end{bmatrix}$	4	$\left\lfloor \frac{(n-2)^2}{4} \right\rfloor$
$t_{i,j}Z_k$	$[2, n-3]$	$[1, n-i-2]$	$\begin{bmatrix} d & & & \\ & ad & & \\ & * & a/d & \\ * & & & a \end{bmatrix}$	2	$\left\lfloor \frac{(n-1)(n-3)(2n-7)}{24} \right\rfloor$
$t_{i,j}X_k$	$[2, n-3]$	$[1, n-i-2]$	$\begin{bmatrix} a & & & \\ * & a & & \\ & & d & \\ * & & * & d \end{bmatrix}$	2	$\left\lfloor \frac{(n-1)(n-3)(2n-7)}{24} \right\rfloor$
$t_{i,j}Y_k$	$[1, \lfloor \frac{n-4}{2} \rfloor]$	$[3, n-2i-1]$	$\begin{bmatrix} a & & & \\ & d & & \\ * & * & a & \\ & * & & d \end{bmatrix}$	2	$\frac{n-3}{6}\left\lfloor \frac{n^2-6n+8}{4} \right\rfloor$
$S(\vec{k})$	-	-	$\begin{bmatrix} a & & & \\ v & a & & \\ * & v & a & \\ * & * & -v & a \end{bmatrix}$	2	3.13

4. Siegel Vectors

Next we consider the case of Siegel vectors. The assumption that $q = 3$, or more precisely that q is odd, will simplify matters more significantly than before. Suppose first that π arises from $\mathrm{K}(\mathfrak{p})$. For $g \in G$ and fixed $n \geq 0$ define the objects

$$\begin{aligned} [g] &= N_G(\mathrm{K}(\mathfrak{p}))g\mathrm{Si}(\mathfrak{p}^n) \\ R_g &= (g\mathrm{Si}(\mathfrak{p}^n)g^{-1} \cap \mathrm{K}(\mathfrak{p}))\mathrm{K}(\mathfrak{p})^+ \\ \mathrm{Supp}(\pi) &= \{[g] : \sigma^{R_g} \neq 0\} = \{[g] : \exists f \in \pi^{\mathrm{Si}(\mathfrak{p}^n)},\ f(g) \neq 0\}. \end{aligned}$$

Then

$$\dim \pi^{\mathrm{Si}(\mathfrak{p}^n)} = \sum_{[g]\in \mathrm{Supp}(\pi)} \dim \sigma^{R_g}.$$

We therefore require a determination of $\mathrm{Supp}(\pi)$ and of $\dim \sigma^{R_g}$ for $[g] \in \mathrm{Supp}(\pi)$. Observe that

$$N_G(\mathrm{K}(\mathfrak{p})) \cap g\mathrm{Si}(\mathfrak{p}^n)g^{-1} = \mathrm{K}(\mathfrak{p}) \cap g\mathrm{Si}(\mathfrak{p}^n)g^{-1}$$

since $\mathrm{K}(\mathfrak{p})$ is the unique maximal compact subgroup of $N_G(\mathrm{K}(\mathfrak{p}))$. In particular, since τ is always an extension of σ (this relies on $q = 3$), we always have $\dim \tau^{R_g} = \dim \sigma^{R_g}$. We first compute $\dim \sigma^R$ for the subgroup $R = \{([{}^{a}\ {}_{b}], [{}^{c}\ {}_{abc^{-1}}]) : a, b, c \in \{\pm 1\}\} \subset \mathrm{GL}_{2,2}(\mathbb{F}_3)$, which will turn out to be the only R_g above.

Remark 4.1: Since q is odd, the image of $Z(\mathrm{K}(\mathfrak{p}))$ has index 2 in $Z(\mathrm{K}(\mathfrak{p})/\mathrm{K}(\mathfrak{p})^+)$ with a representative of the complement given by $(I_2, -I_2)$. A priori, we only know that central characters of representations of $\mathrm{K}(\mathfrak{p})/\mathrm{K}(\mathfrak{p})^+$ must be trivial on the image of $Z(\mathrm{K}(\mathfrak{p}))$ in order for the associated compact induction to have Siegel vectors. As a consequence of the next result, we will see that the relevant representations of $\mathrm{K}(\mathfrak{p})/\mathrm{K}(\mathfrak{p})^+$ are precisely the ones with trivial central character. This will rule out 4 of the 8 representations of G constructed above.

Lemma 4.2: *Let $q = 3$, $\rho_{\pm 1}$ and ρ_0 be the irreducible cuspidal representations of $\mathrm{GL}(2, \mathbb{F}_q)$ defined above.*
a) If $\sigma = [\rho_{\pm 1} \boxtimes \rho_1]$ then $\dim \sigma^R = 0$.
b) If $\sigma = \sigma_\pm$ then $\dim \sigma^R = 1$.

Proof: a) Let $(t_1, t_2) \in R$. Since the ρ_i are cuspidal representations of $\mathrm{GL}(2, \mathbb{F}_q)$, the character table of $\mathrm{GL}(2, \mathbb{F}_3)$ gives that $\dim(\rho_i) = 2$ and $\mathrm{tr}\rho_i(t_j) = 0$ unless t_j is a scalar matrix. So

$$\begin{aligned} |R| \dim \sigma^R &= \sum_{(t_1,t_2)\in R} \mathrm{tr}\rho_{\pm 1}(t_1)\mathrm{tr}\rho_1(t_2) \\ &= 4 \sum_{a=\pm 1} \omega_{\rho_{\pm 1}}(a)\omega_{\rho_1}(a)(1 + \omega_{\rho_1}(-1)) \\ &= 0. \end{aligned}$$

b) Recall $[\rho_0 \boxtimes \rho_0] = \sigma_+ \oplus \sigma_-$. Clearly $(\rho_0 \boxtimes \rho_0)\left(\begin{bmatrix} -1 & \\ & 1 \end{bmatrix}, I_2\right)$ interchanges σ_+ and σ_-. Let $\xi_\pm$ be the irreducible summands of $\rho_0|_{\mathrm{SL}(2,\mathbb{F}_q)}$. It follows that

$$\sigma|_{\mathrm{SL}(2,\mathbb{F}_3)^2} \in \{\xi_+ \boxtimes \xi_+ \oplus \xi_- \boxtimes \xi_-,\ \xi_+ \boxtimes \xi_- \oplus \xi_- \boxtimes \xi_+\}.$$

Let $(t_1, t_2) \in R$. If $\det(t_1) = \det(t_2) = -1$ then $\sigma(t_1, t_2)$ interchanges the two summands, so $\mathrm{tr}\sigma(t_1, t_2) = 0$. Clearly $\mathrm{tr}\sigma(aI_2, \pm aI_2) = \dim\sigma \cdot \omega_\sigma(I_2, \pm I_2) = 2\omega_{\rho_0}(\pm 1) = 2$. The conclusion follows from the computation

$$\begin{aligned}\dim\sigma^R &= |R|^{-1} \sum_{(t_1,t_2)\in R} \mathrm{tr}\sigma(t_1, t_2)\\ &= 8^{-1}(\mathrm{tr}\sigma(I_2, I_2) + \mathrm{tr}\sigma(I_2, -I_2) + \mathrm{tr}\sigma(-I_2, I_2) + \mathrm{tr}\sigma(-I_2, -I_2))\\ &= 1.\end{aligned}$$

This completes the proof. □

Next we obtain a parametrization of $\mathrm{Supp}(\pi)$. Arguing as in the Klingen case, we may take double coset representatives for $N_G(\mathrm{K}(\mathfrak{p}))\backslash G/\mathrm{Si}(\mathfrak{p}^n)$ to have the form $t_{i,j}N(x, y, z)$ where $x, y, z \in \mathfrak{p}$. Since

$$[t_{i,j}N(x, y, z)] = [t_{-1,1}s_1 t_{i,j}N(x, y, z)s_1] = [t_{-i-1,2i+j+1}N(x, z, y)]$$

we can and do assume $i \geq 0$. Since

$$\begin{aligned}[t_{i,j}N(x, y, z)] &= [N(\varpi^{-i-j}x, 0, 0)t_{i,j}N(0, y, z)]\\ &= [N(0, \varpi^{-j}y, 0)t_{i,j}N(x, 0, z)]\\ &= [N(0, 0, \varpi^{-2i-j}z)t_{i,j}N(x, y, 0)]\end{aligned}$$

we may assume $\mathrm{val}(x) \leq i + j$ or $x = 0$, $\mathrm{val}(y) \leq j - 1$ or $y = 0$, and $\mathrm{val}(z) \leq 2i + j$ or $z = 0$, respectively. Define the two subgroups $U_1, U_2 \subset \mathrm{K}(\mathfrak{p})/\mathrm{K}(\mathfrak{p})^+$ by

$$U_1 = \left(\left[\begin{smallmatrix}1 & \\ * & 1\end{smallmatrix}\right], I_2\right), \qquad U_2 = \left(I_2, \left[\begin{smallmatrix}1 & \\ * & 1\end{smallmatrix}\right]\right)$$

and observe that if $R_g/\mathrm{K}(\mathfrak{p})^+ \supset U_1$ or $R_g/\mathrm{K}(\mathfrak{p})^+ \supset U_2$ then $\tau^{R_g} = 0$ by the cuspidality of σ.

Proposition 4.3: *Let $i \geq 0$. If $[t_{i,j}N(x, y, z)] \in \mathrm{Supp}(\pi)$ then we may assume $x = y = z = 0$.*

This will follow from the following lemma. Let $g = t_{i,j}N(x, y, z)$.

Lemma 4.4: *a) We have $1 \leq j \leq n - 2 - 2i$.*
b) We may assume $\mathrm{val}(z) = 2i + 1 + \mathrm{val}(y)$.
c) If $x \neq 0$ we may assume that $0 \leq i + \mathrm{val}(y/x)$.
d) We may assume $xyz = 0$.

e) If $1 \leq j \leq n-2-2i$ *then*

$$R_{t_{i,j}} = \{([{}^{a}{}_{b}], [{}^{c}{}_{abc^{-1}}]) : a, b, c \in \{\pm 1\}\}.$$

f) If $\mathrm{val}(x) \leq i+j$ *then* $[t_{i,j}N(x,0,0)] \notin \mathrm{Supp}(\pi)$.

g) If $\mathrm{val}(y) \leq j-1$ *and* $\mathrm{val}(z) = 2i+1+\mathrm{val}(y)$ *then* $[t_{i,j}N(0,y,z)] \notin \mathrm{Supp}(\pi)$.

Proof: a) If $2i+j \geq n-1$ then take $c \in \mathfrak{p}^{2i+j+1-n}$ and compute

$$g \begin{bmatrix} 1 & & & \\ & 1 & & \\ & & 1 & \\ \varpi^n c & & & 1 \end{bmatrix} g^{-1} = \begin{bmatrix} 1 & & & \\ & 1 & & \\ & & 1 & \\ \varpi^{n-2i-j} c & & & 1 \end{bmatrix}.$$

So $R_g/\mathrm{K}(\mathfrak{p})^+ \supset U_1$. If $j \leq 0$ then take $b \in \mathfrak{o}$ and compute

$$g \begin{bmatrix} 1 & & & \\ xb & 1 & b & \\ & & 1 & \\ & & -xb & 1 \end{bmatrix} g^{-1} = \begin{bmatrix} 1 & & & \\ & & 1-yb & \varpi^j b \\ & & \varpi^{-j}y^2 b & 1-yb \\ \varpi^{-2i-j}x^2 b & & & 1 \end{bmatrix}.$$

Observe that $\varpi^{-j}y^2 b \in \mathfrak{p}$. If $x = 0$ then taking $b \in \mathfrak{p}^{-j}$ shows $R_g/\mathrm{K}(\mathfrak{p})^+$ contains a conjugate of U_2. If $x \neq 0$, then $2i+j-2\mathrm{val}(x)+1 \neq -j$ by parity. So let $M := \max(2i+j-2\mathrm{val}(x)+1, -j) \geq 0$ and take $b \in \mathfrak{p}^M$. Thus $R_g/\mathrm{K}(\mathfrak{p})^+$ contains either U_1 or a conjugate of U_2.

b) Suppose $\mathrm{val}(z) \neq 2i+1+\mathrm{val}(y)$. If $1 \leq \mathrm{val}(y) \leq j-1$ or $1 \leq \mathrm{val}(z) \leq 2i+j+1$ then let $M := \max(j-\mathrm{val}(y), 2i+j+1-\mathrm{val}(z)) \geq 1$ and take $a \in 1+\mathfrak{p}^M$. Then compute

$$g \begin{bmatrix} 1 & & & \\ & a & & \\ & & 1 & \\ & & & a \end{bmatrix} g^{-1} = \begin{bmatrix} 1 & & & \\ & & a & \\ & & \varpi^{-j}y(a-1) & 1 \\ \varpi^{-2i-j}z(1-a) & & & a \end{bmatrix}$$

so either $R_g/\mathrm{K}(\mathfrak{p})^+ \supset U_1$ or $R_g/\mathrm{K}(\mathfrak{p})^+ \supset U_2$. On the other hand, if $\mathrm{val}(y) \geq j$ and $\mathrm{val}(z) \geq 2i+j+1$ then $[g] = [t_{i,j}S(x,0,0)]$.

c), d). Let $A = \begin{bmatrix} 1 & & & \\ c & 1 & & \\ & & 1 & \\ & & -c & 1 \end{bmatrix}$ where $c \in \mathfrak{p}^{i+1}$. Then

$$[g] = [(t_{i,j}At_{i,j}^{-1})t_{i,j}(A^{-1}N(x,y,z)A)] = [t_{i,j}N(x+cy, y, z+2cx+c^2y)].$$

If $\mathrm{val}(x) > i+\mathrm{val}(y)$ take $c = -x/y \in \mathfrak{p}^{i+1}$ so $[g] = [t_{i,j}N(0,y,z-x^2/y)]$.

Suppose $xyz \neq 0$ and let $k = \mathrm{val}(zy/x^2)$. If $0 \leq i + \mathrm{val}(y/x)$, as we may assume by the previous part, then $k = 1 + 2(i + \mathrm{val}(y/x)) \geq 1$ so there exists a unique $s \in 1 + \mathfrak{p}^k$ with $s^2 = 1 - zy/x^2$. Taking $c = -\frac{x}{y}(1-s) \in \mathfrak{p}^{\mathrm{val}(x/y)+k-e} = \mathfrak{p}^{2i+1+\mathrm{val}(y/x)-e} \subset \mathfrak{p}^{i+1}$ shows $[g] = [t_{i,j}N(xs, y, 0)]$.

e) This is clear from the definitions.

f) Suppose $\mathrm{val}(x) \leq i + j$. Let $a \in \mathfrak{p}^{2i+j+1-\mathrm{val}(x)} \subset \mathfrak{p}^{i+1}$ and compute

$$g \begin{bmatrix} 1 & & & \\ a & 1 & & \\ & & 1 & \\ & & -a & 1 \end{bmatrix} g^{-1} = \begin{bmatrix} 1 & & & \\ \varpi^{-i}a & 1 & & \\ & & 1 & \\ 2x\varpi^{-2i-j}a & & -\varpi^{-i}a & 1 \end{bmatrix}.$$

Thus $R_g/\mathrm{K}(\mathfrak{p})^+ \supset U_1$.

g) Set $a = 1 + \frac{v\varpi^j}{2y}$ where $v \in \mathfrak{o}$. Then compute

$$g \begin{bmatrix} 1 & & & \\ & a & & \\ & & a^{-1} & \\ & & & 1 \end{bmatrix} g^{-1} = \begin{bmatrix} 1 & & & \\ & a & & \\ & \varpi^{-j}y(a-a^{-1}) & a^{-1} & \\ & & & 1 \end{bmatrix}$$

and observe that

$$\varpi^{-j}y(a-a^{-1}) = va^{-1}\left(1 + \frac{v\varpi^j}{4y}\right) \equiv v \pmod{\mathfrak{p}}.$$

Thus $R_g/\mathrm{K}(\mathfrak{p})^+ \supset U_2$. □

It is not hard to show that $[t_{i,j}]$ determines i and j when $i \geq 0$ and $1 \leq j \leq n-2-2i$. Thus the number of double cosets in $\mathrm{Supp}(\pi)$ is $\sum_{j=1}^{n-2} \sum_{i=0}^{\lfloor \frac{n-j-2}{2} \rfloor} (1) = \sum_{j=1}^{n-2} \lfloor \frac{n-j}{2} \rfloor = \sum_{j=2}^{n-1} \lfloor \frac{j}{2} \rfloor = \left\lfloor \frac{(n-1)^2}{4} \right\rfloor$. We have proved the following theorem.

Theorem 4.5: *Suppose $q = 3$ and $\pi = c\text{-}\mathrm{Ind}_{N_G(\mathrm{K}(\mathfrak{p}))}^G(\tau)$ is an irreducible depth zero supercuspidal representation with trivial central character. Then*

$$\dim \pi^{\mathrm{Si}(\mathfrak{p}^n)} = \left\lfloor \frac{(n-1)^2}{4} \right\rfloor \cdot \begin{cases} 1 & \text{if } \tau|_{\mathrm{K}(\mathfrak{p})} = \sigma_\pm \\ 0 & \text{otherwise} \end{cases}.$$

Finally, we will now study the case of $\mathrm{Si}(\mathfrak{p}^n)$-fixed vectors in $\pi = c\text{-}\mathrm{Ind}_{ZK}^G(\sigma)$. Since much of the work here is very similar to the case of Klingen vectors, we record the only notable difference, and then the notation necessary to state the final result and the contents of Table 2 below.

As in the Klingen case, one has $\dim \pi^{\mathrm{Si}(\mathfrak{p}^n)} = 0$ if $\sigma = \theta_2$ or $\sigma = \lambda_0\theta_2$. If $\sigma_+ := X_4(\Theta)$ and $\sigma_- := \lambda_0 X_4(\Theta)$ then there is one family of double cosets which will lie in $\mathrm{Supp}(\pi)$ for σ_+ but not σ_-. We refer to Table 2 for the summary. If $g \in G$, define $[g]$, R_g, and $\mathrm{Supp}(\pi)$ as before, with $\mathrm{Si}(\mathfrak{p}^n)$ replacing $\mathrm{Kl}(\mathfrak{p}^n)$. For $k \in \mathbb{Z}$, define $Z_k = \begin{bmatrix} 1 & & & \\ & 1 & & \\ & & 1 & \\ \varpi^k & & & 1 \end{bmatrix}, Y_k = \begin{bmatrix} 1 & & & \\ & 1 & & \\ & \varpi^k & 1 & \\ & & & 1 \end{bmatrix}, X_k = \begin{bmatrix} 1 & & & \\ & 1 & & \\ \varpi^k & & 1 & \\ & \varpi^k & & 1 \end{bmatrix}$.

Lemma 4.6: *A representative for* $[g]$ *may be taken to have the form* $t_{i,j}N(x, y, z)$ *where* $x, y, z \in \mathfrak{p}/\mathfrak{p}^n$ *and* $i \geq 0$. *We may assume* $\mathrm{val}(x) < i+j$, $\mathrm{val}(y) < j$, *or* $\mathrm{val}(z) < 2i + j$ *when* x, y, *or* z, *respectively, are nonzero. If* $x \neq 0$ *then we may assume that* $\mathrm{val}(x) < \mathrm{val}(z)$ *and* $\mathrm{val}(x) < \mathrm{val}(y) + i$. *If* $y \neq 0$ *we may assume* $\mathrm{val}(y) < \mathrm{val}(x)$ *and if* $z \neq 0$ *we may assume* $\mathrm{val}(z) < \mathrm{val}(x) + i$.

Lemma 4.7: *Suppose* $1 \leq j \leq n-1$ *and* $1 \leq i \leq n-1$. *If* $1 \leq \mathrm{val}(y) < j$, $1 \leq \mathrm{val}(z) \leq n-1$, *and* $\mathrm{val}(x) - i < \mathrm{val}(y) < \mathrm{val}(x) < \mathrm{val}(z) < \mathrm{val}(x) + \min(i, \mathrm{val}(x))$.

a) We have $[t_{i,j}N(x, y, z)] \in \mathrm{Supp}(\pi)$ *if and only if the following conditions hold:*

i) $i + j < i + \mathrm{val}(x - \frac{yz}{x}) = j + \mathrm{val}(z/y) < n$

ii) $i < \mathrm{val}(x)$.

b) If conditions i) and ii) hold then R_g *is conjugate to*

$$\left\{ \begin{bmatrix} a & & & \\ v & a & & \\ b & & a & \\ c & b & -v & a \end{bmatrix} \begin{bmatrix} 1 & & & \\ & 1 & & \\ & v & 1 & \\ & & & 1 \end{bmatrix} : a \in \mathbb{F}_q^\times, b, c, v \in \mathbb{F}_q \right\}.$$

c) Let $u \in \mathfrak{o}^\times$. *If conditions i) and ii) hold, then* $[t_{i,j}N(x, y, z)] = [t_{i,j}N(x, uy, z)]$ *if and only if* $u \in 1 + \mathfrak{p}^{j-\mathrm{val}(y)}$.

We can then define $N(\vec{k})$ in an analogous way to $S(\vec{k})$ above.

Lemma 4.8: *a) The number of double cosets of the last type with* $\mathrm{val}(x^2) > \mathrm{val}(yz)$ *is equal to the number with* $\mathrm{val}(x^2) < \mathrm{val}(yz)$.

b) The number of such double cosets with $\mathrm{val}(x^2) > \mathrm{val}(yz)$ *is*

$$3^{\lfloor \frac{n-6}{4} \rfloor} a\left(n - 4\left\lfloor \frac{n-6}{4} \right\rfloor\right) - a(n)$$

where

$$a(n) = \frac{n-3}{6}\left\lfloor \frac{(n-2)(n-4)}{4} \right\rfloor + \frac{1}{2}\left\lfloor \frac{n^2+2n+14}{2} \right\rfloor.$$

c) The number of such double cosets with $\mathrm{val}(x^2) = \mathrm{val}(yz)$ *and* $\mathrm{val}(1-u) = 0$ *is*

$$3^{\lfloor \frac{n-4}{4} \rfloor} b\left(n - 4\left\lfloor \frac{n-4}{4} \right\rfloor\right) - b(n)$$

where

$$b(n) = \frac{1}{2}\left\lfloor \frac{n^2}{4} \right\rfloor + \frac{3}{2}.$$

d) The number of such double cosets with $\mathrm{val}(x^2) = \mathrm{val}(yz)$ *and* $\mathrm{val}(1-u) > 0$ *is*

$$3^{\lfloor \frac{n-5}{4} \rfloor} c\left(n - 4\left\lfloor \frac{n-5}{4} \right\rfloor\right) - c(n)$$

where

$$c(n) = \left\lfloor \frac{(n-1)(n-3)(2n-7)}{24} \right\rfloor + \frac{n^2+3n+15}{2}.$$

Theorem 4.9: *Let* $q = 3$ *and* $n \geq 1$.
a) If $\sigma = \theta_2$ *or* $\sigma = \lambda_0\theta_2$ *and* $\pi = c\text{-}Ind_{ZK}^{G}(\sigma)$*, then* $\dim \pi^{\mathrm{Si}(\mathfrak{p}^n)} = 0$.
b) If $\sigma_+ = X_4(\Theta)$, $\sigma_- = \lambda_0 X_4(\Theta)$, *and* $\pi_\pm = c\text{-}Ind_{ZK}^{G}(\sigma_\pm)$*, then*

$$\dim \pi_\pm^{\mathrm{Si}(\mathfrak{p}^n)} = d(n) * 3^{\lfloor \frac{n-1}{4} \rfloor} + (n-1)(1 \pm 1) + A(n)$$

where

$$d(n) = \begin{cases} 88 & \text{if } n \in 4\mathbb{Z} \\ 38 & \text{if } n \in 1 + 4\mathbb{Z} \\ 51 & \text{if } n \in 2 + 4\mathbb{Z} \\ 66 & \text{if } n \in 3 + 4\mathbb{Z} \end{cases}$$

$$A(n) = \frac{(n-2)}{3}\left\lfloor \frac{(n-1)(n-3)}{4} \right\rfloor - \left\lfloor \frac{n^2+22n+54}{2} \right\rfloor.$$

Table 2. Siegel fixed vector information.

g	j	i	R_g	$\dim \sigma^{R_g}$	$\lvert\{g\}\rvert$
$t_{i,j}$	$[1, n-1]$	0	$\begin{bmatrix} * & * & & \\ * & * & & \\ & & * & * \\ & & * & * \end{bmatrix}$	1 ± 1	$n-1$
$t_{i,j}$	$[3-n, 0]$	$[1-j, \lfloor\frac{n-j-1}{2}\rfloor]$	$\begin{bmatrix} * & & & \\ * & * & * & \\ & & * & \\ & & * & * \end{bmatrix}$	1	$\left\lfloor\frac{n^2-2n+1}{4}\right\rfloor$
$t_{i,j}$	$[1, n-2]$	$[\lceil\frac{n-j}{2}\rceil, n-j-1]$	$\begin{bmatrix} * & & & \\ * & * & & \\ & & * & \\ * & & * & * \end{bmatrix}$	1	$\left\lfloor\frac{n^2-2n+1}{4}\right\rfloor$
$t_{i,j}$	$[1, n-3]$	$[1, \lfloor\frac{n-j-1}{2}\rfloor]$	$\begin{bmatrix} * & & & \\ * & * & & \\ & & * & \\ & & * & * \end{bmatrix}$	4	$\left\lfloor\frac{n^2-4n+4}{4}\right\rfloor$
$t_{i,j}Z_k$	$[1, n-4]$	$[2, n-j-2]$	$\begin{bmatrix} a & & & \\ * & ad & & \\ & & a/d & \\ * & & * & a \end{bmatrix}$	2	$\left\lfloor\frac{(n-1)(n-3)(2n-7)}{24}\right\rfloor$
$t_{i,j}Y_k$	$[2, n-3]$	$[1, \lfloor\frac{n-j-1}{2}\rfloor]$	$\begin{bmatrix} ad & & & \\ & a & & \\ * & * & a & \\ & * & & a/d \end{bmatrix}$	2	$\left\lfloor\frac{(n-1)(n-3)(2n-7)}{24}\right\rfloor$
$t_{i,j}X_k$	$[2, n-i-2]$	$[2, \lfloor\frac{n-2}{2}\rfloor]$	$\begin{bmatrix} a & & & \\ & d & & \\ * & & a & \\ * & * & & d \end{bmatrix}$	2	$\frac{(n-3)}{6}\left\lfloor\frac{(n-2)(n-4)}{4}\right\rfloor$
$N(\vec{k})$	-	-	$\begin{bmatrix} a & & & \\ v & a & & \\ * & v & a & \\ * & * & -v & a \end{bmatrix}$	2	4.8

Acknowledgments

This work was done partially while the author was participating in the virtual program of the Institute for Mathematical Sciences, National University of Singapore, in 2022. We also thank Ralf Schmidt for many useful conversations during the writing of this chapter.

References

1. Enomoto, Hikoe. "The characters of the finite symplectic group $Sp(4, q), q = 2^f$." Osaka J. Math 9.1 (1972): 75–94.
2. Iwahori, Nagayoshi, and Hideya Matsumoto. "On some Bruhat decomposition and the structure of the Hecke rings of p-adic Chevalley groups." Publications Mathématiques de l'IHÉS 25 (1965): 5–48.

3. Johnson-Leung, Jennifer, Brooks Roberts, and Ralf Schmidt. "Stable Klingen Vectors and Paramodular Newforms." arXiv preprint arXiv:2208.08939 (2022).
4. Lust, Jaime. Verifying depth-zero supercuspidal L-packets for inner forms of $GSp(4)$. Diss. UC San Diego, 2010.
5. Roberts, Brooks, and Ralf Schmidt. Local newforms for $GSp(4)$. Lecture Notes in Math., 1918. Springer, Berlin, 2007.
6. Roy, Manami, Ralf Schmidt, and Shaoyun Yi. "Dimension formulas for Siegel modular forms of level 4." Mathematika 69.3 (2023): 795–840.
7. Shinoda, Ken-ichi. "The characters of the finite conformal symplectic group, $CSp(4, q)$." Communications in Algebra 10.13 (1982): 1369–1419.
8. Yi, Shaoyun. "Klingen $\mathfrak{p}^2$ vectors for GSp(4)." The Ramanujan Journal 54, no. 3 (2021): 511–554.

https://doi.org/10.1142/9789819825448_0005

Overview on the Theory of Double Flag Varieties for Symmetric Pairs

Dedicated to the memory of Joe Wolf

Lucas Fresse

Université de Lorraine, CNRS, Institut Élie Cartan de Lorraine, UMR 7502, Vandoeuvre-lès-Nancy, F-54506, France
lucas.fresse@univ-lorraine.fr

Kyo Nishiyama

Department of Mathematics, Aoyama Gakuin University, Fuchinobe 5-10-1, Chuo-ku, Sagamihara 252-5258, Japan
kyo.nishiyama@gmail.com

Let G be a connected reductive algebraic group and K its symmetric subgroup fixed by an involution. The variety $\mathfrak{X} = K/Q \times G/P$ is called a double flag variety, where Q and P are parabolic subgroups of K and G respectively.

In this chapter, we make a survey on the current status of the research and results related to double flag varieties for a symmetric pair (G, K) and report a number of new results and theorems. (By the word "theory", in the title, we mean a series of general results on finiteness and parametrization of orbits under the action of suitable algebraic groups, the geometry related to orbits, and infinitesimal character of the action including relation with nilpotent orbits.)

Most important topic in this chapter is the finiteness of K-orbits on $\mathfrak{X}$. We summarize the classification of $\mathfrak{X}$ of finite type, which are scattered in the literature. In some respects such classifications are complete, and in some cases not. In particular, we get a classification of double flag varieties of finite type when a symmetric pair is of type AIII, using the theorems of Homma who describes "indecomposable" objects of such double flag varieties. Together with these classifications, newly developed embedding theory provides double flag varieties of finite type, which are new. Other ingredients in this chapter are Steinberg theory, generalization of Robinson-Schensted correspondence, and orbit classification via quiver representations.

We hope this chapter is useful for those who want to study the theory of double flag varieties.

Contents

1. Introduction

Let G be a connected reductive algebraic group over the complex number field $\mathbb{C}$ and $\theta \in \operatorname{Aut} G$ an involutive automorphism. There will be no harm in considering an algebraically closed field of characteristic zero, but we use $\mathbb{C}$ for simplicity. The fixed point subgroup $K = G^\theta$ is called a *symmetric subgroup* and (G, K) is called a *symmetric pair*. The subgroup K is also reductive, but it is not connected in general. In this chapter, we assume K is connected, although most of the theory will go fine if we take the connected component of G^θ as K.

We concentrate on the cases where G is classical. Let us list irreducible classical symmetric pairs referring to [21](§10.3).

(AI) $(\mathrm{SL}_n, \mathrm{SO}_n)$ §§6.2 and 6.3; §15.3
(AII) $(\mathrm{SL}_{2n}, \mathrm{Sp}_{2n})$ §§6.2 and 6.3; §15.4
(AIII) $(\mathrm{SL}_n, (\mathrm{GL}_p \times \mathrm{GL}_q) \cap \mathrm{SL}_n)$ $(n = p + q)$ Table 9 in §8.2
(Below, we consider $(\mathrm{GL}_n, \mathrm{GL}_p \times \mathrm{GL}_q)$, which is easier to handle.)

(BDI) $(\mathrm{SO}_n, \mathrm{SO}_p \times \mathrm{SO}_q)$ $(n = p + q)$ §§6.2 and 6.3; §§15.7, 15.8, 15.9
(DIII) $(\mathrm{SO}_{2n}, \mathrm{GL}_n)$ §§6.2 and 6.3; §15.10
(CI) $(\mathrm{Sp}_{2n}, \mathrm{GL}_n)$ §§6.2 and 6.3; §15.5
(CII) $(\mathrm{Sp}_{2n}, \mathrm{Sp}_{2p} \times \mathrm{Sp}_{2q})$ $(n = p + q)$ §§6.2 and 6.3; §15.6

Remark 1.1:

(1) For each type we mention relevant sections, which contain the tables of double flag varieties of finite type (for details, see the explanation below). For the tables of the triple flag varieties of finite type, see the tables in §§6.4 and 6.5.

(2) Sometimes (BDII) refers to $(\mathrm{SO}_{n+1}, \mathrm{SO}_n)$ (see [51](§6.3), for example), but we include this into (BDI) for convenience. Further, in occasions, we separate (BDI) into (BI) and (DI) according as n is odd or even.

Let (G, K) be a symmetric pair. For parabolic subgroups Q of K and P of G, we call $\mathfrak{X} = K/Q \times G/P$ a *double flag variety* on which K acts diagonally. The study of double flag varieties started a decade ago by [55]. Since then, there appeared many works on them; we mention [22, 38, 20, 37, 13, 15, 14, 16, 56, 24] for example. Double flag varieties are intimately related to triple flag varieties (in fact, such triple flag varieties are considered to be special cases of double flag varieties, see §4). There are also many works on these triple/multiple flag varieties, e.g., [44, 45, 79, 82, 12, 47, 48, 49, 3]. The main interests of these papers are proving finiteness of orbits, classifications of multiple flag varieties with finitely many orbits, and generalizations of Steinberg theory including combinatorics such as Robinson-Schensted correspondence. There have been many results including various classification tables of double/triple flag varieties with finitely many orbits and they are scattered over in different places.

In this chapter, we summarize important properties, classification results and tables in the theory of double flag varieties from various resources, and explain general theory of Steinberg maps. We try to keep the description understandable not only for experts but also for readers from different fields. In the course of making summary, we have obtained many entirely new results or improvements on existent theorems. These new results, which advance our understanding of the theory, are in Parts II–IV.

Let us explain the content of the chapter.

In Part I, we introduce basics on double/triple flag varieties. We begin with the well-known theory of flag varieties and Bruhat decompositions. We are interested in double or triple flag varieties with the action of G or K. If

X is a smooth G-variety, then the action of G induces a Hamiltonian action on the cotangent bundle T^*X and it provides a moment map. We introduce the *conormal variety* as the null fiber of the moment map, and describe the irreducible decomposition of the conormal variety using conormal bundles of orbits, if there exist only finitely many orbits. In this respect, the finiteness of orbits plays a central role. We summarize various results related to orbit structures in double/triple flag varieties as well as the classical results on (a single) flag varieties.

In Part II, we discuss finiteness of orbits on double flag varieties, which is crucial in the Steinberg theory developed below. If there are only finitely many K-orbits on $\mathfrak{X} = K/Q \times G/P$, $\mathfrak{X}$ is called of *finite type*. Since $\mathfrak{X}/K \simeq Q\backslash G/P$, it is equivalent to having finitely many Q orbits on G/P. In particular, if $Q = B_K$ is a Borel subgroup of K, classification of the double flag varieties of finite type amounts that of K-spherical partial flag varieties G/P. This classification is achieved by He-N-Ochiai-Oshima [20]. They also classified the double flag varieties of finite type in the case where $P = B_G$ is a Borel subgroup of G. We reproduce the classification in §6 together with the key ideas of the proofs. In fact the classification is reduced to that of multiplicity free action on certain space of nilpotent elements (see §6.1). Together with these results, we also collect the tables of triple flag varieties of finite type in §§6.4 and 6.5. They are quoted from [44, 45], [79], [48, 49]. These classifications are used later to obtain further examples of double flag varieties of finite type.

To classify general double flag varieties of finite type, it is useful to apply representation theory of quivers and consider suitable subcategories of representations. The usage of quiver representations was first introduced by Magyar-Weyman-Zelevinsky [44] for triple flag varieties based on the theory of Gabriel [19], and Kac [27, 29]. We explain core ideas of these theories in §7.

For the classification of double flag varieties of finite type (for type AIII), Homma [24] introduced *joint flag category* and *joint flag varieties* which correspond to the quiver representations associated with the flags in the double flag varieties. Using his idea and results, we classify the double flag varieties of finite type in the case of the symmetric pairs of type AIII. Homma actually classified indecomposable quiver representations of finite type, and it is not immediate to get the explicit form of double flag varieties. Our Table 9 in §8.2 presents precisely when given parabolic subgroups contribute to the double flag varieties of finite type, and this completes the classification in the case of type AIII (Corollary 8.6).

For the symmetric pair $(G, K) = (\mathrm{GL}_n, \mathrm{GL}_p \times \mathrm{GL}_q)$ $(n = p + q)$ of type AIII, we also present a classification of K-orbits on $\mathfrak{X} = K/B_K \times G/P$ in terms of indecomposable quiver representations, where B_K is a Borel subgroup of K and P is a maximal parabolic subgroup of G. See Theorem 9.4 in §9.2.

In Part III, we explain the Steinberg theory for double flag varieties.

For a symmetric pair (G, K), denote by $\mathfrak{g}$ and $\mathfrak{k}$ the Lie algebras of G and K respectively. Let $\mathfrak{g} = \mathfrak{k} + \mathfrak{s}$ be the Cartan decomposition, where $\mathfrak{s}$ is the (-1)-eigenspace of $\theta \in \mathrm{Aut}\,\mathfrak{g}$ (we denote the differential of θ by the same letter). Note that $\mathfrak{s}$ is isomorphic to the tangent space of G/K at the base point eK, on which K acts. We denote by $\mathcal{N}$ the nilpotent variety of $\mathfrak{g}$ and put $\mathcal{N}_{\mathfrak{k}} = \mathcal{N} \cap \mathfrak{k}$ and $\mathcal{N}_{\mathfrak{s}} = \mathcal{N} \cap \mathfrak{s}$. Due to the work of Kostant, Mal'cev and Kostant-Rallis ([46], [39, 41], see also [11]) it is known that there are only finitely many K-orbits in both of the nilpotent varieties $\mathcal{N}_{\mathfrak{k}}$ and $\mathcal{N}_{\mathfrak{s}}$.

If $\mathfrak{X}$ has finitely many K-orbits, then there exists a natural orbit map $\Phi_{\mathfrak{k}} : \mathfrak{X}/K \to \mathcal{N}_{\mathfrak{k}}/K$ by using moment maps. In the classical case where $\mathfrak{X} = G/B \times G/B$ is the product of two full flag varieties on which $K = G$ acts diagonally, this reduces to the classical Steinberg's theory [78]. Namely, there exists a map $W_G \simeq (G/B \times G/B)/G \to \mathcal{N}/G$ called the *Steinberg map*, where W_G denotes the Weyl group of G. In the case of type A, the set of nilpotent orbits $\mathcal{N}/G$ is parameterized by Young diagrams (or partitions) and the fibers of the Steinberg map are identified with the pairs of standard Young tableaux. In such a way, the Steinberg map is identified with the *Robinson-Schensted correspondence* (RS correspondence), a combinatorial procedure. Therefore, in our situation, the orbit map $\Phi_{\mathfrak{k}} : \mathfrak{X}/K \to \mathcal{N}_{\mathfrak{k}}/K$ thus obtained is a generalization of the Steinberg map. In addition, in the case of double flag variety, we get one *extra* map $\Phi_{\mathfrak{s}} : \mathfrak{X}/K \to \mathcal{N}_{\mathfrak{s}}/K$ if certain condition called "nil-projection assumption" holds (see Assumption 10.1 in §10.1 for details). The map $\Phi_{\mathfrak{k}}$ is called the *symmetrized Steinberg map* and $\Phi_{\mathfrak{s}}$ is called *exotic* one. In §11, for symmetric pairs of type AIII, we explain combinatorial procedures to get these two Steinberg maps, which are generalizations of RS correspondence.

In this respect, Nil-projection Assumption is very important. We propose a conjecture that says it always holds whenever the double flag variety is of finite type. For example, this is valid when the parabolic subgroup $P \subset G$ has abelian unipotent radical. See Conjecture 10.7 and Proposition 10.3.

Part IV is devoted to the study of embedding theory of double flag varieties. Since the double flag varieties of finite type are completely classified when the symmetric pair is of type AIII, if we can embed a double flag variety of another type into one of type AIII, so that the structure of orbits is preserved, then we get various examples of double flag varieties of finite type. In [14], this attempt is carried out for the double flag varieties of type CI. In this chapter, we generalize it to a large family of double flag varieties (Theorem 12.3 and Proposition 15.1 in §15.1) and get fairly large, new families of double flag varieties of finite type (see §§15.3–15.10). The embedding theory is equally applicable to triple flag varieties (Theorem 15.2 in §15.2), and we also recover the results in [55].

The advantage of the embedding theory is that it also provides the classification of orbits in terms of the orbits in type AIII. This corresponds to the classical theory of Jordan type of nilpotent orbits for general classical groups. We give such examples in §16, which are taken from [14]. One more advantage of the embedding theory is that it behaves functorially in Steinberg theory. However, we still do not prove the full commutativity between embeddings and Steinberg maps, which are left to Conjecture 14.1.

Finally, let us make some remarks on open questions and related future problems.

Although we get a large portion of double flag varieties of finite type (we believe so), the complete classification of them is still open. Also, in this chapter, we only deal with classical groups, and classification of double flag varieties of finite type for exceptional groups is widely open. Parametrizations of K-orbits on $\mathfrak{X}$ should be interesting and we are expecting new combinatorial structures on orbits and Steinberg maps, which are fruitful subjects.

We only work over algebraically closed field, namely over $\mathbb{C}$, but the same problem over the field of real numbers (or even over p-adic numbers) will be interesting from the view point of infinite dimensional unitary representations (see [37, 38] and [56] for example).

The K-orbit space on $\mathfrak{X}$ admits a natural action of Hecke algebras (see our recent paper [17]), and there should be a rich theory like Springer theory, or categorical study over perverse sheaves since there are finitely many orbits. In this respect, we mention [12, 82] and [67, 68, 69].

As we mention above the tables in this chapter do not exhaust the double flag varieties of finite type (in some cases, it is complete though). However, they give rich examples of them, and we indicate the sections and table numbers at the beginning of this introduction together with the classification of the classical symmetric pairs. We hope they are useful for readers.

Part I. Double Flag Varieties for Symmetric Pairs

2. Flag Varieties and Moment Maps

In this section we introduce (without proofs) basic notions related to flag varieties, which are needed later. For the proofs and more detailed description, see [5, 25, 75] and references therein. Also we will review some facts on moment maps. For this we refer the readers to [9].

2.1. *Flag varieties*

Let G be a connected reductive algebraic group over the complex number field $\mathbb{C}$. Most of the story in this chapter goes through over an algebraically closed field of characteristic zero without changes. (See, for instance, [43] for the general background on Hamiltonian actions and moment maps over algebraically closed fields.) However, we prefer a simpler assumption which will highlight the key points of the theory.

A subgroup $B = B_G$ of G is called a Borel subgroup if it is maximal among connected solvable subgroups of G. It is well-known that all the Borel subgroups are conjugate to each other by inner automorphisms of G, and moreover self-normalizing, i.e., $B = N_G(B) = N_G(\mathfrak{b})$, where $\mathfrak{b}$ denotes the Lie algebra of B.

Theorem 2.1: *The homogeneous space G/B is a projective variety (hence compact). As a set, it can be identified with the set of all Borel subgroups in G, as well as the set of all Borel subalgebras in $\mathfrak{g} = \mathrm{Lie}(G)$.*

The projective variety $\mathscr{B} = G/B$ is called a *complete flag variety* (or a *full* flag variety), and is independent of choice of B. The adjective "complete" is often omitted, but it is sometimes useful to tell a difference from "partial" flag varieties.

A closed subgroup P of G is called *parabolic* if it contains a Borel subgroup. Then P is automatically connected. As well as Borel subgroups, P is self-normalizing: $P = N_G(P) = N_G(\mathfrak{p})$, where $\mathfrak{p}$ is the Lie algebra of P. In the following, we will denote the Lie algebra of a reductive (or "Lie") group G by its corresponding lower case German letter $\mathfrak{g}$.

Theorem 2.2: *A homogeneous space G/P is projective if and only if P is parabolic. As a set, it can be identified with the set of all parabolic subgroups in G that are conjugate to P. Similarly, G/P can be identified with the set of all parabolic subalgebras in $\mathfrak{g}$ that are conjugate to $\mathfrak{p}$.*

If P is a parabolic subgroup, the projective variety $\mathscr{P} = G/P$ is called a (*partial*) *flag variety*. There is a natural projection $G/B \to G/P$ with fiber P/B which is also a flag variety for a smaller group, namely, a Levi subgroup of P. We often omit the word "partial" and simply call flag variety if there is no need to distinguish "complete" ones and "partial" ones.

In the following, the cotangent bundle of flag varieties will play an important role. Since the tangent space at $\mathfrak{b} \in G/B = \mathscr{B}$ is $\mathfrak{g}/\mathfrak{b}$ (note that the stabilizer of $\mathfrak{b}$ is $N_G(\mathfrak{b}) = B$), we get the cotangent space $T^*_{\mathfrak{b}}\mathscr{B} = (\mathfrak{g}/\mathfrak{b})^*$ at $\mathfrak{b}$. By choosing a non-degenerate invariant bilinear form on $\mathfrak{g}$, we can identify $(\mathfrak{g}/\mathfrak{b})^*$ with the nilradical $\mathfrak{n}$ of $\mathfrak{b}$. Thus the cotangent bundle can be described as

$$T^*\mathscr{B} \simeq \{(\mathfrak{b}_1, x) \in \mathscr{B} \times \mathfrak{g} \mid x \in \mathfrak{n}_1\} \simeq G \times_B \mathfrak{n}, \tag{2.1}$$

where in the second expression $\mathscr{B}$ is considered to be the set of Borel subalgebras as explained above and $\mathfrak{n}_1$ is the nilradical of a Borel subalgebra $\mathfrak{b}_1$.

In a very similar way, we can describe the cotangent bundle of a partial flag variety $\mathscr{P} = G/P$. So let $\mathfrak{p}_1 \in \mathscr{P}$ be a parabolic subalgebra conjugate to $\mathfrak{p}$, and denote its nilpotent radical by $\mathfrak{u}_1$ (we will denote a Levi decomposition by $P = LU$, where L is a Levi part and U its unipotent radical in general). Then we get

$$T^*\mathscr{P} \simeq G \times_P \mathfrak{u} \simeq \{(\mathfrak{p}_1, x) \in \mathscr{P} \times \mathfrak{g} \mid x \in \mathfrak{u}_1\}. \tag{2.2}$$

2.2. *Moment maps*

Let X be a smooth variety on which G acts. We call X a smooth G-variety for short. The cotangent bundle T^*X inherits a natural structure of symplectic variety which is G-invariant. Choose $(x, \xi) \in T^*X$, where $x \in X$ and $\xi \in T^*_xX$. The tangent space of T^*X at (x, ξ) is $T_xX \oplus T_\xi(T^*_xX) \simeq T_xX \oplus T^*_xX$, and there is a natural symplectic form ω, for which $T_xX \oplus T^*_xX$ is a polar decomposition by Lagrangian subspaces T_xX and T^*_xX. To get the symplectic form, we adopt the natural pairing between T_xX and T^*_xX, which is the dual of T_xX.

Pick a point $x \in X$ and consider a map $v_x : G \to X$, $v_x(g) = gx$. The differential (at $e \in G$) of this map induces a map from $\mathfrak{g}$ to T_xX, which we denote by the same letter v_x by abuse of notation. For $A \in \mathfrak{g}$, $v_x(A)$ is simply a vector field generated by the action of $\exp tA$ on X. Let us consider a G-equivariant map

$$\mu_X : T^*X \to \mathfrak{g}^*, \qquad \mu_X((x,\xi))(A) = \xi(v_x(A)) \quad \text{for } A \in \mathfrak{g}, \tag{2.3}$$

which is called a "moment map".[a]

Let us apply the framework above to flag varieties. At a point $b \in \mathscr{B}$ which is represented by a Borel subalgebra $\mathfrak{b}_1$, the tangent space is given by $T_b\mathscr{B} = \mathfrak{g}/\mathfrak{b}_1$ and the cotangent space is $T_b^*\mathscr{B} = (\mathfrak{g}/\mathfrak{b}_1)^* \simeq \mathfrak{n}_1$. The vector field generated by $A \in \mathfrak{g}$ is just $A \mod \mathfrak{b}_1$. Thus we get

$$\mu_{\mathscr{B}} : T^*\mathscr{B} \to \mathfrak{g}^*, \qquad \mu_{\mathscr{B}}((\mathfrak{b}_1, x))(A) = \langle x, A\rangle \quad \text{for } x \in \mathfrak{n}_1,\ A \in \mathfrak{g}, \tag{2.4}$$

where $\langle \cdot, \cdot \rangle$ denotes the nondegenerate invariant bilinear form on $\mathfrak{g}$. Note that x kills $\mathfrak{b}_1$ so that the last expression is well defined. By this formula, if we identify $\mathfrak{g}$ and $\mathfrak{g}^*$ by the invariant bilinear form (we always do so below),

$$\mu_{\mathscr{B}}((\mathfrak{b}_1, x)) = x \qquad ((\mathfrak{b}_1, x) \in T^*\mathscr{B}) \tag{2.5}$$

and it is just the second projection. From this formula, since x is nilpotent, the image of the moment map $\operatorname{Im}\mu_{\mathscr{B}}$ is contained in $\mathcal{N}_{\mathfrak{g}}$, the nilpotent variety of $\mathfrak{g}$. It is easy to see that in fact $\operatorname{Im}\mu_{\mathscr{B}} = \mathcal{N}_{\mathfrak{g}}$. Thus we get $\mu_{\mathscr{B}} : T^*\mathscr{B} \to \mathcal{N}_{\mathfrak{g}}$. This map is known to be birational, and gives a G-equivariant resolution of singularity of the nilpotent variety $\mathcal{N}_{\mathfrak{g}}$ called the *Springer resolution* ([73]; see [71] for details). Here we chose a nondegenerate invariant bilinear form on $\mathfrak{g}$. The choice does not affect the structure of orbits or geometric properties of double flag varieties discussed below, although we do not explicitly state it.

Similarly, for the partial flag variety $\mathscr{P} = G/P$, the moment map is also the second projection

$$\mu_{\mathscr{P}}((\mathfrak{p}_1, x)) = x \qquad ((\mathfrak{p}_1, x) \in T^*\mathscr{P};\ \mathfrak{p}_1 \in \mathscr{P},\ x \in \mathfrak{u}_1). \tag{2.6}$$

In this case, the image $\operatorname{Im}\mu_{\mathscr{P}} = \operatorname{Ad}(G)\,\mathfrak{u}$ is the closure of a certain nilpotent orbit. This orbit is called a *Richardson orbit* associated with the parabolic subalgebra $\mathfrak{p}$, and often denoted by $\mathcal{O}_P$. For a parabolic subgroup P, the moment map $\mu_{\mathscr{P}} : T^*\mathscr{P} \to \overline{\mathcal{O}_P}$ is not a resolution in general (see [72], [6](§3.5), [26](§8.8) for example). It is a resolution if the nilpotent orbit is even or $G = \mathrm{GL}_n$.

[a]For any Hamiltonian symplectic G-variety, one can define a moment map. For general definition and properties of moment maps, see [9] and [40].

Let us return to a general G-variety X and the moment map $\mu_X : T^*X \to \mathfrak{g}$ above. We identify $\mathfrak{g}^*$ with $\mathfrak{g}$ as we mentioned.

Take a point $x_0 \in X$ and consider the G-orbit $\mathbb{O} = G \cdot x_0$. We define a Lagrangian subvariety $T^*_{\mathbb{O}}X \subset T^*X$ as follows.

$$T^*_{\mathbb{O}}X = \{(x, \xi) \in T^*X \mid x \in \mathbb{O},\ \xi \in (T_x\mathbb{O})^{\perp}\}, \tag{2.7}$$

where $(T_x\mathbb{O})^{\perp}$ denotes the subspace of T^*_xX consisting of all linear forms ξ killing $T_x\mathbb{O}$. The subvariety $T^*_{\mathbb{O}}X$ is called the *conormal bundle* of the orbit $\mathbb{O}$. Note that a conormal bundle is irreducible since it is a vector bundle over a G-orbit, which is irreducible because G is assumed to be connected.

Lemma 2.3: *Put $Y = \mu_X^{-1}(0) \subset T^*X$, the null fiber of the moment map. Then it is a disjoint union of all conormal bundles.*

$$Y = \coprod_{\mathbb{O} \in X/G} T^*_{\mathbb{O}}X \qquad (disjoint). \tag{2.8}$$

If there are only finitely many G-orbits in X, then Y is an equidimensional variety of dimension $\dim X$, *and its irreducible components are given by the closures of conormal bundles:* $\mathrm{Irr}(Y) = \{\overline{T^*_{\mathbb{O}}X} \mid \mathbb{O} \in X/G\}$.

Proof: The fact that Y is a disjoint union of conormal bundles is straightforward from definitions. Since $\dim T^*_{\mathbb{O}}X = \dim X$, if there appear only finitely many of them, clearly their closures give irreducible components and hence Y is equidimensional of dimension $\dim X$. □

Since Y consists of conormal bundles, we call $Y = \mu_X^{-1}(0)$ a *conormal variety.*

Corollary 2.4: *Assume that there are only finitely many G-orbits in X. There exists a natural bijection between irreducible components of the conormal variety Y and the G-orbits in X: $X/G \simeq \mathrm{Irr}(Y)$, which maps $\mathbb{O} \in X/G$ to the closure of its conormal bundle $\overline{T^*_{\mathbb{O}}X} \subset Y$.*

3. Orbits in Flag Varieties: Known Results, Summary

We give basic facts on the orbits in flag varieties, which should be well-known. See [5, 75, 25] for example.

3.1. *Bruhat decomposition*

Let $B \subset G$ be a Borel subgroup and fix a Levi decomposition $B = TN$, where T is a maximal torus (in G) and N the unipotent radical. The finite group $W = N_G(T)/T$ is called the Weyl group of G, and it coincides with the Weyl group of the root system of G. We often pick representatives for each element in W, and if there is no confusion we simply write $w \in W$ for an element $w \in N_G(T)$.

We have a double coset decomposition

$$G = \coprod_{w \in W} BwB, \tag{3.1}$$

which is called the *Bruhat decomposition*. From this, we know B-orbits on the complete flag variety $\mathscr{B} = G/B$ have representatives in W. Thus $\mathscr{B}/B \simeq W$.

Let P be a parabolic subgroup which contains B. In this case P is called *standard*. We get the standard Levi decomposition $P = LU$, where L is a Levi subgroup containing T and U is the unipotent radical. Since L is reductive, we can consider the Weyl group W_L of L, which is considered to be a subgroup of $W = W_G$. We also denote W_L by W_P.

There is a generalization of the Bruhat decomposition extending to the case of partial flag varieties.

Lemma 3.1: *Let P be a parabolic subgroup. Recall the notation $\mathscr{B} = G/B$ and $\mathscr{P} = G/P$.*

(1) $\mathscr{B}/P \simeq \mathscr{P}/B \simeq W/W_P$.

(2) *For another parabolic subgroup P' of G, the generalized Bruhat decomposition $\mathscr{P}/P' \simeq W_P\backslash W/W_{P'}$ holds. In particular, we get $P\backslash G/P = \mathscr{P}/P \simeq W_P\backslash W/W_P$.*

Proof: This follows from the Bruhat decomposition. See Borel [5](§14.16) and Springer [75](Exercise 8.4.6(3)). □

If X is an irreducible normal G-variety on which a Borel subgroup B acts with finitely many B-orbits, the variety X is called *spherical*. The following theorem is well-known.

Theorem 3.2: (Brion [7], Vinberg [84]) *An irreducible normal G-variety X has finitely many B-orbits if and only if it admits an open B-orbit.*

From above considerations, flag varieties are spherical.

Spherical varieties are ubiquitous in representation theory and bear many nice properties. For this, see Brion [8], Knop and others [33, 36, 34, 35], Perrin [60], Timashev [80] and references therein.

3.2. *Symmetric pairs*

Let θ be an involutive automorphism of G and $K = G^{\theta}$ the subgroup of fixed points of θ. The pair (G, K) is called a *symmetric pair* and K a *symmetric subgroup*. The quotient space G/K is called a *symmetric space*. Note that since the group K is closed and reductive, the quotient G/K is indeed an affine variety.

If G is simply connected, then K is connected. However, without the assumption, it need not be connected in general. In the following, we always assume K is connected.[b]

Example 3.3: For $G = \mathrm{GL}_n(\mathbb{C})$, put $\theta(g) = I_{p,q} g I_{p,q} \quad (g \in G)$, where $I_{p,q} = \mathrm{diag}(\mathbb{1}_p, -\mathbb{1}_q)$ with $\mathbb{1}_m$ being the identity matrix of size m. Then $K = \mathrm{GL}_p(\mathbb{C}) \times \mathrm{GL}_q(\mathbb{C})$, block-diagonally embedded into G. The pair $(\mathrm{GL}_n(\mathbb{C}), \mathrm{GL}_p(\mathbb{C}) \times \mathrm{GL}_q(\mathbb{C}))$ is said to be a symmetric pair of type AIII.

Next, let us consider $G = \mathrm{Sp}_{2n}(\mathbb{C})$, the symplectic group of size $2n$ (and rank n) preserving the standard symplectic form.

$$\mathrm{Sp}_{2n}(\mathbb{C}) = \{g \in \mathrm{GL}_{2n}(\mathbb{C}) \mid {}^t g J_n g = J_n\}, \quad J_n = \begin{pmatrix} 0 & \mathbb{1}_n \\ -\mathbb{1}_n & 0 \end{pmatrix}. \tag{3.2}$$

The involution $\theta(g) = I_{n,n} g I_{n,n} \quad (g \in G)$ defines $K \simeq \mathrm{GL}_n(\mathbb{C})$. This symmetric pair $(\mathrm{Sp}_{2n}(\mathbb{C}), \mathrm{GL}_n(\mathbb{C}))$ is called of type CI.

Finally, let us consider $G = \mathrm{GL}_n(\mathbb{C})$ with the involution $\theta(g) = {}^t g^{-1}$. In this case $K = \mathrm{O}_n(\mathbb{C})$ is an orthogonal group, which is not connected but has 2 connected components.

3.3. *KGB theory: K-orbits in flag varieties*

The following lemma is well-known, but we will give a proof which will clarify some parametrization of K-orbits in G/B.

Lemma 3.4: *G/K is a G-spherical variety, i.e., a Borel subgroup B of G has finitely many orbits in G/K.*

The lemma is equivalent to the claim that the double coset space $B\backslash G/K$ is finite, which is further equivalent to that there are finitely many K-orbits in the complete flag variety G/B.

---

[b]Many of the facts and results below will hold after some modifications even if K is not connected, and for most of our purpose it is sufficient to take a connected component of the identity instead of the whole K.

Proof: We will show there are finitely many K-orbits on $\mathscr{B} = G/B$. For this, we follow the strategy of Millićič [50](§H.2, Theorem 1) (see also [64](Remark 1.8)). We fix a θ-stable Borel subgroup B, which always exists (see [76](Theorem 7.2)). Consider the double flag variety $\mathscr{B}^2 = G/B \times G/B$, on which G acts diagonally. We define another G action on $\mathscr{B}^2$, which is called the θ twisted G action, by

$$\delta_\theta(g)(B_1, B_2) := (gB_1, \theta(g)B_2) \qquad g \in G,\ (B_1, B_2) \in \mathscr{B}^2.$$

We denote the double flag variety with this twisted action by $\mathscr{B}^2_\theta$. As G-varieties, $\mathscr{B}^2$ and $\mathscr{B}^2_\theta$ are isomorphic via

$$\Phi : \mathscr{B}^2 \to \mathscr{B}^2_\theta, \qquad \Phi(B_1, B_2) = (B_1, \theta(B_2)).$$

Thus $\mathscr{B}^2_\theta/\delta_\theta(G) \simeq W$ by the Bruhat decomposition. Let us denote the twisted G-orbit corresponding to $w \in W$ by $\mathcal{O}^\theta_w$. Let $X = \Delta(\mathscr{B}) = \{(B_1, B_1) \mid B_1 \in \mathscr{B}\} \subset \mathscr{B}^2_\theta$ be the (non-twisted) diagonal embedding.

We will show that any irreducible component of $X \cap \mathcal{O}^\theta_w$ is precisely a single K-orbit, which implies there are finitely many orbits in $X \cap \mathcal{O}^\theta_w$. Since $\{\mathcal{O}^\theta_w\}_{w \in W}$ covers $\mathscr{B}^2_\theta$ as well as X, we conclude $X \simeq \mathscr{B}$ has finitely many K-orbits (note that the twisted action and the natural diagonal action of K coincide).

Take $\xi \in X \cap \mathcal{O}^\theta_w$ and consider its tangent space:

$$T_\xi(X \cap \mathcal{O}^\theta_w) \subset T_\xi X \cap T_\xi \mathcal{O}^\theta_w. \tag{3.3}$$

Since $\xi \in X$, it can be written as $\xi = (xB, xB) \in X$ for some $x \in G$. Then the right-hand side of (3.3) is

$$\{(y + \mathfrak{b}^x, y + \mathfrak{b}^x) \mid y \in \mathfrak{g}\} \cap \{(y + \mathfrak{b}^x, \theta(y) + \mathfrak{b}^x) \mid y \in \mathfrak{g}\} \subset \mathfrak{g}/\mathfrak{b}^x \oplus \mathfrak{g}/\mathfrak{b}^x. \tag{3.4}$$

From this, considering $(y + \mathfrak{b}^x, y + \mathfrak{b}^x)$ in the left-hand side of (3.4), we conclude that $y - \theta(y) \in \mathfrak{b}^x$ which implies

$$y = \frac{1}{2}(y + \theta(y)) + \frac{1}{2}(y - \theta(y)) \in \mathfrak{k} + \mathfrak{b}^x.$$

This shows in fact we can take $y \in \mathfrak{k}$ modulo $\mathfrak{b}^x$. We denote the K-orbit through ξ by $\mathbb{O}_K$. The above arguments show $T_\xi X \cap T_\xi \mathcal{O}^\theta_w \subset T_\xi \mathbb{O}_K$. On the other hand, since $\mathbb{O}_K \subset X \cap \mathcal{O}^\theta_w$, it follows $T_\xi \mathbb{O}_K \subset T_\xi(X \cap \mathcal{O}^\theta_w)$. Together with Equation (3.3), we get

$$T_\xi \mathbb{O}_K \subset T_\xi(X \cap \mathcal{O}^\theta_w) \subset T_\xi X \cap T_\xi \mathcal{O}^\theta_w \subset T_\xi \mathbb{O}_K. \tag{3.5}$$

Thus all the terms in the above containments are equal in fact.

Now consider an irreducible component Z of $X \cap \mathcal{O}_w^\theta$. We choose a point ξ from the set of smooth points in Z. Then $T_\xi Z = T_\xi \mathbb{O}_K$ holds by the above consideration, which means $\mathbb{O}_K$ is open dense in Z and we get $Z = \overline{\mathbb{O}_K}$.

If $\xi_0 \in Z \setminus \mathbb{O}_K$, the general theory of algebraic varieties tells $\dim T_{\xi_0} Z \geq \dim Z = \dim \mathbb{O}_K$. While $\xi_0 \in \overline{\mathbb{O}_K} \setminus \mathbb{O}_K$, and in that case, $\dim K \cdot \xi_0 < \dim \mathbb{O}_K$ holds. Thus we get

$$\dim T_{\xi_0} Z \geq \dim Z = \dim \mathbb{O}_K > \dim K \cdot \xi_0 = \dim T_{\xi_0}(X \cap \mathcal{O}_w^\theta) \geq \dim T_{\xi_0} Z,$$

contradiction. Thus we conclude exactly $Z = \mathbb{O}_K$ holds.

So irreducible components of $X \cap \mathcal{O}_w^\theta$ are K-orbits and coincide with connected components. They are finite in number. □

Remark 3.5: By Theorem 3.2, the above lemma is equivalent to the claim that there is an open dense B-orbit in G/K. It is not hard to prove the existence of an open dense B-orbit. In fact, it suffices to show $\mathfrak{g} = \mathfrak{k} + \mathfrak{b}$ for a certain Borel subalgebra.

It is known that there is a θ-split Borel subalgebra $\mathfrak{b}$ (see [85]). Here, the term "θ-split" means $\theta(\mathfrak{b})$ is the opposite Borel subalgebra of $\mathfrak{b}$, so that we have $\mathfrak{g} = \mathfrak{b} + \theta(\mathfrak{b})$. Take an arbitrary $x \in \mathfrak{g}$ and write it as $x = x_1 + x_2$ for $x_1 \in \mathfrak{b}$ and $x_2 \in \theta(\mathfrak{b})$. Then we have $x_2 = \big(x_2 + \theta(x_2)\big) - \theta(x_2)$ and clearly $x_2 + \theta(x_2) \in \mathfrak{k}$, $\theta(x_2) \in \mathfrak{b}$. This proves that $x \in \mathfrak{k} + \mathfrak{b}$, and hence proves $\mathfrak{g} = \mathfrak{k} + \mathfrak{b}$ for a θ-split $\mathfrak{b}$.

Corollary 3.6: *There exists a natural bijection*

$$\mathscr{B}/K = K\backslash G/B \simeq \coprod_{w \in \mathscr{I}} \mathrm{Irr}(X \cap \mathcal{O}_w^\theta),$$

where $\mathscr{I} = \{w \in W \mid \theta(w) = w^{-1}\}$ *is the set of* θ*-twisted involutions and* X *and* $\mathcal{O}_w^\theta$ *are given in the proof of Lemma 3.4.*

Proof: Except for the condition $w \in \mathscr{I}$, the claim has been already established in the proof of the above lemma.

If $X \cap \mathcal{O}_w^\theta$ is not empty, then there exists $g \in G$ which satisfies $(gB, gB) = (gB, \theta(g)wB)$. By [74](Theorem 4.2), a representative of the double coset KvB ($v \in G$) can be chosen so that $v\theta(v)^{-1} \in N_G(T)$, where T is a θ-stable maximal torus in B, which is also assumed to be θ-stable. Thus, if we translate gB by K, we can assume $(vB, vB) = (vB, \theta(v)wB)$ for some v with the property $v\theta(v)^{-1} \in N_G(T)$. Since $vB = \theta(v)wB$, we get $v^{-1}\theta(v)w \in B$. This means $v^{-1}\theta(v)w \in T$ and $w = v\theta(v)^{-1}$ in

$W = N_G(T)/T$. Thus we see $\theta(w) = \theta(v\theta(v)^{-1}) = \theta(v)v^{-1} = w^{-1}$, which concludes that $w \in \mathscr{I}$. □

From the corollary, we get a natural projection

$$\mathscr{B}/K = K\backslash G/B \to \mathscr{I}, \tag{3.6}$$

which is essentially the same as Richardson-Springer ([64](§1.6), [63](Proposition 1.7.1)).

There is another way to parameterize $K\backslash G/B$ by using nilpotent orbits and their Springer fibers. For this, see [10, 81] and references therein.

3.4. *Reductive subgroups acting on flag varieties*

There is a classification, due to Avdeev and Petukhov [2], of the reductive subgroups H in G which act spherically on any given partial flag variety G/P. In particular these subgroups H have finitely many orbits in G/P. There are many examples of such subgroups other than symmetric subgroups, which we already consider (see Lemma 3.4 above).

4. Product of Flag Varieties

For applications of the theory of moment maps introduced in §2, we need a G-variety X with finitely many orbits. For example, if X is a product of partial flag varieties G/P_1 and G/P_2, where P_i $(i = 1, 2)$ are parabolic subgroups, then the diagonal action of G on $X = G/P_1 \times G/P_2$ has finitely many orbits. To see this, we note

$$X/G = (G/P_1 \times G/P_2)/G \simeq P_1\backslash G/P_2 \simeq W_{P_1}\backslash W_G/W_{P_2},$$

where the last isomorphism is given by the Bruhat decomposition (see Lemma 3.1). In this way, we also get a parametrization of G-orbits in X.

It is natural to ask the problem of when a product of flag varieties has finitely many orbits under the diagonal action of G.

4.1. *Triple flag varieties*

Theorem 4.1: *Let G be a classical group of type A, B, C, D, and P_i $(1 \leq i \leq k)$ parabolic subgroups of G, none of which is equal to G. If the product $G/P_1 \times G/P_2 \times \cdots \times G/P_k$ admits finitely many G-orbits under the diagonal G-action then $k \leq 3$ holds.*

The case $k = 1$ is trivial, and as mentioned above, the case $k = 2$ reduces to the Bruhat decomposition. So, in this case, the number of orbits is always finite.

Let us assume $k = 3$, and consider a triple flag variety $G/P_1 \times G/P_2 \times G/P_3$. In general this variety has infinitely many G-orbits, but there are many interesting cases where they admit a finite number of orbits. For classical groups, those cases are classified by Magyar-Weyman-Zelevinsky [44, 45] and Matsuki [48, 49], which gives Theorem 4.1. In [44, 45], they also classified orbits in terms of quiver representations (see §8.1 below for type A). For odd orthogonal groups, Matsuki [47] classified orbits in the case of (I) in [48](Theorem 1.6).

Up to now, there is no simple way to prove the number k should be smaller than 4 for multiple flag varieties of finite type for classical groups, to the knowledge of authors.

The classification of multiple flag varieties of finite type for exceptional groups remains open, except for the cases of type G_2 and F_4 (see [3]). It is known the number k of the flag varieties of finite type is less than or equal to 4 for type E_6, E_7, E_8, that it is 3 for type F_4, and that there is no multiple flag varieties of finite type for type G_2 if k is greater than 2 ([61](Theorem 2)). Actually the result of Popov [61](Theorem 2) specifies the upper bound of the number of flag varieties for which the multiple flag varieties admit an *open* orbit. We reproduce a table in [61] here for the convenience of readers. For the number b_G specified in Table 1, if a multiple flag variety $G/P_1 \times G/P_2 \times \cdots \times G/P_d$ has an open G-orbit then $d \leq b_G$ holds.

Table 1. Number b_G.

G	A_n $(n \geq 1)$	B_n $(n \geq 1)$	C_n $(n \geq 2)$	D_n $(n \geq 4)$	E_6	E_7	E_8	F_4	G_2
b_G	$n+2$	$n+1$	$n+1$	n	4	4	4	3	2

For type A_n, it is widely known that there exists an open orbit in an $(n+2)$-tuple projective spaces. Beyond $(n+2)$ there is no open orbit, which means $b_G = n + 2$ is best possible for type A (cf. Shimamoto [66], which describes families of generic orbits in the product of projective spaces). For other types, these bound seems over estimated. For this, see Theorem 3 in [61].

Even if there is an open orbit, it does not imply the finiteness of orbits. However, if one of the parabolic subgroups is a Borel subgroup, the existence of an open orbit tells that the multiple flag variety is of finite type.

Stembridge [79] gets a complete classification of the triple flag varieties $G/P_1 \times G/P_2 \times G/B$ with B being a Borel subgroup of G, including exceptional cases. For a summary, we refer the readers to [20](Table 1). In this case, there is no such triple flag variety for type E_8, F_4, G_2. If $G = E_6, E_7$, there is no quadruple flag variety of finite type with one of the member G/B because of the dimension reason, i.e., the dimension of such quadruple flag variety is strictly greater than $\dim G$.

4.2. *Double flag varieties for symmetric pairs*

Let us interpret the triple flag varieties in a different way as explained below. Let $\mathbb{G} = G \times G$ and put $\mathbb{P} = P_2 \times P_3$, a parabolic subgroup of $\mathbb{G}$. Choose an involution Θ on $\mathbb{G}$ as a flip of pairs, namely we put $\Theta(g_1, g_2) = (g_2, g_1)$ for $(g_1, g_2) \in \mathbb{G}$. For this involution, the fixed point subgroup is the diagonal embedding of G into $G \times G$, which we denote by $\mathbb{K} \simeq G$, the symmetric subgroup associated with Θ. Hence we can (and do) identify P_1 with a parabolic subgroup of $\mathbb{K}$, and we write $\mathbb{Q} = P_1$. Thus we see

$$\mathbb{K}/\mathbb{Q} \times \mathbb{G}/\mathbb{P} \simeq G/P_1 \times G/P_2 \times G/P_3 \tag{4.1}$$

is the triple flag variety on which $\mathbb{K} \simeq G$ acts. Note that $\mathbb{K}/\mathbb{Q}$ is a flag variety for the symmetric subgroup $\mathbb{K}$ and $\mathbb{G}/\mathbb{P}$ is a flag variety for $\mathbb{G}$.

Following this example, let us define a double flag variety for a symmetric pair. Thus, now we consider G with an involution θ, and denote the corresponding symmetric subgroup by $K = G^\theta$. We pick a parabolic subgroup $Q \subset K$ of K and a parabolic subgroup $P \subset G$ of G. We call the following variety a *double flag variety for the symmetric pair* (G, K), on which K acts diagonally,

$$\mathfrak{X} = K/Q \times G/P. \tag{4.2}$$

The above discussion shows that triple flag varieties are special cases of double flag varieties.

Our main interest in this chapter is the theory of double flag varieties of symmetric pairs.

Example 4.2: We keep the notation in Example 3.3. Let us consider a symmetric pair of type AIII:

$$(G, K) = (\mathrm{GL}_n(\mathbb{C}), \mathrm{GL}_p(\mathbb{C}) \times \mathrm{GL}_q(\mathbb{C})) \qquad (n = p + q),$$

where K is embedded into G block diagonally. A parabolic subgroup P of G is determined by a composition $\mathbf{a} = (a_1, \dots, a_r)$ of n. We denote the parabolic subgroup determined by this composition $\mathbf{a}$ by $P_{\mathbf{a}}$. If we pick a standard basis $\{e_i \mid 1 \leq i \leq n\}$ and define a *standard flag* $\mathscr{F}(\mathbf{a})$ associated with $\mathbf{a}$ by

$$\begin{aligned}\mathscr{F}(\mathbf{a}) : V_0 = \{0\} \subset V_1 \subset V_2 \subset \cdots \subset V_r = \mathbb{C}^n, \\ V_i = \langle e_k \mid 1 \leq k \leq a_1 + a_2 + \cdots + a_i \rangle.\end{aligned} \tag{4.3}$$

The parabolic subgroup $P_{\mathbf{a}}$ is defined to be the stabilizer of the flag $\mathscr{F}(\mathbf{a})$, which has diagonal blocks of size $a_1, a_2, \ldots, a_r$ and blockwise upper triangular.

Similarly, take compositions $\mathbf{b} = (b_1, \ldots, b_{l_1})$ of p and $\mathbf{c} = (c_1, \ldots, c_{l_2})$ of q. Then $Q = P_{\mathbf{b}} \times P_{\mathbf{c}}$ up to K-conjugacy, and $K/Q = \mathrm{GL}_p/P_{\mathbf{b}} \times \mathrm{GL}_q/P_{\mathbf{c}}$ is a product of flag varieties of type A. Thus we get a double flag variety

$$\mathfrak{X} = K/Q \times G/P = \mathrm{GL}_p/P_{\mathbf{b}} \times \mathrm{GL}_q/P_{\mathbf{c}} \times \mathrm{GL}_n/P_{\mathbf{a}} \tag{4.4}$$

on which $K = \mathrm{GL}_p \times \mathrm{GL}_q$ acts.

Example 4.3: We continue examining the examples given in Example 3.3, and consider a symmetric pair of type CI: $(G, K) = (\mathrm{Sp}_{2n}(\mathbb{C}), \mathrm{GL}_n(\mathbb{C}))$.

Let V be a symplectic vector space of dimension $2n$ with a symplectic form $\langle\ ,\ \rangle$, and $V = V^+ \oplus V^-$ a polar decomposition of V by Lagrangian subspaces $V^{\pm}$. For a composition $\mathbf{a}$ of n and $\mathscr{F}(\mathbf{a}) = (V_i^+)_{i=1}^n$ the corresponding flag of isotropic subspaces inside V^+ (cf. Equation (4.3)), we define the subgroup of elements in G which stabilize $\mathscr{F}(\mathbf{a})$ and denote it by $P_{\mathbf{a}}$. Put

$$V_i^- = (V_{n-i}^+)^{\perp} = \{v \in V^- \mid \langle v, V_i^+ \rangle = 0\}. \tag{4.5}$$

It is the dual flag of $\mathscr{F}(\mathbf{a})$ inside V^-. An element $h \in P_{\mathbf{a}}$ preserves both the flag $\mathscr{F}(\mathbf{a})$ and its dual.

Part II. Double Flag Varieties of Finite Type

5. Finiteness Criterions

We will apply the theory explained in §2.2 to the double flag varieties of symmetric pairs. For this purpose, it is important to know if there are finitely many orbits on a double flag variety. Thus we say, a double flag variety $\mathfrak{X} = K/Q \times G/P$ is *of finite type* if there are only finitely many K-orbits on $\mathfrak{X}$. Although the classification of the double flag varieties of finite type is still largely open, there are several different classifications of such varieties.

Roughly speaking, there are three major methods to classify the K-orbits on $\mathfrak{X}$. Let us explain them briefly.

(1) Since $\mathfrak{X}/K \simeq P\backslash G/Q$, the double coset decomposition by (P, Q) will give the classification. This is the strategy used in [20], but explicit decomposition is often very subtle. In [20], the decomposition is reduced to that of unipotent subgroups, and by linearization, it is further reduced to the multiplicity free representations if one of P or Q is a Borel subgroup.

(2) A double flag variety can be considered as configuration spaces of flags (or vector spaces). In this viewpoint, the theory of quiver representations is useful, especially for type A groups. This idea first appeared in [44, 45] for triple flag varieties, and then applied to the double flag variety by Homma in [24]. We will explain this in detail later in §7.

(3) Let $\mathcal{N}_{\mathfrak{k}}$ be the nilpotent variety of $\mathfrak{k}$. Then there is a well defined map between the orbit spaces $\Phi_{\mathfrak{k}} : \mathfrak{X}/K \to \mathcal{N}_{\mathfrak{k}}/K$ (see §10 for the precise definition). We call this map a *symmetrized Steinberg map.* Since the space of nilpotent orbits $\mathcal{N}_{\mathfrak{k}}/K$ is finite and well-understood, if we can determine the fiber of the map $\Phi_{\mathfrak{k}}$, it will produce the classification of $\mathfrak{X}/K$. The fibers often can be described by some combinatorial objects related to Weyl group representations (like Robinson-Schensted correspondence), though there is no general theory yet. The origin of this method is the classical Steinberg theory, and it is generalized to double flag varieties of type AIII in [13, 15, 14, 16]; see also §11 below.

6. Double Flag Varieties Associated to Borel Subgroups

Let us assume that either

(1) G is a *simple* algebraic group, which is simply connected, and $K = G^{\theta}$ is the fixed point subgroup of an involution θ as above; or

(2) $G = G_1 \times G_1$ with $K = \Delta G_1$, which is the fixed point subgroup of the "flip" (see §4.2 around Equation (4.1)). Here G_1 is a simple, simply connected algebraic group.

Note that in both cases K is connected.

For a double flag variety $\mathfrak{X} = K/Q \times G/P$, if one of P or Q is a Borel subgroup of G or K respectively, then there is a complete classification of $\mathfrak{X}$ of finite type in this setting due to He-N-Ochiai-Oshima [20].

6.1. *Ideas of the proofs*

Let us explain the ideas of the proofs in accordance with [20].

We begin with the case where $Q = B_K$ is a Borel subgroup.

Take a θ-stable Borel subgroup B of G and θ-stable maximal torus $T \subset B$, which always exist. Put $B_K = B^\theta = B \cap K$. Since B is θ-stable, B_K is a Borel subgroup of K.

Let P be a parabolic subgroup of G which contains B. Choose a Levi decomposition $P = LU$, where U is the unipotent radical and $T \subset L$, i.e., the decomposition is standard.

Lemma 6.1: *$L \cap K$ is a connected reductive subgroup of K and $L \cap B_K$ is a Borel subgroup of $L \cap K$.*

Proof: Note that $P' := P \cap \theta(P)$ is a minimal θ-stable parabolic subgroup among those that contain $P \cap K$. It is easy to see $L' := L \cap \theta(L)$ is a Levi part of P'. Since P' is θ-stable, $Q' = P' \cap K$ is a parabolic subgroup of K whose Levi component is $L' \cap K$. Now $L \cap K = L \cap \theta(L) \cap K = L' \cap K$ is reductive and connected, since it is a Levi component of a parabolic subgroup Q'. Since $L' \cap B$ is a θ-stable Borel subgroup of L', $L \cap B_K = L \cap \theta(L) \cap B_K = L' \cap B_K = (L' \cap B)^\theta$ is a Borel subgroup of $L \cap K$. □

Let $\mathfrak{u}$ be the Lie algebra of the unipotent radical U of P. We denote

$$\mathfrak{s} = \mathfrak{g}^{-\theta} = \{x \in \mathfrak{g} \mid \theta(x) = -x\}, \tag{6.1}$$

so that $\mathfrak{g} = \mathfrak{k} + \mathfrak{s}$ is the Cartan decomposition with respect to θ.

The following theorem is a key theorem for the classification.

Theorem 6.2: (He-N-Ochiai-Oshima [20](Theorem 4.2)) *Let B_K be a Borel subgroup of K and $P = LU$ a parabolic subgroup of G as explained above. The double flag variety $\mathfrak{X} = K/B_K \times G/P$ is of finite type if and only if the adjoint action of $L \cap K$ on $\mathfrak{u} \cap \mathfrak{s}$ is spherical.*

To prove this theorem, the following theorem due to Panyushev is essential.

Theorem 6.3: (Panyushev [59](Theorem 2.1)) *Let G be a connected reductive algebraic group and X a smooth G-variety. For a smooth locally closed G-stable subvariety M of X, the following* (1)–(3) *are equivalent.*

(1) *The G-variety X is spherical.*

(2) *The normal bundle $T_M X$ is G-spherical.*

(3) *The conormal bundle $T_M^* X$ is G-spherical.*

Proof: [Proof of Theorem 6.2] Let us assume $\mathfrak{u} \cap \mathfrak{s}$ is spherical. We will prove that $X := G/P$ is a spherical K-variety so that it contains finitely many B_K-orbits.

Let $\mathbb{O}$ be a K-orbit through the base point $p = eP$. By Panyushev's theorem, it suffices to prove that the conormal bundle $T^*_{\mathbb{O}}X$ is K-spherical. For that, let us see existence of an open B_K^--orbit, where B_K^- is the opposite Borel subgroup of B_K. Since P contains B_K by assumption, the tangent space of $\mathbb{O}$ at p is $(\mathfrak{k}+\mathfrak{p})/\mathfrak{p} = (\mathrm{Lie}(B_K^-)+\mathfrak{p})/\mathfrak{p}$. Thus we conclude $B_K^- \cdot p \subset \mathbb{O}$ is open dense. The conormal direction of $\mathbb{O}$ at p is $(\mathfrak{k}+\mathfrak{p})^\perp = \mathfrak{s} \cap \mathfrak{p}^\perp = \mathfrak{s} \cap \mathfrak{u}$, which is a $L \cap K$-variety. Note that $L \cap K$ is reductive by Lemma 6.1. The stabilizer $\mathrm{Stab}_{B_K^-}(p) = (P \cap K) \cap B_K^- \supset L \cap B_K^-$, which is a Borel subgroup of $L \cap K$ by the same lemma. Since $\mathfrak{u} \cap \mathfrak{s}$ is spherical, there is a dense open $L \cap B_K^-$-orbit, and we get an open B_K^--orbit on $T^*_{\mathbb{O}}X$.

The converse implication is proved by following the above arguments in reversed order. □

By Theorem 6.2, the classification of the double flag varieties of finite type is reduced to the classification of spherical linear actions (they are often called multiplicity free actions, too). This is already done by Kac [28], and independently by Benson-Ratcliff [4] and Leahy [42].

Next, we consider the case where $P = B$ is a Borel subgroup of G. For this case, we only quote a key theorem in [20].

To do so, we choose a θ-stable parabolic subgroup P' of G which cuts out Q, i.e., $Q = P' \cap K$, which is always possible. By taking conjugation, we can also assume B is θ-stable and P' contains B. Let $P' = L'U'$ be the standard Levi decomposition with respect to a θ-stable maximal torus $T \subset B$, and $P^{L'}_{\mathrm{min}}$ be a minimal theta split parabolic subgroup of L'. Here, "theta split" means $\theta(P^{L'}_{\mathrm{min}})$ becomes the opposite parabolic subgroup of $P^{L'}_{\mathrm{min}}$. Put $M' = P^{L'}_{\mathrm{min}} \cap K$ and denote by M'_0 its connected component.

Theorem 6.4: (He-N-Ochiai-Oshima [20](Theorem 4.4)) *The double flag variety $\mathfrak{X} = K/Q \times G/B$ is of finite type if and only if the adjoint action of M'_0 on $\mathfrak{u}' \cap \mathfrak{s}$ is spherical.*

Thus, again, the classification reduces to that of spherical linear actions.

Let us summarize the classifications of double flag varieties of finite types where one of the parabolic subgroups is a Borel subgroup into several tables in §§6.2–6.5 for the convenience and later use. These tables are all quoted from [20], [79], [44, 45] and [48, 49], with modifications to the present chapter.

6.2. *Classification when* $P = B_G$ *is a Borel subgroup of* G

In the following tables, we separate the conditions by semicolons. If there is a semicolon, then the conditions should be considered independently from the former one(s).

Table 2. Double flag varieties $K/Q \times G/B_G$ of finite type.

$\mathfrak{g}$	$\mathfrak{k}$	$\Pi_K \setminus J_K$ $(Q = Q_{J_K})$
$\mathfrak{sl}_{2n}$ $n \geq 2$	$\mathfrak{sp}_{2n}$	 $\{\alpha_1\}$; $\{\alpha_3\}$ if $n = 3$; any subset of Π_K if $n = 2$
$\mathfrak{sl}_{p+q}$ $p + q \geq 3$	$\mathfrak{sl}_p \oplus \mathfrak{sl}_q \oplus \mathbb{C}$ $1 \leq p \leq q$	  $\{\alpha_1\}$, $\{\alpha_{p-1}\}$, $\{\beta_1\}$, $\{\beta_{q-1}\}$; $\{\beta_i\}(\forall i)$ if $p = 2$; any subset of Π_K if $p = 1$
$\mathfrak{so}_{2n+2}$ $n \geq 3$	$\mathfrak{so}_{2n} \oplus \mathbb{C}$	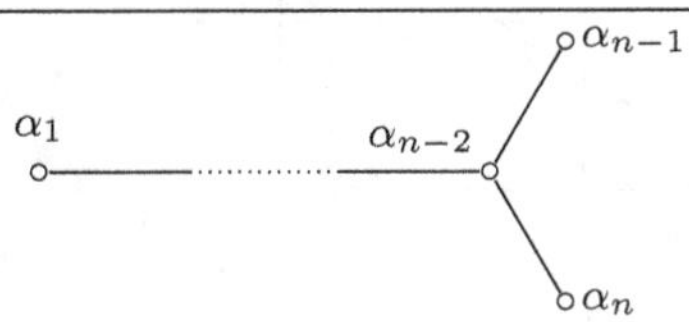 $\{\alpha_{n-1}\}$, $\{\alpha_n\}$
$\mathfrak{so}_{2n+1}$ $n \geq 3$	$\mathfrak{so}_{2n}$	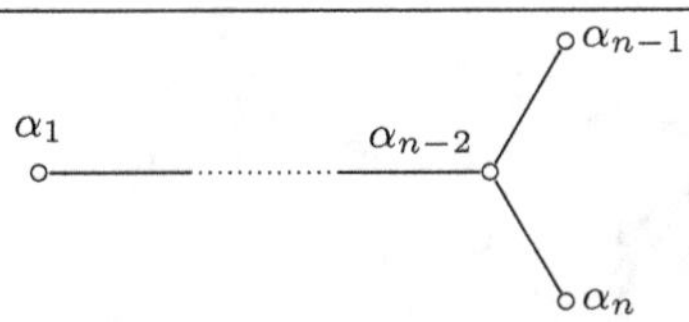 any subset of Π_K
$\mathfrak{so}_{2n+2}$ $n \geq 3$	$\mathfrak{so}_{2n+1}$	 any subset of Π_K
$\mathfrak{so}_{2n+2}$ $n \geq 3$	$\mathfrak{sl}_{n+1} \oplus \mathbb{C}$	 $\{\alpha_1\}$, $\{\alpha_n\}$

Double flag varieties $K/Q \times G/B_G$ of finite type (continued)

$\mathfrak{g}$	$\mathfrak{k}$	$\Pi_K \setminus J_K$ $(Q = Q_{J_K})$
$\mathfrak{sp}_{2(p+q)}$	$\mathfrak{sp}_{2p} \oplus \mathfrak{sp}_{2q}$ $1 \leq p \leq q$	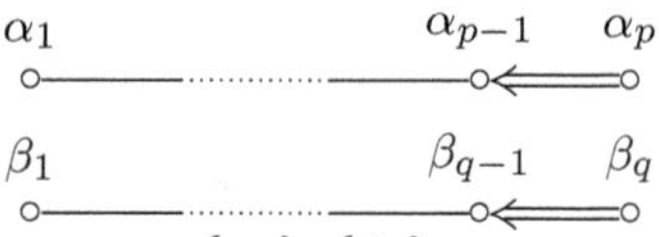 $\{\alpha_1\}$, $\{\beta_1\}$; $\{\alpha_p\}$ if $p \leq 3$; $\{\beta_q\}$ if $p \leq 2$; $\{\beta_q\}$ if $q \leq 3$; $\{\alpha_1, \alpha_2\}$ if $p = 2$; $\{\beta_1, \beta_2\}$ if $q = 2$; $\{\beta_i\}(\forall i)$, $\{\beta_i, \beta_j\}(\forall i, j)$ if $p = 1$
$\mathfrak{f}_4$	$\mathfrak{so}_9$	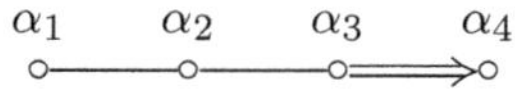 $\{\alpha_i\}(\forall i)$, $\{\alpha_1, \alpha_2\}$
$\mathfrak{e}_6$	$\mathfrak{so}_{10} \oplus \mathbb{C}$	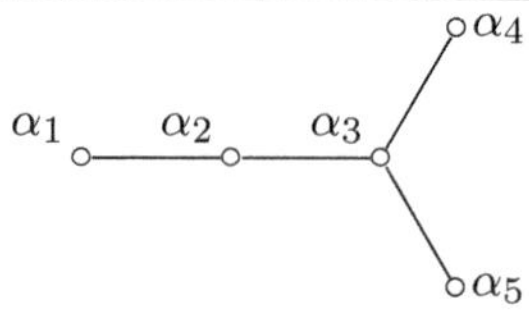 $\{\alpha_1\}$
$\mathfrak{e}_6$	$\mathfrak{f}_4$	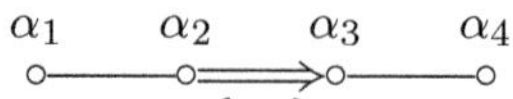 $\{\alpha_1\}$

6.3. *Classification when $Q = B_K$ is a Borel subgroup of K*

Table 3. Double flag varieties $K/B_K \times G/P$ of finite type.

$\mathfrak{g}$	$\mathfrak{k}$	$\Pi \setminus J$ $(P = P_J)$
$\mathfrak{sl}_n$		
$\mathfrak{sl}_n$	$\mathfrak{so}_n$	$\{\alpha_i\}(\forall i)$
$\mathfrak{sl}_{2m}$ $2m = n \geq 4$	$\mathfrak{sp}_{2m}$	$\{\alpha_i\}(\forall i)$, $\{\alpha_i, \alpha_{i+1}\}(\forall i)$, $\{\alpha_1, \alpha_i\}(\forall i)$, $\{\alpha_i, \alpha_{n-1}\}(\forall i)$, $\{\alpha_1, \alpha_2, \alpha_3\}$, $\{\alpha_{n-3}, \alpha_{n-2}, \alpha_{n-1}\}$, $\{\alpha_1, \alpha_2, \alpha_{n-1}\}$, $\{\alpha_1, \alpha_{n-2}, \alpha_{n-1}\}$
$\mathfrak{sl}_{p+q}$ $p + q = n$	$\mathfrak{sl}_p \oplus \mathfrak{sl}_q \oplus \mathbb{C}$ $1 \leq p \leq q$	$\{\alpha_i\}(\forall i)$, $\{\alpha_i, \alpha_{i+1}\}(\forall i)$, $\{\alpha_1, \alpha_i\}(\forall i)$, $\{\alpha_i, \alpha_{n-1}\}(\forall i)$; $\{\alpha_i, \alpha_j\}(\forall i, j)$ if $p = 2$; any subset of Π if $p = 1$
$\mathfrak{so}_{2n+1}$		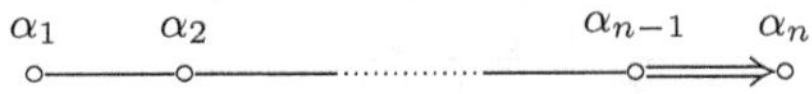
$\mathfrak{so}_{p+q}$ $p+q = 2n+1$	$\mathfrak{so}_p \oplus \mathfrak{so}_q$ $1 \leq p \leq q$	$\{\alpha_1\}$, $\{\alpha_n\}$; $\{\alpha_i\}(\forall i)$ if $p = 2$; any subset of Π if $p = 1$
$\mathfrak{so}_{2n}$		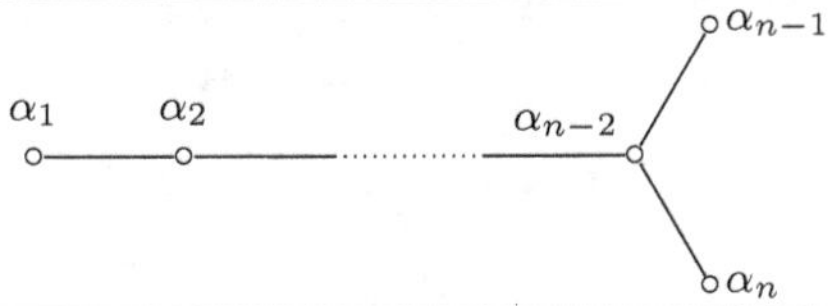
$\mathfrak{so}_{p+q}$ $p + q = 2n$ $n \geq 4$	$\mathfrak{so}_p \oplus \mathfrak{so}_q$ † $1 \leq p \leq q$	$\{\alpha_1\}$, $\{\alpha_{n-1}\}$, $\{\alpha_n\}$; $\{\alpha_i\}(\forall i)$, if $p = 2$, $\{\alpha_i, \alpha_{n-1}\}(\forall i)$, if $p = 2$, $\{\alpha_i, \alpha_n\}(\forall i)$ if $p = 2$; any subset of Π if $p = 1$
$\mathfrak{so}_{2n}$ $n \geq 4$	$\mathfrak{sl}_n \oplus \mathbb{C}$ ‡	$\{\alpha_1\}$, $\{\alpha_2\}$, $\{\alpha_3\}$, $\{\alpha_{n-1}\}$, $\{\alpha_n\}$, $\{\alpha_1, \alpha_2\}$, $\{\alpha_1, \alpha_{n-1}\}$, $\{\alpha_1, \alpha_n\}$, $\{\alpha_{n-1}, \alpha_n\}$, $\{\alpha_2, \alpha_3\}$, $\{\alpha_2, \alpha_4\}$ if $n = 4$

† The subalgebra $\mathfrak{k}$ of $\mathfrak{g}$ is not unique for $(\mathfrak{g}, \mathfrak{k}) \simeq (\mathfrak{so}_8, \mathfrak{so}_7)$, $(\mathfrak{so}_8, \mathfrak{so}_6 \oplus \mathbb{C})$, $(\mathfrak{so}_8, \mathfrak{so}_5 \oplus \mathfrak{so}_3)$. See [20](Remark 5.3) for details.

‡ For $\mathfrak{g} \simeq \mathfrak{so}_{4n}$, a symmetric subalgebra $\mathfrak{k}$ that is isomorphic to $\mathfrak{sl}_{2n} \oplus \mathbb{C}$ is not unique up to inner automorphisms of $\mathfrak{g}$. Here we take K as the Levi part $L_{\Pi \setminus \{\alpha_n\}}$ of the parabolic subgroup $P_{\Pi \setminus \{\alpha_n\}}$.

$\mathfrak{sp}_{2n}$		
$\mathfrak{sp}_{2n}$	$\mathfrak{sl}_n \oplus \mathbb{C}$	$\{\alpha_1\}$, $\{\alpha_n\}$
$\mathfrak{sp}_{2(p+q)}$ $p+q=n$	$\mathfrak{sp}_{2p} \oplus \mathfrak{sp}_{2q}$ $1 \leq p \leq q$	$\{\alpha_1\}$, $\{\alpha_2\}$, $\{\alpha_3\}$, $\{\alpha_n\}$, $\{\alpha_1, \alpha_2\}$; $\{\alpha_i\}(\forall i)$ if $p \leq 2$; $\{\alpha_i, \alpha_j\}(\forall i, j)$ if $p = 1$
$\mathfrak{f}_4$		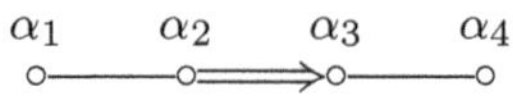
$\mathfrak{f}_4$	$\mathfrak{so}_9$	$\{\alpha_1\}$, $\{\alpha_2\}$, $\{\alpha_3\}$, $\{\alpha_4\}$, $\{\alpha_1, \alpha_4\}$
$\mathfrak{e}_6$		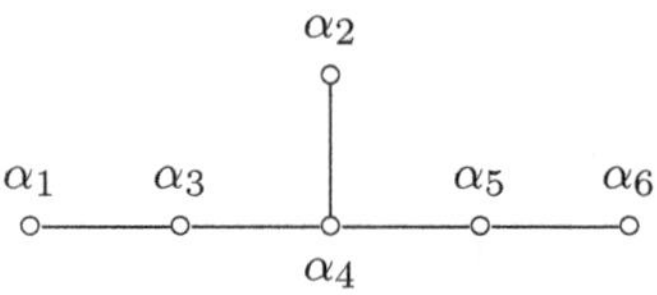
$\mathfrak{e}_6$	$\mathfrak{sp}_8$	$\{\alpha_1\}$, $\{\alpha_6\}$
$\mathfrak{e}_6$	$\mathfrak{sl}_6 \oplus \mathfrak{sl}_2$	$\{\alpha_1\}$, $\{\alpha_6\}$
$\mathfrak{e}_6$	$\mathfrak{so}_{10} \oplus \mathbb{C}$	$\{\alpha_1\}$, $\{\alpha_2\}$, $\{\alpha_3\}$, $\{\alpha_5\}$, $\{\alpha_6\}$, $\{\alpha_1, \alpha_6\}$
$\mathfrak{e}_6$	$\mathfrak{f}_4$	$\{\alpha_1\}$, $\{\alpha_2\}$, $\{\alpha_3\}$, $\{\alpha_5\}$, $\{\alpha_6\}$, $\{\alpha_1, \alpha_2\}$, $\{\alpha_2, \alpha_6\}$, $\{\alpha_1, \alpha_3\}$, $\{\alpha_5, \alpha_6\}$
$\mathfrak{e}_7$		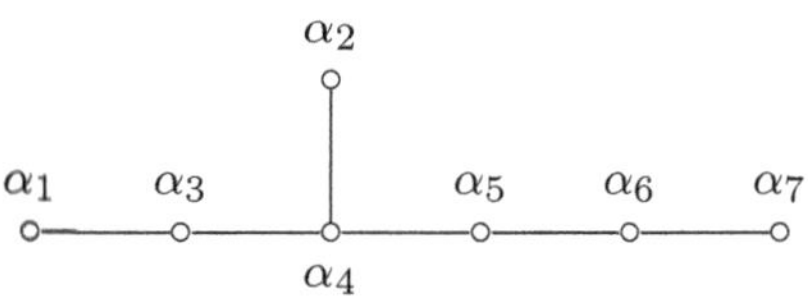
$\mathfrak{e}_7$	$\mathfrak{sl}_8$	$\{\alpha_7\}$
$\mathfrak{e}_7$	$\mathfrak{so}_{12} \oplus \mathfrak{sl}_2$	$\{\alpha_7\}$
$\mathfrak{e}_7$	$\mathfrak{e}_6 \oplus \mathbb{C}$	$\{\alpha_1\}$, $\{\alpha_2\}$, $\{\alpha_7\}$

6.4. *Triple flag variety of finite type, where $P_3 = B_G$*

If one of the parabolic subgroup P_3 is a Borel subgroup B_G of G, the triple flag varieties $\mathfrak{X} = G/P_1 \times G/P_2 \times G/B_G$ are classified by Stembridge [79]. We reproduce his classification here.

Table 4. $G/P_1 \times G/P_2 \times G/B$ of finite type.

$\mathfrak{g}$	$(\Pi \setminus J_1,\ \Pi \setminus J_2)$ up to switching J_1 and J_2
$\mathfrak{sl}_{n+1}$	 $(\{\alpha_i\},\{\alpha_j\})(\forall i,j)$, $(\{\alpha_1,\alpha_i\},\{\alpha_j\})(\forall i,j)$, $(\{\alpha_i,\alpha_n\},\{\alpha_j\})(\forall i,j)$, $(\{\alpha_i,\alpha_{i+1}\},\{\alpha_j\})(\forall i,j)$, $(\{\alpha_i,\alpha_j\},\{\alpha_2\})(\forall i,j)$, $(\{\alpha_i,\alpha_j\},\{\alpha_{n-1}\})(\forall i,j)$, $(\{\alpha_1\},\text{any})$, $(\{\alpha_n\},\text{any})$
$\mathfrak{so}_{2n+1}$	 $(\{\alpha_1\},\{\alpha_i\})(\forall i)$, $(\{\alpha_n\},\{\alpha_n\})$
$\mathfrak{so}_{2n}$ $n \geq 4$	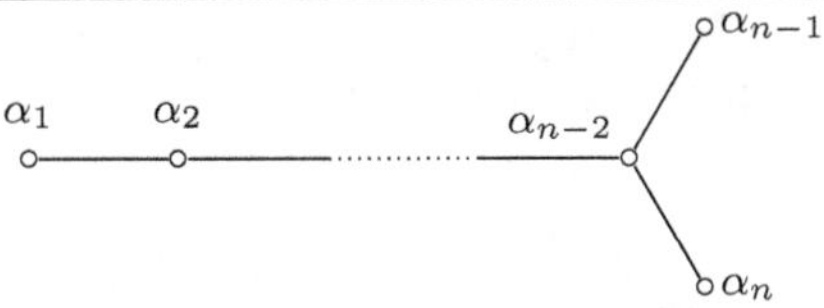 $(\{\alpha_1\},\{\alpha_i\})(\forall i)$, $(\{\alpha_i\},\{\alpha_j\})(i=1,2,3,\ j=n-1,n)$, $(\{\alpha_{n-1}\},\{\alpha_{n-1}\})$, $(\{\alpha_{n-1}\},\{\alpha_n\})$, $(\{\alpha_n\},\{\alpha_n\})$, $(\{\alpha_1\},\{\alpha_i,\alpha_{n-1}\})(\forall i)$, $(\{\alpha_1\},\{\alpha_i,\alpha_n\})(\forall i)$, $(\{\alpha_{n-1}\},\{\alpha_1,\alpha_2\})$, $(\{\alpha_n\},\{\alpha_1,\alpha_2\})$, $(\{\alpha_i\},\{\alpha_j,\alpha_k\})(i=n-1,n,\ \{j,k\}\subset\{1,n-1,n\})$; $(\{\alpha_3\},\{\alpha_2,\alpha_4\})$ if $n=4$; $(\{\alpha_4\},\{\alpha_2,\alpha_3\})$ if $n=4$
$\mathfrak{sp}_{2n}$	 $(\{\alpha_1\},\{\alpha_i\})(\forall i)$, $(\{\alpha_n\},\{\alpha_n\})$
$\mathfrak{e}_6$	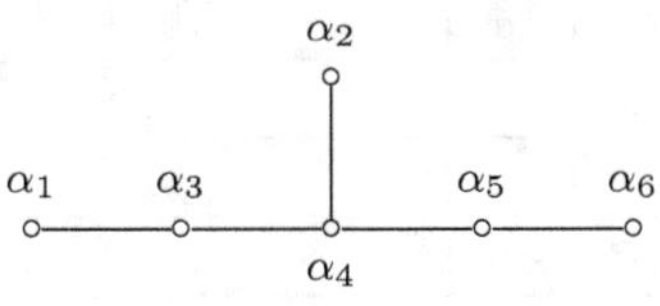 $(\{\alpha_i\},\{\alpha_j\})(i=1,6,\ j\neq 4)$, $(\{\alpha_1\},\{\alpha_1,\alpha_6\})$, $(\{\alpha_6\},\{\alpha_1,\alpha_6\})$

$\mathfrak{e}_7$	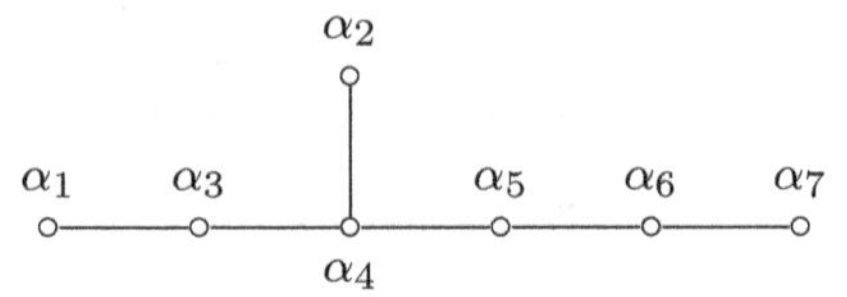 $(\{\alpha_7\}, \{\alpha_1\})$, $(\{\alpha_7\}, \{\alpha_2\})$, $(\{\alpha_7\}, \{\alpha_7\})$

6.5. *Triple flag variety of finite type, where G is classical*

In this subsection, we list triple flag varieties $\mathfrak{X} = G/P_1 \times G/P_2 \times G/P_3$ of finite type, where G is classical. These classifications are quoted from the tables of [44, 45, 48, 49]. We only list the triples when $P_i = P_{J_i}$ $(i = 1, 2, 3)$ are proper parabolic subgroups, where $J_i \subset \Pi$ is a subset of simple roots which generates the Levi part of P_i.

6.5.1. *Type A*

We quote the results from [44](Theorem 2.2) except for the trivial case of double flag variety, which is denoted $(A_{q,r})$. In the following table, $G = \mathrm{GL}_n$ and the numbering of simple roots Π is given as in §6.4. (We use different labeling α, β, γ to indicate the simple roots for P_1, P_2 and P_3 respectively.) In the table, r denotes the number of simple roots in $\Pi \setminus J_3$.

Table 5. Type A triple flag varieties of finite type.

code in [44]	$\Pi \setminus J_1$	$\Pi \setminus J_2$	$\Pi \setminus J_3$
(D_{r+2})	$\{\alpha_i\}$	$\{\beta_j\}$	any subset
(E_6)	$\{\alpha_i\}$	$\{\beta_{j_1}, \beta_{j_2}\}$	$\{\gamma_{k_1}, \gamma_{k_2}\}$
(E_7)	$\{\alpha_i\}$	$\{\beta_{j_1}, \beta_{j_2}\}$	$\{\gamma_{k_1}, \gamma_{k_2}, \gamma_{k_3}\}$
(E_8)	$\{\alpha_i\}$	$\{\beta_{j_1}, \beta_{j_2}\}$	$\{\gamma_{k_1}, \gamma_{k_2}, \gamma_{k_3}, \gamma_{k_4}\}$
$(E^{(a)}_{r+3})$	$\{\alpha_2\}, \{\alpha_{n-2}\}$	$\{\beta_{j_1}, \beta_{j_2}\}$	any subset
$(E^{(b)}_{r+3})$	$\{\alpha_i\}$	$\{\beta_j, \beta_{j+1}\}, \{\beta_1, \beta_j\}, \{\beta_j, \beta_{n-1}\}$	any subset
$(S_{q,r})$	$\{\alpha_1\}, \{\alpha_{n-1}\}$	any subset	any subset

6.5.2. *Type B*

We quote the results from [48](Theorem 1.6). In the following table, $G = \mathrm{SO}_{2n+1}$ and the numbering of simple roots Π is given as in §6.4.

Table 6. Type B triple flag varieties of finite type.

code in [48]	$\Pi \setminus J_1$	$\Pi \setminus J_2$	$\Pi \setminus J_3$
(I)	$\{\alpha_n\}$	$\{\beta_n\}$	any subset
(II)	$\{\alpha_1\}$	$\{\beta_j\}$	any subset
(III)	$\{\alpha_i\}$	$\{\beta_j\}$	$\{\gamma_n\}$
(IV)	$\{\alpha_i\}$	$\{\beta_n\}$	$\{\gamma_i, \gamma_j\}$

6.5.3. *Type C*

We quote the results from [45](Theorem 1.2) except for the trivial case of double flag variety, which is denoted $(\mathrm{Sp}A_{q,r})$. $G = \mathrm{Sp}_{2n}$ and the numbering of simple roots Π is given as in §6.4. (We use different labeling α, β, γ to indicate the simple roots for P_1, P_2 and P_3 respectively.) In the table, r denotes the number of simple roots in $\Pi \setminus J_3$.

Table 7. Type C triple flag varieties of finite type.

code in [45]	$\Pi \setminus J_1$	$\Pi \setminus J_2$	$\Pi \setminus J_3$
$(\mathrm{Sp}D_{r+2})$	$\{\alpha_n\}$	$\{\beta_n\}$	any subset
$(\mathrm{Sp}E_6)$	$\{\alpha_n\}$	$\{\beta_i\}$	$\{\gamma_j\}$
$(\mathrm{Sp}E_7)$	$\{\alpha_n\}$	$\{\beta_i\}$	$\{\gamma_j, \gamma_n\}$
$(\mathrm{Sp}E_8)$	$\{\alpha_n\}$	$\{\beta_i\}$	$\{\gamma_{j_1}, \gamma_{j_2}\}$
$(\mathrm{Sp}E^{(b)}_{r+3})$	$\{\alpha_n\}$	$\{\beta_1\}$	any subset
$(\mathrm{Sp}Y_{r+4})$	$\{\alpha_1\}$	$\{\beta_i\}$	any subset

6.5.4. *Type D*

We quote the results from [49](Theorem 1.7). In the following table, $G = \mathrm{SO}_{2n}$ and the numbering of simple roots Π is given as in §6.4.

Table 8. Type D triple flag varieties of finite type.

code in [49]	$\Pi \setminus J_1$	$\Pi \setminus J_2$	$\Pi \setminus J_3$
(I-1)	$\{\alpha_1\}$	$\{\beta_j\}$	any subset
(I-2)	$\{\alpha_1\}$	$\{\beta_j, \beta_{n-1}\}, \{\beta_j, \beta_n\}$	any subset
(II)	$\{\alpha_{n-1}\}, \{\alpha_n\}$	$\{\beta_1\}, \{\beta_2\}, \{\beta_3\}, \{\beta_1, \beta_2\}$, $\{\beta_1, \beta_{n-1}\}, \{\beta_1, \beta_n\}$, $\{\beta_{n-1}\}, \{\beta_n\}, \{\beta_{n-1}, \beta_n\}$	any subset
(III-1)	$\{\alpha_{n-1}\}, \{\alpha_n\}$	$\{\beta_j\}$ $(4 \leq j \leq n-2)$	$\{\gamma_k\}$
(III-2)	$\{\alpha_{n-1}\}, \{\alpha_n\}$	$\{\beta_j\}$ $(4 \leq j \leq n-2)$	$\{\gamma_{k_1}, \gamma_{k_2}\}$
(III-3)	$\{\alpha_{n-1}\}, \{\alpha_n\}$	$\{\beta_j\}$ $(4 \leq j \leq n-2)$	$\{\gamma_1, \gamma_k, \gamma_\nu\}$ $(\nu \in \{n-1, n\})$, $\{\gamma_k, \gamma_{k+1}, \gamma_\nu\}$ $(\nu \in \{n-1, n\})$, $\{\gamma_k, \gamma_{n-1}, \gamma_n\}$, $\{\gamma_1, \gamma_2, \gamma_{k+2}\}, \{\gamma_1, \gamma_{k+1}, \gamma_{k+2}\}$, $\{\gamma_k, \gamma_{k+1}, \gamma_{k+2}\}$
(III-4)	$\{\alpha_{n-1}\}, \{\alpha_n\}$	$\{\beta_j\}$ $(4 \leq j \leq n-2)$	$\{\gamma_1, \gamma_2, \gamma_3, \gamma_\nu\}$ $(\nu \in \{n-1, n\})$, $\{\gamma_1, \gamma_2, \gamma_{n-1}, \gamma_n\}$, $\{\gamma_1, \gamma_{n-2}, \gamma_{n-1}, \gamma_n\}$, $\{\gamma_{n-3}, \gamma_{n-2}, \gamma_{n-1}, \gamma_n\}$, $\{\gamma_1, \gamma_2, \gamma_3, \gamma_4\}$

7. Quivers and Multiple Flag Varieties

As we explained in §4.1 and §4.2, triple flag varieties are a special case of double flag varieties of symmetric pairs. For classical groups, those multiple flag varieties and their orbits have fruitful descriptions in terms of quiver representations. To see it, let us explain the theory of Kac on quiver representations.

7.1. *Orbits in quiver representations: Theory of Kac*

Let $Q = (Q_0, Q_1)$ be a quiver, where Q_0 is the set of vertices, and Q_1 is the set of arrows. Each $e \in Q_1$ has the unique initial vertex $s(e)$ and the terminal one $t(e)$. So, in general, Q is an oriented graph, which admits multiple edges and loops. However, in this chapter, we only consider *simply laced acyclic quivers* with single edges and no loop. A representation π of the quiver Q is an assignment of a vector space V_i to each vertex $i \in Q_0$ together with an assignment of a linear map $f_e : V_{s(e)} \to V_{t(e)}$ for each arrow $e \in Q_1$. The collection of dimensions $d(\pi) = (\dim V_i)_{i \in Q_0}$ is called the dimension vector. Two representations π and π' are said to be isomorphic if there are isomorphisms $\varphi_i : V_i \to V_i' \ (i \in Q_0)$ such that for each $e \in Q_1$, the diagram

$$\begin{array}{ccc} V_i & \xrightarrow{f_e} & V_j \\ \downarrow{\scriptstyle \varphi_i} & & \downarrow{\scriptstyle \varphi_j} \\ V_i' & \xrightarrow{f_e'} & V_j' \end{array} \qquad \text{(where } i = s(e) \text{ and } j = t(e))$$

is commutative.

Fix a dimension vector $\alpha \in \mathbb{Z}_{\geq 0}^{Q_0}$, and consider all the representations π of Q with the same dimension vector α. Denote the space of such representations by

$$\mathfrak{M}^\alpha(Q) = \{\pi \in \operatorname{Rep} Q \mid d(\pi) = \alpha\}.$$

For this variety, we fix vector spaces V_i for each $i \in Q_0$. Let $G^\alpha = \prod_{i \in Q_0} \mathrm{GL}_{\alpha_i}$, which acts naturally on $\mathfrak{M}^\alpha(Q)$ and it is easy to see that the orbits of G^α represent isomorphism classes of representations of Q with dimension vector α.

Obviously, every finite dimensional representation of a quiver Q decomposes into a direct sum of indecomposable representations. There exists an ingenious description of such indecomposable modules by Kac ([27, 29]), which is originally due to Gabriel for the Dynkin quiver of finite type ([19]). Kac's idea is to relate the quiver to a Cartan matrix A defined by

$$A = 2I_{Q_0} - R, \qquad \text{where } R = (r_{i,j})_{i,j \in Q_0}, \quad r_{i,j} = \#\{e \in Q_1 \mid \{s(e), t(e)\} = \{i, j\}\}.$$

Thus, $r_{i,j}$ denotes the number of edges which connect i and j so that R is a so-called adjacency matrix of the underlying graph of Q. We define a quadratic form in $x = (x_i)_{i \in Q_0}$ by

$$(x|x) = \frac{1}{2}\, {}^t x A x = \|x\|^2 - \frac{1}{2} \sum_{i,j \in Q_0} r_{i,j} x_i x_j,$$

which is called a *Tits form*. The following lemma is easy to prove, but the statement is crucial below.

Lemma 7.1: $\dim \mathfrak{M}^\alpha(Q) - (\dim G^\alpha - 1) = 1 - (\alpha|\alpha)$.

Note that the action of G^α on $\mathfrak{M}^\alpha(Q)$ is not effective and $G^\alpha/\mathbb{C}^\times$ acts effectively.

Let $\mathfrak{g}(A)$ be the Kac-Moody Lie algebra defined by the Cartan matrix $A = (a_{i,j})_{i,j\in Q_0}$ (cf. [30]), and $\Pi = \{\alpha_i \mid i \in Q_0\}$ the set of simple roots so that it satisfies $(\alpha_i|\alpha_j) = \frac{1}{2}a_{i,j}$. Let $\Gamma = \{\sum_{i\in Q_0} k_i\alpha_i \mid k_i \in \mathbb{Z}\}$ be the root lattice and define $\Gamma^+ = \{\sum_i k_i\alpha_i \mid k_i \geq 0\}$. We can (and do) identify a dimension vector $\alpha = (k_i)_{i\in Q_0}$ with an element $\sum_{i\in Q_0} k_i\alpha_i$ in Γ^+. Denote Δ and Δ^+ its root system and the set of positive roots respectively. There are notions of real roots Δ^+_{re} and imaginary roots Δ^+_{im}, the definitions of which we refer the readers to [30].

Theorem 7.2: (Kac [27](Theorem 3))

(1) *For any* $\alpha \in \Gamma^+$, $\mathfrak{M}^\alpha(Q)$ *contains an indecomposable module if and only if* $\alpha \in \Delta^+$.

(2) *There exists a unique isomorphism class of indecomposable modules in* $\mathfrak{M}^\alpha(Q)$ *if and only if* $\alpha \in \Delta^+_{re}$, *which is equivalent to* $(\alpha|\alpha) = 1$.

(3) *If* $\alpha \in \Delta^+_{im}$, *then there are infinitely many isomorphism classes of indecomposable modules in* $\mathfrak{M}^\alpha(Q)$. *There exists a family of orbits of indecomposable modules with* $1 - (\alpha|\alpha)$ *parameters.*

7.2. *Orbits in triple flag varieties*

Let us briefly introduce the classification after Magyar-Weyman-Zelevinsky [44, 45] in the case of type A.

Let $\mathbf{a} = (a_1, \ldots, a_{\ell_\mathbf{a}})$ be a composition of n. We allow zero for several components. For this $\mathbf{a}$, we define a flag variety

$$\mathscr{F}\!\ell_\mathbf{a} = \{F = (F_i)_{i=1}^{\ell_\mathbf{a}} \mid \dim F_i/F_{i-1} = a_i \text{ for } 1 \leq i \leq \ell_\mathbf{a}\},$$

where we put $F_0 = \{0\}$ for convenience. Note that $F_{\ell_\mathbf{a}} = \mathbb{C}^n =: V$.

Consider N-tuple of compositions $(\mathbf{a}_1, \mathbf{a}_2, \ldots, \mathbf{a}_N)$, where each $\mathbf{a}_j = (a_i^{(j)})_{i=1}^{\ell_j}$ is a composition of n. Define a multiple flag variety $\mathscr{X} = \mathscr{F}\!\ell_{\mathbf{a}_1} \times \mathscr{F}\!\ell_{\mathbf{a}_2} \times \cdots \times \mathscr{F}\!\ell_{\mathbf{a}_N} = \prod_{j=1}^N \mathscr{F}\!\ell_{\mathbf{a}_j}$.

For $\mathscr{X}$, consider a star shaped quiver Q. The underlying graph of Q is a tree, which has N-branches of length $\ell_1, \ell_2, \ldots, \ell_N$ with a unique root. The orientation is given from the end vertex of each branch toward the root vertex. We give an example of 3-branched quiver below. Here, each branch represents flags associated to the compositions $\mathbf{a}, \mathbf{b}, \mathbf{c}$, of length p, q and r respectively.

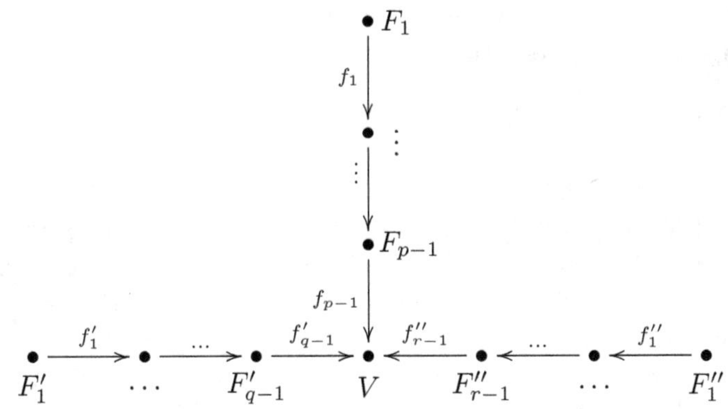

We can associate a point in $\mathscr{X}$

$$(F^{(j)})_{j=1}^{N} \in \prod_{j=1}^{N} \mathscr{F}\!\ell_{\mathbf{a}_j} = \mathscr{X}$$

with a representation of Q. Namely, in the j-th branch of Q, we associate the flag $F^{(j)} = (F_i^{(j)})_{i=1}^{\ell_j} \in \mathscr{F}\!\ell_{\mathbf{a}_j}$ with a representation

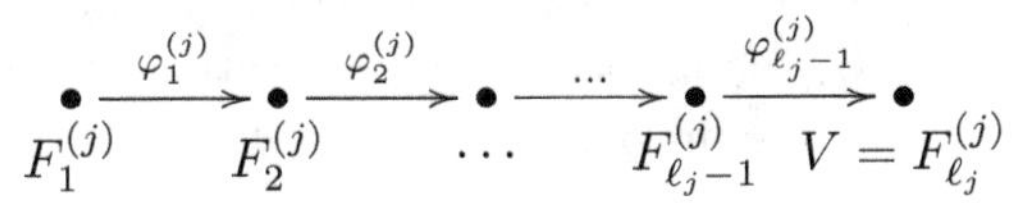

where $\varphi_k^{(j)} : F_k^{(j)} \to F_{k+1}^{(j)}$ is the inclusion map. In this way points in $\mathscr{X}$ are embedded into the representation space $\operatorname{Rep} Q$ of the quiver.

Lemma 7.3: *Let $x, y \in \mathscr{X}$ be two N-tuples of flags and $\xi, \eta \in \operatorname{Rep} Q$ be the corresponding representations of the quiver. Then x and y are in the same* $\mathrm{GL}(V)$*-orbit if and only if ξ and η are isomorphic as representations of Q.*

To make use of this scheme ultimately, we consider a full subcategory of $\operatorname{Rep} Q$ consisting of those objects corresponding to N-tuples of flags of length $\ell_1, \ell_2, \ldots, \ell_N$. Let $\mathscr{MF}(\mathbf{a}_1, \ldots, \mathbf{a}_N)$ be the set of objects in $\operatorname{Rep} Q$ which are isomorphic to representations corresponding to the multiple flags in $\mathscr{X} = \prod_{j=1}^{N} \mathscr{F}\!\ell_{\mathbf{a}_j}$. Then sum up all of these to get a full subcategory

$$\mathscr{MF} = \mathscr{MF}_{\ell_1,\ldots,\ell_N} = \bigcup \mathscr{MF}(\mathbf{a}_1, \ldots, \mathbf{a}_N) \subset \operatorname{Rep} Q,$$

where the union is taken over tuples of compositions of n (of nonnegative integers including zero) with prescribed lengths, and n moves over all nonnegative integers. This subcategory is clearly additive and closed under the decomposition. To state it more precisely, let $\pi \in \mathscr{MF}$ and decompose it into indecomposable representations in $\operatorname{Rep} Q$:

$$\pi = \bigoplus_k \pi_k \qquad (\pi_k \in \operatorname{Rep} Q : \text{indecomposable}).$$

Because the category $\operatorname{Rep} Q$ is equivalent to a category of modules of the path algebra of Q, the Krull-Schmidt theorem applies. Thus a decomposition is always possible and it is indeed unique. It is easy to see the direct summands π_k's are belonging to $\mathscr{MF}$ again. So the decomposition is indeed in the category $\mathscr{MF}$.

By Kac's theorem, the dimension vector $d_k = \dim \pi_k$ is a positive root of the Kac-Moody algebra $\mathfrak{g}(A)$. Moreover if d_k is real, i.e. $(d_k|d_k) = 1$, then π_k is the unique indecomposable object (up to isomorphism) with the dimension vector d_k. If d_k is imaginary, i.e., $(d_k|d_k) \leq 0$, there are infinitely many non-isomorphic classes of indecomposable objects with dimension vector d_k.

Let $\mathscr{X} = \mathscr{F}\ell_{\mathbf{a}_1} \times \mathscr{F}\ell_{\mathbf{a}_2} \times \cdots \times \mathscr{F}\ell_{\mathbf{a}_N}$ be a multiple flag variety. We identify a multiple flag $x \in \mathscr{X}$ with a representation $\pi \in \operatorname{Rep} Q$. Then $d = \dim \pi$ does not depend on the choice of $x \in \mathscr{X}$. The decomposition of π into indecomposables $\pi = \bigoplus_k \pi_k$ induces a decomposition of $d = \sum_k \dim \pi_k$.

Let $\Lambda = \Lambda_{\ell_1,\ell_2,\ldots,\ell_N} = \{\dim \sigma \mid \sigma \in \mathscr{MF}\}$. Then the above considerations read as $d = \sum_k d_k$ where $d_k \in \Lambda$.

Theorem 7.4: *Let d be the dimension vector of any $\pi \in \mathscr{MF}$ corresponding to an N-tuple of flags $x \in \mathscr{X} = \prod_{j=1}^N \mathscr{F}\ell_{\mathbf{a}_j}$. It is independent of the choice of x, but only depending on $(\mathbf{a}_j)_{j=1}^N$.*

(1) *The multiple flag variety $\mathscr{X}$ is of finite type if and only if the following condition* (FT) *holds.*

(FT) *If there is a decomposition $d = d' + d''$ in Λ then $(d'|d'), (d''|d'') \geq 1$.*

(2) *If it is of finite type, N must be less than or equal to 3 (at most a triple flag variety). Moreover,* $\mathrm{GL}(V)$*-orbits in $\mathscr{X}$ are in bijection with decompositions $d = \sum_k d_k$ where $d_k \in \Lambda$ and $(d_k|d_k) = 1$.*

Proof: (1) If there is a summand d' where $(d'|d') \leq 0$, then d' corresponds to an imaginary root, and we know there are infinitely many orbits. Moreover, there are infinitely many such orbits that correspond to representations of the quiver with injective maps. (Indeed, the fact that d' is imaginary combined with Lemma 7.1 implies that every orbit in $\mathfrak{M}^{d'}(Q)$

has positive codimension therein; in addition, the subset of representations with injective maps is dense because of the reason of the rank, thus of the same dimension as $\mathfrak{M}^{d'}(Q)$, hence itself is a union of infinitely many such orbits.) So direct summands should satisfy $(d'|d') \geq 1$. On the other hand if (FT) holds, we consider the unique decomposition by indecomposable representations. Pick a dimension vector d' of any indecomposable summand. Note that d' is a root, which is imaginary or real. By (FT), we get $(d'|d') \geq 1$ and we know d' cannot be imaginary. Therefore it is a real root and there is only one isomorphism class. In this way, any decomposition produces only one isomorphism class and there are only finitely many possibilities of the decompositions of dimension vectors.

(2) Let us rewrite the Tits' form $(d|d)$ in terms of $((\mathbf{a}_j)_{j=1}^N)$. It is easy to see

$$(d|d) = \frac{1}{2}\Big(\sum_{j=1}^{N} \|\mathbf{a}_j\|^2 - (N-2)n^2\Big) \qquad (n = \dim V).$$

Using this formula, we can easily find that

$$(\mathbf{a}_1, \ldots, \mathbf{a}_4) = ((1^2), (1^2), (1^2), (1^2)) \qquad \text{(we omit 0's)}$$

corresponds to an imaginary root. However, if $N \geq 4$, then the dimension vector corresponding to this particular $(\mathbf{a}_1, \ldots, \mathbf{a}_4)$ appears as a summand of *any* dimension vector. Thus multiple flag varieties with $N \geq 4$ factors are not of finite type. □

8. Finite Type Double Flag Varieties of Type AIII

Now we return back to the double flag variety $\mathfrak{X}$ of type AIII.

8.1. *Joint varieties and finiteness criterions*

To find out double flag varieties of finite type, we can also use the same idea along that of [44]. For the double flag varieties of symmetric pairs of type AIII, this is carried out by Hiroki Homma [24]. Let us explain it below.

Consider a triple flag variety

$$\mathscr{X}(\mathbf{a}, \mathbf{b}, \mathbf{c}) = \mathscr{F}\ell_{\mathbf{a}} \times \mathscr{F}\ell_{\mathbf{b}} \times \mathscr{F}\ell_{\mathbf{c}}$$

where $\mathbf{a}, \mathbf{b}, \mathbf{c}$ are compositions of n that satisfy the condition

$$\sum_{i=1}^{\ell_{\mathbf{a}}-1} a_i + \sum_{j=1}^{\ell_{\mathbf{b}}-1} b_j = n.$$

Sometimes we will write $(\sum_{i=1}^{\ell_{\mathbf{a}}-1} a_i, \sum_{j=1}^{\ell_{\mathbf{b}}-1} b_j) = (p, q)$. Thus, if we write $\mathbf{a}' = (a_i)_{i=1}^{\ell_{\mathbf{a}}-1}$ for the sequence obtained from $\mathbf{a}$ after removing the last part of the composition, then $\mathbf{a}'$ is a composition of p and $\mathbf{b}'$ is that of q. Hereafter, we keep this notation $\mathbf{a}'$ for the above meaning. Let us fix (p, q) and $n = p + q$, and denote $K = \mathrm{GL}(V^+) \times \mathrm{GL}(V^-)$, where $V = V^+ \oplus V^-$ and $V^+ = \mathbb{C}^p, V^- = \mathbb{C}^q$. Under this setting, we consider a double flag variety

$$\mathfrak{X} = \mathfrak{X}((\mathbf{a}', \mathbf{b}'); \mathbf{c}) = K/Q \times G/P = \mathrm{GL}_p/Q_1 \times \mathrm{GL}_q/Q_2 \times \mathrm{GL}_n/P, \quad (8.1)$$

where Q_1, Q_2, P are parabolic subgroups of $\mathrm{GL}_p, \mathrm{GL}_q$ and GL_n respectively such that $\mathrm{GL}_p/Q_1 = \mathscr{F}\!\ell_{\mathbf{a}'}$, $\mathrm{GL}_q/Q_2 = \mathscr{F}\!\ell_{\mathbf{b}'}$, and $\mathrm{GL}_n/P = \mathscr{F}\!\ell_{\mathbf{c}}$.

After Homma, we define a *joint variety* $\mathbf{JF}(\mathbf{a}, \mathbf{b}, \mathbf{c})$ by

$$\begin{aligned}\mathbf{JF}(\mathbf{a}, \mathbf{b}, \mathbf{c}) = \{((F_i^a)_{i=1}^{\ell_{\mathbf{a}}}, (F_i^b)_{i=1}^{\ell_{\mathbf{b}}}, (F_i^c)_{i=1}^{\ell_{\mathbf{c}}}) \in \mathscr{F}\!\ell_{\mathbf{a}} \times \mathscr{F}\!\ell_{\mathbf{b}} \times \mathscr{F}\!\ell_{\mathbf{c}} \mid \\ F_{\ell_{\mathbf{c}}}^c = V = F_{\ell_{\mathbf{a}}-1}^a \oplus F_{\ell_{\mathbf{b}}-1}^b\}.\end{aligned}$$

More generally, given a triple (ℓ_1, ℓ_2, ℓ_3) of positive integers, we consider triples of compositions $(\mathbf{a}, \mathbf{b}, \mathbf{c})$ such that

$$(\ell_{\mathbf{a}}, \ell_{\mathbf{b}}, \ell_{\mathbf{c}}) = (\ell_1, \ell_2, \ell_3) \text{ and } \sum_{i=1}^{\ell_{\mathbf{a}}-1} a_i + \sum_{j=1}^{\ell_{\mathbf{b}}-1} b_j = \sum_{k=1}^{\ell_3} c_k \quad (8.2)$$

and the semigroup

$$\Lambda^J = \Lambda^J_{\ell_1, \ell_2, \ell_3} := \{(\mathbf{a}, \mathbf{b}, \mathbf{c}) \text{ satisfying (8.2)}\}$$

formed by such triples. Moreover, we consider the quiver Q of triple branches of length ℓ_1, ℓ_2, ℓ_3. Then any element of $\mathbf{JF}(\mathbf{a}, \mathbf{b}, \mathbf{c})$ determines a representation of Q with the linear maps all injective as explained before. Let $\mathscr{J}(\mathbf{a}, \mathbf{b}, \mathbf{c})$ be the full subcategory of $\operatorname{Rep} Q$ consisting of all the objects obtained in this way. For studying the decompositions by indecomposables we need a larger subcategory. We sum up all $\mathscr{J}(\mathbf{a}, \mathbf{b}, \mathbf{c})$ when $(\mathbf{a}, \mathbf{b}, \mathbf{c}) \in \Lambda^J$ and denote the full subcategory of such objects by $\mathscr{J}$.

The next lemma clarifies the relation of the K-orbits on the double flag varieties and the isomorphism classes of quiver representations. The proof of the lemma is straightforward and we will omit it.

Lemma 8.1: *There are natural bijections between*

(1) *isomorphism classes in* $\mathscr{J}(\mathbf{a}, \mathbf{b}, \mathbf{c})$ *as representations of quiver;*

(2) $\mathrm{GL}(V)$*-orbits on the joint flag variety* $\mathbf{JF}(\mathbf{a}, \mathbf{b}, \mathbf{c})$*;*

(3) K*-orbits in the double flag variety* $\mathfrak{X}((\mathbf{a}', \mathbf{b}'); \mathbf{c})$ *defined in* (8.1).

In the next lemma, we also point out some general properties of the orbits. If $\mathcal{F}, \mathcal{G}$ are objects in $\mathscr{I}$, then we denote by $\mathrm{Hom}(\mathcal{F}, \mathcal{G})$ the complex vector space of morphisms of quiver representations, and we set

$$\langle \mathcal{F}, \mathcal{G} \rangle = \dim \mathrm{Hom}(\mathcal{F}, \mathcal{G}). \tag{8.3}$$

Clearly $\langle \mathcal{F}, \mathcal{G} \rangle$ only depends on $\mathcal{F}, \mathcal{G}$ up to their isomorphism classes. We actually focus on K-orbits of $\mathfrak{X}((\mathbf{a}', \mathbf{b}'); \mathbf{c})$ and, if $\mathcal{F}$ is an element of $\mathscr{I}(\mathbf{a}, \mathbf{b}, \mathbf{c})$, then we denote the corresponding K-orbit by $\mathbb{O}_{\mathcal{F}}$ via Lemma 8.1.

Lemma 8.2:

(1) $\dim \mathbb{O}_{\mathcal{F}} = p^2 + q^2 - \langle \mathcal{F}, \mathcal{F} \rangle$.

(2) *Given* $\mathcal{F}, \mathcal{F}' \in \mathscr{I}(\mathbf{a}, \mathbf{b}, \mathbf{c})$ *the following implication holds:*

$$\mathbb{O}_{\mathcal{F}} \subset \overline{\mathbb{O}_{\mathcal{F}'}} \implies \langle \mathcal{I}, \mathcal{F} \rangle \geq \langle \mathcal{I}, \mathcal{F}' \rangle \text{ for all indecomposable object } \mathcal{I} \in \mathscr{I}.$$

Proof: (1) Since $\mathcal{F} \in \mathscr{I}(\mathbf{a}, \mathbf{b}, \mathbf{c})$, due to the definition of $\mathscr{I}(\mathbf{a}, \mathbf{b}, \mathbf{c})$, every element in $\mathrm{Hom}(\mathcal{F}, \mathcal{F})$ corresponds to an element $f \in \mathrm{End}(V)$ such that $f(F) \subset F$ for all subspace F in $\mathcal{F}$. Moreover, V^+ and V^- themselves are among the subspaces of $\mathcal{F}$ hence we must have $f \in \mathfrak{gl}(V^+) \times \mathfrak{gl}(V^-) = \mathfrak{k}$, and finally

$$\mathrm{Hom}(\mathcal{F}, \mathcal{F}) \cong \{f \in \mathfrak{k} : f(F) \subset F \ \forall F \in \mathcal{F}\} = \mathrm{Lie}(\mathrm{Stab}_K(\mathcal{F})).$$

Thus we get $\langle \mathcal{F}, \mathcal{F} \rangle = \dim \mathrm{Stab}_K(\mathcal{F})$, and consequently $\dim \mathbb{O}_{\mathcal{F}} = \dim K/\mathrm{Stab}_K(\mathcal{F}) = p^2 + q^2 - \langle \mathcal{F}, \mathcal{F} \rangle$.

(2) Follows from a general property of quiver representations; see e.g. [65]. □

Remark 8.3:

(1) Note that

$$p^2 + q^2 = \dim \mathfrak{X}((\mathbf{a}', \mathbf{b}'); \mathbf{c}) + \frac{1}{2}\left(\|\mathbf{a}\|^2 + \|\mathbf{b}\|^2 + \|\mathbf{c}\|^2 - n^2\right)$$

hence the formula in Lemma 8.2(1) is an analogue of [44](Proposition 4.6).

(2) The natural question is whether the converse of the implication in Lemma 8.2(2) is valid. There is a similar discussion in [44](Proposition 4.5). In Theorem 9.4 below we will show the equivalence for one particular double flag variety.

Now we concentrate on the study of the isomorphism classes in $\mathscr{J}$ as representations of the quiver Q.

Lemma 8.4: *Any representation $\pi \in \mathscr{J}$ of Q decomposes into indecomposables in* $\operatorname{Rep} Q$ *uniquely as*

$$\pi = \bigoplus_k \pi_k,$$

where each indecomposable module π_k belongs to $\mathscr{J}$.

This lemma can be obtained in the same way as in the case of multiple flag varieties, and we omit the proof. Note that in a composition $\mathbf{a}$, we allow parts a_i to be zero. Let $\mathbf{a}^+$ be the list of nonzero parts of $\mathbf{a}$ arranged in nonincreasing order. For $(\mathbf{a}, \mathbf{b}, \mathbf{c}) \in \Lambda^J$, let $d(\mathbf{a}, \mathbf{b}, \mathbf{c})$ be the corresponding dimension vector, i.e., the collection of dimensions of the subspaces in the triple of flags.

Theorem 8.5: (Homma [24](Proposition 2.12 and Theorem 2.13)) *Let $\boldsymbol{\lambda} = (\mathbf{a}, \mathbf{b}, \mathbf{c}) \in \Lambda^J$. The following conditions* (1)–(3) *are equivalent.*

(1) *The double flag variety $\mathfrak{X} = \mathfrak{X}((\mathbf{a}', \mathbf{b}'); \mathbf{c})$ is of finite type.*

(2) *Any $\boldsymbol{\mu} = (\mathbf{a}_0, \mathbf{b}_0, \mathbf{c}_0) \in \Lambda^J$ such that $\boldsymbol{\lambda} - \boldsymbol{\mu} \in \Lambda^J$ satisfies $(d(\boldsymbol{\mu})|d(\boldsymbol{\mu})) \geq 1$.*

(3) *There is no $\boldsymbol{\mu} = (\mathbf{a}_0, \mathbf{b}_0, \mathbf{c}_0) \in \Lambda^J$ with $\boldsymbol{\lambda} - \boldsymbol{\mu} \in \Lambda^J$ such that $(\mathbf{a}_0'^{+}, \mathbf{b}_0'^{+}, \mathbf{c}_0)$ is in the following list up to switching the first two partitions.*

$$((1^3),(1^3),(2^3)), \quad ((2^2),(1^5),(3^3)), \quad ((1^2),(1^2),(1^4)),$$
$$((3),(1^5),(2^4)), \quad ((2),(1^3),(1^5)), \quad ((3),(2^2),(1^7)).$$

The equivalence of (1) and (2) is already explained above. For deriving the equivalence with (3), the proof is given in [24](Theorem 2.13) based on case-by-case analysis.

We also point out that Homma's results include the complete list of indecomposable representations of the quiver in the case where $\mathfrak{X}$ is of finite type; see [24](Theorem 2.15). In principle this yields a parametrization of the orbits in terms of quiver representations.

8.2. *Classification of finite type double flag varieties of type AIII*

It is not immediate to read off an actual classification of double flag varieties of finite type from Theorem 8.5. Here we give a complete list describing the parabolic subgroups $Q = Q_1 \times Q_2 \subset K$ and $P \subset G$.

Corollary 8.6: *The double flag variety $\mathfrak{X} = \mathfrak{X}((\mathbf{a}', \mathbf{b}'); \mathbf{c})$ of type AIII is of finite type if and only if it appears (up to switching Q_1 and Q_2) among those in Table 9.*

In the table, we use the above identification of $\mathfrak{X}$ with $\mathrm{GL}_p/Q_1 \times \mathrm{GL}_q/Q_2 \times \mathrm{GL}_n/P$. Moreover we let $|P| = \ell_{\mathbf{c}}$, $|Q_1| = \ell_{\mathbf{a}} - 1$, $|Q_2| = \ell_{\mathbf{b}} - 1$.

Table 9. Double flag varieties of finite type in type AIII.

type	P	Q_1	Q_2
(PA-1)	any	$Q = K$ or mirabolic	
(PA-2)	any	any	GL_1
(PA-3)	any	maximal	GL_2
(P6)	$\|P\| \leq 6$	maximal	GL_q
(P4-1)	$\|P\| \leq 4$	$\|Q_1\| \leq 4$	GL_q
(P4-2)	$\|P\| \leq 4$ & $\min c_i = 1$	any	GL_q
(P4-3)	$\|P\| \leq 4$	any	GL_2
(P3-1)	$\|P\| \leq 3$ & $\min c_i = 1$	any	any
(P3-2)	$\|P\| \leq 3$ & $\min c_i = 2$	any	$\|Q_2\| \leq 2$
(P3-3)	$\|P\| \leq 3$	any	GL_q or mirabolic
(P3-4)	$\|P\| \leq 3$	$\|Q_1\| \leq 4$	maximal
(P2)	$\|P\| \leq 2$ (GL_n or maximal)	any	any

Proof: We first observe that if $\boldsymbol{\lambda} = (\mathbf{a}, \mathbf{b}, \mathbf{c})$ contains some summand $\boldsymbol{\mu}$ which is in the list given in Theorem 8.5(3), then we must have $|P| = \ell_{\mathbf{c}} \geq 3$. Thus, if $|P| = 1$ or 2 (that is, $P = \mathrm{GL}_n$ or maximal), then $\mathfrak{X}$ will be of finite type. This corresponds to the last row (P2) in Table 9.

Next, we consider the case where $|P| = 3$. Up to switching $\mathbf{a}$ and $\mathbf{b}$, we can assume that $|Q_1| = \ell_{\mathbf{a}} - 1 \geq \ell_{\mathbf{b}} - 1 = |Q_2|$.

- If $\min c_i = 1$, then $\boldsymbol{\lambda}$ has no summand arising in the list given in Theorem 8.5(3). This gives (P3-1).

Now assume that $\min c_i \geq 2$. For avoiding summands corresponding to the case $((1^3), (1^3), (2^3))$ in Theorem 8.5(3), it is necessary and sufficient to have $|Q_2| \leq 2$, so let us assume this condition below.

- If $\min c_i = 2$, then the other five triples in Theorem 8.5(3) are automatically avoided, and this yields (P3-2).
- If $\min c_i \geq 3$, then the triple $((2^2), (1^5), (3^3))$ is avoided if and only if $|Q_2| = 1$; or $|Q_2| = 2$ and $\min\{b_1, b_2\} = 1$; or $|Q_1| \leq 4$. (Note

that, since $|Q_2| \leq 2$, (1^5) cannot contribute to $\mathbf{b}$.) Moreover, the remaining four triples of Theorem 8.5(3) are automatically avoided. This gives (P3-3) and (P3-4).

Now suppose that $|P| = 4$. Still we can assume that $|Q_1| \geq |Q_2|$. Then $(\mathbf{a}, \mathbf{b}, \mathbf{c})$ will avoid summands corresponding to $((1^2), (1^2), (1^4))$ if and only if $|Q_2| = 1$, that is $Q_2 = \mathrm{GL}_q$. In such a case, the first two triples in Theorem 8.5(3) are automatically avoided, and to avoid the fourth one $((3), (1^5), (2^4))$, it is necessary and sufficient to have $q \leq 2$; or $|Q_1| \leq 4$; or $\min c_i = 1$. This gives (P4-1), (P4-2), (P4-3), and (PA-2) (in the special case $|P| = 4$).

Next let $|P| \in \{5, 6\}$. Still we assume $|Q_1| \geq |Q_2|$. Then, as above, $(\mathbf{a}, \mathbf{b}, \mathbf{c})$ avoids $((1^2), (1^2), (1^4))$ if and only if $Q_2 = \mathrm{GL}_q$. Now the fifth triple $((2), (1^3), (1^5))$ will be avoided if and only if $q = 1$ or $|Q_1| \leq 2$. In such a case, the other triples in Theorem 8.5(3) are also avoided. This corresponds to (P6) and (PA-2) (in the special case $|P| \in \{5, 6\}$).

Finally, let $|P| \geq 7$. Still we assume $|Q_1| \geq |Q_2|$. First note that we must have $|Q_2| = 1$ (i.e., $Q_2 = \mathrm{GL}_q$), otherwise $((1^2), (1^2), (1^4))$ cannot be avoided. If $q = 1$ or $|Q_1| = 1$, then the six triples are avoided. If $q = 2$, then for avoiding the fifth triple we must have $|Q_1| \leq 2$, and in this case all the triples will be avoided. If $q \geq 3$, then for avoiding the fifth and sixth triples we must have $|Q_1| = 1$ or ($|Q_1| = 2$ and $\min a_i = 1$) (i.e., $Q_1 = \mathrm{GL}_p$ or mirabolic), and then the other triples are also avoided. This yields (PA-1), (PA-2), (PA-3). □

Remark 8.7: The cases (PA-1)–(PA-3), and (P2) are already identified as finite type in [55]. But all the other cases contain new series of finite type double flag varieties.

8.3. *Generalizations*

Theorem 8.5 characterizes double flag varieties of the form $\mathfrak{X} = (\mathrm{GL}_p \times \mathrm{GL}_q)/Q \times \mathrm{GL}_n/P$ with finitely many orbits of $K := \mathrm{GL}_p \times \mathrm{GL}_q$. Actually, the fact that K is a symmetric subgroup of GL_n does not play any role in the argument, and it is natural to consider the following more general problem.

Problem 8.8: *Given a decomposition* $V = \mathbb{C}^n = V_1 \oplus \ldots \oplus V_N$ *with* $p_i := \dim V_i$, *characterize the sequences of compositions* $(\mathbf{a}_1, \ldots, \mathbf{a}_N; \mathbf{c})$ *of*

$p_1, \ldots, p_N$ and n, respectively, such that the multiple flag variety

$$\mathfrak{X} = \left(\mathscr{F}\!\ell_{\mathbf{a}_1}(V_1) \times \cdots \times \mathscr{F}\!\ell_{\mathbf{a}_N}(V_N)\right) \times \mathscr{F}\!\ell_{\mathbf{c}}(V) \tag{8.4}$$

has finitely many orbits for $L := \mathrm{GL}_{p_1} \times \cdots \times \mathrm{GL}_{p_N}$.

It is worth to note that, contrary to the case of usual multiple flag varieties of the form $\mathscr{F}\!\ell_{\mathbf{c}_1}(V) \times \cdots \times \mathscr{F}\!\ell_{\mathbf{c}_N}(V)$, for which it is necessary to have $N \leq 3$ for having a finite number of GL_n-orbits (see [44]), there exist multiple flag varieties $\mathfrak{X}$ of the form (8.4) which are of finite type despite of an arbitrarily large number of factors N as shown in the next example.

Example 8.9: If we have $\mathbf{c} = (1, n-1)$, so that $\mathscr{F}\!\ell_{\mathbf{c}}(V) = \mathbb{P}(V)$, then the variety $\mathfrak{X}$ of (8.4) will have a finite number of L-orbits for any choice of N and $\mathbf{a}_1, \ldots, \mathbf{a}_N$. Indeed, $\mathbb{P}(V)$ has a finite number of T-orbits for any maximal torus $T \subset \mathrm{GL}_n$, hence a fortiori it contains a finite number of Q-orbits, so that $\mathfrak{X} \cong L/Q \times \mathbb{P}(V)$ has a finite number of L-orbits for any parabolic $Q \subset L$ (since Q contains a maximal torus of GL_n).

The next example is less immediate; the justification relies on a straightforward generalization of the criterion in Theorem 8.5(2) (we skip the details).

Example 8.10: Let $V = \mathbb{C}^n = \mathbb{C}^p \oplus \mathbb{C}^q \oplus \mathbb{C}$ with $n = p + q + 1$, $p, q \geq 1$. Then the triple flag variety

$$\begin{aligned}
&\left(\mathbb{P}^{p-1} \times \mathbb{P}^{q-1}\right) \times \mathrm{Gr}_r(\mathbb{C}^n) \\
&\simeq \left(\mathrm{GL}_p/P_{(1,p-1)} \times \mathrm{GL}_q/P_{(1,q-1)} \times \mathrm{GL}_1/\mathrm{GL}_1\right) \times \mathrm{GL}_n/P_{(r,n-r)}
\end{aligned}$$

has a finite number of $\mathrm{GL}_p \times \mathrm{GL}_q \times \mathrm{GL}_1$-orbits. Here $\mathrm{Gr}_r(V)$ denotes the Grassmann variety of r-dimensional subspaces in V.

To check this, it is sufficient to show that every dimension vector of the form $d = (\mathbf{a}_1, \mathbf{a}_2, \mathbf{a}_3, \mathbf{c})$ with $\mathbf{c} = (r, n-r)$, $\mathbf{a}_1 = (1, p-1, q+1)$, $\mathbf{a}_2 = (1, q-1, p+1)$, $\mathbf{a}_3 = (1, n-1)$ satisfies $(d|d) \geq 1$. Now, for seeing this, we compute

$$\begin{aligned}
2(d|d) &= \|\mathbf{c}\|^2 + \|\mathbf{a}_1\|^2 + \|\mathbf{a}_2\|^2 + \|\mathbf{a}_3\|^2 - 2n^2 \\
&= r^2 + (n-r)^2 + 3 + (p-1)^2 + (q+1)^2 + (q-1)^2 + (p+1)^2 \\
&\quad + (n-1)^2 - 2n^2 \\
&= (r-2)^2 + 2\left(p - \frac{r+1}{2}\right)^2 + 2\left(q - \frac{r+1}{2}\right)^2 + 1 \geq 2.
\end{aligned}$$

9. Orbits in the Double Flag Variety of Type AIII

In this section, we show a concrete example of classification of orbits in a double flag variety by using quiver representations. Thus we focus on the category of quiver representations $\mathscr{J}(\mathbf{a},\mathbf{b},\mathbf{c})$ and the double flag variety $\mathfrak{X}((\mathbf{a}',\mathbf{b}');\mathbf{c})$ of type AIII (cf. [15, 16]). We summarize the notation which will be used below:

- $n = p + q$, $r \le n$ are positive integers, and we set
$$\mathbf{a} = (1^p, q), \quad \mathbf{b} = (1^q, p), \quad \mathbf{c} = (r, n-r);$$
- $V = \mathbb{C}^n = V^+ \oplus V^-$ where $V^+ = \mathbb{C}^p \times \{0\}^q$, $V^- = \{0\}^p \times \mathbb{C}^q$, with standard bases $(e_1, \ldots, e_n) = (e_1^+, \ldots, e_p^+, e_1^-, \ldots, e_q^-)$;
- $G = \mathrm{GL}(V) = \mathrm{GL}_n$, $K = \mathrm{GL}(V^+) \times \mathrm{GL}(V^-) = \mathrm{GL}_p \times \mathrm{GL}_q$, $P = P_{(r,n-r)} = \mathrm{Stab}_G(\mathbb{C}^r \times \{0\}^{n-r})$, $B_K = B_p^+ \times B_q^+$, where $B_p^+ \subset \mathrm{GL}_p$, $B_q^+ \subset \mathrm{GL}_q$ are subgroups of upper-triangular matrices corresponding to $\mathbf{a}' = (1^p)$ and $\mathbf{b}' = (1^q)$ respectively;
- $\mathfrak{X} = \mathfrak{X}((\mathbf{a}',\mathbf{b}');\mathbf{c}) = K/B_K \times G/P \simeq (\mathscr{F}\!\ell(V^+) \times \mathscr{F}\!\ell(V^-)) \times \mathrm{Gr}_r(V)$.

9.1. *Category of joint representations*

We consider the quiver $Q = T_{2,p+1,q+1}$ indicated below, which is also denoted as $S(p+1, q+1)$ in [44].

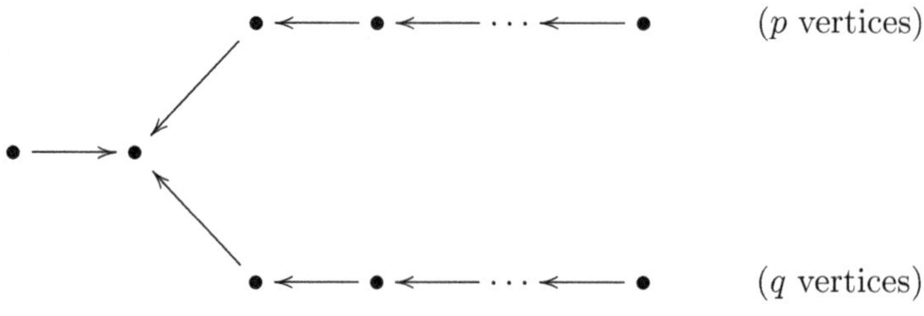

We denote by $\mathscr{J}$ the full subcategory of quiver representations $\mathrm{Rep}\, Q$ consisting of the following representations

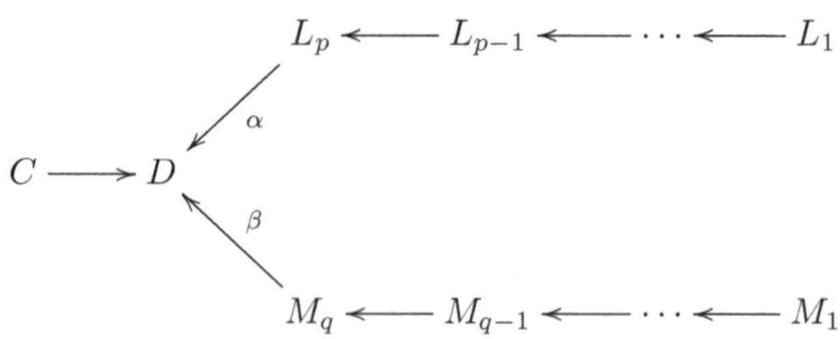

where all maps are injective and $\operatorname{Im}\alpha \oplus \operatorname{Im}\beta = D$. Then $\mathscr{J}(\mathbf{a},\mathbf{b},\mathbf{c})$ is the subset of such representations satisfying $\dim C = r$, $\dim D = n$, $\dim L_i = i$, $\dim M_j = j$ for all i, j, so that the points in the double flag variety $\mathfrak{X} = \mathfrak{X}((\mathbf{a}',\mathbf{b}');\mathbf{c})$ correspond to the objects in $\mathscr{J}(\mathbf{a},\mathbf{b},\mathbf{c})$, for which $D = V$, $L_p = V^+$, and $M_q = V^-$ hold.

In order to classify the K-orbits of $\mathfrak{X}$ by applying Lemma 8.1, we describe the category $\mathscr{J}$ in more detail. From [24], we can describe the indecomposable objects which arise in the decomposition of elements in $\mathscr{J}$. For $i \in \{1, \ldots, p\}$, we define

$$L^i_\bullet = (L^i_1, \ldots, L^i_p) \quad \text{such that} \quad L^i_k = \begin{cases} \langle e_i^+ \rangle & \text{if } k \geq i \\ 0 & \text{if } k < i. \end{cases}$$

Furthermore, we set $L^0_\bullet = (L^0_1, \ldots, L^0_p)$ with $L^0_1 = \ldots = L^0_p = 0$. We define in the same way $M^j_\bullet = (M^j_1, \ldots, M^j_q)$ with e_j^- instead of e_i^+, and $M^0_\bullet = (M^0_1, \ldots, M^0_q) = (0, \ldots, 0)$. We deduce the indecomposable objects $\mathcal{I}_i^+$, $\mathcal{I}_i'^+$, $\mathcal{I}_j^-$, $\mathcal{I}_j'^-$, $\mathcal{I}_{i,j}$ depicted in Figure 1. For each one we include (in parentheses) a graphical representation, which will be explained in §9.2 below.

Recall that if $\mathcal{F}, \mathcal{G}$ are objects in $\mathscr{J}$, we have denoted by $\operatorname{Hom}(\mathcal{F}, \mathcal{G})$ the complex vector space of morphisms of quiver representations, and set $\langle \mathcal{F}, \mathcal{G} \rangle = \dim \operatorname{Hom}(\mathcal{F}, \mathcal{G})$.

Lemma 9.1: *Let $\mathcal{F} \in \mathscr{J}(\mathbf{a},\mathbf{b},\mathbf{c})$ be a quiver representation corresponding to the point $((F_k^+)_{k=1}^p, (F_\ell^-)_{\ell=1}^q, W) \in \mathfrak{X}((\mathbf{a}',\mathbf{b}');\mathbf{c})$. For all $(i,j) \in \{1, \ldots, p\} \times \{1, \ldots, q\}$, we have*

$$\operatorname{Hom}(\mathcal{I}_i^+, \mathcal{F}) = F_i^+ \cap W, \qquad \operatorname{Hom}(\mathcal{I}_i'^+, \mathcal{F}) = F_i^+,$$
$$\operatorname{Hom}(\mathcal{I}_j^-, \mathcal{F}) = F_j^- \cap W, \qquad \operatorname{Hom}(\mathcal{I}_j'^-, \mathcal{F}) = F_j^-,$$
$$\operatorname{Hom}(\mathcal{I}_{i,j}, \mathcal{F}) = (F_i^+ \oplus F_j^-) \cap W.$$

Proof: Straightforward. □

9.2. *Orbit decomposition of $\mathfrak{X}$*

We recall some results and notation from [15, 14, 16]. Since

$$\mathfrak{X} = K/B_K \times G/P \simeq (\mathscr{F\!l}(V^+) \times \mathscr{F\!l}(V^-)) \times \operatorname{Gr}_r(V),$$

it is easy to see $\mathfrak{X}/K \simeq B_K \backslash \operatorname{Gr}_r(V)$. On the other hand, a point in $\operatorname{Gr}_r(V)$ represents an r-dimensional subspace of V, which can be represented as an image of rank r matrix $A \in \mathrm{M}_{n,r}$ ($\dim V = n = p + q$). It is well-known

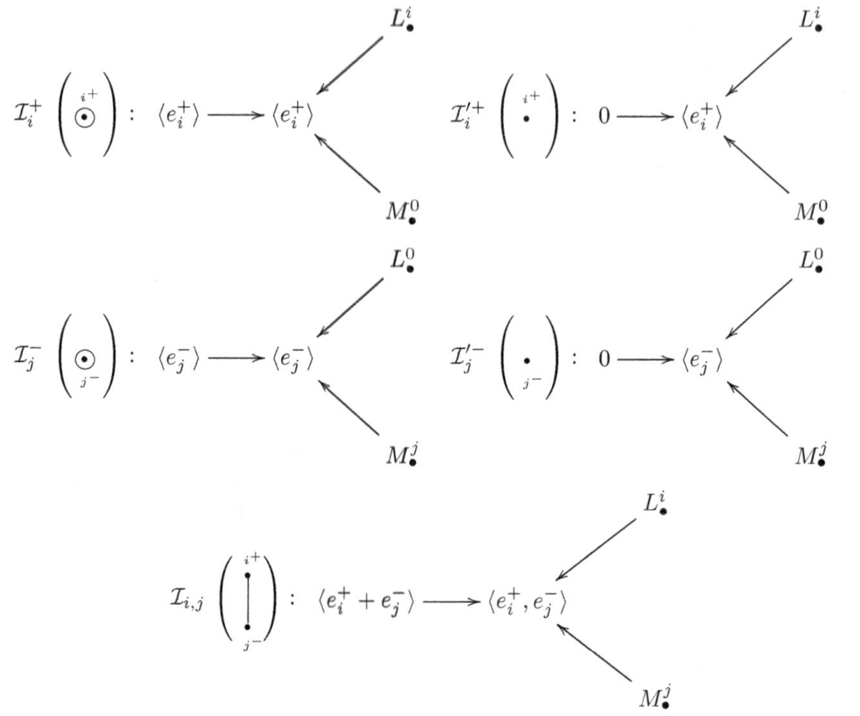

Fig. 1. The indecomposable objects $\mathcal{I}_i^+$, $\mathcal{I}_i'^+$ $\mathcal{I}_j^-$, $\mathcal{I}_j'^-$, $\mathcal{I}_{i,j}$ (for $1 \le i \le p$ and $1 \le j \le q$) of the category $\mathscr{J}$.

that $\mathrm{M}_{n,r}^\circ/\mathrm{GL}_r \simeq \mathrm{Gr}_r(V)$, where $\mathrm{M}_{n,r}^\circ$ denotes the set of full rank (i.e., rank r) matrices. Thus we get $\mathfrak{X}/K \simeq B_K\backslash\mathrm{M}_{p+q,r}^\circ/\mathrm{GL}_r$. In this way, as a representative of an orbit in $\mathfrak{X}/K$, we can take a matrix from $\mathrm{M}_{p+q,r}^\circ$. In fact, we can take "partial permutations" as representatives.

By $\mathfrak{T}_{p,r}$ we denote the set of *partial permutation matrices* of size (p,r), i.e., matrices τ of this shape with coefficients in $\{0,1\}$ and at most one 1 in each row and each column (consequently all the other entries are 0). We put

$$\mathfrak{T} = \left\{\omega = \begin{pmatrix}\tau_1\\ \tau_2\end{pmatrix} \;\middle|\; \tau_1 \in \mathfrak{T}_{p,r}, \tau_2 \in \mathfrak{T}_{q,r}\right\}, \quad \mathfrak{T}^\circ = \{\omega \in \mathfrak{T} \mid \operatorname{rank}\omega = r\} \subset \mathfrak{T}.$$

The latter subset $\mathfrak{T}^\circ$ is equipped with a natural right action of the symmetric group $\mathfrak{S}_r$, and we finally put $\overline{\mathfrak{T}} = \mathfrak{T}^\circ/\mathfrak{S}_r$.

The set $\overline{\mathfrak{T}}$ is in bijection with $\mathfrak{X}/K$ (see [15](Theorem 8.1)). More precisely, we can describe a representative of any K-orbit as follows. The range

$[\omega] := \operatorname{Im}\omega$ is well defined whenever $\omega \in \overline{\mathfrak{T}}$. Let $\mathcal{F}^+$ and $\mathcal{F}^-$ denote the standard flags of V^+ and V^- respectively. Every orbit in the double flag variety $\mathfrak{X}/K$ has a representative of the form

$$\mathcal{F}_\omega := (\mathcal{F}^+, \mathcal{F}^-, [\omega]) \in \mathfrak{X} \qquad (\omega \in \overline{\mathfrak{T}}).$$

Notation 9.2: (a) It is convenient to represent $\omega = \begin{pmatrix}\tau_1 \\ \tau_2\end{pmatrix} \in \overline{\mathfrak{T}}$ by the graph $\Gamma(\omega)$ given below.

- $\Gamma(\omega)$ has $p+q$ vertices labeled with $1^+, \ldots, p^+$ and $1^-, \ldots, q^-$, respectively.
- For every column of ω which contains two 1's, in the i-th and j-th rows of τ_1 and τ_2, respectively, we join the vertices i^+ and j^- of $\Gamma(\omega)$ with an edge.
- For every column of ω which contains only one 1, situated in the i-th row of τ_1 (resp. in the j-th row of τ_2), we circle the corresponding vertex i^+ (resp. j^-) of $\Gamma(\omega)$, and call it a *marked vertex*.
- In particular, the vertex i^+ (resp. j^-) is neither marked nor incident with an edge (we will call this "free point") if and only if the corresponding row of ω contains only 0's.

For instance,

$$\omega = \left(\begin{array}{cccc} 0&0&0&0\\ 1&0&0&0\\ 0&0&0&0\\ 0&1&0&0\\ 0&0&1&0\\ \hline 0&1&0&0\\ 0&0&0&1\\ 1&0&0&0 \end{array}\right) \rightsquigarrow \Gamma(\omega) =$$

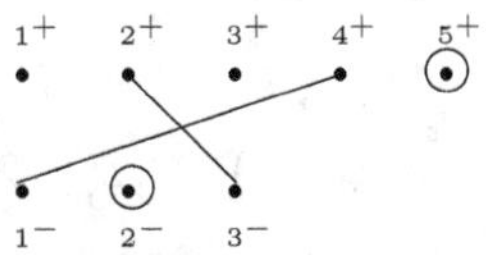

. (9.1)

(b) Note that, according to the representation of the indecomposable objects of the category $\mathscr{I}$ given in Figure 1, the connected components of $\Gamma(\omega)$ give rise to a collection of indecomposable objects which we denote $\operatorname{Indec}(\Gamma(\omega))$. For instance, for the above ω, we have

$$\operatorname{Indec}(\Gamma(\omega)) = \{\mathcal{I}_5^+,\ \mathcal{I}_1'^+,\ \mathcal{I}_3'^+,\ \mathcal{I}_2^-, \mathcal{I}_{2,3},\ \mathcal{I}_{4,1}\}.$$

The next lemma easily follows from the definitions.

Lemma 9.3: *In the category $\mathscr{I}$, we have a multiplicity free decomposition*

$$\mathcal{F}_\omega = \bigoplus\nolimits_{\mathcal{I} \in \operatorname{Indec}(\Gamma(\omega))} \mathcal{I}$$

by indecomposable representations.

Theorem 9.4:

(1) *The map* $\overline{\mathfrak{T}} \to \mathfrak{X}/K$, $\omega \mapsto \mathbb{O}_\omega := K \cdot \mathcal{F}_\omega$, *is a bijection.*

(2) *We have*

$$\dim \mathbb{O}_\omega = p^2 + q^2 - \sum_{x = i^{\pm}\text{: free point}} i - \sum_{x\text{: marked point}} \# \left\{ \begin{array}{l} \text{marked points on the same row} \\ \text{on the left of } x\text{, including } x \end{array} \right\} - \sum_{\alpha\text{: edge}} \# \left\{ \begin{array}{l} \text{marked points and edges} \\ \text{on the left of } \alpha\text{, including } \alpha \end{array} \right\}.$$

(3) *For all* $\omega, \omega' \in \overline{\mathfrak{T}}$, *we have*

$$\mathbb{O}_\omega \subset \overline{\mathbb{O}_{\omega'}} \iff \langle \mathcal{I}, \mathcal{F}_\omega \rangle \geq \langle \mathcal{I}, \mathcal{F}_{\omega'} \rangle \quad \textit{for all indecomposable } \mathcal{I} \in \mathscr{J}.$$

Proof: (1) Follows from [16](Theorem 2.2(1)).

(2) Follows from a combinatorial interpretation of $\langle \mathcal{F}_\omega, \mathcal{F}_\omega \rangle$ by using Lemmas 8.2(1), 9.1, and 9.3.

(3) Following the notation of [16](§2.1), for $(i, j) \in \{0, 1, \ldots, p\} \times \{0, 1, \ldots, q\}$, we denote by $r_{i,j}(\omega)$ the number of marked points or edges of the subgraph of $\Gamma(\omega)$ formed by the vertices k^+, ℓ^- with $1 \leq k \leq i$ and $1 \leq \ell \leq j$. Then, Lemma 9.1 yields the equalities

$$\langle \mathcal{I}_i^+, \mathcal{F}_\omega \rangle = r_{i,0}(\omega), \quad \langle \mathcal{I}_i'^+, \mathcal{F}_\omega \rangle = i,$$
$$\langle \mathcal{I}_j^-, \mathcal{F}_\omega \rangle = r_{0,j}(\omega), \quad \langle \mathcal{I}_j'^+, \mathcal{F}_\omega \rangle = j, \quad \langle \mathcal{I}_{i,j}, \mathcal{F}_\omega \rangle = r_{i,j}(\omega).$$

This observation, combined with [16](Theorem 2.2(4)), yields part (3) of the theorem. □

Remark 9.5: The obtained dimension formula in Theorem 9.4(2) is much simpler than the one given in [16](Theorem 2.2(2)).

Part III. Steinberg Theory for Double Flag Varieties

10. Definition of Steinberg Maps

10.1. *Symmetrized and exotic Steinberg maps for double flag varieties*

Recall the following general situation:

- G is a connected reductive group endowed with an involution $\theta \in \mathrm{Aut}(G)$;
- $K := G^\theta$ is the corresponding symmetric subgroup and we assume that it is connected;
- $P \subset G$ and $Q \subset K$ are parabolic subgroups, and we consider the double flag variety

$$\mathfrak{X} := K/Q \times G/P.$$

By abuse of notation, let $\theta : \mathfrak{g} \to \mathfrak{g}$ denote the differential of the involution θ, which is an involution of the Lie algebra. We also consider the Cartan decomposition

$$\mathfrak{g} = \mathfrak{k} \oplus \mathfrak{s}, \qquad \text{where} \qquad \mathfrak{s} = \{x \in \mathfrak{g} : \theta(x) = -x\}.$$

Let $\mathfrak{g} \to \mathfrak{k}$, $x \mapsto x^\theta$ and $\mathfrak{g} \to \mathfrak{s}$, $x \mapsto x^{-\theta}$ be the projections along this decomposition.

We apply the constructions of §2.2 to the double flag variety $\mathfrak{X}$. In particular, we consider $\mathfrak{X}$ as a K-variety with diagonal K-action. This gives rise to the moment map:

$$\mu : T^*\mathfrak{X} \to \mathfrak{k} \cong \mathfrak{k}^*, \ ((x, \mathfrak{q}_1), (y, \mathfrak{p}_1)) \mapsto x + y^\theta$$

and we get the conormal variety

$$\mathcal{Y} := \mu^{-1}(0) \cong \{(y, \mathfrak{q}_1, \mathfrak{p}_1) \in \mathfrak{g} \times K/Q \times G/P : y \in \mathfrak{n}_{\mathfrak{p}_1}, \ y^\theta \in \mathfrak{n}_{\mathfrak{q}_1}\}.$$

Here we recall that $\mathfrak{n}_{\mathfrak{p}_1}$ and $\mathfrak{n}_{\mathfrak{q}_1}$ respectively denote the nilradicals of the parabolic subalgebras $\mathfrak{p}_1 \subset \mathfrak{g}$ and $\mathfrak{q}_1 \subset \mathfrak{k}$. By Lemma 2.3, $\mathcal{Y}$ is the disjoint union of the conormal bundles

$$\mathcal{Y} = \coprod_{\mathbb{O} \in \mathfrak{X}/K} T^*_{\mathbb{O}}\mathfrak{X}.$$

On the other hand, by using the moment map for the G-action on the second factor, we get a map

$$\phi : \mathcal{Y} \to G\,\mathfrak{n}_{\mathfrak{p}} \subset \mathcal{N}_{\mathfrak{g}}, \qquad (y, \mathfrak{q}_1, \mathfrak{p}_1) \mapsto y,$$

which is simply the projection to the first factor, and it is clearly K-equivariant. By composing with the projections along the Cartan decomposition, this finally yields two K-equivariant maps

$$\phi_{\mathfrak{k}} : \mathcal{Y} \to K\,\mathfrak{n}_{\mathfrak{q}} \subset \mathcal{N}_{\mathfrak{k}}, \qquad (y, \mathfrak{q}_1, \mathfrak{p}_1) \mapsto y^{\theta} \qquad \text{and}$$
$$\phi_{\mathfrak{s}} : \mathcal{Y} \to \mathfrak{s}, \qquad (y, \mathfrak{q}_1, \mathfrak{p}_1) \mapsto y^{-\theta}.$$

The image of $\phi_{\mathfrak{k}}$ is clearly contained in the nilpotent variety $\mathcal{N}_{\mathfrak{k}}$. For $\phi_{\mathfrak{s}}$, we also expect that the image of $\phi_{\mathfrak{s}}$ is contained in the nilpotent variety $\mathcal{N}_{\mathfrak{s}}$ as well, but it is not always the case. See Propositions 10.3, 10.4 and Conjecture 10.7 below.

From now on, we assume the following "nil-projection assumption".

Assumption 10.1: The image of the projection $\phi_{\mathfrak{s}} : \mathcal{Y} \to \mathfrak{s}$ is contained in the nilpotent variety $\mathcal{N}_{\mathfrak{s}}$.

Note that the nilpotent variety $\mathcal{N}_{\mathfrak{s}}$ has a finite number of K-orbits due to [41].

For every orbit $\mathbb{O} \in \mathfrak{X}/K$, the conormal bundle $T^*_{\mathbb{O}}\mathfrak{X}$ is K-stable and irreducible. Hence its images by $\phi_{\mathfrak{k}}$ and $\phi_{\mathfrak{s}}$ are K-stable irreducible subsets of $\mathcal{N}_{\mathfrak{k}}$ and $\mathcal{N}_{\mathfrak{s}}$ respectively. This means the closure of the image of $\mathbb{O} \in \mathfrak{X}/K$ by $\phi_{\mathfrak{k}}$ (respectively $\phi_{\mathfrak{s}}$) coincides with the closure of a single nilpotent K-orbit in $\mathcal{N}_{\mathfrak{k}}$ (respectively $\mathcal{N}_{\mathfrak{s}}$).

Definition 10.2:

(1) The map $\phi_{\mathfrak{k}}$ gives rise to a well-defined map

$$\Phi_{\mathfrak{k}} : \mathfrak{X}/K \to \mathcal{N}_{\mathfrak{k}}/K, \quad \mathbb{O} \mapsto \big(\text{the unique } K\text{-orbit dense in } \phi_{\mathfrak{k}}(T^*_{\mathbb{O}}\mathfrak{X})\big),$$

which we call the *symmetrized Steinberg map.*

(2) Similarly, under Assumption 10.1, we define the *exotic Steinberg map,* using $\phi_{\mathfrak{s}}$:

$$\Phi_{\mathfrak{s}} : \mathfrak{X}/K \to \mathcal{N}_{\mathfrak{s}}/K, \quad \mathbb{O} \mapsto \big(\text{the unique } K\text{-orbit dense in } \phi_{\mathfrak{s}}(T^*_{\mathbb{O}}\mathfrak{X})\big).$$

10.2. *Assumption 10.1 for the exotic Steinberg map*

To define the exotic Steinberg map, we need Assumption 10.1 above. In fact, when the group G is classical and the parabolic subgroup P of G has abelian unipotent radical, this condition always holds. More generally, we have the following proposition.

Proposition 10.3: *Assumption 10.1 holds in the following situations.*

(1) $G = \mathrm{GL}_n$ *and* $P \subset G$ *is a maximal parabolic subgroup;*

(2) $G = \mathrm{Sp}_{2n}$ *and* $P \subset G$ *is the stabilizer of a Lagrangian subspace or of a line;*

(3) $G = \mathrm{SO}_n$ *and* $P \subset G$ *is the stabilizer of an isotropic line or of a Lagrangian subspace (if* n *is even), or of a maximal isotropic subspace if* $n = 5$.

In particular, this implies that Assumption 10.1 holds whenever G *is classical and* P *has an abelian unipotent radical (see [62]).*

Proposition 10.4: *Let* $(G, K) = (\mathrm{GL}_{p+q}, \mathrm{GL}_p \times \mathrm{GL}_q)$ $(p, q \geq 1)$ *be a symmetric pair of type* AIII. *Then, Assumption 10.1 holds whenever we are in one of the following cases that appear in Table 9:*

(1) (PA-1), (PA-2), (PA-3), (P2)*;*

(2) (P3-1), (P3-3) *with* $q = 2$*;*

(3) (P3-4) *with* Q_1 *maximal.*

In particular, this implies that Assumption 10.1 holds whenever the double flag variety $\mathfrak{X} = G/P \times K/Q$ *is of finite type and* P *is a Borel subgroup of* G.

The propositions are shown case-by-case. In fact, they follow from the next technical lemma. (It would be interesting to have a more conceptual proof of the result.)

Lemma 10.5:

(1) *Let* $y \in \mathrm{M}_{p+q}$ *be nilpotent and put*

$$y = \begin{pmatrix} a & b \\ c & d \end{pmatrix} \in \mathcal{N}_{\mathfrak{gl}_{p+q}}, \qquad v := \begin{pmatrix} a & 0 \\ 0 & d \end{pmatrix}, \qquad w := \begin{pmatrix} 0 & b \\ c & 0 \end{pmatrix},$$

with $(a, d) \in \mathfrak{gl}_p \times \mathfrak{gl}_q$, $b \in \mathrm{M}_{p,q}$, $c \in \mathrm{M}_{q,p}$. *Assume that* v *is nilpotent and one of the following conditions* (A)–(G) *occurs. Then* w *is also nilpotent.*

(A) $y^2 = 0$*;*
(B) $y^3 = 0$ *and* $\mathrm{rk}\, y \leq 2$*;*
(C) $y^3 = 0$, $a^2 = 0$, *and* $d^2 = 0$*;*
(D) $y^3 = 0$ *and* $q = 2$*;*
(E) $\mathrm{rk}\, a \leq 1$ *and* $d = 0$*;*
(F) $q = 2$, $a^2 = 0$, *and* $d = 0$*;*
(G) $q = 1$.

(2) Let $y \in \mathfrak{gl}_n$ and let y^* be the adjoint of y with respect to a given orthogonal or symplectic form on $\mathbb{C}^n$. Assume that $v := \frac{1}{2}(y - y^*)$ is nilpotent and $y^2 = 0$. Then $w := \frac{1}{2}(y + y^*)$ is also nilpotent.

Proof: (1) We calculate

$$y^2 = \begin{pmatrix} a^2 + bc & ab + bd \\ ca + dc & d^2 + cb \end{pmatrix}, \tag{10.1}$$

and get $\operatorname{Tr} y^2 = \operatorname{Tr} a^2 + \operatorname{Tr} d^2 + 2 \operatorname{Tr} bc$. Since y, a, and d are all nilpotent, we obtain

$$\operatorname{Tr} cb = 0. \tag{10.2}$$

(G) When $q = 1$, (10.2) already implies that $cb = 0$, thus $(bc)^2 = b(cb)c = 0$, and so w is nilpotent.

(F) When $d = 0$ and $a^2 = 0$, we see that

$$y^4 = \begin{pmatrix} (bc)^2 + abca & bcab + abcb \\ cabc + cbca & (cb)^2 \end{pmatrix}.$$

Whence $0 = \operatorname{Tr} y^4 = \operatorname{Tr}(bc)^2 + \operatorname{Tr} abca + \operatorname{Tr}(cb)^2 = 2 \operatorname{Tr}(cb)^2$ because $\operatorname{Tr} abca = \operatorname{Tr} bca^2 = 0$. If $q = 2$, since cb is a matrix of size 2 such that $\operatorname{Tr}(cb) = \operatorname{Tr}(cb)^2 = 0$, we conclude that $(cb)^2 = 0$, so that $(bc)^2 b = b(cb)^2 = 0$ holds. Therefore, $w^5 = 0$.

(D) If $y^3 = 0$ and $d^2 = 0$ (which holds in particular if $q = 2$), then we have

$$0 = y^3 = \begin{pmatrix} a^3 + abc + bca + bdc & a^2 b + abd + bcb \\ ca^2 + cbc + dca & cab + cbd + dcb \end{pmatrix}. \tag{10.3}$$

Multiplying the upper right coefficient by c on the left, we get

$$(cb)^2 = -ca^2 b - cabd. \tag{10.4}$$

Multiplying the lower diagonal coefficient by d on the right, we obtain

$$\operatorname{Tr}(cabd) = -\operatorname{Tr}(cbd^2) - \operatorname{Tr}(dcbd) = -\operatorname{Tr}(cbd^2) = 0, \tag{10.5}$$

since $d^2 = 0$. Finally, multiplying the upper diagonal coefficient of (10.3) by a and using $\operatorname{Tr} a^4 = 0$, we get

$$2 \operatorname{Tr}(ca^2 b) = -\operatorname{Tr}(bdca) = 0 \tag{10.6}$$

because of (10.5). Altogether, (10.4), (10.5), and (10.6) imply that $\operatorname{Tr}(cb)^2 = 0$. When $q = 2$, this fact combined with (10.2) is sufficient for ensuring that $(cb)^2 = 0$, whence $(bc)^2 b = 0$. Therefore, $w^5 = 0$.

(C) Here, in addition to $y^3 = 0$ and $d^2 = 0$, we assume that $a^2 = 0$. Hence we can replace a^3, a^2b, and ca^2 by zero in (10.3). Multiplying the lower left (resp. upper right) block of the above matrix by b on the left (resp. by c on the right), we obtain

$$-bdca = (bc)^2 = -abdc,$$

whence

$$(bc)^4 = bdca^2bdc = 0$$

(since $a^2 = 0$). Similarly it can be shown that $(cb)^4 = 0$, hence w is nilpotent.

(A) If $y^2 = 0$, then from (10.1) we have

$$w^2 = \begin{pmatrix} bc & 0 \\ 0 & cb \end{pmatrix} = -\begin{pmatrix} a^2 & 0 \\ 0 & d^2 \end{pmatrix} = -v^2.$$

Since v is nilpotent, we conclude that w is also nilpotent in this case.

(E) Up to enlarging the size of the matrices by adding rows/columns of zeros, we can assume that $p = q \geq 2$. By $E_{i,j}^{(p)}$ we will denote the elementary matrix of size (p, p) with 1 at the position (i, j). Also, we can assume that $\operatorname{rk} a = 1$ and $d = 0$ and, up to conjugation by $\mathrm{GL}_p \times \mathrm{GL}_p$, we can assume that $a = E_{1,2}^{(p)}$. Fix $\epsilon \in \{0, 1\}$. Then, we have

$$\begin{aligned}\det(\epsilon v + w - t\,\mathbb{1}_{2p}) &= \det\begin{pmatrix} \epsilon a - t\,\mathbb{1}_p & b \\ c & -t\,\mathbb{1}_p \end{pmatrix} = \det((\epsilon a - t\,\mathbb{1}_p)(-t\,\mathbb{1}_p) - bc) \\ &= \det(t^2\,\mathbb{1}_p - bc - \epsilon t\,E_{1,2}^{(p)}) = f(t^2) + \epsilon t g(t^2)\end{aligned}$$

with certain polynomials f, g. Since $t^{2p} = \det(v+w-t\,\mathbb{1}_{2p}) = f(t^2)+t\,g(t^2)$ (because $y = v + w$ is nilpotent), we conclude that $g \equiv 0$, hence $\det(w - t\mathbb{1}_{2p}) = t^{2p}$, and so w is nilpotent.

(B) Finally we assume that $y^3 = 0$ and $\operatorname{rk} y \leq 2$. For this remaining case, we reason by induction on $p+q$, with immediate initialization if $p = 0$ or $q = 0$. We have $2 \geq \operatorname{rk} y \geq \max\{\operatorname{rk} a, \operatorname{rk} d\}$. Since we have already dealt with (A) and (C), we can assume that $\operatorname{rk} y = 2$ and one of a^2, d^2 is nonzero, say $a^2 \neq 0$, and so $\operatorname{rk} a = 2$. Then up to conjugating under $\mathrm{GL}_p \times \mathrm{GL}_q$, we can assume that

$$a = E_{1,2}^{(p)} + E_{2,3}^{(p)}.$$

Since $\operatorname{rk} y = 2$, this implies in particular that the first column of y is zero. Let y', v', w' be the submatrices of y, v, w obtained by removing the first row and column. We can apply the induction hypothesis to y', and we get that w' is nilpotent. Since the first column of w (like the one of y) is zero, we conclude that w is nilpotent.

(2) We have

$$0 = y^2 = (v + w)^2 = (v^2 + w^2) + (vw + wv).$$

Since $v^* = -v$ and $w^* = w$, we get that $(v^2 + w^2)^* = v^2 + w^2$ and $(vw + wv)^* = -(vw + wv)$. Due to the uniqueness of the decomposition of a matrix as sum of self-adjoint and skew-adjoint elements, we derive that $v^2 + w^2 = vw + wv = 0$. In particular, knowing that v is nilpotent, we conclude that w is nilpotent. □

Example 10.6: In general, Assumption 10.1 will be valid if and only if the implication

$$(y \in G\mathfrak{n}_{\mathfrak{p}} \quad \text{and} \quad y^{\theta} \in K\mathfrak{n}_{\mathfrak{q}}) \quad \Longrightarrow \quad y^{-\theta} \in \mathcal{N}_{\mathfrak{s}} \tag{10.7}$$

holds. When P and Q are Borel subgroups, the left-hand side of (10.7) just means that y and y^{θ} are nilpotent, and the implication of (10.7) is *not true* in general. Consider for instance the symmetric pair $(G, K) = (\mathrm{GL}_6, \mathrm{GL}_3 \times \mathrm{GL}_3)$ of type AIII, and

$$y = \begin{pmatrix} a & \mathbb{1}_3 \\ b & -a \end{pmatrix} \quad \text{with} \quad a = \begin{pmatrix} 0 & 1 & 0 \\ 0 & 0 & -1 \\ 0 & 0 & 0 \end{pmatrix} \quad \text{and} \quad b = \begin{pmatrix} 0 & 0 & 1 \\ 0 & 0 & 0 \\ 1 & 0 & 0 \end{pmatrix}.$$

Then we have $y^4 = 0$, $(y^{\theta})^3 = 0$, but $y^{-\theta} \notin \mathcal{N}_{\mathfrak{s}}$. However, note that the corresponding double flag variety is *not* of finite type.

We do not have a counterexample where a double flag variety is *of finite type* and the image of the projection $\phi_{\mathfrak{s}} : \mathcal{Y} \to \mathfrak{s}$ is not contained in the nilpotent variety. Thus, based on the cases already solved by Propositions 10.3 and 10.4, we propose a conjecture here.

Conjecture 10.7: (Nil-projection Conjecture) If a double flag variety $\mathfrak{X}$ is of finite type, then the image of the projection $\phi_{\mathfrak{s}} : \mathcal{Y} \to \mathfrak{s}$ is contained in the nilpotent variety $\mathcal{N}_{\mathfrak{s}}$.

11. Generalization of Robinson-Schensted Correspondence

We keep the setting and notation of the previous section. Let us assume

$$\mathfrak{X} \text{ has finitely many } K\text{-orbits, that is, } \mathfrak{X} \text{ is of finite type,} \tag{11.1}$$

and we further assume Assumption 10.1, i.e., the image of the projection $\phi_{\mathfrak{s}}$ is contained in the nilpotent variety so that we have $\phi_{\mathfrak{s}} : \mathcal{Y} \to \mathcal{N}_{\mathfrak{s}}$.

Because $\mathfrak{X}$ is of finite type, as already pointed out in §2.2, the conormal variety $\mathcal{Y}$ is equidimensional and the closures of the conormal bundles $T^*_{\mathbb{O}}\mathfrak{X}$, for $\mathbb{O} \in \mathfrak{X}/K$, are the irreducible components of $\mathcal{Y}$. Moreover, in this case, the K-orbits in $\mathfrak{X}$ can be described combinatorially, whereas nilpotent orbits have a well-known combinatorial description (see for example, [11]), hence the maps $\Phi_{\mathfrak{k}}$ and $\Phi_{\mathfrak{s}}$ can be explicitly described by the combinatorial data.

In the previous section, we have extended the setting of the classical work of Steinberg [77], which deals with the standard double flag variety $\mathfrak{X} = G/B \times G/B$ on which G acts diagonally. This standard theory yields the classical *Steinberg map* $\Phi : W \to \mathcal{N}_{\mathfrak{g}}/G$, also considered in [23]. In the next subsections, first we explain how classical Steinberg theory can be obtained as a particular case of the construction given above and recall from [77, 78] the calculation of Φ in the case where $G = K = \mathrm{GL}_n$. Then we calculate $\Phi_{\mathfrak{k}}$ and $\Phi_{\mathfrak{s}}$ for the symmetric pair $(G, K) = (\mathrm{GL}_n, \mathrm{GL}_p \times \mathrm{GL}_q)$ of type AIII and the double flag variety $\mathfrak{X}$ considered in §9.

11.1. *Classical Steinberg map and Robinson–Schensted correspondence*

Let us consider $\mathbb{G} = G \times G$ and define $\theta \in \operatorname{Aut}\mathbb{G}$ as $\theta(g_1, g_2) = (g_2, g_1)$ $((g_1, g_2) \in \mathbb{G})$, which is the "flip" of two components. Put $\mathbb{K} = \mathbb{G}^\theta = \Delta G$, which is the diagonal embedding $G \overset{\Delta}{\hookrightarrow} \mathbb{G}$. We identify $\mathbb{K}$ and G sometimes. Now consider the parabolic subgroup $\mathbb{Q} = \mathbb{K}$ and a Borel subgroup $\mathbb{B} = B \times B \subset G \times G = \mathbb{G}$. Then the double flag variety associated with these data is

$$\mathfrak{X} = \mathbb{K}/\mathbb{Q} \times \mathbb{G}/\mathbb{B} = G/B \times G/B,$$

on which $G(= \mathbb{K})$ acts diagonally.

By virtue of Bruhat decomposition, $\mathfrak{X}$ has finitely many G-orbits ($G = \mathbb{K}$ as noted above), parametrized by the elements of the Weyl group $W = N_G(T)/T$, where T is a maximal torus in B. So let us write the decomposition as

$$\mathfrak{X} = G/B \times G/B = \coprod_{w \in W} \mathbb{O}_w \quad \text{where} \quad \mathbb{O}_w = G \cdot (B, wB).$$

Note that the conormal direction at $(B, wB) \in \mathfrak{X}$ is

$$\{(x + \mathfrak{b}, x + {}^w\mathfrak{b}) : x \in \mathfrak{g}\}^\perp = \mathfrak{n} \cap {}^w\mathfrak{n},$$

where $\mathfrak{n} \subset \mathfrak{b} = \mathrm{Lie}(B)$ stands for the nilpotent radical of our Borel subalgebra, and the notation ${}^w\mathfrak{b}$ or ${}^w\mathfrak{n}$ stands for the conjugate by $w \in W$. Therefore the conormal bundles of orbits are given by

$$T^*_{\mathbb{O}_w}\mathfrak{X} = G \cdot \{(\mathfrak{b}, {}^w\mathfrak{b}, x) : x \in \mathfrak{n} \cap {}^w\mathfrak{n}\}.$$

In the present situation, the nilpotent varieties are

$$\mathcal{N}_{\mathfrak{k}} = \mathcal{N}^{\theta} = \{(x,x) : x \in \mathcal{N}_{\mathfrak{g}}\} \quad \text{and} \quad \mathcal{N}_{\mathfrak{s}} = \mathcal{N}^{-\theta} = \{(x,-x) : x \in \mathcal{N}_{\mathfrak{g}}\}$$

and both are identified with $\mathcal{N}_{\mathfrak{g}}$. Since we have chosen $\mathbb{Q} = \mathbb{K}$, the symmetrized Steinberg map $\Phi_{\mathfrak{k}}$ is the trivial map which assigns every orbit $\mathbb{O}_w$ to the trivial orbit, and Assumption 10.1 is trivially satisfied. Then we focus on the (exotic) Steinberg map

$$\Phi = \Phi_{\mathfrak{s}} : \mathfrak{X}/G \cong W \to \mathcal{N}_{\mathfrak{g}}/G$$

which is explicitly given as

$\Phi : w \mapsto$ (the unique nilpotent G-orbit $\mathcal{O}$ which intersects $\mathfrak{n} \cap {}^w\mathfrak{n}$ densely).

Let us further assume $G = \mathrm{GL}_n$. In this case, the Steinberg map Φ can be computed explicitly, in terms of the *Robinson-Schensted correspondence* (RS correspondence)

$$\mathrm{RS} : \mathfrak{S}_n \xrightarrow{\sim} \coprod\nolimits_{\lambda \vdash n} \mathrm{STab}(\lambda) \times \mathrm{STab}(\lambda), \quad w \mapsto (\mathrm{RS}_1(w), \mathrm{RS}_2(w)).$$

Here, the union on the right-hand side is over partitions of n (equivalently, Young diagrams of size n) and $\mathrm{STab}(\lambda)$ denotes the set of *standard Young tableaux* of shape λ (i.e., the boxes of λ are filled with integers $1, 2, \ldots, n$ and the filling is increasing in both directions, horizontal and vertical). Let us recall the definition and some properties of the RS correspondence briefly (see [18] for more details).

Definition 11.1:

(1) If T is a Young tableau and a an integer, let $T \leftarrow a$ be the tableau obtained by inserting a into T as follows:

- If a is greater than or equal to every entry in the first row of T, then we insert a at the end of the first row of T.
- Otherwise, let b be the biggest entry of the first row which is $> a$; we substitute b by a, and insert b into the subtableau formed by the remaining rows of T, according to the same rule.

(2) If $a_1, \ldots, a_k$ is a list of integers, let $\mathrm{Rowinsert}(a_1, \ldots, a_k)$ be the tableau obtained by inserting successively $a_1, \ldots, a_k$ according to the rule described above, that is,

$$\mathrm{Rowinsert}(a_1, \ldots, a_k) = (\emptyset \leftarrow a_1 \leftarrow \cdots \leftarrow a_k).$$

(3) The tableau $\mathrm{RS}_1(w)$ is defined by

$$\mathrm{RS}_1(w) = \mathrm{Rowinsert}(w_1, \ldots, w_n)$$

while we define $\mathrm{RS}_2(w)$ as the unique tableau such that the shape of the subtableau of $\mathrm{RS}_2(w)$ of entries $1, \ldots, k$ coincides with the shape of $\mathrm{Rowinsert}(w_1, \ldots, w_k)$.

For later use, we mention some properties of the Robinson–Schensted correspondence:

Proposition 11.2: (See, e.g., [18, 83])

(1) *The map*

$$\mathrm{RS} = (\mathrm{RS}_1, \mathrm{RS}_2) : \mathfrak{S}_n \to \coprod\nolimits_{\lambda \vdash n} \mathrm{STab}(\lambda) \times \mathrm{STab}(\lambda)$$

so obtained is a bijection.

(2) *If* $w \in \mathfrak{S}_n$, *then* $(\mathrm{RS}_1(w^{-1}), \mathrm{RS}_2(w^{-1})) = (\mathrm{RS}_2(w), \mathrm{RS}_1(w))$.

(3) *If* $w_0 : k \mapsto n - k + 1$ *is the longest element of* $\mathfrak{S}_n$, *then* $\mathrm{RS}_1(ww_0) = {}^t\mathrm{RS}_1(w)$, *where* ${}^tT$ *denotes the transposition of a tableau* T, *and* $\mathrm{RS}_1(w_0 w) = {}^t\mathbf{S}(\mathrm{RS}_1(w))$, *where* $\mathbf{S}$ *stands for the* Schützenberger involution (*see* [83](§2)).

Remark 11.3: More generally, if $\sigma : \{j_1 < \ldots < j_n\} \to \{i_1 < \ldots < i_n\}$ is a bijection, then we define

$$\mathrm{RS}_1(\sigma) = \mathrm{Rowinsert}(\sigma(j_1), \ldots, \sigma(j_n)),$$

which is a Young tableau with entries $i_1, \ldots, i_n$, and we can define $\mathrm{RS}_2(\sigma)$ as the unique Young tableau of entries $j_1, \ldots, j_n$ such that

$$\mathbf{sh}\,\mathrm{RS}_2(\sigma)|_{\{j_1,\ldots,j_k\}} = \mathbf{sh}\,\mathrm{RS}_1(\sigma|_{\{j_1,\ldots,j_k\}}) \quad \text{for all } k = 1, \ldots, n,$$

where the notation on the left hand side stands for the shape of the subtableau of entries $j_1, \ldots, j_k$. Moreover, from Proposition 11.2(2) we easily have $\mathrm{RS}_2(\sigma) = \mathrm{RS}_1(\sigma^{-1})$.

Finally, we recall that nilpotent orbits of $\mathfrak{g} = \mathfrak{gl}_n$ are parametrized by the set $\mathcal{P}(n) = \{\lambda \vdash n\}$ of partitions of n through Jordan normal form:

$$\mathcal{N}_{\mathfrak{gl}_n} = \coprod_{\lambda \vdash n} \mathcal{O}_\lambda.$$

Let us denote the shape of a tableau T by $\mathbf{sh}\, T$, which is also interpreted as a partition (or a Young diagram).

Theorem 11.4: (Steinberg, [77, 78]) *For $G = \mathrm{GL}_n$ and $\mathfrak{X} = G/B \times G/B$, the Steinberg map $\Phi : \mathfrak{X}/G \cong W \to \mathcal{N}_{\mathfrak{g}}/G$ is given by, for $w \in W$,*

$$\Phi : \mathbb{O}_w \mapsto \mathcal{O}_\lambda \quad \textit{such that } \lambda = \mathbf{sh}(\mathrm{RS}_1(w)) = \mathbf{sh}(\mathrm{RS}_2(w)).$$

11.2. *Description of Steinberg maps for a double flag variety of type AIII*

In this section, we use the same notation as in the beginning of §9. Thus our double flag variety is

$$\begin{aligned} \mathfrak{X} = K/B_K \times G/P &= \left(\mathrm{GL}_p/B_p^+ \times \mathrm{GL}_q/B_q^+\right) \times \mathrm{GL}_n/P_{(r,n-r)} \\ &\simeq \left(\mathscr{F}\!\ell(V^+) \times \mathscr{F}\!\ell(V^-)\right) \times \mathrm{Gr}_r(V). \end{aligned}$$

The purpose of this section is to give a combinatorial description of the Steinberg maps

$$\Phi_{\mathfrak{k}} : \mathfrak{X}/K \to \mathcal{N}_{\mathfrak{k}}/K \quad \text{and} \quad \Phi_{\mathfrak{s}} : \mathfrak{X}/K \to \mathcal{N}_{\mathfrak{s}}/K.$$

To this end, we first need to review the combinatorial parametrization of the various orbit sets that appear in the picture.

11.2.1. *Parametrization of $\mathcal{N}_{\mathfrak{k}}/K$*

In the present case, the Cartan decomposition is $\mathfrak{g} = \mathfrak{k} \oplus \mathfrak{s}$ with

$$\mathfrak{k} = \left\{ \begin{pmatrix} a & 0 \\ 0 & d \end{pmatrix} : a \in \mathfrak{gl}_p,\ d \in \mathfrak{gl}_q \right\},$$

$$\mathfrak{s} = \left\{ \begin{pmatrix} 0 & b \\ c & 0 \end{pmatrix} : b \in \mathrm{M}_{p,q},\ c \in \mathrm{M}_{q,p} \right\}.$$

We have natural bijections $\mathcal{N}_{\mathfrak{k}} \cong \mathcal{N}_{\mathfrak{gl}_p} \times \mathcal{N}_{\mathfrak{gl}_q}$ and so $\mathcal{N}_{\mathfrak{k}}/K \cong \mathcal{N}_{\mathfrak{gl}_p}/\mathrm{GL}_p \times \mathcal{N}_{\mathfrak{gl}_q}/\mathrm{GL}_q \cong \mathcal{P}(p) \times \mathcal{P}(q)$ where the last identification is through Jordan normal form.

11.2.2. *Parametrization of* $\mathcal{N}_{\mathfrak{s}}/K$

Nilpotent orbits of $\mathfrak{s}$ are parametrized by a set of signed Young diagrams:

Definition 11.5: (a) A *signed Young diagram* (of signature (p, q)) is a Young diagram of size $p+q$ whose boxes are filled in with p symbols $+$ and q symbols $-$ so that:

- two consecutive boxes of the same row have opposite signs, so that each row is a sequence of alternating signs;
- we identify two such fillings up to permutation of rows, in particular we can standardize the filling in such a way that, among rows which have the same length, the rows starting with a $+$ are above those starting with a $-$ (if there is any).

Let $\mathcal{SYD}(p, q)$ denote the set of signed Young diagrams of signature (p, q). For $\Lambda \in \mathcal{SYD}(p, q)$, the shape of Λ is an element of $\mathcal{P}(p+q)$, also denoted by $\mathbf{sh}(\Lambda)$. By $\#\Lambda_{\leq k}(\pm)$ we will denote the number of $\pm$'s within the first k columns of Λ.

Proposition 11.6: (see, e.g., [11, 57]) *We have a bijection*

$\mathcal{SYD}(p, q) \to \mathcal{N}_{\mathfrak{s}}/K,\ \Lambda \mapsto \{x \in \mathcal{N}_{\mathfrak{s}} : \forall k \geq 1,\ \dim\ker x^k \cap V^{\pm} = \#\Lambda_{\leq k}(\pm)\}.$

11.2.3. *Parametrization of* $\mathfrak{X}/K$ *by partial permutations*

We recall from §9.2 that the K-orbits of $\mathfrak{X}$ are parametrized by the set

$$\overline{\mathfrak{T}} = \left\{\omega = \begin{pmatrix}\tau_1\\ \tau_2\end{pmatrix} : \tau_1 \in \mathfrak{T}_{p,r},\ \tau_2 \in \mathfrak{T}_{q,r},\ \operatorname{rk}\omega = r\right\} / \mathfrak{S}_r.$$

In fact, we have a bijection

$$\overline{\mathfrak{T}} \xrightarrow{\sim} \mathfrak{X}/K,\ \omega \mapsto K(\mathcal{F}^+, \mathcal{F}^-, [\omega])$$

where $[\omega] = \operatorname{Im}\omega$ and $\mathcal{F}^+, \mathcal{F}^-$ are the standard flags of V^+, V^-, respectively (see Theorem 9.4). Note also that we have a graphical expression $\Gamma(\omega)$ of the elements $\omega \in \overline{\mathfrak{T}}$ (see Notation 9.2).

11.2.4. *Combinatorial Steinberg maps*

Now, according to the parametrization of orbits from the previous subsections, the symmetrized and exotic Steinberg maps $\Phi_{\mathfrak{k}}$ and $\Phi_{\mathfrak{s}}$ can be considered as combinatorial maps

$$\Phi_{\mathfrak{k}} : \overline{\mathfrak{T}} \to \{(\lambda, \mu) : \lambda \vdash p,\ \mu \vdash q\} \quad \text{and} \quad \Phi_{\mathfrak{s}} : \overline{\mathfrak{T}} \to \mathcal{SYD}(p, q). \tag{11.2}$$

By using these maps, we will finally reformulate another parametrization of $\mathfrak{X}/K$ in terms of standard tableaux in Theorem 11.10, which clarifies the irreducible components of the conormal fibers of Steinberg maps.

Let us describe these maps $\Phi_{\mathfrak{k}}$ and $\Phi_{\mathfrak{s}}$ explicitly.

11.3. *Generalized Robinson-Schensted correspondence related to* $\Phi_{\mathfrak{k}}$

The following theorem gives a complete algorithm to calculate the map $\Phi_{\mathfrak{k}} : \overline{\mathfrak{T}} \to \{(\lambda, \mu) : \lambda \vdash p,\ \mu \vdash q\}$.

Theorem 11.7: *Let $\omega \in \overline{\mathfrak{T}}$. We use the following notation, relative to the graph $\Gamma(\omega)$:*

- *Let $i_1 < \ldots < i_k$, $\ell_1 < \ldots < \ell_s$, and $\ell'_1 < \ldots < \ell'_{s'}$ be respectively the numbers of the end points, marked points, and free points, within the line of positive vertices of $\Gamma(\omega)$. Let $j_1 < \ldots < j_k$, $m_1 < \ldots < m_t$, and $m'_1 < \ldots < m'_{t'}$ be similarly the numbers of the end points, marked points, and free points, within the negative vertices of $\Gamma(\omega)$.*
- *Let L, L', M, M' be the vertical Young tableaux with entries $\{\ell_i\}_{i=1}^{s}$, $\{\ell'_i\}_{i=1}^{s'}$, $\{m_i\}_{i=1}^{t}$, $\{m'_i\}_{i=1}^{t'}$, respectively. Let $\sigma : \{j_1, \ldots, j_k\} \to \{i_1, \ldots, i_k\}$ be the bijection such that $\sigma(j) = i$ whenever (i^+, j^-) is an edge of $\Gamma(\omega)$.*

Then, the image of ω by the symmetrized Steinberg map is the pair of partitions

$$\Phi_{\mathfrak{k}}(\omega) = (\lambda, \mu) = \left(\mathbf{sh}(L \cdot \mathrm{RS}_1(\sigma) \cdot L'), \mathbf{sh}(M \cdot \mathrm{RS}_2(\sigma) \cdot M') \right) \tag{11.3}$$

where $\cdot$ is the sliding operation of the plactic monoid (see [18] *(§1.2–2.1)).*

Example 11.8: For ω as in (9.1), we obtain

$$L \cdot \mathrm{RS}_1(\sigma) \cdot L' = \ytableaushort{5} \cdot \ytableaushort{2,4} \cdot \ytableaushort{1,3} = \ytableaushort{13,2,4,5}, \quad M \cdot \mathrm{RS}_2(\sigma) \cdot M' = \ytableaushort{2} \cdot \ytableaushort{1,3} = \ytableaushort{13,2},$$

hence

$$\Phi_{\mathfrak{k}}(\omega) = (\lambda, \mu) \quad \text{with} \quad \lambda = \ydiagram{2,1,1,1}, \ \mu = \ydiagram{2,1}.$$

Remark 11.9: The properties of the row insertion algorithm also yield the following rephrasing of the formula giving $\Phi_{\mathfrak{k}}(\omega) = (\lambda, \mu)$ in Theorem 11.7. Specifically, if we introduce the permutations

$$w_{\mathfrak{k},+} = \begin{pmatrix} 1 & \cdots & s & s+1 & \cdots & s+k & s+k+1 & \cdots & p \\ \ell_s & \cdots & \ell_1 & \sigma(j_1) & \cdots & \sigma(j_k) & \ell'_{s'} & \cdots & \ell'_1 \end{pmatrix} \in \mathfrak{S}_p,$$

$$w_{\mathfrak{k},-} = \begin{pmatrix} 1 & \cdots & t & t+1 & \cdots & t+k & t+k+1 & \cdots & q \\ m_t & \cdots & m_1 & \sigma^{-1}(i_1) & \cdots & \sigma^{-1}(i_k) & m'_{t'} & \cdots & m'_1 \end{pmatrix} \in \mathfrak{S}_q,$$

then we obtain

$$(\lambda, \mu) = \big(\mathbf{sh}(\mathrm{RS}_1(w_{\mathfrak{k},+})), \mathbf{sh}(\mathrm{RS}_1(w_{\mathfrak{k},-}))\big).$$

Proof: [Sketch of proof of Theorem 11.7] According to Definition 10.2, we need to consider the conormal bundle

$$T^*_{\mathbb{O}_\omega}\mathfrak{X} = K \cdot \{(\mathcal{F}_p^+, \mathcal{F}_q^+, [\omega], y) : y \in \mathcal{D}_\omega\}$$

with the conormal direction

$$\mathcal{D}_\omega = \{y \in \mathfrak{n}_\omega : y^\theta \in \mathfrak{n}_p^+ \times \mathfrak{n}_q^+\}.$$

Here $\mathfrak{n}_\omega = \{y : \operatorname{Im} y \subset [\omega] \subset \ker y\}$ is the nilradical of the parabolic subalgebra determined by $[\omega]$ and $\mathfrak{n}_p^+, \mathfrak{n}_q^+$ stand for the subalgebras of strictly upper triangular matrices of $\mathfrak{gl}_p$ and $\mathfrak{gl}_q$, respectively.

For $a \in \mathfrak{gl}_p$, it can be shown that

$$\exists b, c, d \text{ such that } \begin{pmatrix} a & b \\ c & d \end{pmatrix} \in \mathcal{D}_\omega \quad \Longleftrightarrow \quad a \in \mathfrak{n}_p^+ \cap \mathrm{Ad}(w_{\mathfrak{k},+})\mathfrak{n}_p^+$$

where $w_{\mathfrak{k},+}$ is as in Remark 11.9 above. A reformulation of Steinberg's Theorem 11.4 gives that the unique nilpotent GL_p-orbit which intersects the linear space $\mathfrak{n}_p^+ \cap \mathrm{Ad}(w_{\mathfrak{k},+})\mathfrak{n}_p^+$ is the one corresponding to the partition

$$\mathbf{sh}\,\mathrm{RS}_1(w_{\mathfrak{k},+}) = \mathbf{sh}\,\mathrm{RS}_2(w_{\mathfrak{k},+}).$$

Arguing similarly with the block d and the permutation $w_{\mathfrak{k},-}$, we finally get that the product of nilpotent orbits that intersects the projection $\phi_{\mathfrak{k}}(T^*_{\mathbb{O}_\omega}\mathfrak{X}) = K \cdot \{y^\theta : y \in \mathcal{D}_\omega\}$ densely is $\mathcal{O}_\lambda \times \mathcal{O}_\mu$ with

$$(\lambda, \mu) = (\mathbf{sh}\,\mathrm{RS}_1(w_{\mathfrak{k},+}), \mathbf{sh}\,\mathrm{RS}_1(w_{\mathfrak{k},-})).$$

This (combined with Remark 11.9) yields the claimed assertion. See [16](§4.3) for a complete proof. □

As a byproduct of Theorem 11.7, we can determine the fibers of $\Phi_{\mathfrak{k}}$ in terms of a combinatorial correspondence which extends the Robinson–Schensted correspondence. It involves the following sets of 5-tuples: for $(\lambda, \mu) \in \mathcal{P}(p) \times \mathcal{P}(q)$, let

$$\mathcal{T}_{\lambda,\mu} = \{(T_1, T_2; \lambda', \mu'; \nu) \mid T_1 \in \mathrm{STab}(\lambda),\ T_2 \in \mathrm{STab}(\mu), \text{satisfying } (\star)\} \tag{11.4}$$

where

$$(\star)\ \ \nu \Subset \lambda' \Subset \lambda,\ \nu \Subset \mu' \Subset \mu, \text{ and } |\lambda'| + |\mu'| = |\nu| + r.$$

Hereafter, $|\lambda|$ stands for the size of a partition λ, and the notation $\lambda' \Subset \lambda$ means that λ' is a Young subdiagram of λ such that the skew diagram $\lambda \setminus \lambda'$ is *column strip* (i.e., it contains at most one box in each row).

Theorem 11.10: *We use the notation of Theorem 11.7 and Remark 11.9. We have a commutative diagram*

$$\begin{array}{ccc}
\overline{\mathfrak{X}} & \xrightarrow[\text{gRS}]{\sim} & \displaystyle\coprod_{(\lambda,\mu)\in\mathcal{P}(p)\times\mathcal{P}(q)} \mathcal{T}_{\lambda,\mu} \\
& {\scriptstyle \Phi_{\mathfrak{k}}} \searrow \quad \swarrow {\scriptstyle \pi} & \\
& \mathcal{P}(p) \times \mathcal{P}(q) &
\end{array}$$

where π is the natural map and gRS *is a bijection defined by*

$$\begin{aligned}\mathrm{gRS}(\omega) = \big(&L \cdot \mathrm{RS}_1(\sigma) \cdot L', M \cdot \mathrm{RS}_2(\sigma) \cdot M'; \\ &\mathbf{sh}(L \cdot \mathrm{RS}_1(\sigma)), \mathbf{sh}(M \cdot \mathrm{RS}_2(\sigma));\ \mathbf{sh}(\mathrm{RS}_1(\sigma))\big).\end{aligned} \tag{11.5}$$

In particular, gRS *restricts to a bijection*

$$\Phi_{\mathfrak{k}}^{-1}(\lambda, \mu) \xrightarrow{\sim} \mathcal{T}_{\lambda,\mu}$$

for all $(\lambda, \mu) \in \mathcal{P}(p) \times \mathcal{P}(q)$.

Proof: We only need to justify that the map gRS is a bijection; the rest of the statement will then follow from Theorem 11.7.

First, we note that the considered map is well defined. Let $(T_1, T_2; \lambda', \mu'; \nu)$ denote the right-hand side of (11.5). The fact that $\lambda \setminus \lambda'$, $\lambda' \setminus \nu$, $\mu \setminus \mu'$, and $\mu' \setminus \nu$ are column strips follows from [18](Proposition in §1.1), and by construction we have $|\lambda'| + |\mu'| = 2k + s + t = |\nu| + r$.

For showing the bijectivity, let $(T_1, T_2; \lambda', \mu'; \nu) \in \mathcal{T}_{\lambda,\mu}$. Two applications of [18](Proposition in §1.1) yield a unique 6-tuple (S_1, S_2, L, L', M, M'), where S_1, S_2 are Young tableaux of shape ν and L, L', M, M' are vertical Young tableaux, all contents being disjoint, such that $T_1 = L \cdot S_1 \cdot L'$, $T_2 = M \cdot S_2 \cdot M'$, $\lambda' = \mathbf{sh}(L \cdot S_1)$, and $\mu' = \mathbf{sh}(M \cdot S_2)$. Letting $\sigma : J \to I$ be the bijection associated with the pair (S_1, S_2) via Robinson–Schensted correspondence, the data $I, L, L', J, M, M', \sigma$ determine a unique element $\omega \in \overline{\mathfrak{T}}$ fitting with the statement of Theorem 11.7, and by construction the map $(T_1, T_2; \lambda', \mu'; \nu) \mapsto \omega$ so obtained is the inverse procedure of gRS. □

Example 11.11:

(1) Let $(p, q, r) = (5, 3, 4)$. For ω as in (9.1) and Example 11.8, we obtain

$$\omega = \begin{pmatrix} e_2 & e_4 & e_5 & 0 \\ \hline e_3 & e_1 & 0 & e_2 \end{pmatrix} \rightsquigarrow \Gamma(\omega) =$$

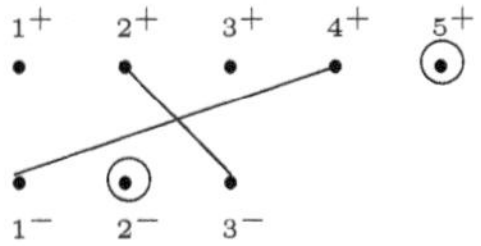

$$\mathrm{gRS}(\omega) = \left(\begin{array}{|c|c|}\hline 1 & 3 \\ \hline 2 \\ \cline{1-1} 4 \\ \cline{1-1} 5 \\ \cline{1-1}\end{array}, \begin{array}{|c|c|}\hline 1 & 3 \\ \hline 2 \\ \cline{1-1}\end{array}; \begin{array}{|c|}\hline \ \\ \hline \ \\ \hline \ \\ \hline\end{array}, \begin{array}{|c|c|}\hline \ & \ \\ \hline \ \\ \cline{1-1}\end{array}; \begin{array}{|c|}\hline \ \\ \hline \ \\ \hline\end{array} \right).$$

Here, to describe ω as a pair of partial permutations, we use e_k for the k-th elementary basis vector, i.e., a column vector whose k-th entry is 1 and any other entries are 0 (we do not distinguish whether it is belonging to $\mathbb{C}^p$ or $\mathbb{C}^q$ for simplicity).

(2) We give another example for $(p, q, r) = (5, 4, 4)$.

$$\omega = \begin{pmatrix} e_1 & e_2 & e_3 & e_5 \\ \hline e_2 & e_4 & 0 & e_3 \end{pmatrix} \rightsquigarrow \Gamma(\omega) =$$

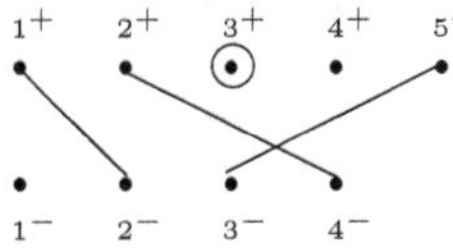

$$\mathrm{gRS}(\omega) = \left(\begin{array}{|c|c|c|}\hline 1 & 2 & 4 \\ \hline 3 & 5 \\ \cline{1-2}\end{array}, \begin{array}{|c|c|}\hline 1 & 3 \\ \hline 2 \\ \cline{1-1} 4 \\ \cline{1-1}\end{array}; \begin{array}{|c|c|}\hline \ & \ \\ \hline \ & \ \\ \hline\end{array}, \begin{array}{|c|c|}\hline \ & \ \\ \hline \ \\ \cline{1-1}\end{array}; \begin{array}{|c|c|}\hline \ & \ \\ \hline \ \\ \cline{1-1}\end{array} \right).$$

(3) Let $p = q = r = n$. In this case, $\overline{\mathfrak{T}}$ contains the subset

$$\overline{\mathfrak{T}}_0 := \left\{ \omega_\sigma = \begin{pmatrix} \sigma \\ \hline \mathbb{1}_n \end{pmatrix} : \sigma \in \mathfrak{S}_n \right\}.$$

Then we have

$$\mathrm{gRS}(\omega_\sigma) = (\mathrm{RS}_1(\sigma), \mathrm{RS}_2(\sigma); \lambda, \mu; \nu)$$

with $\lambda = \mu = \nu$ is the common shape of $\mathrm{RS}_1(\sigma)$ and $\mathrm{RS}_2(\sigma)$. In this way, gRS is indeed an extension of Robinson–Schensted correspondence.

(4) Let $p = r \geq 2$ and $q = 1$. Then $\overline{\mathfrak{T}}$ contains three families of elements $\{\omega_i^{\mathrm{I}}\}_{i=1}^p$, $\{\omega_i^{\mathrm{II}}\}_{i=1}^p$, and $\{\omega^{\mathrm{III}}\}$, with respective graphical representations as below.

$\Gamma(\omega_i^{\mathrm{I}}) =$ 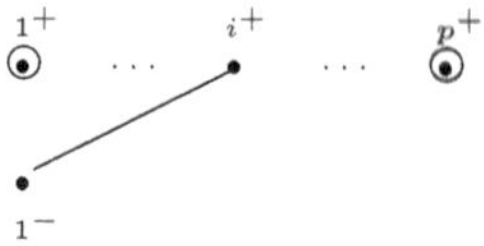 $\Gamma(\omega_i^{\mathrm{II}}) =$ 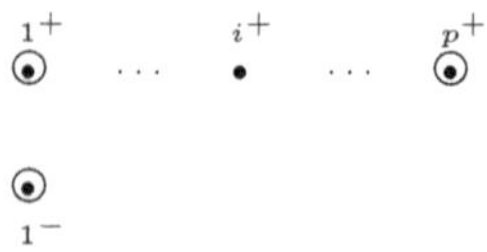

$\Gamma(\omega^{\mathrm{III}}) =$ 

Their images by gRS are as follows.

ω	ω_1^{I}	$\omega_i^{\mathrm{I}},\ i > 1$
gRS(ω)	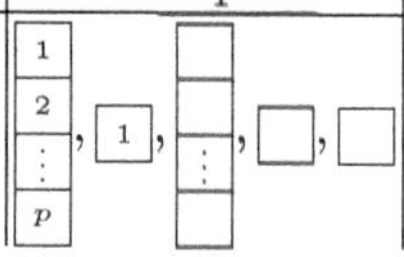	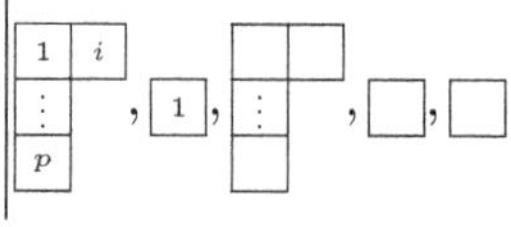

ω	ω_1^{II}	$\omega_i^{\mathrm{II}},\ i > 1$	ω^{III}
gRS(ω)	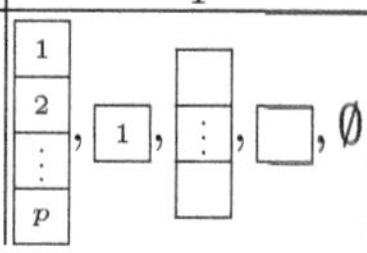	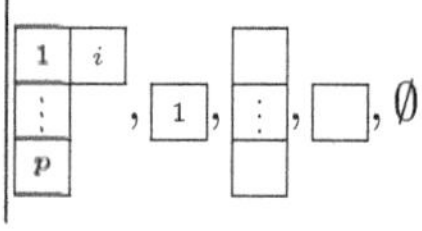	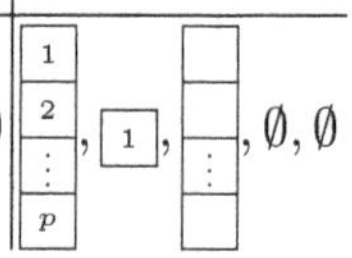

11.3.1. *Relation to the Travkin's mirabolic Robinson-Schensted correspondence*

In [82], Travkin considers the variety

$$X = \mathscr{F}\ell(V) \times \mathscr{F}\ell(V) \times V,$$

with diagonal action of $\mathrm{GL}(V)$, which is a slight variation of the triple flag variety

$$\mathscr{F}\ell(V) \times \mathscr{F}\ell(V) \times \mathbb{P}(V) = \mathrm{GL}_n/B_n^+ \times \mathrm{GL}_n/B_n^+ \times \mathrm{GL}_n/P_{\mathrm{mir}},$$

where $P_{\mathrm{mir}} \subset \mathrm{GL}_n$ is mirabolic — the stabilizer of a line in $V = \mathbb{C}^n$. Then, using an explicit description of the conormal bundles to the orbits, he derives a *mirabolic* Robinson–Schensted correspondence ([82](§3.3))

$$\begin{aligned} X/\mathrm{GL}_n &\simeq \{(w, \sigma) \mid w \in \mathfrak{S}_n,\ \sigma \text{ a descent of } w\} \\ &\xrightarrow{\sim} \coprod_{(\lambda,\mu)\in\mathcal{P}(n)^2} \{(T_1, T_2, \nu) \mid T_1 \in \mathrm{STab}(\lambda),\ T_2 \in \mathrm{STab}(\mu),\ \text{satisfying } (\star\star)\}, \end{aligned} \tag{11.6}$$

where a *descent* of w means a (possibly empty) sequence $i_1 < \ldots < i_s$ with $w(i_1) > \ldots > w(i_s)$, and the property $(\star\star)$ is given below (compare it with the condition $(\star)$ in §11.3).

$(\star\star)$ $\lambda \setminus \nu$ and $\mu \setminus \nu$ are column strip, that is, $\nu \Subset \lambda$ and $\nu \Subset \mu$.

On the other hand, in [15], we consider an "open big cell" $\mathfrak{X}^\circ$ in the double flag variety $\mathfrak{X}$ of type AIII below,

$$\mathfrak{X} = (\mathrm{GL}_n/B_n^+ \times \mathrm{GL}_n/B_n^+) \times \mathrm{GL}_{2n}/P_{(n,n)} = K/B_K \times G/P_S \quad (P_S := P_{(n,n)}).$$

If we denote by U_S° the unipotent radical of the opposite parabolic subgroup P_S°, the open big cell is defined as a K-stable subvariety

$$\mathfrak{X}^\circ = K/B_K \times U_S^\circ P_S/P_S \simeq (\mathrm{GL}_n/B_n^+ \times \mathrm{GL}_n/B_n^+) \times \mathrm{M}_n,$$

where $(k_1, k_2) \in K = \mathrm{GL}_n \times \mathrm{GL}_n$ acts on M_n by $k_1 A k_2^{-1}$ $(A \in \mathrm{M}_n)$. It is obvious that $\mathfrak{X}^\circ/K \simeq B_n^+ \backslash \mathrm{M}_n / B_n^+$, and the representatives of $B_n^+ \times B_n^+$-orbits in M_n are precisely given by partial permutations $\mathfrak{T}_{n,n}$ (see §9.2); for the proof, we refer to [18], and also [16](Lemma 3.4). In [16], the representatives are described as $\omega = \begin{pmatrix} \tau_1 \\ \mathbb{1}_n \end{pmatrix}$ with $\tau_1 \in \mathfrak{T}_{n,n}$. If we apply the algorithm for the generalized Robinson–Schensted correspondence explained in §11.3, we precisely obtain a triple (T_1, T_2, ν) belonging to the set in the right-hand side of the Travkin's mirabolic RS correspondence (11.6) (note that $L = M' = \emptyset$ in Theorem 11.10 so that we do not need intermediate λ' and μ' because $\lambda' = \nu$ and $\mu' = \mu$). Thus we get a bijection

$$\begin{aligned} &\mathfrak{X}^\circ/K \simeq \mathfrak{T}_{n,n} \\ &\xrightarrow{\sim} \coprod_{(\lambda,\mu)\in\mathcal{P}(n)^2} \{(T_1, T_2, \nu) \mid T_1 \in \mathrm{STab}(\lambda),\ T_2 \in \mathrm{STab}(\mu),\ \text{satisfying } (\star\star)\}. \end{aligned} \tag{11.7}$$

Consequently all of these five sets that appear in (11.6) and (11.7) are in bijection. In particular we get

Proposition 11.12: *There is a bijection between orbits in* $(\mathrm{GL}_n/B_n^+ \times \mathrm{GL}_n/B_n^+ \times \mathbb{C}^n)/\mathrm{GL}_n$ *and* $(\mathrm{GL}_n/B_n^+ \times \mathrm{GL}_n/B_n^+ \times \mathrm{M}_n)/(\mathrm{GL}_n \times \mathrm{GL}_n)$.

We get this proposition by comparing mirabolic/generalized RS correspondence. However, constructing bijection between Travkin's set $\{(w, \sigma) \mid w \in S_n, \sigma \subset \{1, \ldots, n\}$ is a descent in$w\}$ and the partial permutations $\mathfrak{T}_{n,n}$ is somewhat straightforward. In fact, for $\tau \in \mathfrak{T}_{n,n}$, we get a permutation matrix $w \in \mathfrak{S}_n$ by adding 1's in some columns $i_1, \ldots, i_k$ in such a way that $i_1 < i_2 < \cdots < i_k$ and $w(i_1) > w(i_2) > \cdots > w(i_k)$, where $k = n - \operatorname{rank} \tau$. Then $\sigma = \{w(i_1), w(i_2), \ldots, w(i_k)\}$ is a descent for w, and this actually gives a bijection between orbital representatives.

We thank Anthony Henderson for pointing out the bijection to us.

Remark 11.13:

(1) In the mirabolic case, there is a natural map from the conormal variety to the *enhanced nilpotent cone* $\mathcal{N}_{\mathfrak{gl}_n} \times V$ (see [1]), which is GL_n-equivariant. It is interesting that the orbits $(\mathcal{N}_{\mathfrak{gl}_n} \times V)/\mathrm{GL}_n$ are classified by the bipartitions $Q_n = \{(\mu, \nu) \mid |\mu| + |\nu| = n\}$ which is further in bijection with $\mathfrak{P} = \{(\nu, \lambda) \mid |\lambda| = n, \nu \Subset \lambda\}$. Note that the latter set $\mathfrak{P}$ also classifies the pairs (U, W), where U is an irreducible unipotent representation of $\mathrm{GL}_n(\mathbb{F}_q)$, and W is an irreducible constituent of the restriction of U to the mirabolic subgroup $P_n(\mathbb{F}_q)$ ([86]; see also [82](§2.2)).

(2) For the symplectic group, Syu Kato developed a generalization of Deligne-Langlands theory over the exotic nilpotent cone ([31]). In this case, Henderson-Trapa [22] and Nandakumar-Rosso-Saunders [53] establish a kind of Robinson–Schensted correspondence called the exotic RS correspondence. They are also related to the set of bipartitions Q_n or $\mathfrak{P}$.

(3) There is yet another bijection for $\mathfrak{T}_{n,n}$ by using so-called "admissible signed Young diagrams". For this, see [70].

It would be interesting to develop a general theory which explains all of these related results.

11.4. *Combinatorial description of the exotic Steinberg map $\Phi_{\mathfrak{s}}$*

The description of the exotic Steinberg map in terms of signed Young diagrams is more involved.

Theorem 11.14: *Let $\omega \in \overline{\mathfrak{T}}$. With the notation of Theorem 11.7, we consider the bijections*

$$w_{\mathfrak{s},+} = \begin{pmatrix} m_1 & \cdots & m_t & j_1 & \cdots & j_k & q+1 & \cdots & q+s' \\ -1 & \cdots & -t & \sigma(j_1) & \cdots & \sigma(j_k) & \ell'_{s'} & \cdots & \ell'_1 \end{pmatrix}, \tag{11.8}$$

$$w_{\mathfrak{s},-} = \begin{pmatrix} \ell_1 & \cdots & \ell_s & i_1 & \cdots & i_k & p+1 & \cdots & p+t' \\ -1 & \cdots & -s & \sigma^{-1}(i_1) & \cdots & \sigma^{-1}(i_k) & m'_{t'} & \cdots & m'_1 \end{pmatrix}. \tag{11.9}$$

Then, the image of ω *by the exotic Steinberg map is the signed Young diagram* $\Phi_{\mathfrak{s}}(\omega) = \Lambda$ *determined as follows. Here,* $\#\Lambda_{\leq c}(+)$, *resp.* $\#\Lambda_{\leq c}(-)$ *(resp.* $\#\lambda_{\leq c}$*) denotes the number of* $+$*'s, resp.* $-$*'s (resp. boxes) in the first* c *columns of* Λ *(respectively* λ*).*

(1) *For every* $c \geq 1$ *even,*

$$\#\Lambda_{\leq c}(+) = \#\lambda_{\leq c} \quad \textit{and} \quad \#\Lambda_{\leq c}(-) = \#\mu_{\leq c},$$

where $(\lambda, \mu) = \Phi_{\mathfrak{k}}(\omega)$.

(2) *For every* $c \geq 1$ *odd,*

$$\#\Lambda_{\leq c}(+) = s - t + \#\lambda'_{\leq c} \quad \textit{and} \quad \#\Lambda_{\leq c}(-) = t - s + \#\mu'_{\leq c},$$

where (λ', μ') *is the pair of Young diagrams given by*

$$(\lambda', \mu') = \big(\mathbf{sh}(\mathrm{RS}_1(w_{\mathfrak{s},+})), \mathbf{sh}(\mathrm{RS}_1(w_{\mathfrak{s},-}))\big).$$

Proof: [Sketch of proof] In view of the definition of $\Phi_{\mathfrak{s}}$ in Definition 10, we need to determine the nilpotent $\mathrm{GL}_p \times \mathrm{GL}_q$-orbit of $\mathfrak{s}$ which intersects the projection $\{y^{-\theta} : y \in \mathcal{D}_\omega\}$ densely, where $\mathcal{D}_\omega$ is the conormal direction already considered in the proof of Theorem 11.7.

Since every $y \in \mathcal{D}_\omega$ satisfies $y^2 = 0$, we already have

$$(y^{-\theta})^{2m} = (-1)^m (y^\theta)^{2m} \quad \text{for all } m \geq 1, \text{ all } y \in \mathcal{D}_\omega.$$

This, combined with Theorem 11.7, proves (1).

For showing (2), we need to characterize the blocks $c \in \mathrm{M}_{q,p}$ satisfying the condition

$$\exists a, b, d \text{ such that } \begin{pmatrix} a & b \\ c & d \end{pmatrix} \in \mathcal{D}_\omega.$$

This is done, again, by relying on Steinberg's result (Theorem 11.4), though the construction here is less direct than in Theorem 11.7; see [16](§4.4) for a complete proof. □

Example 11.15:

(1) Taking again ω as in (9.1) and Example 11.8, we have

$$w_{\mathfrak{s},+} = \begin{pmatrix} 1 & 2 & 3 & 4 & 5 \\ 4 & -1 & 2 & 3 & 1 \end{pmatrix} \quad \text{and} \quad w_{\mathfrak{s},-} = \begin{pmatrix} 2 & 4 & 5 \\ 3 & 1 & -1 \end{pmatrix},$$

which (combined with the calculation of $\Phi_{\mathfrak{k}}(\omega)$ in Example 11.8) yields

$\lambda' =$ 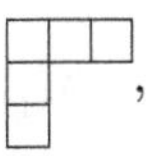, $\mu' =$ , and $\Lambda =$ 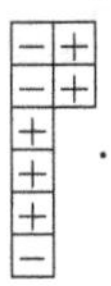.

(2) Assume $p = r \geq 2$ and $q = 1$ as in Example 11.11(4). With the same notation as in Example 11.11, we obtain the following images by the exotic Steinberg map $\Phi_{\mathfrak{S}}$.

ω	ω_1^{I}	$\omega_i^{\mathrm{I}},\ i \geq 2$	$\omega_i^{\mathrm{II}},\ i \geq 1$	ω^{III}
$\Phi_{\mathfrak{S}}(\omega)$	$\begin{array}{\|c\|}\hline + \\ \hline \vdots \\ \hline + \\ \hline - \\ \hline\end{array}$	$\begin{array}{\|c\|c\|}\hline + & - \\ \hline + \\ \cline{1-1} \vdots \\ \cline{1-1}\end{array}$	$\begin{array}{\|c\|c\|}\hline - & + \\ \hline + \\ \cline{1-1} \vdots \\ \cline{1-1}\end{array}$	$\begin{array}{\|c\|c\|}\hline + & - \\ \hline + \\ \cline{1-1} \vdots \\ \cline{1-1}\end{array}$

Unlike the case of the symmetrized Steinberg map, we have almost no information about the fibers of the exotic Steinberg map $\Phi_{\mathfrak{S}}$. It would be interesting to identify the fibers, which will give different parametrization of the orbits $\mathfrak{X}/K$ in terms of a generalized Robinson–Schensted correspondence analogous to Theorem 11.10.

Part IV. Embedding Theory for Double Flag Varieties

Let us return back to a general double flag variety associated with a symmetric pair. Thus (G, K) is a symmetric pair, P, Q being parabolic subgroups of G and K respectively, and we consider a double flag variety $\mathfrak{X} = K/Q \times G/P$.

Here by "embedding theory" we mean general results which guarantee an embedding between orbit sets, and general consequences of these results. There exist general results on orbit embedding (see [58], [54]), and one can apply them to the case of double flag varieties ([14]). If $\mathfrak{X}$ can be embedded into a larger, well-understood double flag variety, we can deduce finiteness of orbits or even orbit classifications from those of the larger double flag variety.

12. Embedding of Double Flag Varieties and their Orbits

Let $\mathbb{G}$ be a (larger) connected reductive algebraic group over $\mathbb{C}$, and let $\theta, \sigma \in \operatorname{Aut} \mathbb{G}$ be two involutions which are commutative. Let $\mathbb{K} = \mathbb{G}^\theta$ be the fixed point subgroup of θ, and assume that our G is the fixed point subgroup of σ, i.e., $G = \mathbb{G}^\sigma$. Then $(\mathbb{G}, \mathbb{K})$ is a symmetric pair, and $K = G^\theta = \mathbb{K}^\sigma = G \cap \mathbb{K}$ is a symmetric subgroup of both G and $\mathbb{K}$. We assume all the subgroups G, K and $\mathbb{K}$ are connected.

$$
\begin{array}{ccccc}
 & & \mathbb{G} & & \\
 & \overset{\sigma}{\swarrow} & & \overset{\theta}{\searrow} & \\
\mathbb{G}^\sigma = G & & & & \mathbb{K} = \mathbb{G}^\theta \\
 & \underset{\theta}{\searrow} & & \underset{\sigma}{\swarrow} & \\
 & & K = G^\theta = G \cap \mathbb{K} = \mathbb{G}^{\sigma,\theta} & &
\end{array}
$$

For any parabolic subgroup P of G, there exists a σ-stable parabolic subgroup $\mathbb{P}$ of $\mathbb{G}$ satisfying $P = \mathbb{P}^\sigma = \mathbb{P} \cap G$. Similarly, for a parabolic subgroup Q of K, we choose a σ-stable parabolic subgroup $\mathbb{Q}$ of $\mathbb{K}$ such that $Q = \mathbb{Q}^\sigma = \mathbb{Q} \cap K$. We denote $\mathbb{X} = \mathbb{G}/\mathbb{P} \times \mathbb{K}/\mathbb{Q}$ a double flag variety for $(\mathbb{G}, \mathbb{K})$, and $\mathfrak{X} = G/P \times K/Q$ as before. Then we get a natural embedding

$$\mathfrak{X} \hookrightarrow \mathbb{X}, \qquad (gP, kQ) \mapsto (g\mathbb{P}, k\mathbb{Q}) \quad (g \in G, k \in K)$$

and we will use the same notation $\mathfrak{X}$ for its embedded image in $\mathbb{X}$.

By abuse of notation, let σ also denote the automorphism of $\mathbb{X}$ defined by $\sigma(\boldsymbol{g} \cdot \mathbb{P}, \boldsymbol{k} \cdot \mathbb{Q}) = (\sigma(\boldsymbol{g}) \cdot \mathbb{P}, \sigma(\boldsymbol{k}) \cdot \mathbb{Q})$ so that σ is a $\mathbb{K}$-equivariant automorphism on $\mathbb{X}$, i.e., $\sigma(h \cdot \boldsymbol{x}) = \sigma(h) \cdot \sigma(\boldsymbol{x})$ holds for any $h \in \mathbb{K}$ and $\boldsymbol{x} \in \mathbb{X}$.

Definition 12.1: For a σ-stable subgroup $\mathbb{H}$ of $\mathbb{G}$, let us denote

$$\mathbb{H}^{-\sigma} = \{h \in \mathbb{H} \mid \sigma(h) = h^{-1}\}.$$

We say that the subgroup $\mathbb{H}$ *admits $(-\sigma)$-square roots* if for any $h \in \mathbb{H}^{-\sigma}$ there exists an $f \in \mathbb{H}^{-\sigma}$ which satisfies $h = f^2$. In this case f is called a $(-\sigma)$-square root of h.

Lemma 12.2: *If both $\mathbb{P}$ and $\mathbb{Q}$ admit $(-\sigma)$-square roots, then $\mathfrak{X}$ coincides with the fixed point set $\mathbb{X}^\sigma := \{\boldsymbol{x} \in \mathbb{X} \mid \sigma(\boldsymbol{x}) = \boldsymbol{x}\}$ of σ.*

Proof: Take $\boldsymbol{g}\mathbb{P} \in (\mathbb{G}/\mathbb{P})^\sigma$. Then we get $\boldsymbol{g}^{-1}\sigma(\boldsymbol{g}) \in \mathbb{P}^{-\sigma}$ since $\mathbb{P}$ is σ-stable. Therefore there exists a $(-\sigma)$-square root $f \in \mathbb{P}^{-\sigma}$ of $\boldsymbol{g}^{-1}\sigma(\boldsymbol{g})$. By this definition $\boldsymbol{g}f = \sigma(\boldsymbol{g})f^{-1} = \sigma(\boldsymbol{g}f)$ holds, and hence $\boldsymbol{g}f \in G$, which in turn implies $\boldsymbol{g}\mathbb{P} = (\boldsymbol{g}f)\mathbb{P}$ belongs to $G/P = G \cdot \mathbb{P} \subset \mathbb{G}/\mathbb{P}$. The proof of $\boldsymbol{k}\mathbb{Q} \in K/Q$ in $\mathbb{K}/\mathbb{Q}$ is the same. $\square$

The natural K-equivariant embedding $\mathfrak{X} \hookrightarrow \mathbb{X}$ induces an orbit map

$$
\begin{array}{rcl}
\iota : \mathfrak{X}/K & \longrightarrow & \mathbb{X}/\mathbb{K}, \\
\Psi & & \Psi \\
\mathcal{O} & \longmapsto & \mathbb{O} = \mathbb{K} \cdot \mathcal{O}
\end{array}
$$

which is not injective in general. (In the previous sections, we often used the notation $\mathbb{O}$ to denote a K-orbit in $\mathfrak{X}$, but here we change it to $\mathcal{O}$.) However, in a nice situation (like the following theorem), the map ι becomes an injection and we get $\mathcal{O} = \mathbb{O} \cap \mathfrak{X}$ is a single K-orbit.

Theorem 12.3: *Let us consider the double flag varieties $\mathfrak{X} = K/Q \times G/P$ and $\mathbb{X} = \mathbb{K}/\mathbb{Q} \times \mathbb{G}/\mathbb{P}$ above. We assume*

> (E) *The parabolic subgroups $\mathbb{P}$ and $\mathbb{Q}$ admit $(-\sigma)$-square roots, and, for any σ-stable parabolic subgroups $\mathbb{P}_1 \subset \mathbb{G}$ and $\mathbb{Q}_1 \subset \mathbb{K}$ which are conjugate to $\mathbb{P}$ and $\mathbb{Q}$ respectively, the intersection $\mathbb{P}_1 \cap \mathbb{Q}_1$ also admits $(-\sigma)$-square roots.*

Under the condition (E), *the following statements hold.*

(1) *The natural orbit map $\iota : \mathfrak{X}/K \to \mathbb{X}/\mathbb{K}$ defined by $\iota(\mathcal{O}) = \mathbb{K} \cdot \mathcal{O}$ for $\mathcal{O} \in \mathfrak{X}/K$ is injective, i.e., if we put $\mathbb{O} = \iota(\mathcal{O})$, then it holds $\mathcal{O} = \mathbb{O} \cap \mathfrak{X} = \mathbb{O}^{\sigma}$.*

(2) *For any $\mathbb{K}$-orbit $\mathbb{O}$ in $\mathbb{X}$, the intersection $\mathbb{O} \cap \mathfrak{X}$ is either empty or a single K-orbit.*

For the proof, we refer the readers to [14](Theorem 3.3), where Theorem 3.2 in [54] is used in crucial way. Note that we do not need finiteness of the orbits for this theorem.

Corollary 12.4: *Under the same notation and the condition* (E) *above, if the double flag variety $\mathbb{X} = \mathbb{K}/\mathbb{Q} \times \mathbb{G}/\mathbb{P}$ is of finite type, then $\mathfrak{X} = K/Q \times G/P$ is also of finite type.*

13. Embedding of Conormal Bundles

From now on, we always assume the condition (E) in Theorem 12.3.

Let us denote by $\mathfrak{G}$ the Lie algebra of $\mathbb{G}$. Similarly we will use German capital letters for the Lie algebras of the groups denoted by black board bold letters. We fix a nondegenerate invariant bilinear form on $\mathfrak{G}$ which is also invariant under the involutions σ and θ once and for all, and we identify the duals of $\mathfrak{g}, \mathfrak{k}$ and $\mathfrak{K}$ with themselves using this bilinear form. Let us consider the cotangent bundles $T^*\mathfrak{X}$ and $T^*\mathbb{X}$. Since $T^*\mathbb{X} \simeq (\mathbb{G} \times_{\mathbb{P}} \mathfrak{u}_{\mathfrak{P}}) \times (\mathbb{K} \times_{\mathbb{Q}} \mathfrak{u}_{\mathfrak{Q}})$, and $\mathfrak{u}_{\mathfrak{P}}$ and $\mathfrak{u}_{\mathfrak{Q}}$ are σ-stable, we have a natural extension of $\sigma \in \operatorname{Aut} \mathbb{X}$ to the whole $T^*\mathbb{X}$. We prove

Lemma 13.1: *$T^*\mathfrak{X}$ is the set of σ-fixed points in $T^*\mathbb{X}$, i.e., $(T^*\mathbb{X})^{\sigma} = T^*\mathfrak{X}$ holds.*

Proof: Let us prove $(\mathbb{G} \times_{\mathbb{P}} \mathfrak{u}_{\mathfrak{P}})^\sigma = G \times_P \mathfrak{u}_{\mathfrak{p}}$. Take $[\boldsymbol{g}, \boldsymbol{u}] \in \mathbb{G} \times_{\mathbb{P}} \mathfrak{u}_{\mathfrak{P}}$. Then $\sigma([\boldsymbol{g}, \boldsymbol{u}]) = [\sigma(\boldsymbol{g}), \sigma(\boldsymbol{u})] = [\boldsymbol{g}, \boldsymbol{u}]$ if and only if there exists $\boldsymbol{p} \in \mathbb{P}$ such that $\sigma(\boldsymbol{g}) = \boldsymbol{g}\boldsymbol{p}$ and $\sigma(\boldsymbol{u}) = \boldsymbol{p}^{-1}\boldsymbol{u}$. So we get $\boldsymbol{p} = \boldsymbol{g}^{-1}\sigma(\boldsymbol{g}) \in \mathbb{P}^{-\sigma}$, and it has a $(-\sigma)$-square root $f \in \mathbb{P}^{-\sigma}$. From this, we know $\sigma(\boldsymbol{g}f) = \boldsymbol{g}f$, which means $\boldsymbol{g}f \in G = \mathbb{G}^\sigma$. Also we see

$$\sigma(f^{-1}\boldsymbol{u}) = f\sigma(\boldsymbol{u}) = f\boldsymbol{p}^{-1}\boldsymbol{u} = ff^{-2}\boldsymbol{u} = f^{-1}\boldsymbol{u},$$

which proves $f^{-1}\boldsymbol{u} \in (\mathfrak{u}_{\mathfrak{P}})^\sigma = \mathfrak{u}_{\mathfrak{p}}$. Thus $[\boldsymbol{g}, \boldsymbol{u}] = [\boldsymbol{g}f, f^{-1}\boldsymbol{u}] \in G \times_P \mathfrak{u}_{\mathfrak{p}}$.

The proof for $(\mathbb{K} \times_{\mathbb{Q}} \mathfrak{u}_{\mathfrak{Q}})^\sigma = K \times_Q \mathfrak{u}_{\mathfrak{q}}$ is the same. □

Let us denote the moment maps by

$$\mu_{\mathbb{X}} : T^*\mathbb{X} \to \mathfrak{K}^* \simeq \mathfrak{K}, \qquad \mu_{\mathfrak{X}} : T^*\mathfrak{X} \to \mathfrak{k}^* \simeq \mathfrak{k}. \tag{13.1}$$

Clearly $\mu_{\mathbb{X}}$ commutes with σ and $\mu_{\mathbb{X}}\big|_{T^*\mathfrak{X}} = \mu_{\mathfrak{X}}$. Let $\mathcal{Y}_{\mathbb{X}} = \mu_{\mathbb{X}}^{-1}(0)$ and $\mathcal{Y}_{\mathfrak{X}} = \mu_{\mathfrak{X}}^{-1}(0)$ be the corresponding conormal varieties.

Theorem 13.2: *Assume the condition* (E) *in Theorem* 12.3.

(1) *The set of fixed points in $\mathcal{Y}_{\mathbb{X}}$ is equal to $\mathcal{Y}_{\mathfrak{X}}$. Thus we get $(\mathcal{Y}_{\mathbb{X}})^\sigma = \mathcal{Y}_{\mathbb{X}} \cap T^*\mathfrak{X} = \mathcal{Y}_{\mathfrak{X}}$.*

(2) *By Theorem 12.3, there is an embedding $\iota : \mathfrak{X}/K \hookrightarrow \mathbb{X}/\mathbb{K}$. If we denote $\mathbb{O} = \iota(\mathcal{O})$ for $\mathcal{O} \in \mathfrak{X}/K$, we have $(T^*_{\mathbb{O}}\mathbb{X})^\sigma = T^*_{\mathcal{O}}\mathfrak{X}$.*

Proof: (1) $(\mathcal{Y}_{\mathbb{X}})^\sigma = (T^*\mathbb{X})^\sigma \cap \mu_{\mathbb{X}}^{-1}(0) = T^*\mathfrak{X} \cap \mu_{\mathbb{X}}^{-1}(0) = T^*\mathfrak{X} \cap \mu_{\mathfrak{X}}^{-1}(0) = \mathcal{Y}_{\mathfrak{X}}$.

(2) We have

$$\begin{aligned}(T^*_{\mathbb{O}}\mathbb{X})^\sigma &= \{([\mathbb{P}_1, u], [\mathbb{Q}_1, v]) \in \mathcal{Y}_{\mathbb{X}}^\sigma : (\mathbb{P}_1, \mathbb{Q}_1) \in \mathbb{O}\} \\ &= \{([P_1, u], [Q_1, v]) \in \mathcal{Y}_{\mathfrak{X}} : (P_1, Q_1) \in \mathbb{O} \cap \mathfrak{X}\} \\ &= T^*_{\mathcal{O}}\mathfrak{X}\end{aligned}$$

where we use part (1) and the fact that $\mathbb{O} \cap \mathfrak{X} = \mathcal{O}$. □

14. Compatibility with Steinberg Maps

In this section we keep the assumption (E) in Theorem 12.3 so that we have an orbit embedding $\iota : \mathfrak{X}/K \hookrightarrow \mathbb{X}/\mathbb{K}$, which induces the embedding of the conormal bundle $T^*_{\mathcal{O}}\mathfrak{X} \hookrightarrow T^*_{\mathbb{O}}\mathbb{X}$.

Recall the nilpotent varieties $\mathcal{N}_{\mathfrak{k}} = \mathcal{N}_{\mathfrak{g}}^\theta$ and $\mathcal{N}_{\mathfrak{s}} = \mathcal{N}_{\mathfrak{g}}^{-\theta}$. Similarly, we get the nilpotent varieties $\mathcal{N}_{\mathfrak{K}}$ and $\mathcal{N}_{\mathfrak{S}}$ for $\mathfrak{G} = \mathrm{Lie}(\mathbb{G})$, where $\mathcal{N}_{\mathfrak{S}} = \mathcal{N}_{\mathfrak{G}}^{-\theta}$.

Since $\mathcal{N}_{\mathfrak{g}}^{\pm\theta}$ are varieties of σ-fixed points in $\mathcal{N}_{\mathfrak{G}}^{\pm\theta}$, we have natural orbit maps

$$\iota^{\pm\theta} : \mathcal{N}_{\mathfrak{g}}^{\pm\theta}/K \to \mathcal{N}_{\mathfrak{G}}^{\pm\theta}/\mathbb{K}, \qquad \iota^{\pm\theta}(\mathcal{O}) = \mathbb{K} \cdot \mathcal{O} \quad (\mathcal{O} \in \mathcal{N}_{\mathfrak{g}}^{\pm\theta}/K).$$

The maps $\iota^{\pm\theta}$ are not injective in general.

The Assumption 10.1 on the image of the Steinberg map clearly holds for $\mathfrak{X}$ whenever it holds for $\mathbb{X}$, which we assume below. Under this assumption, we get the Steinberg maps $\Phi^{\pm\theta} : \mathfrak{X}/K \to \mathcal{N}_{\mathfrak{g}}^{\pm\theta}/K$ relative to $\mathfrak{X}$ and also $\mathbb{\Phi}^{\pm\theta} : \mathbb{X}/\mathbb{K} \to \mathcal{N}_{\mathfrak{G}}^{\pm\theta}/\mathbb{K}$ relative to $\mathbb{X}$ (see Definition 10.2; here we use $\Phi^{\pm\theta}$ instead of $\Phi_{\mathfrak{k}}$ or $\Phi_{\mathfrak{s}}$, since we want to have notation which keep track of the involution). It is natural to expect the following.

Conjecture 14.1: The orbit embedding $\iota : X/K \hookrightarrow \mathbb{X}/\mathbb{K}$ is compatible with the Steinberg maps $\Phi^{\pm\theta}$ and $\mathbb{\Phi}^{\pm\theta}$, i.e., the diagram below is commutative.

$$\begin{array}{ccc} \mathfrak{X}/K & \xrightarrow{\quad \Phi^{\pm\theta} \quad} & \mathcal{N}_{\mathfrak{g}}^{\pm\theta}/K \\ \Big\downarrow{\scriptstyle \iota} & & \Big\downarrow{\scriptstyle \iota^{\pm\theta}} \\ \mathbb{X}/\mathbb{K} & \xrightarrow{\quad \mathbb{\Phi}^{\pm\theta} \quad} & \mathcal{N}_{\mathfrak{G}}^{\pm\theta}/\mathbb{K} \end{array} \tag{14.1}$$

Unfortunately, the formalism of functoriality of taking σ-invariants does not imply Conjecture 14.1. Let $\mathcal{O} \subset \mathfrak{X}$ be a K-orbit corresponding to the $\mathbb{K}$-orbit $\mathbb{O} = \iota(\mathcal{O})$ in $\mathbb{X}$. By the definition of $\mathbb{\Phi}^{\pm\theta}$, there exists a non-empty $\mathbb{K}$- and σ-stable dense open subset $\mathbb{U}^{\pm} \subset T^*_{\mathbb{O}}\mathbb{X}$ which satisfies

$$\mathbb{U}^{\pm} \subset (\phi_{\mathbb{X}}^{\pm\theta})^{-1}\big(\mathbb{\Phi}^{\pm\theta}(\mathbb{O})\big) \cap T^*_{\mathbb{O}}\mathbb{X}, \tag{14.2}$$

where we denote the Steinberg maps by $\phi_{\mathbb{X}}^{\pm\theta} : \mathcal{Y}_{\mathbb{X}} \to \mathcal{N}_{\mathfrak{G}}^{\pm\theta}$. Note that $T^*_{\mathcal{O}}\mathfrak{X} = (T^*_{\mathbb{O}}\mathbb{X})^{\sigma}$ by Theorem 13.2.

Lemma 14.2: *The equality $\mathbb{\Phi}^{\pm\theta}(\iota(\mathcal{O})) = \iota^{\pm\theta}(\Phi^{\pm\theta}(\mathcal{O}))$ holds (so that the diagram* (14.1) *commutes) if and only if $\mathbb{U}^{\pm}$ intersects with $T^*_{\mathcal{O}}\mathfrak{X}$.*

Note that, in Lemma 14.2, the commutativity of the diagram for $+\theta$ follows from the condition that $\mathbb{U}^{+} \cap T^*_{\mathcal{O}}\mathfrak{X} \neq \emptyset$ (and similarly for $-\theta$). If the condition holds for both of $\pm\theta$, the open set $\mathbb{U} := \mathbb{U}^{+} \cap \mathbb{U}^{-}$ intersects with $T^*_{\mathcal{O}}\mathfrak{X}$, since $T^*_{\mathcal{O}}\mathfrak{X}$ is irreducible.

15. Finiteness of Orbits via Embedding Theory

In this section, we provide useful examples of the embedding theory explained above. In fact, it turns out to be a powerful tool for finding out new double flag varieties of finite type. We use this embedding theory in two different ways. One is for the embedding to triple flag varieties, and the other is that into double flag varieties of type AIII. We begin with the embedding into type AIII, and then come back to the triple flag varieties.

15.1. *Embedding of double flag varieties into type AIII*

Let $\mathbb{G} = \mathrm{GL}_n$ be a general linear group, and put $\mathrm{M}_n = \mathrm{M}_n(\mathbb{C})$, which is an ambient vector space of $\mathbb{G}$.[c] Let us consider two commuting involutions $\sigma, \theta \in \mathrm{Aut}\,\mathbb{G}$ defined by

$$\theta(g) = I\,g\,I^{-1}, \qquad \sigma(g) = J\,{}^t g^{-1} J^{-1}, \tag{15.1}$$

where $I, J \in \mathrm{GL}_n$ are certain regular matrices. Since σ and θ are involutions, we must have $I^2 = J^2 = \mathbb{1}_n$ up to nonzero constant multiple. Put

$$\mathbb{K} = \mathbb{G}^\theta = \{g \in \mathrm{GL}_n \mid IgI^{-1} = g\},$$
$$G = \mathbb{G}^\sigma, \qquad K = G^\theta = \mathbb{K}^\sigma$$

as before. Then the commutativity of θ and σ enforces $J\,{}^t I^{-1} = \alpha\, IJ$, i.e., $\sigma(I) = \alpha I$, for some $\alpha \in \mathbb{C}^\times$. We allow $I = \mathbb{1}_n$ (the identity matrix), and in that case we have $\mathbb{K} = \mathbb{G}$ and $K = G$.

Proposition 15.1: *Under the above setting, the condition* (E) *in Theorem 12.3 is satisfied for any choice of σ-stable parabolic subgroups $\mathbb{P} \subset \mathbb{G}$ and $\mathbb{Q} \subset \mathbb{K}$. In particular, we get an embedding of orbits $\mathfrak{X}/K \hookrightarrow \mathbb{X}/\mathbb{K}$ of double flag varieties.*

Proof: Let us prove that $\mathbb{P}$ admits $(-\sigma)$-square roots (Definition 12.1). To prove it, we follow the arguments in [58](Theorem 1). So, we define an anti-involution $\tau : \mathrm{M}_n \to \mathrm{M}_n$ by $\tau(g) = J\,{}^t g J^{-1} = \sigma(g^{-1})$. This is a linear anti-algebra homomorphism.

Take $g \in \mathbb{P}^{-\sigma}$, which means $\sigma(g) = g^{-1}$. Note that $\sigma(g) = g^{-1}$ is equivalent to $\tau(g) = g$.

By an easy linear algebra argument, we get a polynomial $f(T) \in \mathbb{C}[T]$ in a variable T, which satisfies $f(g)^2 = g$. Since g is invertible, $f(g)$ is

[c] Of course $\mathrm{M}_n(\mathbb{C})$ is the Lie algebra of $\mathbb{G}$, but its structure of a finite dimensional associative algebra is more important.

also invertible and hence in $\mathbb{G}$. Since $\tau(f(g)) = f(\tau(g)) = f(g)$, we get $f(g) \in \mathbb{G}^{-\sigma}$. To prove that $f(g) \in \mathbb{P}$, we use a characterization of parabolic subgroups due to Kempf [32](p. 305) and Mumford [52](§2.2). Namely, there exists a one parameter subgroup $\lambda : \mathbb{C}^\times \to \mathbb{G}$ such that

$$\mathbb{P} = \{g \in \mathbb{G} \mid \text{the limit } \lim_{t\to 0} \lambda(t) g \lambda(t)^{-1} =: g_0 \text{ exists}\}. \tag{15.2}$$

Using this, we see $\lambda(t) f(g) \lambda(t)^{-1} = f(\lambda(t) g \lambda(t)^{-1}) \to f(g_0)$ has a limit, and conclude that $f(g) \in \mathbb{P} \cap \mathbb{G}^{-\sigma} = \mathbb{P}^{-\sigma}$. So $f(g)$ is a desired $(-\sigma)$-square root of g.

For the parabolic subgroup $\mathbb{Q}$, the arguments are similar, and we note $f(g) \in \mathbb{K} = \mathbb{G}^\theta$ holds because θ is just a conjugation by I.

We did not use any particular properties of $\mathbb{P}$ and $\mathbb{Q}$, so the same technique also proves that $\mathbb{P}_1 \cap \mathbb{Q}_1$ admits $(-\sigma)$-square roots. □

15.2. *Embedding of double flag varieties into triple flag varieties*

Now let us consider the case of triple flag varieties.

Let G be a connected reductive group and $K = G^\theta$ a symmetric subgroup of G fixed by a certain involution $\theta \in \operatorname{Aut} G$ as above. We will embed a double flag variety $\mathfrak{X} = K/Q \times G/P$ into a triple flag variety $\mathbb{X}$ in K-equivariant way (cf. §4). For that purpose, choose parabolic subgroups P_1, P_2, P_3 of G in the following way.

We take $P_1 := P$ and $P_2 := \theta(P)$. For the parabolic subgroup Q of K, we can choose a θ-stable parabolic subgroup $P_3 \subset G$ which cuts out Q from K, i.e., $P_3 = \theta(P_3)$ and $Q = P_3 \cap K$. Thus our triple flag variety is $\mathbb{X} = G/P_1 \times G/P_2 \times G/P_3$ as in §4.

We consider the following setting:

$$\begin{aligned}
&\mathbb{G} = G \times G, \quad \mathbb{P} = P \times \theta(P) \subset \mathbb{G}\\
&\quad \sigma, \tau \in \operatorname{Aut}\mathbb{G} : \text{involutions}, \quad \sigma(g_1, g_2) := (\theta(g_2), \theta(g_1))\\
&\qquad\qquad\qquad\qquad\qquad\qquad\quad \tau(g_1, g_2) := (g_2, g_1)\\
&\mathbb{K} = \mathbb{G}^\tau = \{(g, g) \mid g \in G\} \simeq G, \qquad \mathbb{Q} = \{(g, g) \mid g \in P_3\} \simeq P_3\\
&G = \mathbb{G}^\sigma = \{(g, \theta(g)) \mid g \in G\} \simeq G, \quad K = \mathbb{G}^{\sigma,\tau} = \{(g, g) \mid g = \theta(g)\} \simeq G^\theta.
\end{aligned}$$

Here we use the involution τ instead of θ, which is used in the former sections, because we want to keep the setting $K = G^\theta$.

Now notice

$$\mathbb{X} = \mathbb{K}/\mathbb{Q} \times \mathbb{G}/\mathbb{P} \simeq G/P_3 \times \big(G/P \times G/\theta(P)\big) \tag{15.3}$$

is the triple flag variety.

Theorem 15.2: *Let $G = \mathrm{GL}_n$ and consider $\theta(g) = J\,{}^t g^{-1} J^{-1}$ for J which satisfies $J^2 = \mathbb{1}_n$. Apply the setting explained above in this subsection. Then we get an orbit embedding $\mathfrak{X} = K/Q \times G/P \hookrightarrow \mathbb{K}/\mathbb{Q} \times \mathbb{G}/\mathbb{P} = \mathbb{X}$, where $\mathbb{X}$ is the triple flag variety given in* (15.3)*. In particular, if the triple flag variety $\mathbb{X}$ is of finite type, the double flag variety $\mathfrak{X} = K/Q \times G/P$ is of finite type.*

Proof: We use Theorem 12.3. What we must prove is the condition (E). Let us see

$$\mathbb{P}^{-\sigma} = \{(g, \theta(g^{-1})) \mid g \in P\} \simeq P,$$
$$\mathbb{Q}^{-\sigma} = \{(g, g) \mid g \in P_3,\ g^{-1} = \theta(g)\} \simeq P_3^{-\theta}.$$

Since $\theta(g) = J\,{}^t g^{-1} J^{-1}$, the same argument of the proof of Proposition 15.1 can be applied. It is literally the same but after exchanging θ here and σ there. □

Remark 15.3: If we are interested in just the finiteness of orbits in a double flag variety, there is a simpler comparison with triple flag varieties. See Theorem 3.4 in [55]. The above Theorem 15.2 is a refinement of Theorem 3.1 in [55] for the above mentioned involution θ.

In the following, we list the tables of double flag varieties of finite type which are obtained by embedding into triple flag varieties of finite type as well as double flag varieties of type AIII. The readers are warned that the lists are far from complete, and they overlap the already given lists in §6. Note also that, in these lists, we omit the trivial cases for which $G = P$ or $Q = K$.

Since type AII is logically simpler than type AI, let us begin with type AII. We will discuss on type AI in the second subsection (§15.4).

15.3. *Type AII*

We use Theorem 15.2 above. Let us consider GL_{2n} instead of GL_n and the involution $\theta(g) = J\,{}^t g^{-1} J^{-1}$ defined as follows:

$$J = \begin{pmatrix} 0 & -\Sigma_n \\ \Sigma_n & 0 \end{pmatrix}, \quad \Sigma_n = \begin{pmatrix} & & 1 \\ & \cdot^{\cdot^{\cdot}} & \\ 1 & & \end{pmatrix}.$$

Then our symmetric pair becomes $(\mathrm{GL}_{2n}, \mathrm{Sp}_{2n})$. We compare the double flag variety $\mathfrak{X}$ with the triple flag variety for type A, the table of finite type in §6.5.1 gives the following table of finite type double flag variety.

To indicate parabolic subgroups, we use the subset of simple roots $J \subset \Pi$ as in §5, and the numbering of simple roots is indicated below.

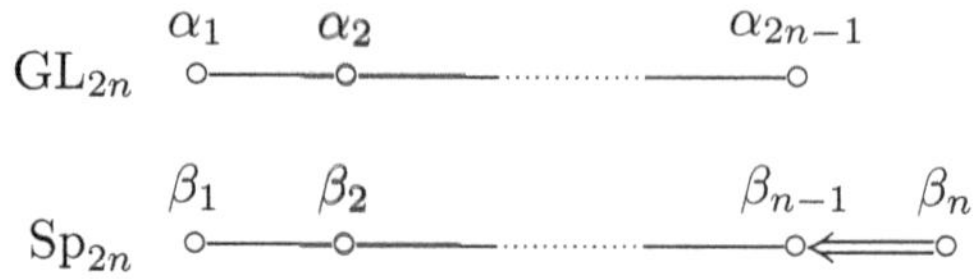

(G, K)	$\Pi_G \setminus J_G$ $(P = P_{J_G})$	$\Pi_K \setminus J_K$ $(Q = Q_{J_K})$
$(\mathrm{GL}_{2n}, \mathrm{Sp}_{2n})$	$\{\alpha_i\}$	any subset
	$\{\alpha_i, \alpha_j\}$	$\{\beta_n\}$

15.4. *Type AI*

Again we use Theorem 15.2 above. Let us consider the involution $\theta(g) = {}^t g^{-1}$ so that $J = \mathbb{1}_n$ in this case. Then our symmetric pair becomes $(\mathrm{SL}_n, \mathrm{SO}_n)$ (we prefer SL_n rather than GL_n). We compare $\mathfrak{X}$ with the triple flag variety of type A as in the former subsection and get the following table of finite type double flag varieties.

The numbering of simple roots is indicated below.

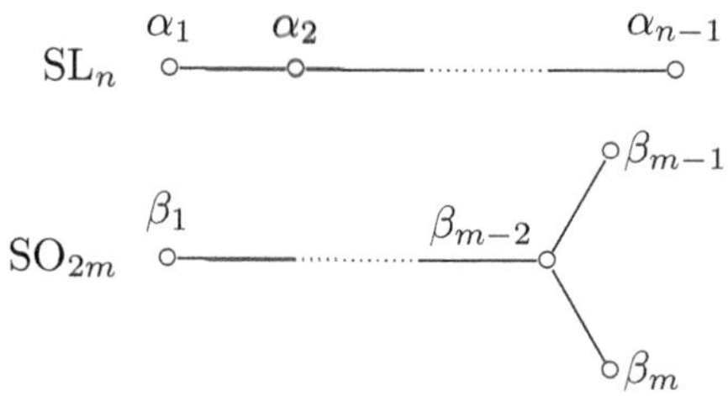

(G, K)	$\Pi_G \setminus J_G$ $(P = P_{J_G})$	$\Pi_K \setminus J_K$ $(Q = Q_{J_K})$	
$(\mathrm{SL}_n, \mathrm{SO}_n)$	$\{\alpha_i\}$	any subset	
	$\{\alpha_i, \alpha_j\}$	$\{\beta_{m-1}\}, \{\beta_m\}$	if $n = 2m$

15.5. *Type CI*

In the setting of §15.1, we take $2n$ instead of n, and choose

$$J = \begin{pmatrix} 0 & -\Sigma_n \\ \Sigma_n & 0 \end{pmatrix}, \quad \Sigma_n = \begin{pmatrix} & & 1 \\ & \cdot^{\cdot^{\cdot}} & \\ 1 & & \end{pmatrix}, \quad \text{and} \quad I = \begin{pmatrix} \mathbb{1}_n & \\ & -\mathbb{1}_n \end{pmatrix}.$$

Thus we get an embedding

$$\begin{aligned}\mathfrak{X} &= K/Q \times G/P = \mathrm{GL}_n/Q \times \mathrm{Sp}_{2n}/P \hookrightarrow \\ &(\mathbb{K}_1/\mathbb{Q}_1 \times \mathbb{K}_2/\mathbb{Q}_2) \times \mathbb{G}/\mathbb{P} = (\mathrm{GL}_n/\mathbb{Q}_1 \times \mathrm{GL}_n/\mathbb{Q}_2) \times \mathrm{GL}_{2n}/\mathbb{P} = \mathbb{X}.\end{aligned}$$

With this embedding, a σ-stable parabolic subgroup $\mathbb{P}$ of $\mathbb{G} = \mathrm{GL}_{2n}$ corresponds to a composition $(c_1, c_2, \ldots, c_\ell)$ of $2n$ with $c_i = c_{\ell-i+1}$, i.e., central symmetric. Moreover, since $\mathbb{Q}_1 \times \mathbb{Q}_2$ is σ-stable, the parabolic subgroups $\mathbb{Q}_1 \subset \mathbb{K}_1 \cong \mathrm{GL}_n$ and $\mathbb{Q}_2 \subset \mathbb{K}_2 \cong \mathrm{GL}_n$ are conjugated by σ, and hence correspond to the same composition of n; then $Q = (\mathbb{Q}_1 \times \mathbb{Q}_2)^\sigma \subset K \cong \mathrm{GL}_n$ also corresponds to this common composition of n. Taking this into account, Table 9 of the finite-type double flag varieties of type AIII tells us the following type CI double flag varieties of finite type. Each label of the following item indicates the labels in Table 9.

- (P2) $P = P_{(n,n)}$, and Q is any parabolic subgroup. We call the parabolic subgroup $P = P_{(n,n)}$ a Siegel parabolic subgroup, which stabilizes a Lagrangian subspace.
- (P3-1) $P = P_{(1,2n-2,1)}$, and any Q. We call the parabolic subgroup $P = P_{(1,2n-2,1)}$ a mirabolic parabolic subgroup of Sp_{2n}, which stabilizes a line.
- (P3-2) $P = P_{(2,2n-4,2)}$ or $P_{(n-1,2,n-1)}$, and Q is a maximal parabolic subgroup.
- (P3-3) P is a maximal parabolic subgroup, which stabilizes an isotropic subspace, and $Q = Q_{(1,n-1)}$ or $Q_{(n-1,1)}$ is a mirabolic parabolic subgroup, which stabilizes a line (or a hyperplane).
- (P3-4) P is a maximal parabolic subgroup, and Q is a maximal parabolic subgroup.

  - In the rest of the cases (P4-1)–(P4-3), (P6), (PA-1)–(PA-3), the only possibility for having $Q = (\mathbb{Q}_1 \times \mathbb{Q}_2) \cap K$ is that $Q = K = \mathrm{GL}_n$, but we disregard this (trivial) situation in this section.

From this, we get the following type CI double flag varieties of finite type. To indicate parabolic subgroups, we use the subset of simple roots $J \subset \Pi$, and the numbering of simple roots is indicated below.

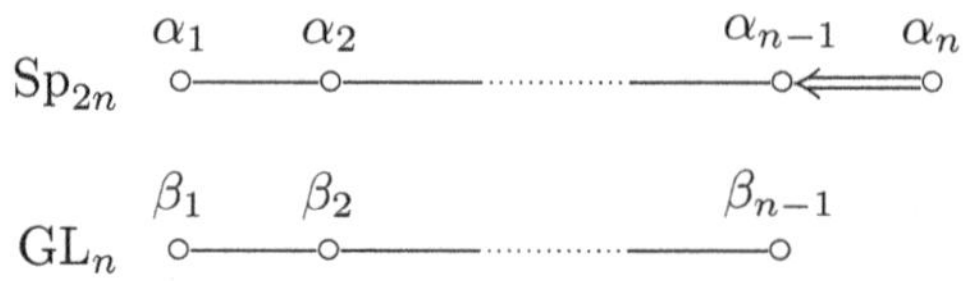

(G, K)	$\Pi_G \setminus J_G$ $(P = P_{J_G})$	$\Pi_K \setminus J_K$ $(Q = Q_{J_K})$
$(\mathrm{Sp}_{2n}, \mathrm{GL}_n)$	$\{\alpha_1\}, \{\alpha_n\}$	any subset
	$\{\alpha_i\}$	$\{\beta_j\}$

Note that, in the first case, Q can be a Borel subgroup B_K of K, and it already appeared in §6.3.

15.6. *Type CII*

In the setting of §15.1, we take $2n$ instead of n, $2p$ (respectively $2q$) instead of p (respectively q), where $n = p + q$. Choose

$$J = \begin{pmatrix} 0 & -\Sigma_n \\ \Sigma_n & 0 \end{pmatrix}, \quad \Sigma_n = \begin{pmatrix} & & 1 \\ & \iddots & \\ 1 & & \end{pmatrix},$$

$$I = \begin{pmatrix} -I_{p,q} & \\ & I_{q,p} \end{pmatrix}, \quad I_{p,q} = \begin{pmatrix} \mathbb{1}_p & \\ & -\mathbb{1}_q \end{pmatrix}.$$

Thus our embedding becomes

$$\begin{aligned} \mathfrak{X} &= (K_1/Q_1 \times K_2/Q_2) \times G/P = (\mathrm{Sp}_{2p}/Q_1 \times \mathrm{Sp}_{2q}/Q_2) \times \mathrm{Sp}_{2n}/P \\ &\hookrightarrow (\mathbb{K}_1/\mathbb{Q}_1 \times \mathbb{K}_2/\mathbb{Q}_2) \times \mathbb{G}/\mathbb{P} \\ &\qquad = (\mathrm{GL}_{2p}/\mathbb{Q}_1 \times \mathrm{GL}_{2q}/\mathbb{Q}_2) \times \mathrm{GL}_{2n}/\mathbb{P} = \mathbb{X}. \end{aligned}$$

We do not repeat the consideration for each cases of finite type double flag varieties of type AIII, but only list the table of type CII double flag varieties of finite type.

The numbering of simple roots for $G = \mathrm{Sp}_{2n}$ is the same as type CI case, and those for $K_1 = \mathrm{Sp}_{2p}$ and $K_2 = \mathrm{Sp}_{2q}$ are also the same except for that we use $\{\beta_1, \ldots, \beta_p\}$ for K_1 and $\{\gamma_1, \ldots, \gamma_q\}$ for K_2 instead of α_i's.

Note that the roles of K_1 and K_2 are interchangeable. In this table, there are obtained 4 new cases other than those which have been already listed in the case of $P = B_G$ or $Q = B_K$ (see §§6.2 and 6.3).

(G,K)	$\Pi_G \setminus J_G$ $(P = P_{J_G})$	$\Pi_{K_1} \setminus J_{K_1}$ $(Q_1 = Q_{J_{K_1}})$	$\Pi_{K_2} \setminus J_{K_2}$ $(Q_2 = Q_{J_{K_2}})$
$(\mathrm{Sp}_{2n}, \mathrm{Sp}_{2p}\times\mathrm{Sp}_{2q})$	$\{\alpha_1\}, \{\alpha_n\}$	any subset	any subset
	$\{\alpha_2\}, \{\alpha_{n-1}\}$	any subset	$\{\gamma_q\}$
	$\{\alpha_i\}$	any subset	$\emptyset$ $(Q_2 = K_2)$ $\{\gamma_1\}$ if $q = 1$
	$\{\alpha_i\}$	$\{\beta_j\}, \{\beta_j, \beta_p\}$	$\{\gamma_q\}$
	$\{\alpha_1, \alpha_n\}$	any subset	$\emptyset$ $(Q_2 = K_2)$
	$\{\alpha_i, \alpha_n\}$	$\{\beta_j\}, \{\beta_j, \beta_p\}$	$\emptyset$ $(Q_2 = K_2)$
	$\{\alpha_i, \alpha_n\}$	any subset	$\emptyset$ if $q = 1$
	$\{\alpha_i, \alpha_j\}$, $\{\alpha_i, \alpha_j, \alpha_n\}$	$\{\beta_p\}$	$\emptyset$ $(Q_2 = K_2)$
	any subset	$\{\beta_p\}$	$Q_2 = K_2 = \mathrm{Sp}_2$ $(q = 1)$
	any subset	$\{\beta_1\}$ if $p = 1$	$\emptyset$ $(Q_2 = K_2)$

15.7. *Type BI (BDI)*

In the setting of §15.1, we take $2n+1$ instead of n, $2p+1$ (respectively $2q$) instead of p (respectively q), where $n = p + q$. Choose

$$J = \begin{pmatrix} \Sigma_{2p+1} & 0 \\ 0 & \Sigma_{2q} \end{pmatrix}, \quad \Sigma_m = \begin{pmatrix} & & 1 \\ & \cdot^{\cdot^{\cdot}} & \\ 1 & & \end{pmatrix}, \quad I = \begin{pmatrix} \mathbb{1}_{2p+1} & \\ & -\mathbb{1}_{2q} \end{pmatrix}.$$

Thus our embedding becomes

$$\begin{aligned} \mathfrak{X} = (K_1/Q_1 \times K_2/Q_2) \times G/P &= (\mathrm{SO}_{2p+1}/Q_1 \times \mathrm{SO}_{2q}/Q_2) \times \mathrm{SO}_{2n+1}/P \\ \hookrightarrow (\mathbb{K}_1/\mathbb{Q}_1 \times \mathbb{K}_2/\mathbb{Q}_2) \times \mathbb{G}/\mathbb{P} & \\ &= (\mathrm{GL}_{2p+1}/\mathbb{Q}_1 \times \mathrm{GL}_{2q}/\mathbb{Q}_2) \times \mathrm{GL}_{2n+1}/\mathbb{P} = \mathbb{X}. \end{aligned}$$

Below, we list the table of type BI (BDI) double flag varieties of finite type.

Here is the numbering of simple roots for $G = \mathrm{SO}_{2n+1}$ and $K_2 = \mathrm{SO}_{2q}$. The numbering for K_1 is similar as G and the difference is that we use $\{\beta_1, \ldots, \beta_p\}$ instead of α_i's.

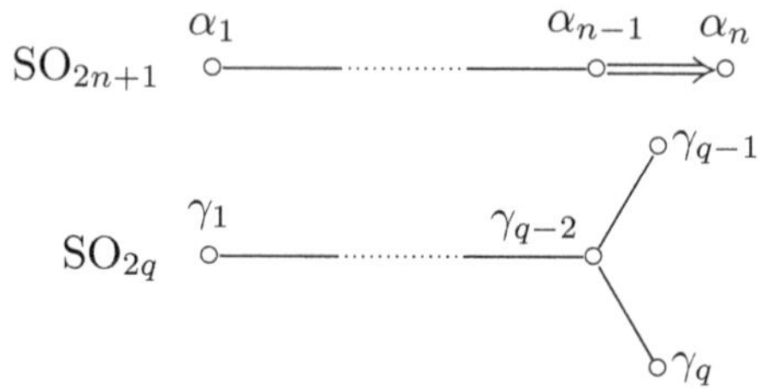

(G, K)	$\Pi_G \setminus J_G$ $(P = P_{J_G})$	$\Pi_{K_1} \setminus J_{K_1}$ $(Q_1 = Q_{J_{K_1}})$	$\Pi_{K_2} \setminus J_{K_2}$ $(Q_2 = Q_{J_{K_2}})$
$(\mathrm{SO}_{2n+1}, \mathrm{SO}_{2p+1} \times \mathrm{SO}_{2q})$	$\{\alpha_1\}, \{\alpha_n\}$	any subset	any subset
	$\{\alpha_2\}$	any subset	$\{\gamma_{q-1}\}, \{\gamma_q\}$
	$\{\alpha_i\}$	any subset	$\emptyset$ $(Q_2 = K_2)$ $\{\gamma_1\}$ if $q = 1$
	$\{\alpha_i\}$	$\emptyset$ $(Q_1 = K_1)$	any subset
	$\{\alpha_i\}$	$\{\beta_j\}$	$\{\gamma_{q-1}\}, \{\gamma_q\}$
	$\{\alpha_i, \alpha_j\}$	$\emptyset$ $(Q_1 = K_1)$	$\{\gamma_{q-1}\}, \{\gamma_q\}$
	any subset	$\emptyset$ $(Q_1 = K_1)$	$\{\gamma_1\}$ if $q = 1$
	any subset	$Q_1 = \mathrm{SO}_1$ if $p = 0$	any subset

15.8. *Type DI (BDI), even case*

In the setting of §15.1, we take $2n$ instead of n, $2p$ (respectively $2q$) instead of p (respectively q), where $n = p + q$. Choose

$$J = \begin{pmatrix} \Sigma_{2p} & 0 \\ 0 & \Sigma_{2q} \end{pmatrix}, \quad \Sigma_m = \begin{pmatrix} & & 1 \\ & \cdot^{\cdot^{\cdot}} & \\ 1 & & \end{pmatrix}, \quad I = \begin{pmatrix} \mathbb{1}_{2p} & \\ & -\mathbb{1}_{2q} \end{pmatrix}.$$

Thus our embedding becomes

$$\begin{aligned} \mathfrak{X} &= (K_1/Q_1 \times K_2/Q_2) \times G/P = (\mathrm{SO}_{2p}/Q_1 \times \mathrm{SO}_{2q}/Q_2) \times \mathrm{SO}_{2n}/P \\ &\hookrightarrow (\mathbb{K}_1/\mathbb{Q}_1 \times \mathbb{K}_2/\mathbb{Q}_2) \times \mathbb{G}/\mathbb{P} \\ &\qquad = (\mathrm{GL}_{2p}/\mathbb{Q}_1 \times \mathrm{GL}_{2q}/\mathbb{Q}_2) \times \mathrm{GL}_{2n}/\mathbb{P} = \mathbb{X}. \end{aligned}$$

Below, we list the table of type DI (BDI) double flag varieties of finite type (even case).

Here is the numbering of simple roots for $G = \mathrm{SO}_{2n}$. The numbering for K_1 and K_2 is similar as G and the difference is that we use $\{\beta_1, \dots, \beta_p\}$ and $\{\gamma_1, \dots, \gamma_q\}$ instead of α_i's.

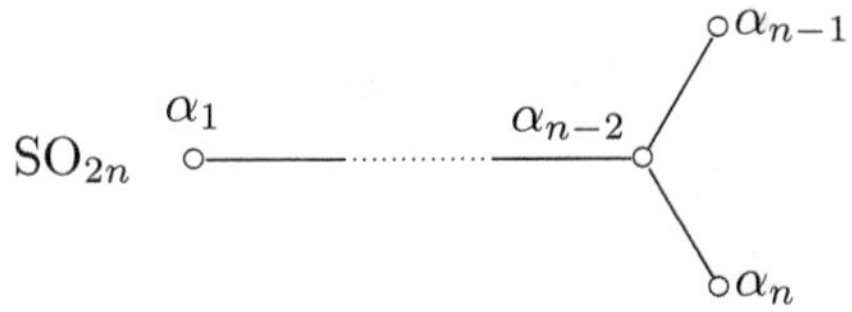

(G,K)	$\Pi_G \setminus J_G$ $(P = P_{J_G})$	$\Pi_{K_1} \setminus J_{K_1}$ $(Q_1 = Q_{J_{K_1}})$	$\Pi_{K_2} \setminus J_{K_2}$ $(Q_2 = Q_{J_{K_2}})$
$(\mathrm{SO}_{2n},$	$\{\alpha_1\}, \{\alpha_{n-1}\}, \{\alpha_n\}$	any subset	any subset
$\mathrm{SO}_{2p}\times\mathrm{SO}_{2q})$	$\{\alpha_2\}, \{\alpha_{n-1}, \alpha_n\}$	any subset	$\{\gamma_{q-1}\}, \{\gamma_q\}$
	$\{\alpha_i\}$	any subset	$\{\gamma_1\}$ if $q = 1$
	$\{\alpha_i\}, \{\alpha_1, \alpha_{n-1}\},$ $\{\alpha_1, \alpha_n\}, \{\alpha_{n-1}, \alpha_n\}$	any subset	$\emptyset$ $(Q_2{=}K_2)$
	$\{\alpha_i, \alpha_{n-1}\}, \{\alpha_i, \alpha_n\}$	any subset	$\emptyset$ $(Q_2{=}K_2)$ if $q{=}1$
	$\{\alpha_i\}$	$\{\beta_j\},$ $\{\beta_j, \beta_{p-1}\}, \{\beta_j, \beta_p\}$	$\{\gamma_{q-1}\}, \{\gamma_q\}$
	$\{\alpha_i, \alpha_{n-1}\}, \{\alpha_i, \alpha_n\}$	$\{\beta_j\},$ $\{\beta_j, \beta_{p-1}\}, \{\beta_j, \beta_p\}$	$\emptyset$ $(Q_2{=}K_2)$
	$\{\alpha_i, \alpha_j\}, \{\alpha_i, \alpha_j, \alpha_{n-1}\},$ $\{\alpha_i, \alpha_j, \alpha_n\}, \{\alpha_i, \alpha_{n-1}, \alpha_n\}$	$\{\beta_{p-1}\}, \{\beta_p\}$	$\emptyset$ $(Q_2{=}K_2)$
	any subset	$\emptyset$ $(Q_1{=}K_1)$	$\{\gamma_1\}$ if $q{=}1$
	any subset	$\{\beta_{p-1}\}, \{\beta_p\}$	$\emptyset$ $(Q_2{=}K_2)$ if $q{=}1$

15.9. *Type DI (BDI), odd case*

In the setting of §15.1, we take $2n$ instead of n, $2p+1$ (respectively $2q+1$) instead of p (respectively q), where $n = p + q + 1$. Choose

$$J = \begin{pmatrix} \Sigma_{2p+1} & 0 \\ 0 & \Sigma_{2q+1} \end{pmatrix}, \quad \Sigma_m = \begin{pmatrix} & & 1 \\ & \cdot^{\cdot^{\cdot}} & \\ 1 & & \end{pmatrix}, \quad I = \begin{pmatrix} \mathbb{1}_{2p+1} & \\ & -\mathbb{1}_{2q+1} \end{pmatrix}.$$

Thus our embedding becomes

$$\mathfrak{X} = (K_1/Q_1 \times K_2/Q_2) \times G/P = (\mathrm{SO}_{2p+1}/Q_1 \times \mathrm{SO}_{2q+1}/Q_2) \times \mathrm{SO}_{2n}/P \hookrightarrow (\mathbb{K}_1/\mathbb{Q}_1 \times \mathbb{K}_2/\mathbb{Q}_2) \times \mathbb{G}/\mathbb{P} = (\mathrm{GL}_{2p+1}/\mathbb{Q}_1 \times \mathrm{GL}_{2q+1}/\mathbb{Q}_2) \times \mathrm{GL}_{2n}/\mathbb{P} = \mathbb{X}.$$

Below, we list the table of type DI (BDI) double flag varieties of finite type (odd case).

The numbering of simple roots for $G = \mathrm{SO}_{2n}$ and $K_1 = \mathrm{SO}_{2p+1}$ are the same as that of type BI and DI (even case) except for the labeling Greek letters.

(G, K)	$\Pi_G \setminus J_G$ $(P = P_{J_G})$	$\Pi_{K_1} \setminus J_{K_1}$ $(Q_1 = Q_{J_{K_1}})$	$\Pi_{K_2} \setminus J_{K_2}$ $(Q_2 = Q_{J_{K_2}})$
$(\mathrm{SO}_{2n}, \mathrm{SO}_{2p+1} \times \mathrm{SO}_{2q+1})$	$\{\alpha_1\}, \{\alpha_{n-1}\}, \{\alpha_n\}$	any subset	any subset
	$\{\alpha_i\}, \{\alpha_{n-1}, \alpha_n\}$ $\{\alpha_1, \alpha_{n-1}\}, \{\alpha_1, \alpha_n\}$	any subset	$\emptyset$ $(Q_2 = K_2)$
	$\{\alpha_i, \alpha_{n-1}\}, \{\alpha_i, \alpha_n\}$	$\{\beta_j\}$	$\emptyset$ $(Q_2 = K_2)$
	any subset	any subset	$K_2 = SO_1$ if $q = 0$

15.10. *Type DIII*

In the setting of §15.1, we take $2n$ instead of n, n instead of p (respectively q). Choose

$$J = \Sigma_{2n} = \begin{pmatrix} & & 1 \\ & \cdot^{\cdot^{\cdot}} & \\ 1 & & \end{pmatrix}, \quad I = \begin{pmatrix} \mathbb{1}_n & \\ & -\mathbb{1}_n \end{pmatrix}.$$

Thus our embedding becomes

$$\begin{aligned} \mathfrak{X} &= K/Q \times G/P = \mathrm{GL}_n/Q \times \mathrm{SO}_{2n}/P \hookrightarrow \\ &(\mathbb{K}_1/\mathbb{Q}_1 \times \mathbb{K}_2/\mathbb{Q}_2) \times \mathbb{G}/\mathbb{P} = (\mathrm{GL}_n/\mathbb{Q}_1 \times \mathrm{GL}_n/\mathbb{Q}_2) \times \mathrm{GL}_{2n}/\mathbb{P} = \mathbb{X}. \end{aligned}$$

Below, we list the table of type DIII double flag varieties of finite type.

The numbering of simple roots for $G = \mathrm{SO}_{2n}$ and $K = \mathrm{GL}_n$ are the same as that of type DI and type CI respectively.

(G, K)	$\Pi_G \setminus J_G$ $(P = P_{J_G})$	$\Pi_K \setminus J_K$ $(Q = Q_{J_K})$
$(\mathrm{SO}_{2n}, \mathrm{GL}_n)$	$\{\alpha_1\}, \{\alpha_{n-1}\}, \{\alpha_n\}$	any subset
	$\{\alpha_i\}$	$\{\beta_j\}$

Here in this table, the embedding theory does not give the cases where P is a Borel subgroup (hence can be any parabolic subgroup) and $\Pi_K \setminus J_K = \{\beta_1\}$ or $\{\beta_{n-1}\}$ (mirabolic case), which is of finite type and listed in §6.2.

16. An Example of Embedding Theory for Type CI into Type AIII

In this section, we give an example of the embedding of the case of type CI into type AIII based on [14]. As in §15.5, let us begin with $\mathbb{G} = \mathrm{GL}_{2n}$ and commuting involutions $\sigma, \theta \in \mathrm{Aut}\,\mathbb{G}$ defined by

$$\theta(g) = I^{-1}\, g\, I \qquad\qquad I = I_{n,n} = \mathrm{diag}(\mathbb{1}_n, -\mathbb{1}_n),$$

$$\sigma(g) = J^{-1}\, {}^t g^{-1} J \qquad\qquad J = J_n = \begin{pmatrix} & -\mathbb{1}_n \\ \mathbb{1}_n & \end{pmatrix}.$$

Note that we use a convention for J which is different from the one in §12. We define a symplectic form on $V = \mathbb{C}^{2n}$ by $(u, v) = {}^t u J v$. Note that $V^+ = \langle \boldsymbol{e}_1, \boldsymbol{e}_2, \ldots, \boldsymbol{e}_n \rangle \subset V$ is a Lagrangian subspace with respect to this symplectic form, where $\boldsymbol{e}_k$ denotes the k-th elementary basis vector as usual. Then

$$\mathbb{K} = \mathbb{G}^\theta = \mathrm{GL}_n \times \mathrm{GL}_n, \quad G = \mathbb{G}^\sigma = \mathrm{Sp}_{2n}, \quad K = G^\theta \simeq \mathrm{GL}_n \tag{16.1}$$

are all connected, and $(G, K) = (\mathrm{Sp}_{2n}, \mathrm{GL}_n)$ is a symmetric pair of type CI and $(\mathbb{G}, \mathbb{K}) = (\mathrm{GL}_{2n}, \mathrm{GL}_n \times \mathrm{GL}_n)$ is that of type AIII.

Let us denote by $B_n^\pm$ the Borel subgroup of GL_n consisting of upper/lower triangular matrices. Define

$$\mathbb{B}_\mathbb{K} = B_n^+ \times B_n^- \subset \mathbb{K} = \mathrm{GL}_n \times \mathrm{GL}_n, \qquad \mathbb{P} = \mathrm{Stab}_\mathbb{G}(V^+) \subset \mathbb{G} = \mathrm{GL}_{2n}$$

both of which are σ-stable. Then $Q = B_K = \mathbb{B}_\mathbb{K}^\sigma$ is a Borel subgroup in K and $P = P_S = \mathbb{P}^\sigma$ is the Siegel parabolic subgroup of $G = \mathrm{Sp}_{2n}$ stabilizing a Lagrangian subspace V^+. Thus our double flag varieties are

$$\mathfrak{X} = K/B_K \times G/P = \mathrm{GL}_n/B_n^+ \times \mathrm{Sp}_{2n}/P_S \simeq \mathscr{F}\!\ell_n \times \mathrm{LGr}(\mathbb{C}^{2n}) \tag{16.2}$$

and

$$\begin{aligned} \mathbb{X} = \mathbb{K}/\mathbb{B}_\mathbb{K} \times \mathbb{G}/\mathbb{P} &= \big(\mathrm{GL}_n/B_n^+ \times \mathrm{GL}_n/B_n^-\big) \times \mathrm{GL}_{2n}/P_{(n,n)} \\ &\simeq \big(\mathscr{F}\!\ell_n \times \mathscr{F}\!\ell_n\big) \times \mathrm{Gr}_n(\mathbb{C}^{2n}), \end{aligned} \tag{16.3}$$

where $\mathrm{LGr}(V)$ denotes the Lagrangian Grassmannian, the variety of all the Lagrangian subspaces in $V = \mathbb{C}^{2n}$, and $\mathscr{F}\!\ell_n$ is the variety of complete flags in $\mathbb{C}^n$. We use the notation $P_{(n,n)}$ for the parabolic subgroup in GL_{2n} determined by the partition (n, n) of $2n$.

Proposition 15.1 enables orbit embedding in this case.

Theorem 16.1: *Let $\mathfrak{X}$ and $\mathbb{X}$ be the double flag varieties defined in* (16.2) *and* (16.3) *respectively. Then the orbit map $\mathfrak{X}/K \to \mathbb{X}/\mathbb{K}$ is injective, i.e.,* *for any $\mathbb{K}$-orbit $\mathbb{O}$ in $\mathbb{X}$, the intersection $X \cap \mathbb{O}$ is either empty or a single K-orbit.*

Let us describe the explicit embedding of orbits in this case. First, we clarify the explicit embedding map $\mathfrak{X} \to \mathbb{X}$ and the involutive automorphism on $\mathbb{X}$. Note the following isomorphism from §§9.2 and 11.2.3.

$$\begin{aligned} \mathbb{X}/\mathbb{K} &\simeq B_n^+ \times B_n^- \backslash \mathrm{GL}_{2n}/P_{(n,n)} \simeq B_n^+ \times B_n^- \backslash \mathrm{Gr}_n(\mathbb{C}^{2n}) \\ &\simeq B_n^+ \times B_n^- \backslash \mathrm{M}_{2n,n}^\circ/\mathrm{GL}_n \simeq \mathfrak{T}_n^\circ/S_n, \end{aligned} \tag{16.4}$$

where $\mathrm{M}_{2n,n}^\circ$ denotes the set of $2n$ by n matrices of full rank (i.e., rank n), and $\mathfrak{T}^\circ = \mathfrak{T}_n^\circ$ is the set of pairs of partial permutation matrices which are full rank (see §9.2). Note that we put the subscript n.

Let us explain the above isomorphism more precisely. Take $\omega = \begin{pmatrix} \tau_1 \\ \tau_2 \end{pmatrix} \in \mathfrak{T}_n^\circ$. Then we can find matrices $\xi_1, \xi_2 \in \mathrm{M}_n$ such that

$$\boldsymbol{g} = \begin{pmatrix} \tau_1 & \xi_1 \\ \tau_2 & -\xi_2 \end{pmatrix} \in \mathrm{GL}_{2n} = \mathbb{G}$$

is a regular matrix. Note that the coset $\boldsymbol{g}\mathbb{P}$ does not depend on the choice of ξ_1, ξ_2 but only depending on ω. The $\mathbb{K}$-orbit $\mathbb{O}$ which goes through $(\mathbb{B}_\mathbb{K}, \boldsymbol{g}\mathbb{P}) \in \mathbb{X}$ is denoted by $\mathbb{O}_\omega$.

There is a great possibility for the choice of ξ_1, ξ_2. In fact, we can choose ξ_1, ξ_2 in such a way that $\begin{pmatrix} \xi_1 \\ \xi_2 \end{pmatrix} \in \mathfrak{T}_n^\circ$ is also a pair of partial permutations of full rank and it satisfies

$${}^t\xi_1\tau_1 - {}^t\xi_2\tau_2 = 0. \tag{16.5}$$

For this, see [14](Lemma 4.4). In the following, we always assume this.

We define an involution on $\mathfrak{T}_n^\circ/S_n \simeq \mathbb{X}/\mathbb{K}$ by $\sigma(\mathbb{O}_\omega) = \mathbb{O}_{\sigma(\omega)}$ $(\omega \in \mathfrak{T}_n^\circ)$, which is denoted by the same letter σ by abuse of notation. Note that $\sigma(\omega)$ is determined modulo the right multiplication by S_n.

Proposition 16.2: *For $\omega = \begin{pmatrix} \tau_1 \\ \tau_2 \end{pmatrix} \in \mathfrak{T}_n^\circ$ and $\boldsymbol{g} = \begin{pmatrix} \tau_1 & \xi_1 \\ \tau_2 & -\xi_2 \end{pmatrix}$ chosen above, let $\mathbb{O}_\omega$ be the $\mathbb{K}$-orbit in $\mathbb{X}$ through the point $(\mathbb{B}_\mathbb{K}, \boldsymbol{g}\mathbb{P})$. Then $\sigma(\mathbb{O}_\omega) = \mathbb{O}_{\sigma(\omega)}$ is given by*

$$\sigma(\omega) = \sigma\left(\begin{pmatrix} \tau_1 \\ \tau_2 \end{pmatrix}\right) = \begin{pmatrix} \xi_2 \\ \xi_1 \end{pmatrix} \pmod{S_n}.$$

Proof: We have $\sigma((\mathbb{B}_{\mathbb{K}}, \boldsymbol{g}\mathbb{P})) = (\mathbb{B}_{\mathbb{K}}, \sigma(\boldsymbol{g})\mathbb{P})$ and $\sigma(\boldsymbol{g}) = J_n^{-1}\,{}^t\boldsymbol{g}^{-1}J_n$. Let us compute ${}^t\boldsymbol{g}^{-1}$. We get

$$\begin{aligned}{}^t\boldsymbol{g}\boldsymbol{g} &= \begin{pmatrix} {}^t\tau_1 & {}^t\tau_2 \\ {}^t\xi_1 & -{}^t\xi_2 \end{pmatrix}\begin{pmatrix} \tau_1 & \xi_1 \\ \tau_2 & -\xi_2 \end{pmatrix} \\ &= \begin{pmatrix} {}^t\tau_1\tau_1 + {}^t\tau_2\tau_2 & {}^t\tau_1\xi_1 - {}^t\tau_2\xi_2 \\ {}^t\xi_1\tau_1 - {}^t\xi_2\tau_2 & {}^t\xi_1\xi_1 + {}^t\xi_2\xi_2 \end{pmatrix} =: \begin{pmatrix} d_1 & 0 \\ 0 & d_2 \end{pmatrix},\end{aligned}$$

by the property ${}^t\xi_1\tau_1 - {}^t\xi_2\tau_2 = 0$. An easy calculation tells that $d_1 = {}^t\tau_1\tau_1 + {}^t\tau_2\tau_2$ is a diagonal matrix with diagonal entries 1 or 2, and so is $d_2 = {}^t\xi_1\xi_1 + {}^t\xi_2\xi_2$. Thus we get ${}^t\boldsymbol{g}^{-1} = \boldsymbol{g}\boldsymbol{d}^{-1}$ with $\boldsymbol{d} = \mathrm{diag}(d_1, d_2)$. From this, we compute

$$\sigma(\boldsymbol{g}) = J_n^{-1}\,{}^t\boldsymbol{g}^{-1}J_n = J_n^{-1}\boldsymbol{g}\boldsymbol{d}^{-1}J_n = \begin{pmatrix} \xi_2 & \tau_2 \\ \xi_1 & -\tau_1 \end{pmatrix} \cdot \mathrm{diag}(-d_2^{-1}, -d_1^{-1}),$$

and get $\sigma(\boldsymbol{g})\mathbb{P} = \begin{pmatrix} \xi_2 & \tau_2 \\ \xi_1 & -\tau_1 \end{pmatrix}\mathbb{P}$. This implies $\sigma\left(\begin{pmatrix} \tau_1 \\ \tau_2 \end{pmatrix}\right) = \begin{pmatrix} \xi_2 \\ \xi_1 \end{pmatrix}$. □

Theorem 16.3: *For* $\omega = \begin{pmatrix} \tau_1 \\ \tau_2 \end{pmatrix} \in \mathfrak{T}_n^\circ/S_n$*, let* $\mathbb{O}_\omega$ *be the corresponding* $\mathbb{K}$*-orbit in* $\mathbb{X}$*. Then the following* (1)–(4) *are all equivalent.*

(1) $\mathbb{O}_\omega \cap \mathfrak{X} \neq \emptyset$*, hence it is a single* K*-orbit.*

(2) $\sigma(\mathbb{O}_\omega) = \mathbb{O}_\omega$*, i.e., the* $\mathbb{K}$*-orbit is* σ*-stable.*

(3) ${}^t\tau_1\tau_2 \in \mathrm{Sym}_n$.

(4) ${}^t\tau_2\tau_1 \in \mathrm{Sym}_n$.

In particular, the set of K*-orbits in the double flag variety* $\mathfrak{X} = \mathrm{GL}_n/B_n^+ \times \mathrm{Sp}_{2n}/P_S$ *of type* CI *is parametrized by* $\mathscr{C}_n^\circ/S_n$*, where*

$$\mathscr{C}_n^\circ := \left\{\omega = \begin{pmatrix} \tau_1 \\ \tau_2 \end{pmatrix} \in \mathfrak{T}_n^\circ \;\middle|\; {}^t\tau_1\tau_2 = {}^t\tau_2\tau_1 \in \mathrm{Sym}_n\right\}.$$

Proof: For (2)⇒(3) we use that $\sigma(\omega) = \omega \mod S_n$, which means

$$\sigma(\omega) = \begin{pmatrix} \xi_2 \\ \xi_1 \end{pmatrix} = \begin{pmatrix} \tau_1 s \\ \tau_2 s \end{pmatrix}$$

for some $s \in S_n$. The other implications are straightforward. □

As in the case of $\mathbb{X}/\mathbb{K}$ (see (16.4)), there are natural bijections

$$\begin{aligned}\mathfrak{X}/K &\simeq (\mathscr{F}\!\ell_n \times \mathrm{LGr}(\mathbb{C}^{2n}))/\mathrm{GL}_n \simeq B_n^+\backslash\mathrm{Sp}_{2n}/P_S \simeq B_n^+\backslash\,\mathrm{LGr}(\mathbb{C}^{2n}) \\ &\simeq B_n^+\backslash\mathscr{S}_{2n,n}^\circ/\mathrm{GL}_n \simeq \mathscr{C}_n^\circ/S_n,\end{aligned} \tag{16.6}$$

where

$$\begin{aligned}\mathscr{S}^{\circ}_{2n,n} &= \{A \in \mathrm{M}^{\circ}_{2n,n} \mid {}^{t}AJ_nA = 0\} \\ &= \{A = \begin{pmatrix} A_1 \\ A_2 \end{pmatrix} \in \mathrm{M}^{\circ}_{2n,n} \mid {}^{t}A_1A_2 = {}^{t}A_2A_1 \in \mathrm{Sym}_n\}.\end{aligned} \tag{16.7}$$

Note that the actions of $b \in B_n^+$ are all defined by the left multiplications by $\begin{pmatrix} b & 0 \\ 0 & {}^{t}b^{-1} \end{pmatrix}$. The above theorem tells that all these coset spaces are in bijection with $\mathscr{C}^{\circ}_n / S_n$.

Acknowledgments

The authors thank the referee for his/her precise reading, and useful suggestions. The second author (K. N.) is supported by JSPS KAKENHI Grant Number #21K03184.

References

1. P. Achar and A. Henderson. Orbit closures in the enhanced nilpotent cone. *Advances in Mathematics*, **219**(1):27–62, 2008.
2. R. Avdeev and A. Petukhov. Spherical actions on isotropic flag varieties and related branching rules. *Transform. Groups*, **26**(3):719–774, 2021.
3. D. Barbasch, S. Da Silva, B. Elek, and G. G. Krishnan. Finite type multiple flag varieties of exceptional groups. arXiv:1708.06341, (2017).
4. C. Benson and G. Ratcliff. A classification of multiplicity free actions. *J. Algebra*, **181**(1):152–186, 1996.
5. A. Borel. *Linear algebraic groups*, second edn. vol. 126, *Graduate Texts in Mathematics*, Springer-Verlag, New York, 1991.
6. W. Borho and R. MacPherson. Partial resolutions of nilpotent varieties. In *Analysis and topology on singular spaces, II, III (Luminy, 1981)*, vols. 101–102, *Astérisque*, pp. 23–74. Soc. Math. France, Paris, 1983.
7. M. Brion. Quelques propriétés des espaces homogènes sphériques, *Manuscripta Math.*, **55**(2):191–198, 1986.
8. M. Brion. Spherical varieties. In *Highlights in Lie algebraic methods*, vol. 295, *Progr. Math.*, pp. 3–24. Birkhäuser/Springer, New York, 2012.
9. N. Chriss and V. Ginzburg. *Representation theory and complex geometry.* Birkhäuser Boston Inc., Boston, MA, 1997.
10. D. Ciubotaru, K. Nishiyama, and P. E. Trapa. Regular orbits of symmetric subgroups on partial flag varieties. In *Representation theory, complex analysis, and integral geometry*, pp. 61–86. Birkhäuser/Springer, New York, 2012.
11. D. H. Collingwood and W. M. McGovern. *Nilpotent orbits in semisimple Lie algebras.* Van Nostrand Reinhold Mathematics Series, Van Nostrand Reinhold Co., New York, 1993.
12. M. Finkelberg, V. Ginzburg, and R. Travkin. Mirabolic affine Grassmannian and character sheaves, *Selecta Math. (N.S.)*, **14**(3-4):607–628, 2009.

13. L. Fresse and K. Nishiyama. On the exotic Grassmannian and its nilpotent variety. *Represent. Theory*, **20**:451–481, 2016.
14. L. Fresse and K. Nishiyama. Orbit embedding for double flag varieties and Steinberg maps. In *Lie groups, number theory, and vertex algebras*, vol. 768, *Contemp. Math.*, pp. 21–42. Amer. Math. Soc., Providence, RI, 2021.
15. L. Fresse and K. Nishiyama. A generalization of Steinberg theory and an exotic moment map. *Int. Math. Res. Not. IMRN*, (1):1–62, 2022.
16. L. Fresse and K. Nishiyama. On generalized Steinberg theory for type AIII. *Algebr. Comb.*, **6**(1):165–195, 2023.
17. L. Fresse and K. Nishiyama. Action of Hecke algebra on the double flag variety of type AIII. *Adv. in Appl. Math.*, **153**:Paper No. 102614,19, 2024.
18. W. Fulton. *Young tableaux. With applications to representation theory and geometry.* vol. 35, *London Mathematical Society Student Texts*, Cambridge University Press, Cambridge, 1997.
19. P. Gabriel. Unzerlegbare Darstellungen. I, *Manuscripta Math.*, **6**:71–103; correction, ibid. 6 (1972), 309, 1972.
20. X. He, K. Nishiyama, H. Ochiai, and Y. Oshima. On orbits in double flag varieties for symmetric pairs. *Transf. Groups*, **18**1091–1136, 2013.
21. S. Helgason, *Differential geometry, Lie groups, and symmetric spaces.* vol. 80, *Pure and Applied Mathematics*, Academic Press Inc. [Harcourt Brace Jovanovich Publishers], New York, 1978.
22. A. Henderson and P. E. Trapa. The exotic Robinson-Schensted correspondence. *J. Algebra*, **370**:32–45, 2012.
23. V. Hinich and A. Joseph. Orbital variety closures and the convolution product in Borel-Moore homology. *Selecta Math. (N.S.)*, **11**(1):9–36, 2005.
24. H. Homma. Double flag varieties and representations of quivers. arXiv: 2103.14509, 2021.
25. J. E. Humphreys. *Linear algebraic groups.* Graduate Texts in Mathematics, No. 21, Springer-Verlag, New York-Heidelberg, 1975.
26. J. C. Jantzen. Nilpotent orbits in representation theory. In *Lie theory*, vol. 228, *Progr. Math.*, pp. 1–211. Birkhäuser Boston, Boston, MA, 2004.
27. V. G. Kac, Infinite root systems, representations of graphs and invariant theory, *Invent. Math.*, **56**(1):57–92, 1980.
28. V. G. Kac. Some remarks on nilpotent orbits. *J. Algebra*, **64**(1):190–213, 1980.
29. V. G. Kac. Infinite root systems, representations of graphs and invariant theory. II. *J. Algebra*, **78**(1):141–162, 1982.
30. V. G. Kac. *Infinite-dimensional Lie algebras*, 3rd edn. Cambridge University Press, Cambridge, 1990.
31. S. Kato. An exotic Deligne-Langlands correspondence for symplectic groups. *Duke Math. J.*, **148**(2):305–371, 2009.
32. G. R. Kempf. Instability in invariant theory. *Ann. of Math. (2)*. **108**(2):299–316, 1978.
33. F. Knop. Some remarks on multiplicity free spaces. In *Representation theories and algebraic geometry (Montreal, PQ, 1997)*, vol. 514, *NATO Adv. Sci. Inst. Ser. C Math. Phys. Sci.*, pp. 301–317. Kluwer Acad. Publ., Dordrecht, 1998.

34. F. Knop, B. Krötz, T. Pecher, and H. Schlichtkrull. Classification of reductive real spherical pairs I. The simple case. *Transform. Groups*, **24**(1):67–114, 2019.
35. F. Knop, B. Krötz, T. Pecher, and H. Schlichtkrull. Classification of reductive real spherical pairs II. The semisimple case. *Transform. Groups*, **24**(2):467–510, 2019.
36. F. Knop and B. Van Steirteghem. Classification of smooth affine spherical varieties. *Transform. Groups*, **11**(3):495–516, 2006.
37. T. Kobayashi and T. Matsuki. Classification of finite-multiplicity symmetric pairs. *Transform. Groups*, **19**(2):457–493, 2014.
38. T. Kobayashi and T. Oshima. Finite multiplicity theorems for induction and restriction. *Adv. Math.*, **248**:921–944, 2013.
39. B. Kostant. Lie group representations on polynomial rings. *Amer. J. Math.*, **85**:327–404, 1963.
40. B. Kostant. Quantization and representation theory. In *Representation theory of Lie groups*, pp. 287–316. Cambridge Univ. Press, Cambridge, 1979.
41. B. Kostant and S. Rallis. Orbits and representations associated with symmetric spaces. *Amer. J. Math.*, **93**:753–809, 1971.
42. A. S. Leahy. A classification of multiplicity free representations. *J. Lie Theory*, **8**(2):367–391, 1998.
43. I. Losev. Combinatorial invariants of algebraic Hamiltonian actions. *Mosc. Math. J.*, **8**(3):493–519, 616, 2008.
44. P. Magyar, J. Weyman, and A. Zelevinsky. Multiple flag varieties of finite type. *Adv. Math.*, **141**(1):97–118, 1999.
45. P. Magyar, J. Weyman, and A. Zelevinsky. Symplectic multiple flag varieties of finite type. *J. Algebra*, **230**(1):245–265, 2000.
46. A. Malcev. On semi-simple subgroups of Lie groups. *Bull. Acad. Sci. URSS. Sér. Math. [Izvestia Akad. Nauk SSSR]*. **8**:143–174, 1944.
47. T. Matsuki. An example of orthogonal triple flag variety of finite type. *J. Algebra*, **375**:148–187, 2013.
48. T. Matsuki. Orthogonal multiple flag varieties of finite type I: Odd degree case. *J. Algebra*, **425**:450–523, 2015.
49. T. Matsuki. Orthogonal multiple flag varieties of finite type II: Even degree case. arXiv:1903.06335, 2019.
50. D. Miličić. Localization and Representation Theory of Reductive Lie Groups. `http://www.math.utah.edu/~milicic/Eprints/book.pdf`, 1993.
51. M. Mimura and H. Toda. *Topology of Lie groups. I, II.* Japanese edn. vol. 91, *Translations of Mathematical Monographs*, American Mathematical Society, Providence, RI, 1991.
52. D. Mumford, J. Fogarty, and F. Kirwan. *Geometric invariant theory*, third edn. vol. 34, *Ergebnisse der Mathematik und ihrer Grenzgebiete*, Springer-Verlag, Berlin, 1994.
53. V. Nandakumar, D. Rosso, and N. Saunders. Irreducible components of exotic Springer fibres. *J. Lond. Math. Soc. (2)*. **98**(3):609–637, 2018.
54. K. Nishiyama. Enhanced orbit embedding. *Comment. Math. Univ. St. Pauli*, **63**(1-2):223–232, 2014.

55. K. Nishiyama and H. Ochiai. Double flag varieties for a symmetric pair and finiteness of orbits. *J. Lie Theory*, **21**(1):79–99, 2011.
56. K. Nishiyama and B. Ørsted. Real double flag varieties for the symplectic group. *J. Funct. Anal.*, **274**(2):573–604, 20108.
57. T. Ohta. The closures of nilpotent orbits in the classical symmetric pairs and their singularities. *Tohoku Math. J. (2)*. **43**(2):161–211, 1991.
58. T. Ohta. An inclusion between sets of orbits and surjectivity of the restriction map of rings of invariants. *Hokkaido Math. J.*, **37**(3):437–454, 2008.
59. D. I. Panyushev. On the conormal bundle of a G-stable subvariety. *Manuscripta Math.*, **99**(2):185–202, 1999.
60. N. Perrin. On the geometry of spherical varieties. *Transform. Groups*, **19**(1):171–223, 2014.
61. V. L. Popov. Tensor product decompositions and open orbits in multiple flag varieties. *J. Algebra*, **313**(1):392–416, 2007.
62. R. Richardson, G. Röhrle, and R. Steinberg. Parabolic subgroups with abelian unipotent radical. *Invent. Math.*, **110**(3):649–671, 1992.
63. R. W. Richardson and T. A. Springer. Combinatorics and geometry of K-orbits on the flag manifold. In *Linear algebraic groups and their representations (Los Angeles, CA, 1992)*, vol. 153, *Contemp. Math.*, pp. 109–142. Amer. Math. Soc., Providence, RI, 1993.
64. R. W. Richardson and T. A. Springer. Complements to: "The Bruhat order on symmetric varieties" [Geom. Dedicata **35** (1990), no. 1-3, 389–436; MR1066573 (92e:20032)]. *Geom. Dedicata*, **49**(2):231–238, 1994.
65. C. Riedtmann. Degenerations for representations of quivers with relations. *Ann. Sci. École Norm. Sup. (4)*, **19**(2):275–301, 1986.
66. N. Shimamoto. Description of infinite orbits on multiple projective spaces. *J. Algebra*, **563**:404–425, 2020.
67. T. Shoji and K. Sorlin. Exotic symmetric space over a finite field, I. *Transform. Groups*, **18**(3):877–929, 2013.
68. T. Shoji and K. Sorlin. Exotic symmetric space over a finite field, II. *Transform. Groups*, **19**(3):887–926, 2014.
69. T. Shoji and K. Sorlin. Exotic symmetric space over a finite field, III. *Transform. Groups*, **19**(4):1149–1198, 2014.
70. R. Singh. A Robinson–Schensted correspondence for partial permutations. arXiv:2010.13918, 2020.
71. P. Slodowy. *Simple singularities and simple algebraic groups.* vol. 815, *Lecture Notes in Mathematics*, Springer, Berlin, 1980.
72. N. Spaltenstein. The fixed point set of a unipotent transformation on the flag manifold. *Nederl. Akad. Wetensch. Proc. Ser. A Indag. Math.*, **79**(5):452–456, 1976.
73. T. A. Springer. The unipotent variety of a semi-simple group. In *Algebraic Geometry (Internat. Colloq., Tata Inst. Fund. Res., Bombay, 1968)*, pp. 373–391. Oxford Univ. Press, London, 1969.
74. T. A. Springer. Some results on algebraic groups with involutions. In *Algebraic groups and related topics (Kyoto/Nagoya, 1983)*, vol. 6, *Adv. Stud. Pure Math.*, pp. 525–543. North-Holland, Amsterdam, 1985.

75. T. A. Springer. *Linear algebraic groups*, second edn. vol. 9, *Progress in Mathematics*, Birkhäuser Boston Inc., Boston, MA, 1998.
76. R. Steinberg. *Endomorphisms of linear algebraic groups.* Memoirs of the American Mathematical Society, No. 80, American Mathematical Society, Providence, R.I., 1968.
77. R. Steinberg. On the desingularization of the unipotent variety. *Invent. Math.*, **36**:209–224, 1976.
78. R. Steinberg. An occurrence of the Robinson-Schensted correspondence. *J. Algebra*, **113**(2):523–528, 1988.
79. J. R. Stembridge. Multiplicity-free products and restrictions of Weyl characters. *Represent. Theory*, **7**:404–439 (electronic), 2003.
80. D. A. Timashev. *Homogeneous spaces and equivariant embeddings.* vol. 138, *Encyclopaedia of Mathematical Sciences*, Springer, Heidelberg, 2011. Invariant Theory and Algebraic Transformation Groups, 8.
81. P. E. Trapa. Generalized Robinson-Schensted algorithms for real groups. *Internat. Math. Res. Notices*, (15):803–834, 1999.
82. R. Travkin. Mirabolic Robinson-Schensted-Knuth correspondence. *Selecta Math. (N.S.)*, **14**(3-4):727–758, 2009.
83. M. A. A. van Leeuwen. The Robinson-Schensted and Schützenberger algorithms, an elementary approach. *Electron. J. Combin.*, **3**(2):Research Paper 15, approx. 32, 1996.
84. È. B. Vinberg. Complexity of actions of reductive groups. *Funktsional. Anal. i Prilozhen.*, **20**(1):1–13, 96, 1986.
85. T. Vust. Opération de groupes réductifs dans un type de cônes presque homogènes. *Bull. Soc. Math. France*, **102**:317–333, 1974.
86. A. V. Zelevinsky. *Representations of finite classical groups.* vol. 869, *Lecture Notes in Mathematics*, Springer-Verlag, Berlin-New York, 1981. A Hopf algebra approach.

https://doi.org/10.1142/9789819825448_0006

On Local Character Expansions for Principal Series Representations of General Linear Groups

Maxim Gurevich
Department of Mathematics
Technion – Israel Institute of Technology
Haifa, Israel.
maxg@technion.ac.il

We obtain two explicit formulas for the full local character expansion of any irreducible representation of a p-adic general linear group in principal blocks. The first, generalizing previous work of the author on the Iwahori-spherical case, expresses the expansion in terms of dimensions of degenerate Whittaker models. The second gives a closed expression in terms of values of Kazhdan-Lusztig polynomials of a suitable permutation group.

1. Introduction

Two areas of study are typical focal points of character theory of representations of reductive p-adic groups. One examines how standard representations break down into irreducible subquotients. The second delves into asymptotic properties of irreducible representations, as encoded in microlocal invariants that are associated with adjoint nilpotent orbits of the group. The precise links between these two branches of study, particularly the role of the Kazhdan-Lusztig theory, which is known to govern the former problem, have often remained a subject of folklore.

In this chapter, we explicate two formulas that regard the Harish-Chandra–Howe local character expansion for principal series of general linear groups, by leveraging the well-developed Zelevinsky theory applicable to this context.

Let $G_n = GL_n(F)$ be the general linear locally compact group, defined over a p-adic field F. Let $\mathfrak{g}_n = \mathfrak{gl}_n(F)$ be its Lie algebra. The set $P(n)$, consisting of all integer partitions of n, naturally parameterizes the nilpotent $\mathrm{Ad}(G_n)$-orbits in $\mathfrak{g}_n$.

For a partition $\alpha \in P(n)$, the orbit $\mathcal{O}_\alpha$ is taken to be the set of nilpotent matrices, whose sizes of Jordan blocks are described by the parts of α.

Given a finite-length smooth complex G_n-representation π, its trace is known to provide a distribution on the group, which is represented by a locally constant integrable function Θ_π on regular elements. For a regular element $X \in \mathfrak{g}_n$ close enough to zero, the celebrated Harish-Chandra–Howe local character expansion [10, 8] takes the form

$$\Theta_\pi(1+X) = \sum_{\alpha \in P(n)} c_\alpha(\pi)\widehat{\mu}_{\mathcal{O}_\alpha}(X), \quad c_\alpha(\pi) \in \mathbb{Z}, \tag{1.1}$$

where $\widehat{\mu}_{\mathcal{O}_\alpha}$, $\alpha \in P(n)$, are suitably normalized functions representing the $\mathfrak{g}_n$-Fourier transform of the orbital integral distribution coming from the orbit $\mathcal{O}_\alpha$.

In [7], with an assumption of a good residual characteristic for F, a formula was pointed out, which relates the set of integers $\{c_\alpha(\pi)\}_{\alpha \in P(n)}$, for an irreducible Iwahori-spherical π, with another well-studied set of integer invariants.

The second set is given by the dimensions of certain degenerate Whittaker models attached to the same G_n-representation π and a nilpotent $\mathrm{Ad}(G_n)$-orbit.

Namely, for $\alpha \in P(n)$, an exact normalized Jacquet functor $\mathbf{r}_\alpha$ is defined, which produces smooth M_α-representations out of smooth G_n-representations, where $M_\alpha < G_n$ is a corresponding Levi subgroup.

We denote the integers

$$d_\alpha(\pi) = \dim_{\mathbb{C}} \mathrm{Wh}(\mathbf{r}_\alpha(\pi)), \ \alpha \in P(n),$$

where Wh stands for the Whittaker functor on M_α-representations.

Our first goal is to give a generalization of this formula that will encompass all representations in Bernstein blocks that arise from a minimal Levi subgroup (maximal torus), remove the residual characteristic constraint, and offer an alternative method of proof.

Let us make the stated notions more precise.

For a partition $\alpha = (\alpha_1, \ldots, \alpha_k)$ in

$$P(n) = \{(\alpha_1 \geq \ldots \geq \alpha_k) \,:\, \alpha_i \in \mathbb{Z}_{>0},\ \alpha_1 + \ldots + \alpha_k = n\},$$

the standard Levi subgroup $M_\alpha < G_n$ consists of block-diagonal matrices, with block sizes $\alpha_1, \ldots, \alpha_k$.

We also write $P_\alpha < G_n$ for the standard parabolic subgroup generated by M_α and the upper-triangular matrices in G_n, $N_\alpha < P_\alpha$ for its unipotent radical, and $\mathfrak{n}_\alpha = \mathrm{Lie}(N_\alpha) < \mathfrak{g}_n$.

Now, $\mathcal{O}_\alpha$ becomes the unique nilpotent $\mathrm{Ad}(G_n)$-orbit in $\mathfrak{g}_n$, for which $\mathfrak{n}_{\alpha^t} \cap \mathcal{O}_\alpha$ is dense in $\mathfrak{n}_{\alpha^t}$. Here, $\alpha \mapsto \alpha^t$ is the combinatorial transposition involution on $P(n)$.

We say that an irreducible G_n-representation π is *principal*, when $\mathbf{r}_{(1,1,\ldots,1)}(\pi) \neq 0$. In other words, there exists a character χ of the maximal torus $(F^\times)^n \cong T = M_{(1,\ldots,1)} < G_n$, for which π is a sub-representation[a] of $\mathbf{i}_{(1,\ldots,1)}(\chi)$, where $\mathbf{i}_\alpha$ is the normalized parabolic induction (adjoint to Jacquet) functor.

Theorem 1.1: *For each principal irreducible representation π of G_n and each partition $\alpha \in P(n)$, the formula*

$$d_\alpha(\pi) = \sum_{\beta \in P(n)} s(\alpha, \beta^t) c_\beta(\pi)$$

holds, where the coefficients $s(\alpha, \beta) \in \mathbb{Z}_{\geq 0}$ are the familiar combinatorial invariants given as

$$s(\alpha,\beta) = \\ \#\left\{ A \subseteq \{1,\ldots,k\} \times \{1,\ldots,l\} : \begin{matrix} \alpha_i = \#\{j : (i,j) \in A\},\ \forall 1 \leq i \leq k \\ \beta_j = \#\{i : (i,j) \in A\},\ \forall 1 \leq j \leq l \end{matrix} \right\}, \tag{1.2}$$

for $\alpha = (\alpha_1, \ldots, \alpha_k)$, $\beta = (\beta_1, \ldots, \beta_l)$.

We invite the reader to the discussion in the introduction section of [7] on how the unitriangular (with respect to the topological closure order on nilpotent orbits) formula of Theorem 1.1 naturally refines the renowned result from [16] concerning the relationship between character expansions and generalized Whittaker models.

Back in [7], we utilized Hecke algebra techniques to reduce the problem, for the case of the Iwahori-spherical Bernstein block, to a matter involving the ring of symmetric functions. In Section 2, we instead highlight an alternative computation method for both sets of invariants, employing the basis of Zelevinsky-standard representations for the Grothendieck groups of G_n.

[a]Note, that in other common terminology a *principal series* representation may refer to situations where $\pi = \mathbf{i}_{(1,\ldots,1)}(\chi)$ is irreducible. This notion is more restrictive than ours.

1.1. *Kazhdan-Lusztig theory*

The second goal of this note is to highlight a formula for producing the explicit values of the expansion $\{c_\alpha(\pi)\}_{\alpha\in P(n)}$, for a principal representation π, in terms of Kazhdan-Lusztig polynomials of a relevant permutations group.

We say that an irreducible G_n-representation π is *integral principal*, when it is principal and the irreducible subquotients of $\mathbf{r}_{(1,\dots,1)}(\pi)$ are (unramified) characters χ of $(F^\times)^n$ of the form $\chi(a_1,\dots,a_n) = \prod_{i=1}^n |a_i|_F^{t_i}$ with $t_i \in \mathbb{Z}$ ($|\,|_F$ being the p-adic norm).

We write $\mathrm{Irr}_{\mathrm{int}}(G_n)$ for the set of isomorphism classes of irreducible integral principal G_n-representations. As detailed at the head of Section 3, computations of local character expansions for principal representations are easily reduced to the integral case.

Let us now restate the Zelevinsky classification for $\mathrm{Irr}_{\mathrm{int}}(G_n)$.

For $k \geq 1$, we consider the integer lattice $\mathbb{Z}^k$ with its cone $D_k = \{(\alpha_1 \geq \dots \geq \alpha_k)\} \subseteq \mathbb{Z}^k$.

Considering the natural action of S_k on $\mathbb{Z}^k$, the cone D_k becomes a full set of orbit representatives. We denote the projection $p : \mathbb{Z}^k \to D_k$ that takes a tuple to its representative.

For $x_0 \in \mathbb{Z}^k$, we would also like to consider a shifted S_k-action on $\mathbb{Z}^k$ given as

$$\omega *_{x_0} x := \omega \cdot (x - x_0) + x_0, \quad \text{for } x \in \mathbb{Z}^k.$$

We further denote the set

$$C(n,k) = \{(\alpha_1,\dots,\alpha_k) : \alpha_i \in \mathbb{Z}_{>0}, \alpha_1 + \dots + \alpha_k = n\} \subseteq \mathbb{Z}^k$$

of compositions of n into k parts, the set of pairs

$$M_{n,k} = \{(\lambda,\eta) \in \mathbb{Z}^k \times \mathbb{Z}^k : \eta - \lambda \in C(n,k)\},$$

and the set $\mathrm{Mult}_{n,k} = M_{n,k}/S_k$ of orbits of the diagonal S_k-action on $M_{n,k}$. An element of $\mathrm{Mult}_{n,k}$, represented by a pair $((\lambda_1,\dots,\lambda_k),(\eta_1,\dots,\eta_k)) \in M_{n,k}$, can be thought of as a multiset of k intervals $\{[\lambda_i,\eta_i]\}_{i=1}^k$ on the integer line, whose sum of lengths equals to n.

The Zelevinsky classification ([17]) sets up a bijection

$$Z : \bigcup_{k\geq 1} \mathrm{Mult}_{n,k} \to \mathrm{Irr}_{\mathrm{int}}(G_n). \tag{1.3}$$

For a partition $\mu = (\mu_1,\dots,\mu_t) \in P(n)$ with $t \leq k$, we consider μ as an element of D_k by padding its tail with 0's if necessary.

Theorem 1.2: *Let $\pi = Z(\mathfrak{m}) \in \mathrm{Irr}_{\mathrm{int}}(G_n)$ be an irreducible integral principal representation, parameterized by the S_k-orbit $\mathfrak{m}$ of a pair $(\lambda, \eta) \in M_{n,k}$.*

Let $\alpha = \eta - \lambda \in C(n,k)$ be the associated composition. Let $\omega_1, \omega_2 \in S_k$ be the permutations for which the product $\omega_1\omega_2$ is longest possible while subject to

$$\omega_1 \cdot \lambda,\ \omega_2^{-1} \cdot \eta \in D_k.$$

Then, for a partition $\mu \in P(n)$ with $\mu_1 \leq k$, the identity

$$c_\mu(\pi) = \sum_{\tau \in S_k \,:\, p(\tau *_{-\lambda} \alpha) = \mu^t} \epsilon(\tau) P_{\omega_0\omega_1\tau\omega_2,\, \omega_0\omega_1\omega_2}(1)$$

holds, where P_{ζ_1,ζ_2} stands for the Kazhdan-Lusztig polynomial attached to a pair of permutations $\zeta_1, \zeta_2 \in S_k$, $\epsilon : S_k \to \{\pm 1\}$ is the sign map, and $\omega_0 \in S_k$ denotes the longest permutation in the group.

For partitions $\mu \in P(n)$ with $\mu_1 > k$, we have $c_\mu(\pi) = 0$.

Given that Bernstein blocks of G_n-representations are inherently equivalent to module categories over type A affine Hecke algebras, it is reasonable to speculate that the formula presented in Theorem 1.2 is expressible using Kazhdan-Lusztig polynomials associated with the affine Weyl group of type A. However, the author has not encountered a documentation of this particular approach in literature. We suggest further investigation into the role of Kazhdan-Lusztig theory in the context of local character expansion coefficients.

We should also comment on the formula from the perspective of the Langlands reciprocity. While it is clear that the Zelevinsky classification is an early instance of the reciprocity for general linear groups, it involves additional computational issues. Specifically, the Langlands parameter corresponding to a given representation $\pi \in \mathrm{Irr}_{\mathrm{int}}(G_n)$ can be naturally associated with a multisegment $\mathfrak{m}_\pi \in \mathrm{Mult}_{k,n}$ through a correspondence of the form $\pi = Z(\widehat{\mathfrak{m}_\pi})$, where the mapping $\mathfrak{m} \mapsto \widehat{\mathfrak{m}}$ is the Zelevinsky involution on $\bigcup_{k \geq 1} \mathrm{Mult}_{n,k}$, which was algorithmically explicated in [15].

Thus, in terms of complexity, the formula of Theorem 1.2 yields the local character expansion of principal representations when provided with algorithms for S_n-Kazhdan-Lusztig polynomials, complemented by an algorithm for the Zelevinsky involution. This observation aligns with recent insights presented in [2, 3], where it was demonstrated that the wavefront set invariant for further cases of unipotent representations of p-adic reductive groups can be directly extracted from the Aubert-Zelevinsky transform of its Langlands parameter.

Acknowledgements

The chapter has been shaped by insights gained from engaging discussions I had during participation in the July 2022 IMS workshop. I am particularly grateful to Raphaël Beuzart-Plessis, Monica Nevins, Emile Okada, and Marie-France Vignéras for sharing their perspectives on this matter. I extend my appreciation to the organizers of this wonderful workshop in Singapore.

This research is supported by the Israel Science Foundation (Grant Number: 737/20).

2. Zelevinsky-Standard Representations

Let $\widehat{F^\times}$ be the collection of smooth complex one-dimensional $GL_1(F)$-representations (characters). For $\chi \in \hat{F^\times}$, let $\Re(\chi) \in \mathbb{R}$ be the number for which $|\chi(a)| = |a|_F^{\Re(\chi)}$ holds, for all $a \in F^\times$.

One-dimensional G_n-representations are of the form $\chi^{(n)} := \chi \circ \det$, for $\chi \in \widehat{F^\times}$.

Given a composition $\alpha = (\alpha_1, \ldots, \alpha_k) \in C(n,k)$, in similarity with the partition case, we let $P_\alpha = M_\alpha N_\alpha < G_n$ be the corresponding standard parabolic subgroup. As customary, we treat the corresponding normalized parabolic induction functor $\mathbf{i}_\alpha$ as a product functor

$$\pi_1 \times \cdots \times \pi_k = \mathbf{i}_\alpha(\pi_1 \boxtimes \cdots \boxtimes \pi_k)$$

that takes $G_{\alpha_1} \times \cdots \times G_{\alpha_k}$-representations to G_n-representations.

For forthcoming needs, we would like to consider the larger set

$$C_0(n,k) = \{(\alpha_1, \ldots, \alpha_k) \ : \alpha_i \in \mathbb{Z}_{\geq 0}\ , \alpha_1 + \ldots + \alpha_k = n\} \subseteq \mathbb{Z}^k$$

of compositions with possible zero parts. The definitions of the functors $\mathbf{r}_\alpha$ and $\mathbf{i}_\alpha$ are extended to $\alpha \in C_0(n,k)$ by considering G_0 as a trivial group and its representations as neutral to the parabolic induction product. In particular, for $\chi \in \widehat{F^\times}$, we may take $\chi^{(0)}$ as a trivial representation of G_0.

For $\alpha \in C_0(n,k)$ and characters $\chi_1, \ldots, \chi_k \in \widehat{F^\times}$, a G_n-representation

$$\zeta = \zeta(\alpha, \chi_1, \ldots, \chi_k) := \chi_1^{(\alpha_1)} \times \cdots \times \chi_k^{(\alpha_k)} \tag{2.1}$$

is constructed. We say that ζ is a *principal Zelevinsky-standard representation* whenever $\Re(\chi_i) \geq \Re(\chi_j)$ holds, for all $1 \leq i < j \leq k$.

In [17] it is shown that each irreducible G_n-representation π corresponds to a unique Zelevinsky-standard representation $\zeta(\pi)$, in which π appears as a sub-representation. It is also a standard corollary that the collection of Zelevinsky-standard representations forms a basis for the Gorthendieck group of the category of finite-length G_n-representations.

Finally, it is a standard outcome of the theory of Bernstein blocks that the collection of principal Zelevinsky-standard representations forms a basis for the Gorthendieck group of the direct summand of the category that is generated by the principal irreducible representations.

Note now, that due to trace additivity and exactness of Whittaker and Jacquet functors, both $c_\mu(\cdot)$ and $d_\mu(\cdot)$, for $\mu \in P(n)$, are additive functionals on the above mentioned Grothendieck groups.

This short discussion sums up to the following reduction.

Proposition 2.1: *Theorem 1.1 will follow, if the identity in its statement is proved with π being a principal Zelevinsky-standard representation, rather than an irreducible representation.*

Indeed, Theorem 1.1 now clearly follows from the previous proposition and the following two lemmas.

Lemma 2.2: *For a principal Zelevinsky-standard representation $\zeta = \zeta(\widetilde{\beta}, \chi_1, \ldots, \chi_k)$ with $\widetilde{\beta} \in C(n,k)$, and a given partition $\alpha \in P(n)$, we have*[b]

$$d_\alpha(\zeta) = s(\alpha, p(\widetilde{\beta})).$$

Proof: We apply the basic Mackey theory for p-adic general linear groups, as stated for example in [12, 2.2] (a case of the Bernstein-Zelevinsky geometric lemma). It follows that the $M_\alpha = M_{(\alpha_1,\ldots,\alpha_l)}$-representation $\mathbf{r}_\alpha(\zeta)$ is decomposed into sub-quotients of the form

$$(\chi_{1,1} \times \cdots \times \chi_{k,1}) \boxtimes \cdots \boxtimes (\chi_{1,l} \times \cdots \times \chi_{k,l}),$$

for all possible $k \times l$ matrices $A = (\chi_{i,j})$ of representations, such that $\chi_{i,1} \boxtimes \cdots \boxtimes \chi_{i,l}$ is an irreducible sub-quotient of $\mathbf{r}_{\gamma_i}(\chi_i^{(\beta_i)})$, for all $i = 1, \ldots, k$, and choices of $\gamma_i \in C_0(\beta_i, l)$ for which the group ranks fit.

Let $\mathcal{A}$ denote the set of such matrices A, up to isomorphism of their entries.

By exactness properties of the Whittaker functor, $d_\alpha(\zeta)$ will count the cardinality of the subset $\mathcal{A}_{\text{gen}} \subseteq \mathcal{A}$, consisting of matrices all of whose entries are generic representations (assuming G_0-representations are counted as generic).

[b]Note that the application of p in the formula is somewhat redundant, since $s(\alpha, \widetilde{\beta})$ may be defined directly for compositions that are not necessarily partitions.

Now, since all $\chi_i^{(\beta_i)}$ are one-dimensional, so are all possible entries of matrices in $\mathcal{A}$. Since one-dimensional G_m-representations are generic, if and only if, $m \leq 1$, the set $\mathcal{A}_{\text{gen}}$ is identified with the set of $k \times l$ matrices with entries in $\{0,1\}$, for which rows sum up to the values of β_i while columns sum up to the values of $\alpha_{\tilde{j}}$. The cardinality of this set is given by the number $s(p(\widetilde{\beta}), \alpha) = s(\alpha, p(\widetilde{\beta}))$. □

Lemma 2.3: *For a principal Zelevinsky-standard representation $\zeta = \zeta(\alpha, \chi_1, \ldots, \chi_k)$ with $\alpha \in C(n,k)$, and a given partition $\beta \in P(n)$, we have*

$$c_\beta(\zeta) = \begin{cases} 1 \; p(\alpha) = \beta^t \\ 0 \; p(\alpha) \neq \beta^t \end{cases}.$$

The proof of Lemma 2.3 will follow readily from the multiplicative nature of local character expansions, as was recently explicated by Henniart-Vignéras. We recall it here in a form convenient for our needs.

Proposition 2.4: *(Theorem 1.5 [11])*[c] *Given a tuple of finite-length G_{α_i}-representations π_i, $i = 1, \ldots, k$, with $(\alpha_1, \ldots, \alpha_k) \in C(n,k)$, and a partition $\mu \in P(n)$, we have*

$$c_\mu(\pi_1 \times \cdots \times \pi_k) = \sum c_{\beta^1}(\pi_1) \cdots c_{\beta^k}(\pi_k),$$

where the sum is over all tuples of partitions $\beta^i \in P(\alpha_i)$, $i = 1, \ldots, k$, that satisfy

$$p(\beta^1 + \ldots + \beta^k) = \mu.$$

Here, partitions are considered as elements of the lattice $\mathbb{Z}^n$ through 0 padding as before.

Remark 2.5: We have until now left the coefficients $c_\mu(\cdot)$ defined only up to a constant factor, without specifying a particular normalization. Indeed, Proposition 2.4 remains valid when a specific normalization is imposed, as outlined in [11]. However, instead of delving into the details of this normalization, which would require us to deviate somewhat from the main discussion, we opt to interpret Lemma 2.3 as a convenient way to pin down the constants.

[c]In [11], notation for c_μ corresponds to our c_{μ^t}, which results in a visible difference between sources in descriptions of the multiplicative property.

In essence, it suffices to avoid fixing a normalization while regarding the lemma as a statement on vanishing behavior of coefficients, under the assumption that Proposition 2.4 is valid.

Proof: [Proof of Lemma 2.3] Since all $\chi_i^{(\alpha_i)}$ are one-dimensional representations, the only contribution for their local character expansions comes from the zero orbit. In other words, we have $c_\beta(\chi_i^{(\alpha_i)}) \neq 0$, if and only if, $\beta = \kappa(\alpha_i)$, where $\kappa(r) = (1, \ldots, 1) \in P(r)$ stands for the minimal partition of an integer $r \geq 1$.

By Proposition 2.4, it now follows that the only non-vanishing local character expansion coefficient of the Zelevinsky-standard representation ζ is parameterized by the partition

$$\mu = p(\kappa(\alpha_1) + \ldots + \kappa(\alpha_k)) \in P(n).$$

It is easy to verify that $\mu = \alpha^t$. □

3. Explicit Expansions

The theory of [17] stipulates that any principal irreducible G_n-representation π can be constructed in the form

$$\pi = (\chi_1^{(\gamma_1)} \otimes \pi_1) \times \cdots \times (\chi_r^{(\gamma_r)} \otimes \pi_r),$$

for $\pi_i \in \mathrm{Irr}_{\mathrm{int}}(G_{\gamma_i})$, $i = 1, \ldots, r$, and characters $\chi_1, \ldots, \chi_r \in \widehat{F^\times}$.

Since tensoring with one-dimensional representations does not affect local character expansions, it follows from the above decomposition and Proposition 2.4 that the formula of Theorem 1.2 suffices to give an explicit local character expansions for all principal irreducible G_n-representations, in terms of their Zelevinsky parameters.

Let us now recall the details of the construction of the Zelevinsky bijection (1.3).

For $t \in \mathbb{Z}$, we write $\nu^t \in \widehat{F^\times}$ for the unramified character given by $\nu^t(a) = |a|_F^t$.

We consider the set of pairs

$$\overline{M}_{n,k} = \{(\lambda, \eta) \in D_k \times \mathbb{Z}^k \ : \ \eta - \lambda \in C_0(n, k)\}.$$

For a pair $(\lambda, \eta) \in \overline{M}_{n,k}$, a G_n-representation is constructed by parabolic induction of the form

$$\zeta(\lambda, \eta) = \zeta(\eta - \lambda, \nu^{\frac{1}{2}(\lambda_1+\eta_1+1)}, \ldots, \nu^{\frac{1}{2}(\lambda_k+\eta_k+1)}). \tag{3.1}$$

We will say that such $\zeta(\lambda, \eta)$ is an *integral Zelevinsky-standard representation.*

For an orbit $\mathfrak{m} \in \mathrm{Mult}_{n,k}$, it is evident that there are unique tuples $\lambda(\mathfrak{m}), \eta(\mathfrak{m}) \in D_k$, for which

$$(\lambda(\mathfrak{m}), \omega \cdot \eta(\mathfrak{m})) \in \overline{M}_{n,k}$$

is satisfied, for a permutation $\omega \in S_k$. We set $\omega = \omega(\mathfrak{m}) \in S_k$ to be the longest such permutation.

We set

$$\zeta(\mathfrak{m}) = \zeta(\lambda(\mathfrak{m}), \omega(\mathfrak{m}) \cdot \eta(\mathfrak{m})).$$

The bijection (1.3) is now constructed by taking $Z(\mathfrak{m})$ to be the unique irreducible sub-representation of $\zeta(\mathfrak{m})$, for each $\mathfrak{m} \in \mathrm{Mult}_{n,k}$.

Remark 3.1: Each integral Zelevinsky-standard representation $\zeta(\mathfrak{m})$ is isomorphic to a principal Zelevinsky-standard representation.

This fact is not tautological, since the condition in (2.1) requires $\frac{1}{2}(\lambda(\mathfrak{m}) + \omega(\mathfrak{m})\eta(\mathfrak{m})) \in D_k$ to hold, while the construction of $\zeta(\mathfrak{m})$ implies only the non-equivalent condition $\lambda(\mathfrak{m}) \in D_k$. Yet, basic commutation relations on the parabolic induction product provide this fact.

We now recall the role of Kazhdan-Lusztig polynomials in the character theory of finite-length G_n-representations. This perspective was initially shaped by the conjectures of [18], which were subsequently proved in the affine Hecke algebra setting in [1], by means of geometric realization of standard modules. An account of this theory can be found in the survey [14], 2.3, while a precise translation of these findings into the p-adic group context can be read in [5], Section 3. We also recommend the introduction section of [9] for an effective overview of this theory.

An alternative approach may employ the KLR categorification of the dual canonical basis for quantum groups, coupled with Lusztig's description of the transition matrix between PBW and canonical bases and categorical equivalences with the representation theory of affine Hecke algebras. For explorations of this approach, the reader may delve into the details presented in [6], Section 3, or [4].

With this background in hand, we limit ourselves to a presentation of an established corollary of the aforementioned discussion.

For a finite-length G_n-representations π, let $[\pi]$ denote its projection into the Grothendieck group of G_n.

Proposition 3.2: *For a multisegment* $\mathfrak{m} \in \mathrm{Mult}_{n,k}$, *we have an expansion*

$$[Z(\mathfrak{m})] = \sum_{\sigma \in S_k \,:\, (\lambda(\mathfrak{m}),\, \sigma\cdot\eta(\mathfrak{m})) \in \overline{M}_{n,k}} \epsilon(\sigma\omega) P_{\omega_0\sigma, \omega_0\omega(\mathfrak{m})}(1)[\zeta(\lambda(\mathfrak{m}),\, \sigma \cdot \eta(\mathfrak{m}))],$$

of the corresponding irreducible G_n*-representation on the basis of Zelevinsky-standard representations.*

Proof: The formula is given explicitly, for example, in [13, Corollary 10.1]. □

Proof: [Proof of Theorem 1.2] Note, that $\omega(\mathfrak{m}) = \omega_1\omega_2$, $\omega_1 \cdot \lambda = \lambda(\mathfrak{m})$ and $\eta = \omega_2 \cdot \eta(\mathfrak{m})$ hold.

Let us set the composition

$$\alpha(\mathfrak{m}) = \omega(\mathfrak{m}) \cdot \eta(\mathfrak{m}) - \lambda(\mathfrak{m}) = \omega_1 \cdot \alpha \in C(n, k).$$

Since $C_0(n, k)$ is a S_k-invariant set, the equivalence

$$\tau *_{-\lambda} \alpha \in C_0(n, k) \;\Leftrightarrow\; (\omega_1\tau\omega_1^{-1}) *_{-\lambda(\mathfrak{m})} \alpha(\mathfrak{m}) \in C_0(n, k)$$

holds, for all $\tau \in S_k$, with both compositions belonging to the same S_k-orbit.

Hence, it suffices to prove the identity

$$c_\mu(\pi) = \sum_{\tau \in S_k \,:\, p(\tau *_{-\lambda(\mathfrak{m})} \alpha(\mathfrak{m})) = \mu^t} \epsilon(\tau) P_{\omega_0\tau\omega(\mathfrak{m}),\, \omega_0\omega(\mathfrak{m})}(1).$$

Clearly, for $\tau \in S_k$, the conditions $\tau *_{-\lambda(\mathfrak{m})} \alpha(\mathfrak{m}) \in C_0(n, k)$ and $(\lambda(\mathfrak{m}),\, \tau\omega(\mathfrak{m}) \cdot \eta(\mathfrak{m})) \in \overline{M}_{n,k}$ are equivalent.

Moreover, by Lemma 2.3, for $\tau \in S_k$ that satisfies the above condition, we have

$$c_\mu(\zeta(\lambda(\mathfrak{m}),\, \tau\omega(\mathfrak{m}) \cdot \eta(\mathfrak{m}))) = \begin{cases} 1 & p(\tau *_{-\lambda(\mathfrak{m})} \alpha(\mathfrak{m})) = \mu^t \\ 0 & \text{otherwise} \end{cases}.$$

The identity now follows by exactness from Proposition 3.2. □

References

1. Neil Chriss and Victor Ginzburg. *Representation theory and complex geometry.* Birkhäuser Boston, Inc., Boston, MA, 1997.
2. Dan Ciubotaru, Lucas Mason-Brown, and Emile Okada. The wavefront sets of Iwahori-spherical representations of reductive p-adic groups. *arXiv preprint arXiv:2112.14354*, 2021.

3. Dan Ciubotaru, Lucas Mason-Brown, and Emile Okada. Wavefront sets of unipotent representations of reductive p-adic groups II. *J. Reine Angew. Math.*, 823:191–253, 2025.
4. Maxim Gurevich. An identity of parabolic Kazhdan-Lusztig polynomials arising from square-irreducible modules. *Journal of the Australian Mathematical Society*, 2019.
5. Maxim Gurevich. On restriction of unitarizable representations of general linear groups and the non-generic local Gan-Gross-Prasad conjecture. *J. Eur. Math. Soc. (JEMS)*, 24(1):265–302, 2022.
6. Maxim Gurevich. Simple modules for quiver Hecke algebras and the Robinson-Schensted-Knuth correspondence. *J. Lond. Math. Soc. (2)*, 107(2):704–749, 2023.
7. Maxim Gurevich. A triangular system for local character expansions of Iwahori-spherical representations of general linear groups. *Comptes Rendus. Mathématique*, 361:21–30, 2023.
8. Harish-Chandra. *Admissible invariant distributions on reductive p-adic groups*, volume 16 of *University Lecture Series*. American Mathematical Society, Providence, RI, 1999.
9. Anthony Henderson. Nilpotent orbits of linear and cyclic quivers and Kazhdan-Lusztig polynomials of type A. *Represent. Theory*, 11:95–121 (electronic), 2007.
10. Roger Howe. The Fourier transform and germs of characters (case of Gl_n over a p-adic field). *Math. Ann.*, 208:305–322, 1974.
11. Guy Henniart and Marie-France Vignéras. Representations of $GL_n(D)$ near the identity. *Proc. Lond. Math. Soc. (3)*, 129(6):art.id.e70000, 2024.
12. Erez Lapid and Alberto Mínguez. On parabolic induction on inner forms of the general linear group over a non-archimedean local field. *Selecta Math. (N.S.)*, 22(4):2347–2400, 2016.
13. Erez Lapid and Alberto Mínguez. Geometric conditions for $\square$-irreducibility of certain representations of the general linear group over a non-archimedean local field. *Adv. Math.*, 339:113–190, 2018.
14. Bernard Leclerc, Maxim Nazarov, and Jean-Yves Thibon. Induced representations of affine Hecke algebras and canonical bases of quantum groups. In *Studies in memory of Issai Schur (Chevaleret/Rehovot, 2000)*, volume 210 of *Progr. Math.*, pages 115–153. Birkhäuser Boston, Boston, MA, 2003.
15. C. Mœglin and J.-L. Waldspurger. Sur l'involution de Zelevinski. *J. Reine Angew. Math.*, 372:136–177, 1986.
16. C. Mœglin and J.-L. Waldspurger. Modèles de Whittaker dégénérés pour des groupes p-adiques. *Math. Z.*, 196(3):427–452, 1987.
17. A. V. Zelevinsky. Induced representations of reductive p-adic groups. II. On irreducible representations of $\mathrm{GL}(n)$. *Ann. Sci. École Norm. Sup. (4)*, 13(2):165–210, 1980.
18. A. V. Zelevinskiĭ. The p-adic analogue of the Kazhdan-Lusztig conjecture. *Funktsional. Anal. i Prilozhen.*, 15(2):9–21, 96, 1981.

https://doi.org/10.1142/9789819825448_0007

Projection of Elliptic Orbits and Branching Laws

Hongyu He
Department of Mathematics, Louisiana State University, Baton Rouge, Louisiana 70803, USA
livingstone@alum.mit.edu

Let G be a Lie group, and $H \subset G$ a closed Lie subgroup. Let π be an irreducible unitary representation of G. In this chapter, we briefly discuss the orbit method and its application to the branching problem $\pi|_H$. We use the Gan-Gross-Prasad branching law for the pair $(G, H) = (U(p, q), U(p, q-1))$ as an example to illustrate the relation between $\mathrm{Pro}^{\mathfrak{u}(p,q)}_{\mathfrak{u}(p,q-1)}\mathcal{O}(\lambda)$ and the branching law of the discrete series $D_\lambda|_{U(p,q-1)}$ for λ a regular elliptic element. We also discuss some results regarding branching laws and wave front sets. The presentation of this chapter does not follow the historical timeline of development.

1. Coadjoint Orbits and Their Projections

Let G be a connected Lie group. Let $\mathfrak{g}$ be its Lie algebra, and $\mathfrak{g}^*$ its dual space (over $\mathbb{R}$). The Lie group G acts on $\mathfrak{g}$ and $\mathfrak{g}^*$ respectively. Each orbit is called an adjoint orbit or a coadjoint orbit respectively. If $\mathfrak{g}$ is reductive, then $\mathfrak{g}$ can be identified with $\mathfrak{g}^*$ such that the adjoint action coincides with coadjoint action. Then adjoint orbits can be identified with coadjoint orbits. We denote the set of adjoint orbits by $\mathfrak{g}//G$ and the set of coadjoint orbits by $\mathfrak{g}^*//G$. For each $\lambda \in \mathfrak{g}^*$, let $\mathcal{O}(\lambda)$ be the corresponding orbit generated by λ. Since our discussion will not involve the topology on $\mathfrak{g}//G$ or $\mathfrak{g}^*//G$, we will not address this issue in this chapter.

One classical example is the unitary Lie algebra $\mathfrak{u}(n)$, consisting of the $n \times n$ skew-Hermitian matrices. The unitary group $U(n)$ acts on $\mathfrak{u}(n)$ adjointly. The adjoint orbits in this case are in one-to-one correspondence with $i\lambda$ with $\lambda \in \mathbb{R}^n$ and

$$\lambda_1 \geq \lambda_2 \ldots \geq \lambda_n.$$

Let H be a Lie subgroup of G. Let $\mathcal{O}(\lambda)$ be a coadjoint orbit of G. Let $\mathrm{Pro}_{\mathfrak{h}}^{\mathfrak{g}} : \mathfrak{g}^* \to \mathfrak{h}^*$ be defined as

$$(\mathrm{Pro}_{\mathfrak{h}}^{\mathfrak{g}}\phi)(h) = \phi(h), \qquad (\forall\ \phi \in \mathfrak{g}^*,\ \ h \in \mathfrak{h}).$$

Clearly the set $\mathrm{Pro}_{\mathfrak{h}}^{\mathfrak{g}}\mathcal{O}(\lambda)$ is H-invariant under the coadjoint action. Hence $\mathrm{Pro}_{\mathfrak{h}}^{\mathfrak{g}}\mathcal{O}(\lambda)$ is a union of coadjoint orbits of H. For each $\lambda \in \mathfrak{g}^*//G$, we define $\mathrm{Pro}_{\mathfrak{h}}^{\mathfrak{g}}(\lambda)$ to be the subset of $\mathfrak{h}^*//H$,

$$\{\mu \in \mathfrak{h}^*//H : \mathcal{O}(\mu) \in \mathrm{Pro}_{\mathfrak{h}}^{\mathfrak{g}}(\mathcal{O}(\lambda))\}.$$

Notice $\mathrm{Pro}_{\mathfrak{h}}^{\mathfrak{g}}\mathcal{O}(\lambda)$ is a subset of $\mathfrak{h}^*$ and $\mathrm{Pro}_{\mathfrak{h}}^{\mathfrak{g}}\lambda$ is a subset of $\mathfrak{h}^*//H$.

It may have been more reasonable to use $\mathrm{Res}_{\mathfrak{h}}^{\mathfrak{g}}$, instead of $\mathrm{Pro}_{\mathfrak{h}}^{\mathfrak{g}}$. However, in most applications in the literature, $\mathfrak{g}$ and $\mathfrak{h}$ are both reductive. In this situation, by identifying $\mathfrak{g}^*$ with $\mathfrak{g}$ and $\mathfrak{h}^*$ with $\mathfrak{h}$, $\mathrm{Pro}_{\mathfrak{h}}^{\mathfrak{g}}$ is indeed a projection from $\mathfrak{g}$ to $\mathfrak{h}$. Since we focus on the case both G and H are reductive, we retain the notation $\mathrm{Pro}_{\mathfrak{h}}^{\mathfrak{g}}$. One advantage we gain is that $\mathfrak{g}$ and $\mathfrak{h}$ can both be treated as linear Lie algebras.

The most well-known case of orbital projection is the following

Theorem 1.1: [Cauchy Interlacing Relation] *Let $i\mathfrak{u}(n)$ be the space of $n\times n$ Hermitian matrices. Define $\mathrm{Pro}_{i\mathfrak{u}(n-1)}^{i\mathfrak{u}(n)}$ to be the projection of a Hermitian matrix to its upper left $(n-1)\times(n-1)$ submatrix. Then $\mu \in \mathrm{Pro}_{i\mathfrak{u}(n-1)}^{i\mathfrak{u}(n)}(\lambda)$ if and only if (λ, μ) satisfies the Cauchy interlacing relations:*

$$\lambda_1 \geq \mu_1 \geq \lambda_2 \geq \mu_2 \geq \ldots \geq \lambda_{n-1} \geq \mu_{n-1} \geq \lambda_n.$$

The coadjoint orbits become homogeneous symplectic manifolds when they are equipped with the canonical Kirillov-Kostant-Souriau symplectic form. The orbital projection $\mathrm{Pro}_{\mathfrak{h}}^{\mathfrak{g}}\mathcal{O}(\lambda)$ can then be interpreted as the moment map of the Hamiltonian H-action on $\mathcal{O}(\lambda)$. The case that both G and H are compact was intensively studied in the past. In this case, the $\mathrm{Pro}_{\mathfrak{h}}^{\mathfrak{g}}(\lambda)$ is a convex polytope, often known as the Kirwan polytope, denoted by $\Delta_H(\lambda)$. We shall refer the reader to [18] for references and historical account.

In this chapter, we will focus on the case where both G and H are noncompact. When G is semisimple and noncompact, $\mathrm{Pro}_{\mathfrak{h}}^{\mathfrak{g}}(\mathcal{O}(\lambda))$ is a lot more complex. First of all, there are often nonconjugate Cartan subalgebras and each Cartan subalgebra yields a class of semisimple adjoint orbits. Each class of semisimple adjoint orbit must be treated differently. Secondly, semisimple adjoint orbits do not exhaust all adjoint orbit and there are nilpotent orbits which behave quite differently. Thirdly to treat all adjoint orbits, an induction process will be needed and it will involve the nilpotent orbits of certain smaller subalgebra. Generally speaking, $\mathrm{Pro}^{\mathfrak{g}}_{\mathfrak{h}}(\lambda)$ will not be convex unless H is compact.

2. Branching Laws

Let G be a Lie group and π be a unitary representation of G. Let H be a Lie subgroup of type I and $\hat{H}$ the unitary dual of H ([7]). Then $\pi|_H$ decomposes into a direct integral of irreducible unitary representations of H with multiplicities

$$\int_{\mu \in \hat{H}} \oplus^{m_\pi(\mu)} \mu d_\pi(\mu).$$

Here $m_\pi(\mu)$ is the multiplicity of μ in π, which can assume the value ∞. Perhaps, the most important part of this direct integral decomposition is the discrete spectrum, namely, those $\mu \in \hat{H}$ that occur as subrepresentations of $\pi|_H$. We denote the discrete spectrum by $\pi|_H^{dis}$.

The fundamental case to study is the decomposition of $\pi|_H$ when π is irreducible. For irreducible π, the direct integral decomposition of $\pi|_H$ is often called a branching law. One of the most well-known branching laws is the following

Theorem 2.1: [Weyl] *Let $\lambda \in \mathbb{Z}^n$ be arranged in descending order. Let π_λ be the irreducible unitary representation of $U(n)$ with highest weight λ. Let $U(n-1)$ be any subgroup of $U(n)$ preserving a nonzero vector in $\mathbb{C}^n$. Then*

$$\pi_\lambda|_{U(n-1)} \cong \bigoplus_{\mu \in \mathbb{Z}^{n-1}, \lambda_1 \geq \mu_1 \geq \lambda_2 \geq \mu_2 \geq \ldots \geq \lambda_{n-1} \geq \mu_{n-1} \geq \lambda_n} \tau_\mu.$$

Here τ_μ is an irreducible unitary representation of $U(n-1)$ corresponding to the highest weight μ.

We observe here that Cauchy interlacing relation is manifested here in the branching law.

Here we list a few important cases of branching laws.

(1) Let (G, H) be a symmetric pair, i.e., the subgroup H is the set of fixed points of a certain involution of G. Let π be an irreducible unitary representation of G. If π is an induced representation, then the problem of $\pi|_H$ can be reduced to a direct integral decomposition of a Hilbert space on a homogeneous vector bundle. However, when π is a discrete series representation, $\pi|_H$ can be difficult to study. Some important results were obtained by Kobayashi in a series of

papers ([41], [42], [43]) regarding the discrete decomposability of a larger class of unitary representation $A(\mathfrak{q}, \lambda)$.

(2) Let (G, H) be both compact. Then every $\pi \in \hat{G}$ is finite dimensional and is in the discrete series. This case has been intensely studied. A particular example is $G = U(n) \times U(n)$ and $H = U(n)$ diagonally embedded in G ([40]). The branching law in this case is exactly the decomposition of tensor product of two irreducible unitary representations, given by Littlewood-Richardson rule.

(3) Let (H_1, H_2) be a dual reductive pair in a symplectic group $Sp_{2n}(\mathbb{R})$ and ω the Weil representation of $\widetilde{Sp}_{2n}(\mathbb{R})$. Then branching law $\omega|_{\widetilde{H_1 \times H_2}}$ is given by the L^2 Howe's correspondence ([32]). We understand well the cases that (H_1, H_2) are of the same size ([53], [1]) or (H_1, H_2) in the stable range ([44]) or one of (H_1, H_2) is compact ([33]). There are many other cases that we do not have a complete description of the branching law ([45]).

(4) There are also the problems of studying the multiplicities in the branching laws. A great deal of effort was made to understand the multiplicity one branching laws, both in compact cases and noncompact cases. For the compact case, see [16] and the references therein.

The literature for the branching laws is vast. It is impossible to include all the references here. The main point is that many of these branching problems have "counterpart" problems for coadjoint orbits, as we shall discuss in the next section.

3. Orbit Method

From Theorem 1.1 and Theorem 2.1, we already see that orbit projection and the branching law are related. In fact, this kind of relation is exactly suggested by the orbit method. The orbit method, pioneered by Kirillov and Kostant, is a method that produces a correspondence between coadjoint orbits and irreducible unitary representations, by which both the structure of coadjoint orbits and irreducible unitary representations can be better understood.

In the early 1960s, Kirillov proved that there is a one-to-one correspondence between irreducible unitary representations of a simply connected nilpotent group N and the coadjoint orbits in $\mathfrak{n}^*$ ([35]). Later Auslander and Kostant extended this correspondence to type I solvable groups ([2]). Auslander and Kostant proved that the correspondence between coadjoint orbits and irreducible unitary representations still holds if one imposes an admissible condition on the coadjoint orbits. For G noncompact and semisimple, the correspondence between $\hat{G}$ and $\mathfrak{g}^*//G$ runs into several problems. For example, not every coadjoint orbit will have a corresponding irreducible unitary representation. Hence there is no one-to-one correspondence between $\mathfrak{g}^*//G$ and $\hat{G}$. In [9], Duflo defined an admissible condition on coadjoint orbits for all semisimple Lie groups. This produces an remedy—one should use the admissible orbit datum, instead of coadjoint orbit alone, to associate to irreducible unitary representations ([9], [65]). Since then, a great deal of progress has been made to establish a correspondence between the unitary dual and admissible coadjoint orbit datum ([65]).

The most well-understood part of the orbit method is the correspondence between the tempered dual and admissible coadjoint orbits ([10]). In the case that G is real reductive, a theorem due to Harish-Chandra and Rossmann provides a good foundation to discuss irreducible tempered representations and coadjoint orbits ([21], [23], [55], [61]). We remark that the correspondence between tempered representations and admissible coadjoint orbits have been extended to all type I real Lie groups by Duflo and his school, namely the decomposition of $L^2(G)$ can be parametrized by admissible coadjoint orbits in the sense of Duflo ([10], [63]).

Theorem 3.1: [Rossmann] *Let G be a real reductive group of Harish-Chandra class. Let π be an irreducible tempered representation of G with regular infinitesimal character ([38]). Then there exists a unique coadjoint orbit $\mathcal{O}(\lambda)$ such that $\Theta_\pi(g)$, the Harish-Chandra character of π satisfies*

$$\Theta_\pi(\exp x) = C_\pi p^{-1}(x)\mathcal{F}(\mu(\mathcal{O}(\lambda)))(x), \qquad (x \in \Omega)$$

as distributions on $\mathfrak{g}$. Here C_π is a constant; $\mu(\mathcal{O}(\lambda))$ is the canonical invariant measure on $\mathcal{O}(\lambda)$; $p^{-1}(x)$ is a factor related to $\exp : \mathfrak{g} \to G$; Ω is an open neighborhood of zero in $\mathfrak{g}$ and $\mathcal{F}$ is the Fourier transform from tempered distributions $\mathcal{S}'(\mathfrak{g}^*)$ to $\mathcal{S}'(\mathfrak{g})$.

Essentially, the pull back of the Harish-Chandra character $\exp^*(\Theta_\pi)$, as a distribution on Ω, can be identified locally with the Fourier transform of the invariant measure $d\mathcal{O}_\lambda$ multiplied by $C_\pi p^{-1}(x)$. Alternatively, for any ϕ a smooth function with compact support in a small open neighborhood of the identity in G,

$$\int \phi(\exp x)\Theta_\pi(\exp x)dx = C_\pi \int_{\mathcal{O}_\lambda} \mathcal{F}(p^{-1}(x)\phi(\exp x))(\xi)d\xi.$$

Rossmann's character formula only relates the Harish-Chandra character to the Fourier transform of an admissible orbit in a small neighborhood of the identity. M. Vergne established the following elegant result, known as the Poisson-Plancherel formula for all semisimple Lie groups ([62]), in which the complete Fourier transforms of generic admissible orbits all appear.

Theorem 3.2: [Vergne] *Let G be a semisimple Lie group. Let*

$$E(G) = \{x \in \mathfrak{g} \mid \exp x = e\}.$$

Then there exist a G-invariant measure d_E on $E(G)$ and a G-invariant measure d_v on $\mathfrak{g}^*$ such that for any smooth function ϕ with compact support

$$\int_{E(G)} \phi(x) d_E(x) = \int_{\mathfrak{g}^*} \mathcal{F}(p(x)\phi(x))(\xi) d_v(\xi).$$

Here $E(G)$ is a (lattice-like) subset of elliptic elements in $\mathfrak{g}$ and $d_v(g)$ can be regarded as a Lie algebra analog of the Plancherel measure on $\hat{G}$. Recall that the classical Plancherel formula related $\phi(e)$ to an integral of $Tr(\pi(\phi))$ with respect to the Plancherel measure on $\pi \in \hat{G}$.

There are many other important results concerning coadjoint orbits and irreducible unitary representations, which are beyond the scope of this chapter. What is important for the purpose of this chapter, is that, associated to each discrete series representation of a semisimple group, there is a unique elliptic coadjoint orbit, generated by its Harish-Chandra parameter, from which many analytic properties of the representation can be obtained.

To understand what the orbit method says about branching law, let us recall the theorem of Kirillov regarding branching laws of nilpotent groups (page 81, [35]).

Theorem 3.3: [Kirillov] *Let N be a simply connected nilpotent Lie group and H a closed connected Lie subgroup. For each $\lambda \in n^*//N$, let π_λ be the irreducible unitary representation constructed by polarization of the coadjoint orbit $\mathcal{O}(\lambda)$. Then*

$$\pi_\lambda|_H = \int_{\mu \in \mathrm{Pro}_{\mathfrak{h}}^{\mathfrak{n}}(\lambda)} \oplus^{m(\mu)} \tau_\mu d_\lambda(\mu),$$

with τ_μ the irreducible unitary representation of H corresponding to $\mu \in \mathfrak{h}^*//H$.

This theorem can be characterized by the following commutative diagram

$$\begin{array}{ccc} \mathcal{O}(\lambda) \subset \mathfrak{n}^* & \longrightarrow & \pi_\lambda \\ \big\downarrow \mathrm{Pro}^{\mathfrak{n}}_{\mathfrak{h}} & & \big\downarrow Res|_H \\ \sqcup_{\mu \in \mathrm{Pro}^{\mathfrak{n}}_{\mathfrak{h}}(\lambda)} \mathcal{O}(\mu) & \longrightarrow & \int_{\mu \in \mathrm{Pro}^{\mathfrak{n}}_{\mathfrak{h}}(\lambda)} \oplus^{m(\mu)} \tau_\mu d_\lambda(\mu) \end{array} \tag{3.1}$$

The remaining question is whether Kirillov's theorem can be extended to other groups.

When G is compact, H is a closed Lie subgroup, π an irreducible unitary representation, $\pi|_H$ decomposes discretely into a direct sum of irreducible representations of H with multiplicities:

$$\pi|_H = \oplus_{\sigma \in \hat{H}} \sigma \otimes M(\pi, \sigma).$$

Here $M(\pi, \sigma)$ records the multiplicity. On the geometric side, there is the moment map, which can be identified with $\mathrm{Pro}^{\mathfrak{g}}_{\mathfrak{h}}$, maps any coadjoint orbit $\mathcal{O}$ into $\mathfrak{h}^*$. The famous Guillemin-Sternberg conjecture, proved by Meinrenken, established a precise relation between $M(\pi, \sigma)$ and the geometry of the preimage of $\mathcal{O}(\sigma)$ ([19], [29], [47], [48]). Here the correspondence between coadjoint orbits and unitary dual is readily provided by the Borel-Weil theorem.

When G is semisimple noncompact and H is compact, π a tempered representation of G, Guillemin-Sternberg conjecture remains to be true with some modifications ([51], [13]). As pointed out by us by Prof. Vergne, what holds true for H compact, still applies to the case H noncompact under the assumption that the projection $\mathrm{Pro}^{\mathfrak{g}}_{\mathfrak{k}_H} : \mathcal{O} \to \mathfrak{k}^*_H$ is proper where K_H is a maximal compact subgroup of H. In such case, the representation corresponding to $\mathcal{O}$ should decompose discretely under K_H, therefore decompose discretely under H ([52]).

Recently, Liu, Oshima and Yu established a branching law for $G = Spin(n, 1)$ and H the minimal parabolic subgroup. They proved that for any irreducible unitary representation π of G, $\pi|_H$ decomposes into a finite direct sum. In this case, they confirmed a conjecture of Duflo, relating the branching law of the discrete series to $\mathrm{Pro}^{\mathfrak{g}}_{\mathfrak{h}}$ of the elliptic orbits ([46]).

For (G, H) noncompact, the relation between the branching laws and orbit projection is much more delicate. Nevertheless, many techniques for compact H still make sense. For instance, when $\mathrm{Pro}^{\mathfrak{g}}_{\mathfrak{h}} : \mathcal{O} \to \mathfrak{h}^*$ is proper, the push-forward of the Louiville measure, often known as the Duistermaat–Heckman measure, still exists ([14], [12]). In some sense, the Duistermaat–Heckman measure is still closely related to the multiplicity space $M(\pi|_H, \sigma)$

even though a few subtleties, like ρ-shift, continuous spectrum and nontemperedness, may complicate the relation. In any case, as we shall see later, the study of orbit projection problem can often provide new insights and perspectives to the study of branching problem and vice versa.

4. Discrete Series and Elliptic Orbits

Let G be a semisimple Lie group. Let $G \times G$ act on $L^2(G)$ from the left and right. Then $L^2(G)$ becomes a unitary representation of $G\times G$. The decomposition of $L^2(G)$ into the direct integral of irreducible unitary representations of $G \times G$ is known as the Plancherel formula. In the 1960s, Harish-Chandra successfully carried out the determination of the Plancherel formula. One critical step is the determination of the discrete spectrum $L^2(G)^{dis}$. Harish-Chandra proved that discrete spectrum exists if and only if the group G has a compact Cartan subgroup. He also parametrized the discrete spectrum of $L^2(G)$. We now state Harish-Chandra's parametrization of discrete series ([22], [24]).

Theorem 4.1: [Harish-Chandra] *Let G be a connected reductive Lie group of Harish-Chandra class. Let K be a maximal compact subgroup of G. Then G has discrete series representation if and only if $rank(G) = rank(K)$. Let T be a maximal torus of K. Let L be the integral lattice dual to T and ρ be half of the sum of positive roots of $\mathfrak{g}$ with respect to $\mathfrak{t}$. Let $W(K, \mathfrak{t})$ be the Weyl group of K with respect to $\mathfrak{t}$. Let $(L + \rho)'$ be the regular elements in $L + \rho$ with respect to the action of $W(\mathfrak{g}_{\mathbb{C}}, \mathfrak{t})$. Then discrete series of G is in one-to-one correspondence with $(L+\rho)'//W(K, \mathfrak{t})$, the $W(K, \mathfrak{t})$ orbits of $(L + \rho)'$.*

The element $\lambda \in (L + \rho)'//W(K, \mathfrak{t})$, or simply $\lambda \in (L + \rho)'$ is called the Harish-Chandra parameter of the corresponding discrete series D_λ.

Recall that the semisimple adjoint orbits can be represented by elements in the Cartan subalgebras of $\mathfrak{g}$. Hence they are parametrized by the related Weyl group orbit in the related Cartan subalgebra. The elliptic adjoint orbits are simply orbits generated by elements in the compact Cartan subalgebra. Identify $\mathfrak{g}$ with $\mathfrak{g}^*$ under the Killing form. $L + \rho$, as a subset of $\mathfrak{t}^*$, becomes a subset in $\mathfrak{g}^*$. The orbits in $\mathfrak{t}^*//W(K)$ are then in one-to-one correspondence with the elliptic coadjoint orbits in $\mathfrak{g}^*$. Now we can attach the discrete series D_λ to the elliptic coadjoint orbit $\mathcal{O}(\lambda)$.

Despite Harish-Chandra's great success of classifying discrete series representation, the structure of the discrete series representation is still not well-understood. In particular, we do not have a good understanding of K-types of discrete series representations and their multiplicities. The branching laws of discrete series $\pi|_K$ are provided by the Blattner's formula ([30]). However, the coefficients in Blattner's formula are difficult to compute and it is not easy to know which ones are nonzero. The related problem on the orbit side, $\mathrm{Pro}_{\mathfrak{k}}^{\mathfrak{g}}$ of the elliptic orbit $\mathcal{O}(\lambda)$, is equally difficult. For some recent progresses see [51], [26], [4] and Atlas of Lie groups project.

As to restrictions to noncompact subgroups, even though discrete series representation can be constructed geometrically as cohomology classes ([57]), the restrictions of cohomolgy classes to submanifolds may not even make sense. There is no general construction of discrete series by which branching laws can be easily derived. One exception here is the holomorphic discrete series. A range of branching laws can be established for discrete series by Duflo-Vargas ([11]), Vargas ([59, 60]), Paradan ([51]), Liu-Oshima-Yu ([46]), Baldoni-Vergne ([5]) among others. Most efforts have been focused on the case that π is H-admissible in the sense of Kobayashi. To this end, Duflo made a precise conjecture relating the H-admissibility to the properness of the moment map $m : \mathcal{O}(\lambda) \to \mathfrak{h}^*$. See [46] for the precise statement of Duflo's conjecture.

5. The Gan-Gross-Prasad Branching problem for unitary groups

Let us now consider the discrete series of $G = U(p, q)$. The compact Cartan subalgebra can be identified with $i\mathbb{R}^p \times i\mathbb{R}^q$. In linear algebra terms, an elliptic element in $\mathfrak{u}(p, q)$ is a diagonalizable matrix in $\mathfrak{u}(p, q)$ with purely imaginary eigenvalues

$$(i\chi_1, i\chi_2, \dots, i\chi_{p+q}).$$

The regular elements are just those $i\chi$ with distinct χ_i. Suppose that χ_i's are all distinct. The defining Hermitian form of signature (p, q) restricted onto each eigenspace $E(i\chi_k)$ is either positive definite or negative definite. If we attach a $+$ sign $(+1)$ or $-$ sign (-1) to $i\lambda_k$, we obtain a sequence of sign in $\{\pm 1\}^{p+q}$ with p $+1$'s and q -1's. We denote it by z. Now we see that the regular elements in a compact Cartan subalgebra can be parametrized by (χ, z) with $\chi \in \mathbb{R}^{p+q}$ and $z \in \{\pm 1\}^{p+q}$ and z has signature (p, q). The Weyl group $W(K, \mathfrak{t})$ acts on $(i\chi, z)$ by permuting the $(i\chi, z)$ simultaneously and preserving z. In other words, $W(K, \mathfrak{t})$ permutes those $i\chi_i$ with the same sign z_i. For convenience, one may think of $(i\chi, z)$ as signed eigenvalues, with each eigenvalue $i\chi_i$ a sign z_i attached to it. Now an regular elliptic orbit

$\mathcal{O}(i\chi, z)$ is simply the conjugacy class of matrices in $\mathfrak{u}(p,q)$ with the signed eigenvalue $(i\chi, z)$.

We make a quick remark here. If χ is not regular, $E(i\lambda_k)$ may be more than one dimensional. In this case, we will have λ_k appear $\dim(E(i\lambda_k))$ times. Suppose that the defining Hermitian form restricted onto $E(i\lambda_k)$, which is necessarily nondegenerate, has signature (r,s). Then we assign r $+1$'s and s -1's to λ_k's. There are $\frac{(r+s)!}{r!s!}$ ways to do this. Hence $(i\chi, z)$ will no longer be unique. In any case, elements in a compact Cartan subalgebra can still be represented by its signed eigenvalues $(i\chi, z)$. There is no ambiguity what $\mathcal{O}(i\chi, z)$ is even when $(i\chi, z)$ is not regular.

Definition 5.1: *[*[17], [15]*]* We say that two signed elliptic element $(i\chi, z)$ and $(i\eta, t)$, of $U(p,q)$ and $U(p-1,q)$ respectively, satisfy the Gan-Gross-Prasad interlacing relation, if one can line up χ and η in the descending ordering such that the corresponding sequence of signs from z and t only has the following eight adjacent pairs

$$(\oplus +), (+\oplus), (-\ominus), (\ominus -), (+-), (-+), (\oplus\ominus), (\ominus\oplus).$$

Here $\oplus$ and $\ominus$ represent $+1$ and -1 in t, and $+$ and $-$ represent $+1$ and -1 in z. We call such a sign sequence the (interlacing) sign pattern of (χ, z) and (η, t). Clearly, when there is neither $-$ nor $\ominus$, this interlacing relation is exactly the classical Cauchy interlacing relation.

Now we have the following theorem regarding the projection of elliptic orbits in $\mathfrak{u}(p,q)$.

Theorem 5.2: *Let $(i\chi, z)$ be a regular elliptic element in a compact Cartan subalgebra in $\mathfrak{u}(p,q)$. Suppose that $q \geq 1$. Then an elliptic orbit $\mathcal{O}(i\eta, t)$ appears in $\mathrm{Pro}^{\mathfrak{u}(p,q)}_{\mathfrak{u}(p,q-1)}\mathcal{O}(i\chi, z)$ if and only if $(i\chi, z)$ and $(i\eta, t)$ satisfy the Gan-Gross-Prasad interlacing relation.*

The proof of this theorem is based on linear algebra. We omit it here.

Let us look at the elliptic orbit corresponding to holomorphic discrete series. Consider $(i\chi, z)$ with

$$\chi_1 > \chi_2 \ldots > \chi_{p+q},$$

$$z_1 = z_2 = \ldots z_p = 1, \qquad z_{p+1} = z_{p+2} = \ldots = z_{p+q} = -1.$$

We have the sign pattern of $(i\chi, z)$:

$$\overbrace{+ + \ldots +}^{p}\overbrace{- - \ldots -}^{q}.$$

We now insert $(i\eta, t)$ into $(i\chi, z)$ in descending order. The only sign pattern allowed by GGP interlacing relation is

$$\oplus + \oplus + \oplus + \ldots \oplus + - \ominus - \ominus - \ldots \ominus -.$$

Hence the elliptic element appears in the projection must have the sign pattern

$$\overbrace{\oplus \oplus \ldots \oplus}^{p} \overbrace{\ominus \ominus \ldots \ominus}^{q-1}.$$

We see that these elliptic elements again correspond to holomorphic discrete series. In addition, we can show that these elliptic elements exhaust all regular semisimple orbits in $\mathrm{Pro}^{\mathfrak{u}(p,q)}_{\mathfrak{u}(p,q-1)}(\mathcal{O}(\xi, z))$.

A similar statement holds on the branching law side, as conjectured by Gan, Gross and Prasad ([17], [15]). Let $(i\chi, z)$ be a Harish-Chandra parameter for a discrete series representation $D(i\chi, z)$ for $U(p, q)$. Here $\chi \in \mathbb{R}^{p+q}$ is a sequence of distinct integers if $p+q$ is odd, of half integers if $p+q$ is even. Let $D(i\eta, t)$ be a discrete series representation of $U(p-1, q)$.

Theorem 5.3: [[27]] *Suppose $q \geq 1$. The discrete spectrum*

$$D(i\chi, z)|^{dis}_{U(p,q-1)} = \hat{\oplus}_{(\eta,t)} D(i\eta, t)$$

where the direct sum is taking over all those Harish-Chandra parameters $(i\eta, t)$ such that $(i\chi, z)$ and $(i\eta, t)$ satisfy the GGP interlacing relation.

The main idea of the proof is to relate the branching law of $D(i\chi, z)$ to the branching laws of the Weil representation ω restricted to a noncompact dual pair $(U(n), U(n+1))$. One critical ingredient is the determination of discrete spectrum of ω, due to Jian-Shu Li ([45]). This allows us to apply Howe's correspondences in a see-saw fashion to obtain the discrete spectrum of $D(i\chi, z)|_{U(p,q-1)}$ inductively.

As we can see, in the case of $(U(p, q), U(p, q-1))$, the projection of elliptic coadjoint orbits $\mathrm{Pro}^{\mathfrak{g}}_{\mathfrak{h}}$ does tell correctly which discrete series representations occurs in D_H. However, one cannot use this fact to prove the branching law. The orbit method is a good way to suggest the branching laws. But one can hardly use it to prove the branching laws when G is semisimple ([37]). Kirillov's branching law is an exception since all irreducible unitary representations can be obtained through polarization.

6. Wave Front Set

The projection of coadjoint orbits $\mathrm{Pro}^{\mathfrak{g}}_{\mathfrak{h}}$, in the most general cases, will not tell exactly which representations occurs in $\pi|_H$. In addition, the correspondence between coadjoint orbits and unitary dual may not be available. One remedy is to study the wave front set. Let π be a unitary representation of G, not necessarily irreducible. The wave front set $WF(\pi)$, defined by Howe, is the union of the wave front sets of all matrix coefficients of π ([31]). It is a conic subset of the cotangent bundle T^*G. Due to the action of G, it is enough to consider T^*_eG which can then be identified with $\mathfrak{g}^*$. Without loss of generality, we assume $WF(\pi)$ is in $\mathfrak{g}^*$. Due to the adjoint action of G, $WF(\pi)$ is invariant under the coadjoint action of G.

When π is an irreducible unitary representation of a semisimple Lie group, $WF(\pi)$ lies in the zero set of all homogeneous G-invariant functions on $\mathfrak{g}^*$. Hence $WF(\pi)$ lies in $\mathcal{N}(\mathfrak{g}^*)$, the nilcone of $\mathfrak{g}^*$. In particular, $WF(\pi)$ can be defined as the wave front set of the Harish-Chandra character $\Theta_\pi(g)$. It is closed related to other invariants of π, the asymptotic cycle, characteristic cycle, and associated variety ([3], [58], [64]). Since a series of representations may share the same wave front set, one usually does not gain precise information about the branching laws. The issue is less about G, since many irreducible representations of G with the same wave front set also have the similar branching laws. It is more about the group H. More precisely, let us consider the following diagram

$$\begin{array}{ccc} \pi \subset \hat{G} & \xrightarrow{WF} & WF(\pi) \\ \Big\downarrow {\scriptstyle Res} & & \Big\downarrow {\scriptstyle \mathrm{Pro}} \\ \pi|_H = \int_{\mu \in \hat{H}} \oplus^{m(\mu)} \mu d_\lambda(\mu) & \xrightarrow{???} & \mathrm{Pro}^{\mathfrak{n}}_{\mathfrak{h}} WF(\pi) \end{array} \tag{6.1}$$

One important result, due to Howe, asserts that

$$WF(\pi|_H) \supseteq \mathrm{Pro}^{\mathfrak{g}}_{\mathfrak{h}} WF(\pi).$$

But we do not know whether $WF(\pi|_H) = \mathrm{Pro}^{\mathfrak{g}}_{\mathfrak{h}} WF(\pi)$. Even if we assume the equality, we still cannot read $\pi|_H$ off the $WF(\pi|_H)$. Nevertheless, once we know $\mathrm{Pro}^{\mathfrak{g}}_{\mathfrak{h}} WF(\pi)$, we gain some information about $\pi|_H$. There are many important results regarding branching laws based on wave front sets.

Theorem 6.1: [Kobayashi, Cor. 3.4 [43]] *Let π be an irreducible unitary representation of a reductive group G. Let H be a reductive subgroup. Suppose that $\pi|_H$ is infinitesimally discretely decomposable, then*

$$\mathrm{Pro}^{\mathfrak{g}}_{\mathfrak{h}}(WF(\pi)) \subseteq \mathcal{N}(\mathfrak{h}^*).$$

Here reductive subgroup means H is reductive in G. Infinitesimally discretely decomposable means that the underlying $(\mathfrak{g}, K)$ module can be decomposed as a direct sum of irreducible $(\mathfrak{h}, K_H)$ submodules.

Clearly, this theorem gives us a criterion for $\pi|_H$ not infinitesimally discretely decomposable.

For the branching law $\omega|_{\tilde{G}}$ in the dual reductive setting, we have

Theorem 6.2: [Przebinda] *Let (G, G') be a dual pair $Sp_{2N}(\mathbb{R})$. Let ω be the Weil representation of $\widetilde{Sp}_{2N}(\mathbb{R})$. If π is in the discrete spectrum of $\omega|_{\tilde{G}}$ then*

$$WF(\pi) \subseteq \mathrm{Pro}_{\mathfrak{g}}^{\mathfrak{sp}_{2N}(\mathbb{R})} WF(\omega).$$

Here the $WF(\omega)$ is the minimal nilpotent orbit of $\mathfrak{sp}_{2N}(\mathbb{R})$ consisting of all rank 1 and 0 matrices in $\mathfrak{sp}_{2N}(\mathbb{R})$.

In fact, Przebinda proved this theorem for all $\pi \in \mathcal{R}(\tilde{G}, \omega)$ (Cor. 2.8, page 557 [54]). This theorem can be applied to detect those representations that are not in $\mathcal{R}(\tilde{G}, \omega)$ ([28]). Important results concerning $\mathrm{Pro}_{\mathfrak{g}}^{\mathfrak{sp}_{2N}(\mathbb{R})} WF(\omega)$ can be found in [6].

Finally, we shall mention a result due to Harris, Olafsson and myself that describes the wave front set for any unitary representation weakly contained in the L^2 space of a reductive group of Harish-Chandra class ([25]). Let G be a reductive Lie group of Harish-Chandra class. Let $\widehat{G}_{\mathrm{temp}}$ be the tempered dual ([39]), the part of unitary dual appearing in the Plancherel formula of $L^2(G)$. To each irreducible tempered representation σ of G, as we have seen for the discrete series, Duflo and Rossmann associated a finite union of coadjoint orbits $\mathcal{O}_\sigma \subset \mathfrak{g}^*$ ([8], [55], [56]). In the generic case, $\mathcal{O}_\sigma$ is a single coadjoint orbit.

For each π weakly contained in the regular representation, we define the orbital support of π by

$$\mathcal{O}\text{-}\operatorname{supp}\pi = \bigcup_{\sigma \in \operatorname{supp}\pi} \mathcal{O}_\sigma.$$

Here $\operatorname{supp}\pi$ is a closed subset of $\hat{G}_{\mathrm{temp}}$.

Theorem 6.3: *If G is a noncompact reductive Lie group of Harish-Chandra class and π is weakly contained in the regular representation of G, then*

$$WF(\pi) = AC(\mathcal{O}\text{-}\operatorname{supp}\pi),$$

where $AC(S)$ *for any* S *in a linear space* V *is defined to be*

$$AC(S) = \{\xi \in V \mid \Gamma \text{ an open cone containing } \xi \Rightarrow \Gamma \cap S \text{ is unbounded}\} \cup \{0\}.$$

When G is compact and connected, this result is known. See Cor 5.10 of [34], Proposition 2.3 of [31] and [20].

Corollary 6.4: *Let* π *be a unitary representation of a reductive Lie group* G *in the Harish-Chandra class. Suppose that* $WF(\pi)$ *contains a regular elliptic element in* $\mathfrak{g}^*$. *Then there are infinitely many discrete series representations in the discrete spectrum of* π.

Combined with Howe's result that $WF(\pi|_H) \supseteq \mathrm{Pro}_{\mathfrak{h}}^{\mathfrak{g}} WF(\pi)$, we have

Corollary 6.5: *Let* G *be a Lie group and* H *be a Lie subgroup that is reductive of the Harish-Chandra class. Let* π *be an irreducible unitary representation of* G *such that* $\pi|_H$ *is weakly contained in* $L^2(H)$. *Suppose that* $\mathrm{Pro}_{\mathfrak{h}}^{\mathfrak{g}} WF(\pi)$ *contains a regular elliptic element in* $\mathfrak{h}^*$. *Then there are infinitely many discrete series representations of* H *in the discrete spectrum of* $\pi|_H$.

Acknowledgements

I would like to thank Prof. R. Howe for getting me interested in discrete series, Prof. M. Vergne for her very helpful comments and her interests in this work, and J. Yu for bringing their work ([46]) to my attention. I would also like to thank the editors for allowing me to revise Section 3 after the original manuscript was accepted.

References

1. J. Adams and D. Barbasch, "Genuine Representations of the Metaplectic Group", *Compositio Mathematica*, Volume 113 Issue 1, 23–66, 1998.
2. M. Auslander and B. Kostant, "Quantization and representations of solvable Lie groups", *Bull. Amer. Math. Soc.* 73, 692–695, 1967.
3. D. Barbasch and D. Vogan, "The Local Structure of Characters", *Journal of Functional Analysis*, (37), 27–55, 1980.
4. V. Baldoni and M. Vergne, "Discrete series representations and K multiplicities for U (p, q). User's guide", arXiv:1008.5360 (2010).
5. V. Baldoni and M. Vergne, "Tensor Product of Holomorphic Discrete Series Representations of and Quivers", Symmetry in Geometry and Analysis, Volume 1: Festschrift in Honor of Toshiyuki Kobayashi, 103–139, 2024.
6. A. Daszkiewicz, W. Kraskiewicz and T. Przebinda, "Nilpotent Orbits and Complex Dual Pairs", *Journal of Algebra* (V. 190), (518–539), 1997.

7. J. Dixmier, *C^*-algebras*, North Holland Press, Amsterdam (1977).
8. M. Duflo, "Fundamental Series Representations of a Semisimple Lie Group", *Functional Analysis Applications*, Volume 4, 122–126, 1970.
9. M. Duflo, "Théorie de Mackey pour les groupes de Lie algébriques", Acta Math. 149, 153–213, 1982.
10. M. Duflo, "On the Plancherel formula for almost algebraic real Lie groups", Lie group representations, III (College Park, Md., 1982/1983), Lecture Notes in Math., vol. 1077, Springer, Berlin, 101–165, 1984.
11. M. Duflo and J. Vargas, "Branching Laws for Square Integrable Representations", Proceedings of the Japan Academy **86**, Series A, 49–54, 2010.
12. M. Duflo, G. Heckman, and M. Vergne, "Projection d'orbites, formule de Kirillov et formule de Blattner", Harmonic analysis on Lie groups and symmetric spaces (Kleebach, 1983), M´em. Soc. Math. France (N.S.), No. 15, 65–128, 1984.
13. M. Duflo and M. Vergne, "Kirillov's formula and Guillemin–Sternberg conjecture", Comptes Rendus Math. Vol. 349, 1213–1217, 2011.
14. J. Duistermaat and G. J. Heckman, "On the variation in the cohomology of the symplectic form of the reduced phase space", Invent. Math. 69, 259–268, 1982; Addendum, 72 153–158, 1983.
15. W.-T. Gan, B. Gross and D. Prasad, "Symplectic local root numbers, central critical L values, and restriction problems in the representation theory of classical groups. Sur les conjectures de Gross et Prasad. I", Asterisque No. 346, 1–109, 2012.
16. R. Goodman and N. Wallach, *Representations and Invariants of the Classical Groups*, Cambridge University Press, 2003.
17. B. Gross and D. Prasad, "On the decomposition of a representation of SO_n when restricted to SO_{n-1}", *Canad. J. Math.* No. 5 Vol. 44, 974–1002, 1992.
18. V. Guillemin and R. Sjamaar, *Convexity Properties of Hamiltonian Group Actions*, CRM Monograph Series, American Mathematical Society, Providence. 2005.
19. V. Guillemin and S. Sternberg, "Geometric quantization and multiplicities of group representations", Invent. Math. 67, 515–538, 1982.
20. S. Hansen, J. Hilgert and S. Keliny, *Asymptotic K-support and Restrictions of Representations*, Representation Theory, Volume 13, 460–469, 2009.
21. Harish-Chandra, "Invariant Eigendistributions on a semisimple Lie Group", *Transactions of American Mathematical Society*, (V. 119), 457–508, 1965.
22. Harish-Chandra, "The Discrete Series for Semisimple Lie Groups II: Explicit Determination of the Characters", Acta Mathematica **116**, 1–111, 1966.
23. Harish-Chandra, Invariant eigendistributions on a semisimple Lie algebra, Inst. Hautes Etudes Sci. Publ. Math. No. 27, 5–54, 1965.
24. Harish-Chandra, Discrete series for semisimple Lie groups I, II, Acta Math. 113, 241–318, 1965: 116, 1–111, 1966.
25. B. Harris, H. He and G. Olafsson, "Wave front sets of reductive Lie group representations", Duke Math. J. 165(5), 793–846, 2016.
26. M. Havlickova, *Boundary of K-types in Discrete series*, MIT Thesis 2008.

27. H. He, "On the Gan–Gross–Prasad conjecture for U(p, q)", *Inventiones Mathematicae*, Vol. 209 (3), 837–884, 2017.
28. H. He, "Unipotent Representations and Quantum Induction", Memoirs of A. M. S. (1496) 2024.
29. G. Heckman, "Projections of orbits and asymptotic behavior of multiplicities for compact connected Lie groups", Inventiones Mathematicae 67 (2), 333–356, 1982.
30. H. Hecht and W. Schmid, *A Proof of Blattner's Conjecture*, Inventiones Mathematicae, Volume 31, 129–154, 1975.
31. R. Howe, *Wave Front Sets of Representations of Lie Groups*, In *Automorphic Forms, Representation Theory, and Arithmetic* (Bombay 1979), Tata Inst. Fund. Res. Studies in Math., 10, 117–140, 1981.
32. R. Howe, "Transcending Classical Invariant Theory", *Journal of American Mathematical Society* (Vol. 2), 535–552, 1989.
33. M. Kashiwara and M. Vergne, "On the Segal-Shale-Weil Representations and Harmonic Polynomials", *Invent. Math.* (44), 1–47, 1978.
34. M. Kashiwara and M. Vergne, *K-types and Singular Spectrum*, Noncommutative Harmonic Analysis (Proc. Third Colloq., Marseilles-Luminy, 1978), Lecture Notes in Mathematics, Volume 728, Springer Verlag, Berlin, 177–200, 1979.
35. A. Kirillov, "Unitary representations of nilpotent Lie groups", *Uspehi Mat. Nauk 17*, no. 4 (106), 57–110, 1962.
36. A. Kirillov, "The characters of unitary representations of Lie groups", Functional Anal. Appl. 2, 90–93, 133–146 (1968), 3, 29–38, 1969.
37. A. Kirillov, *Lectures on the Orbit Method*, American Mathematical Society, Providence, 2004.
38. A. Knapp, *Representation Theory on Semisimple Groups: An Overview Based on Examples*, Princeton University Press, 1986.
39. A. Knapp and G. Zuckerman, *Classification of Irreducible Tempered Representations of Semisimple Lie Groups*, Annals of Mathematics Series 2, Volume 118, 389–501, 1982.
40. A. Knutson and T. Tao, "The Honeycomb Model of GLn(C) Tensor Products I: Proof of the Saturation Conjecture", J. A. M. S. V. 12 No. 4, 1055–1090, 1999.
41. T. Kobayashi, *Discrete decomposability of the restriction of $A_{\mathfrak{q}}(\lambda)$ with respect to reductive subgroups and its applications*, Inventiones Mathematicae **117**, no. 2, 181–205, 1994.
42. T. Kobayashi, *Discrete decomposability of the restriction of $A_{\mathfrak{q}}(\lambda)$ with respect to reductive subgroups. II. Micro-local analysis and asymptotic K-support*, Annals of Mathematics Series 2, Volume 147, Number 3, 709–729, 1998.
43. T. Kobayashi, *Discrete decomposability of the restriction of $A_{\mathfrak{q}}(\lambda)$ with respect to reductive subgroups. III. Restriction of Harish-Chandra modules and associated varieties*, Inventiones Mathematicae, Volume 131, Number 2, 229–256, 1998.

44. J.-S. Li, "Singular Unitary Representation of Classical Groups", Inventiones Mathematicae (V. 97), 237–255, 1989.
45. J.-S. Li, "Theta lifting for unitary representations with nonzero cohomology", Duke Mathematical Journal. 61 (V3), 913–937, 1990.
46. G. Liu, T. Oshima and J. Yu, "Restriction of irreducible Unitary Representations of $Spin(N,1)$ to parabolic subgroups", *Representation Theory* V. 27, 887–932, 2023.
47. E. Meinrenken, *On Riemann-Roch formulas for multiplicities*, J. A. M. S. 9, 373–389, 1996.
48. E. Meinrenken, "Symplectic surgery and the $Spin^c$-Dirac operator", Adv. Math. 134, 240–277, 1998.
49. E. Meinrenken and R. Sjamaar, "Singular reduction and quantization", Topology 38, 699–763, 1999.
50. B. Ørsted and J. Vargas, "Restriction of Square Integrable Representations: Discrete Spectrum", Duke Mathematical Journal, Volume 123, Number 3, 609–633, 2004.
51. P.E. Paradan, *$Spin^c$-quantization and the K-multiplicities of the discrete series*, Ann. Sci. ´Ecole Norm. Sup. (4), Vol. 36, No. 5, 805–845, 2003.
52. P.E. Paradan, "Kirillov's orbit method: the case of Discrete Series representations", Duke Math. J. 168, No. 16, 3103–3134, 2019.
53. A. Paul, "Howe Correspondence for Real Unitary Groups", Journal of Functional Analysis 159, 384–431, 1998.
54. T. Przebinda, "Characters, Dual Pairs, and Unitary Representations", Duke Journal of Mathematics (V. 69), 547–592, 1993.
55. W. Rossmann, "Kirillov's Character Formula for Reductive Lie Groups", Inventiones Mathematicae, Volume 48, Number 3, 207–220, 1978.
56. W. Rossmann, *Limit Characters of Reductive Lie Groups*, Inventiones Mathematicae, Volume 61, Number 1, 53–66, 1980.
57. W. Schmid, "L^2 cohomology and discrete series", Annals of Math. 103, 375–394, 1976.
58. W. Schmid and K. Vilonen, "Characteristic Cycles and Wave Front Cycles of Representations of Reductive Lie Groups", Annals of Mathematics Series 2, Volume 151, Number 3, 1071–1118, 2000.
59. J. Vargas, "Restriction of square integrable representations, a review", Symposium in Representation Theory, edited by Yamashita, J., Hokadate, Japan, 62–79, 2003.
60. J. Vargas, "Associated symmetric pair and multiplicities of admissible restriction of Discrete Series", Int. J. Math. 27, 12, 1–29, 2016.
61. M. Vergne, "On Rossmann's Character Formula for Discrete Series", Inventiones Math. 54, 11–14, 1979.
62. M. Vergne, "A Poisson-Plancherel Formula for Semisimple Lie groups", Annals of Mathematics 115, 639–666, 1982.
63. M. Vergne, *Representations of Lie groups and the orbit method*, Emmy Noether Colloquium, Birkhauser, Boston 1983.

64. D. Vogan, "Associated Varieties and Unipotent Representations", Harmonic Analysis on Reductive Groups (Brunswick, ME, 1989), Progress in Mathematics 101, 315–388, Birkhauser, Boston, MA 1991.
65. D. Vogan, "The Orbit Method and Unitary Representations for Reductive Lie Groups", *Algebraic and Analytic Methods in Representation Theory*, 243–339, 1994.
66. A. Weil, "Sur certains groupes d'opérateurs unitaries", Acta Math., (113), 143–211, 1965.

https://doi.org/10.1142/9789819825448_0008

Mathematics around the Heisenberg Group

Roger Howe
Department of Mathematics
Yale University
New Haven, CT 06520-8283, USA
roger.howe@yale.edu

Contents

Introduction

This chapter can be regarded as a sequel or supplement to [33]. It expands on some of the topics that were treated briefly or not at all in that paper, and goes more deeply into others, most notably applications to partial differential equations. There is some overlap, but this chapter does not try to incorporate the discussion of [33], so the author hopes it will be worth the reader's while to read both. Briefly, as indicated by the title, [33] deals mainly with topics in harmonic analysis. This chapter examines dual pairs and the theta correspondence, with their applications to representation theory and automorphic forms, which have been understood with much greater clarity in the time since the publication of [33]. Regarding harmonic analysis, it focuses on the oscillator semigroup, introduced in [35], which provides substantially deeper connections to PDE than were appreciated at the time of [33].

1. The Heisenberg Group

The Heisenberg group is actually a family of groups H_n, one for each odd dimension $2n+1$. H_n can be characterized as the simply connected, two step nilpotent Lie group with one-dimensional center, and of dimension $2n+1$. This description may not make evident to the reader why it could be a central object in mathematics. This chapter will try to provide some rationale for this status.

1.1. *The Heisenberg Lie algebra*

Introducing the Lie algebra h_n of the Heisenberg group H_n helps in understanding its structure. Both h_n and H_n have straightforward realizations in terms of matrices. The Lie algebra h_n can be realized as the $(n+2)\times(n+2)$ matrices of the form

$$h_n = \left\{ \begin{bmatrix} 0 & x_1 & x_2 & \cdots & x_n & z \\ 0 & 0 & 0 & \cdots & 0 & y_1 \\ 0 & 0 & 0 & \cdots & 0 & y_2 \\ \vdots & \vdots & \vdots & \ddots & \vdots & \vdots \\ 0 & 0 & 0 & \dots & 0 & y_n \\ 0 & 0 & 0 & \dots & 0 & 0 \end{bmatrix} \right\}. \tag{1.1.1}$$

For the group H_n, you take the exponentials of these matrices. (The series terminates at the quadratic term.) In terms of sets, this just involves putting 1s down the diagonal. It is straightforward from this realization that the exponential map from h_n to H_n is a diffeomorphism.

We can write

$$h_n \simeq W \oplus Z(h_n), \tag{1.1.2}$$

where $Z(h_n) = Z \simeq \mathbf{R}$ is the one-dimensional center (all entries except z in the above description should equal 0), and W (for which instead $z = 0$, and the x_j and y_j are arbitrary) is a linear complement (of dimension $2n$) to Z in h_n. The Lie bracket $[w_1, w_2]$ takes pairs of elements of W to $Z \simeq \mathbf{R}$, so it constitutes a skew-symmetric bilinear form on W. Also, it is non-degenerate, since W is complementary to Z, so every element must have a non-zero commutator with some other one. A non-degenerate skew-symmetric bilinear form is known in linear algebra as a *symplectic* form. Thus the Lie bracket on h_n is effectively defined by a symplectic form $\langle \cdot, \cdot \rangle$ on W, and vice versa. The matrices of $\exp(W)$ are those with coordinates

$$\left(\vec{x}, \vec{y}, \frac{\vec{x} \cdot \vec{y}}{2}\right), \tag{1.1.3}$$

where here $\vec{x} \cdot \vec{y}$ denotes the standard inner product on $\mathbf{R}^n$.

In these coordinates, the group law is

$$(\vec{x}, \vec{y}, t) \times (\vec{x}', \vec{y}', t') = \left(\vec{x} + \vec{x}', \vec{y} + \vec{y}', t + t' + \frac{\vec{x} \cdot \vec{y}'}{2}\right). \tag{1.1.4}$$

1.2. *Automorphisms*

It is a standard fact of linear algebra ([40], or [20] B.2.4) that up to linear isomorphism, there is a unique symplectic form on a given $2n$-dimensional vector space. Moreover, it is not hard to argue that any two possible choices for W are conjugate by an inner automorphism of H. Thus, the decomposition (1.1.2) is essentially unique.

Also, the symplectic group — the group of linear isomorphisms of W that preserve the form $\langle \cdot, \cdot \rangle$ — can be extended to h_n by letting it act trivially on Z. Thus, it constitutes a subgroup of the automorphism group of h_n, *mutatis mutandem* of H_n. We will denote this group by $Sp(\langle \cdot, \cdot \rangle)$, or by $Sp(W)$, or by $Sp_{2n}(\mathbf{R})$, or just by Sp_{2n}, according to context.

We note that the dimension of Sp_{2n} is $2n^2 + n$ — more than double the dimension of $GL_n(\mathbf{R}) = Aut(\mathbf{R}^n)$. (Also, we can think of $GL_n(\mathbf{R})$ as a subgroup of $Sp_{2n}(\mathbf{R})$ — the subgroup that preserves the direct sum decomposition $W \simeq X \oplus X^* = X \oplus Y$ described below in §1.3.) Thus, the Heisenberg Lie algebra/group is a highly symmetric object.

In addition to $Sp(W)$, h_n allows the automorphisms

$$\gamma_a : (w, t) \to (aw, a^2 t). \tag{1.2.1}$$

Here a can be any non-zero (real) number. So arbitrary dilations of W are also allowed as automorphisms. (Note that these γ_a commute with $Sp(W)$.) The γ_a only produce positive multiplications of Z, but we will see in the next subsection that in fact we can multiply Z by any scalar; but the automorphisms that do this will not commute with $Sp(W)$.

1.3. *Connections with Heisenberg's Canonical Commutation Relations*

It is also well known that for a given symplectic form $\langle \cdot, \cdot \rangle$ on a vector space W, we can find a decomposition

$$W \simeq X \oplus Y \tag{1.3.1}$$

where X and Y are *isotropic*, that is, the restriction of the form $\langle \cdot, \cdot \rangle$ to either X or Y is identically zero. In the matrix realization (1.1.1) above, X can be taken to be the set of elements for which the last column vanishes, and Y the set for which the first row vanishes. Further, for any basis $\{e_j : 1 \leq j \leq n\}$ for X, we can find a dual basis $\{f_j\}$ for Y, so that

$$\langle e_j, f_k \rangle = \delta_{jk}, \tag{1.3.2}$$

where δ_{jk} is Kronecker's delta. Again, such decompositions and the related bases are essentially unique, up to action by the symplectic group $Sp(W) \simeq Sp_{2n}(\mathbf{R})$. The spaces X and Y are clearly maximal with respect to the property of being isotropic. Such maximal isotropic subspaces are sometimes called *Lagrangian* subspaces.

The relation between the e_j and f_j essentially duplicates Werner Heisenberg's Canonical Commutation Relations

$$[P_j, Q_k] = \hbar \delta_{jk}, \tag{1.3.3}$$

where

$$Q_k = (\text{multiplication by}) x_k \tag{1.3.3a}$$

is the position operator for the k-th spatial coordinate of a family of particles, and

$$P_j = \hbar \frac{\partial}{\partial x_j} \tag{1.3.3b}$$

is the corresponding momentum operator, and $\hbar$ is Planck's constant, which for mathematical convenience, we here normalize to be 1.

These relations are considered foundational to quantum mechanics. Heisenberg proposed them in 1925–26 ([6], [8], [9]), in papers joint with Max Born and Pascual Jordan. In the following year, Hermann Weyl [79] noted that they essentially defined a Lie algebra, and that there was an associated Lie group, which would be represented by unitary operators generated by exponentiating the P_j and iQ_j. Weyl felt that assuming that the P_j and iQ_j generated one-parameter groups was a good way to avoid the possible pathologies associated with unbounded operators (which the P_j and Q_j would necessarily be).

Physicists took to referring to the Lie group associated to the CCR as the "Weyl group". However, since mathematicians understand "Weyl group" as referring to quite a different kind of group, the name "Heisenberg group" is used in this chapter.

Not long after Heisenberg's proposal, Marshall Stone [66] and John von Neumann [69], independently, showed that the concrete representation of the CCR by (multiples of) $P_j = \frac{\partial}{\partial x_j}$ and $Q_k =$ (multiplication by) x_k, are, up to isomorphism, essentially the only technically reasonable realization of the CCR.

Thus, we can think of the CCR and their standard realization as defining a unique group, and a unique (up to unitary equivalence) unitary representation of that group.

In fact, it is not hard to describe all the unitary representations of H_n. By standard arguments, one knows that, in an irreducible unitary representation ρ of H_n, the center $Z(H_n)$ will be represented by scalar operators — multiples of the identity. The specific multiple for each element of Z will define a character ξ_ρ of Z:

$$\rho(0, z) = \xi_\rho(z)I, \tag{1.3.4}$$

where I is the identity operator on the Hilbert space of ρ. If ξ_ρ is the trivial character, then ρ factors to $H_n/Z \simeq W \simeq \mathbf{R}^{2n}$, so ρ itself will be a character of H_n/Z. The Stone-von Neumann Theorem can be interpreted to say

Theorem 1.3.5:

(*a*) *For each non-trivial character of $Z(H_n)$, there is a unique, up to unitary equivalence, irreducible unitary representation of H_n with that central character. For the character $\xi(t) = e^{it}$, it may be realized on*

$L^2(\mathbf{R}^n)$, *with a standard basis for* W *being realized by the operators* P_j *and* iQ_j *of the CCR (Equations* (1.3.3a), (1.3.3b)).

(*b*) *Any representation* ρ_ξ *for* ξ *non-trivial is equivalent to any other under an automorphism of* H_n.

Stone-von Neumann implies that our particular ρ is unique up to unitary equivalence, and that the Lie algebra h_n can be represented by appropriate multiples of the P_j and Q_j. The irreducible representations of an arbitrary connected nilpotent Lie group have been described very elegantly by A. A. Kirillov [43], but the representations of H_n were known long before this.

When representation theory had developed much more, it became appreciated that these infinite dimensional representations of H_n have another valuable property — that of being square integrable. We will describe what this means slightly later. As a final remark, to supplement §1.2, note that the linear map $\delta_{a(X,Y)}$ of h_n defined by

$$\delta_{a(X,Y)}(e_j) = ae_j; \quad \delta_{a(X,Y)}(f_j) = f_j; \quad \text{and} \quad \delta_{a(X,Y)}(z) = az, \quad \text{for } a \in \mathbf{R}^\times \tag{1.3.6}$$

is an automorphism of h. Thus $Aut(H)$ acts transitively on the non-zero elements of $Z(H)$, and likewise on the non-trivial characters of Z. So all of the infinite dimensional representations of H_n can be converted to each other by means of automorphisms of H_n.

1.4. *Oscillator representation*

Since the symplectic group, acting by automorphisms of H_n, as described just above in §1.2, preserves any of the representations ρ_ξ of H_n, it follows more or less obviously that for any g in Sp_{2n}, there is a unitary operator $\omega(g)$ on the Hilbert space $\mathcal{H}_\xi$ where ρ_ξ acts, such that

$$\omega(g)(\rho_\xi(h))\omega(g)^{-1} = \rho_\xi(g(h)). \tag{1.4.1}$$

Because ρ_ξ is irreducible, $\omega(g)$ is determined up to scalar multiples (which should have absolute value 1, since $\omega(g)$ is unitary). This in turn implies that ω is multiplicative, modulo scalars:

$$\omega(g_1)\omega(g_2) = \alpha(g_1, g_2)\omega(g_1g_2) \tag{1.4.2}$$

for some complex number $\alpha(g_1, g_2)$, of absolute value 1.

It is of interest to determine the extent to which the scalar factors can be eliminated. If they can be eliminated entirely, ω will define an ordinary unitary representation of $Sp(W)$. The general situation embodied here was studied by George Mackey [48]. This particular situation was first studied carefully by Irving Segal and David Shale [59], [60], [61]. They determined that the scalar factors could be adjusted so that α always takes the values ± 1. Thus, ω can be defined as a genuine representation of a two-fold cover $\tilde{S}p(W) = Mp(W)$, often called the *metaplectic group.* This representation is known in the literature by several names: Segal-Shale-Weil representation, Weil representation, metaplectic representation, oscillator representation.... Here we use "oscillator representation".

Although $Mp(W)$ is much larger than $H(W)$, the oscillator representation ω is not irreducible. However, it almost is: it has two irreducible summands, which are the eigenspaces for the center of $Mp(W)$. This center has order 4.

Also, strictly speaking, we should have included the central character ξ in the notation for ω and written ω_ξ. However, the dependence of ω on ξ is very weak. Due to the automorphisms γ_a of Equation (1.2.1), one can see that ω can be taken to be independent of ξ for all $\xi \circ \gamma_a$. Also, one can take $\omega_{\overline{\xi}} = \overline{\omega}_\xi$. We probably should also note that the "positive" and "negative" characters of H_n, which give rise to two distinct versions of ω and that these are interchanged by the automorphism $\delta_{-1(X,Y)}$ defined in §1.3. So if you are willing to ignore complex conjugation, or take the automorphisms $\delta_{a(X,Y)}$ of Equation (1.3.6) as given, there is essentially only one ω. We also remark that ω is a representation of holomorphic type, as discussed in §2.3 below. (Its dual will then be antiholomorphic.)

The oscillator representation has many remarkable properties. It is an important aspect of the relevance of the Heisenberg group to mathematics more broadly. One indication of the special nature of ω is its "size". In various senses, it's components are the smallest possible non-trivial irreducible unitary representations of Mp, and by a good margin. There are several measures of the size of representations of reductive Lie groups. One of the most widely used is *Gelfand-Kirillov dimension* [68]. The GK dimension of the components of ω is only about half that of the next smallest irreducible unitary representations of $Mp(W)$. The components of the oscillator representation are also distinguished from other unitary representations by their restrictions to the unipotent radical of the stabilizer of a Lagrangian subspace ([24], [45]).

Some of the phenomena governed by the oscillator representation, in invariant theory, representation theory and the theory of automorphic forms are sketched in §§2 and 3 below. In §4, relations with analysis and PDE are described. There are also many notable connections to mathematical physics, but we will only describe a few. The papers [34] and [2] discuss connections with physics in greater detail.

1.5. *Extension to more general fields*

In 1949, George Mackey published an article [47], in which he pointed out that the SvN Theorem could be interpreted in such a way as to apply to a much larger class of groups. In the SvN Theorem, one copy of $\mathbf{R}^n$ (say the one spanned by the Q_j) functions as the Pontrjagin dual of the other copy (in this context, the one spanned by the P_j). In the associated unitary representation characterized by SvN, the first copy acts on $L^2(\mathbf{R}^n)$ by the regular representation, i.e., by translation operators. The other copy of $\mathbf{R}^n$ acts by means of the dual action — it multiplies functions by characters of $\mathbf{R}^n$.

Mackey observed that this construction could be extended to any locally compact abelian group $\mathbf{A}$. One can let $\mathbf{A}$ act on $L^2(\mathbf{A})$ by translation operators, and let $\hat{\mathbf{A}}$ act by multiplication by characters. Then the commutator of any translation with any multiplication-by-character will be a scalar, so, if $\mathbf{T}$ denotes the unit circle in $\mathbf{C}$, which is further identified with the scalar operators on the relevant Hilbert space, the set

$$H(\mathbf{A}) = \mathbf{A} \times \hat{\mathbf{A}} \times \mathbf{T} \tag{1.5.1}$$

acquires the structure of a two-step nilpotent group, with center $\mathbf{T}$ and commutator quotient $\mathbf{A} \times \hat{\mathbf{A}}$. Moreover, $H(\mathbf{A})$ comes with a ready-made unitary representation on which $\mathbf{T}$ acts by scalar operators in the standard way. Mackey showed that this representation was unique — up to unitary equivalence, it is the only representation of $H(\mathbf{A})$ with the specified action of $\mathbf{T}$. We will refer to this generalization of SvN as MSvN.

Among the groups covered by MSvN are the ones where $\mathbf{A}$ is $\mathbf{F}^n$, where $\mathbf{F}$ is a locally compact field, such as a p-adic field — a finite extension of $\mathbf{Q}_p$, the p-adic numbers, or the field of quotients of power series over a finite field; or also, simply a finite field. In all these situations, the Pontrjagin dual of $\mathbf{A}$ is isomorphic to $\mathbf{A}$, and the commutator operation can be lifted to a symplectic form on $W(\mathbf{A}) = \mathbf{A} \oplus \hat{\mathbf{A}} \simeq \mathbf{A} \oplus \mathbf{A}^*$, where $\mathbf{A}^*$ indicates the vector space dual of $\mathbf{A}$. Thus there is an analog of $H_n = H(\mathbf{R}^n)$ for every local field, and also for finite fields. Indeed, in the explicit matrix realization (1.1.1) of h_n, and the parallel one for H_n, we can clearly let the matrix entries x_j, y_j and z belong to any field, or indeed, any commutative ring. In particular, we can consider Heisenberg groups defined over any local field, and over finite fields.

In the first half of the 20th century, number theorists came to see the value in working with what they called the *adeles* [75]. The adeles for $\mathbf{Q}$ are the locally compact ring formed by taking the *restricted direct product* of $\mathbf{R}$ with all the $\mathbf{Q}_p$. This is the union (technically, inductive limit) of the direct product of a finite number of these fields (always including $\mathbf{R}$), together with the integers of the remaining ones. The p-adic integers $\mathbf{Z}_p$ are the closure of the integers $\mathbf{Z}$ in $\mathbf{Q}_p$. They are a compact, open subring of $\mathbf{Q}_p$. The construction of the adeles allows also the construction of the Heisenberg group over the adeles, and the corresponding symplectic group. These are likewise the restricted direct products of the associated local groups. Mackey's theory applies also to them.

1.6. *Oscillator representation for local fields*

In 1964, André Weil, a colleague of Harish-Chandra at the Institute for Advanced Study in Princeton, NJ, having studied the work of Carl Ludwig Siegel, had the insight that Siegel's construction of "Siegel modular forms" — (see [65]) involved the (projective) representation of the adelic symplectic group acting on the space of the representation of the adelic Heisenberg group [74]. He computed the obstruction to its being an actual representation, and showed that at each local place, it was of order 2, and that it was trivial on the subgroup of rational points of the group. Besides its main consequences, he showed that these results could be used to give another proof of the law of quadratic reciprocity. (This is actually a bit of overkill. You can deduce quadratic reciprocity from the oscillator representation of groups over finite fields, where no central extension is required, [23].)

The number of distinct oscillator representations for $Mp_{2n}(\mathbf{F})$, for $\mathbf{F}$ a locally compact field, is $\#\left(\frac{\mathbf{F}^\times}{(\mathbf{F}^\times)^2}\right)$. Again, they are all interchanged by the automorphisms of the type $\delta_{a(X,Y)}$ of Equation (1.3.4).

1.7. *Square integrability*

As representation theory developed, and especially through the work of Harish-Chandra, the significance of the "square integrability" of a representation became appreciated. It is especially relevant for the Heisenberg group. A representation is called *square integrable* if it appears as a discrete summand in the L^2 space of the group. The representations ρ_ξ of H are square integrable *modulo the center*. This can be formulated in the following way.

For a given character ξ of $Z(H)$, let $L^2(H, \xi)$ begine the space of functions f on H that transform by ξ under $Z(H)$:

$$f(hz) = \xi(z)f(h) \tag{1.7.1}$$

for $h \in H$ and $z \in Z(H)$. Restriction to W defines an isometric isomorphism

$$\tau : L^2(H, \xi) \longrightarrow L^2(W). \tag{1.7.2}$$

The space $L^2(H, \xi)$ has a convolution product formed by taking the natural quotient of convolution on H. (Technically, in general, although convolution can be defined for two functions in L^2, this would not be a Banach algebra, because the result might not again be in L^2; but because of square integrability, $L^2(H, \xi)$ will be a Banach algebra, as specified in Theorem 1.7.4 below.) We can transfer this to $L^2(W)$ by the isomorphism (1.7.2). This results in what is called *twisted convolution*: given two functions f_1 and f_2 on W, then

$$f_1 \natural_2(w') = \int_W f_1(w) f_2(w' - w) \xi(\tfrac{1}{2}\langle w', w \rangle)\, dw. \tag{1.7.3}$$

(Cf. formula (62) of [33], using the character $\xi(z) = e^{2\pi i z}$.)

Theorem 1.7.4: *The representation mapping $\rho : h \to \rho(h)$ is an isometric algebra isomorphism from $L^2(W)$ with its twisted convolution structure to* $\mathbf{HS}(\mathcal{H})$*, the algebra of Hilbert-Schmidt operators on the Hilbert space $\mathcal{H}$ of ρ_ξ.*

Remark 1.7.5:

(a) The basic ingredient in the proof of Theorem 1.7.4 is the Plancherel Theorem — unitarity of Fourier transform.

(b) It can be shown that this theory is valid also for groups defined over finite fields. For finite fields of odd characteristic, the oscillator representation is a bonafide representation of the symplectic group — no extension is necessary. (This is fairly easy to see using the standard cohomological interpretation of the scalar factors $\alpha(g_1, g_2)$ of Equation (1.4.2).)

1.8. *Timeline*

The developments described above took place over several decades. The approximate timeline is as follows:

Werner Heisenberg states the CCR 1925	refinement and amplification with Max Born and Pascual Jordan 1926	Hermann Weyl connection to Lie theory 1927

Stone-von Neumann Theorem 1929–30	Mackey generalization 1949

Segal and Shale explicit oscillator representation over $\mathbf{R}$ 1959–60	Andre Weil oscillator representation for all local fields 1964

2. Interaction with Classical Invariant Theory

2.1. *Weyl's FFT*

In the 1930s, Weyl concentrated on invariant theory. The results of his studies are described in his book [80]. A major result was his *First Fundamental Theorem of Invariant Theory* (FFT) for standard classical actions. Let G be a group, and $\gamma : G \to GL(V)$ be a representation of G on a vector space V. By the FFT for γ, Weyl means the description of a generating set for the space $\mathbf{P}(V^k \oplus (V^*)^\ell)^G$ of G-invariant polynomials for the action of G on any number k of copies of V, together with ℓ copies of the dual action on V^*.

Weyl gives such a description for the *standard classical actions*, which are:

(a) $GL(V)$ acting on V, for any vector space V;
(b) the orthogonal group (the isometry group of a non-degenerate symmetric bilinear form (aka *inner product*)) on a space V; and
(c) the symplectic group (the isometry group of a symplectic (non-degenerate, skew symmetric bilinear) form) on a space V (which is necessarily even-dimensional).

Since the examples (b) and (c) are self-dual, the dual action does not need to be explicitly considered in these cases.

Weyl's FFT can be summarized by the statement that, for these classical actions, the algebra of invariants is generated by the obvious quadratic invariants: the pairings between various copies of V and V^* for GL_n, and between various copies of V for O_n or Sp_{2n}.

In the 3/4 century since Weyl's work, an FFT has been established for very few other examples. See [20], §4.6, for more information.

2.2. *The Weyl algebra*

In his discussion of the CCR, Weyl introduced what are now called *Weyl algebras.* The n-th Weyl algebra can be described as the associative algebra with identity, $\mathbf{W}_n$, generated by elements P_j and Q_j, for $1 \leq j \leq n$, where the P_j and Q_j satisfy the CCR (see (1.3.3)). Using the realization (1.3.3a), (1.3.3b) of the CCR, it can also be described as the ring of polynomial coefficient differential operators attached to a given vector space. It acts in an obvious way on the algebra of polynomials on that vector space. It can also be described as the universal enveloping algebra for the Heisenberg Lie algebra h_n, modulo the relation that identifies a generator of $Z(h_n)$ with a constant.

Polynomial rings are well-known to be *graded* rings: the monomials of length d in the generators span the space $\mathbf{P}^d$ of polynomials of degree d, and it is a standard fact that $\mathbf{P}^d \cdot \mathbf{P}^e = \mathbf{P}^{d+e}$. The Weyl algebra, however, is not graded. Indeed the CCR imply that the difference $P_jQ_j - Q_jP_j$, does not vanish, but has lower degree than either P_j or Q_j.

Instead, the Weyl algebra is *filtered.* If we let $\mathbf{W}^{(d)}$ denote the span of all products of the P_j and Q_k of length at most d, then it is clear that

$$\mathbf{W}^{(d)} \cdot \mathbf{W}^{(e)} \subset \mathbf{W}^{(d+e)}. \tag{2.2.1}$$

Taking commutators tends to reduce degree. The CCR say that

$$[\mathbf{W}^{(1)}, \mathbf{W}^{(1)}] = \mathbf{W}^{(0)}.$$

From the fact that taking commutator with a fixed element is a derivation, it follows that, for positive d and e, we have

$$[\mathbf{W}^{(d)}, \mathbf{W}^{(e)}] \subset \mathbf{W}^{(d+e-2)}. \tag{2.2.2}$$

Thus in particular,

$$[\mathbf{W}^{(2)}, \mathbf{W}^{(k)}] \subset \mathbf{W}^{(k)}.$$

Therefore $\mathbf{W}^{(2)}$ is a Lie algebra, and it acts by commutator on $\mathbf{W}^{(1)}/\mathbf{W}^{(0)} = W$, and preserves the symplectic commutator form on W. It is not hard to check that this action induces an isomorphism

$$\mathbf{W}^{(2)}/\mathbf{W}^{(1)} \simeq sp(W), \tag{2.2.3}$$

where $sp(W)$ is the Lie algebra of infinitesimal symplectic isometries of W.

It can also be checked without great difficulty that the span of the symmetrized products of elements of W provides a Lie algebra complement to $\mathbf{W}^{(1)}$ in $\mathbf{W}^{(2)}$. In terms of P and Qs, this is the span of

$$Q_jQ_k = x_jx_k, \qquad P_jP_k = \frac{\partial^2}{\partial x_j \partial x_k} \qquad \text{and}$$
$$\frac{1}{2}\left(x_j\frac{\partial}{\partial x_k} + \frac{\partial}{\partial x_k}x_j\right) = x_j\frac{\partial}{\partial x_k} + \frac{1}{2}\delta_{jk}. \tag{2.2.4}$$

The additive term $\frac{1}{2}\delta_{jk}$ is related to the need to use the metaplectic covering group of Sp_{2n} to realize the oscillator representation.

It also follows from (2.2.2) that the associated graded algebra

$$\tilde{\mathbf{W}} = Gr\mathbf{W} = \Sigma_{d=0}^{\infty}\mathbf{W}^{(d)}/\mathbf{W}^{(d-1)}, \tag{2.2.5}$$

is isomorphic to the polynomial ring $\mathbf{P}(V \oplus V^*)$, where $V \simeq \mathbf{R}^n$ is the span of the $Q_j = x_j$ and $V^* \simeq (\mathbf{R}^n)^*$ is the span of the $P_j = \frac{\partial}{\partial x_j}$.

For us, an important consequence of this is

Corollary 2.2.6: *If V is the space of a classical action for the classical group G, then the action of G on $\mathbf{W} = \mathbf{W}(V)$, the Weyl algebra associated to V, by conjugation is linearly isomorphic to the polynomial ring of a classical action; precisely, the action on $V \oplus V^*$.*

Note also that since the action of $GL(V)$ on $\mathbf{W}(V)$ is by conjugation, the subalgebra $\mathbf{W}(V)^G$ of G-invariants consists of the operators in $\mathbf{W}(V)$ that commute with G. Thus the FFT implies that, if the action of G on V is classical, then the commuting ring $\mathbf{W}(V)^G$ is generated by its order two elements, which constitute the centralizer of G inside $\tilde{\mathbf{W}}^{(2)} = sp(V \oplus V^*)$.

From this, taking advantage of the rich algebra structure of $\mathbf{W}(V)$, it is not hard to argue (see [36]) that

Proposition 2.2.7: *Each isotypic component of $\mathbf{P}(V)$ under the action of G is an irreducible module for the joint action of G and the centralizer Lie algebra $g' = sp(V \oplus V^*)^G$.*

This picture can be further refined. The Lie algebra $sp(V \oplus V^*)$ has a grading:

$$\begin{aligned} sp(V \oplus V^*) &\simeq S^2(V \oplus V^*) \simeq S^2(V) \oplus (V \otimes V^*) \oplus S^2(V^*) \\ &= sp(V \oplus V^*)^{(2,0)} \oplus sp(V \oplus V^*)^{(1,1)} \oplus sp(V \oplus V^*)^{(0,2)}. \quad (2.2.8) \end{aligned}$$

The operators listed in Equation (2.2.4) provide bases for each summand of this decomposition.

Since $G \subseteq GL(V)$, the centralizer algebra g' is compatible with this decomposition:

$$\begin{aligned} g' &= sp(V \oplus V^*)^G \\ &= (sp(V \oplus V^*)^{(2,0)})^G \oplus (sp(V \oplus V^*)^{(1,1)})^G \oplus (sp(V \oplus V^*)^{(0,2)})^G \\ &= g'^{(2,0)} \oplus g'^{(1,1)} \oplus g'^{(0,2)}. \quad (2.2.9) \end{aligned}$$

Since the action of G on $\mathbf{P}(V)$ preserves degree, the G-isotypic components in $\mathbf{P}(V)$ are graded: each one is the sum of its homogeneous subspaces. The operators in $g'^{(0,2)}$, being pure second order differentiations, reduce the degree of a polynomial by 2. This means that they will annihilate the polynomials of lowest degree in any G-isotypic component. We will call the sum of all these lowest-degree subspaces the *G-harmonics*, since the elements of $g'^{(0,2)}$ bear some resemblance to the Laplace operator. (Indeed, in the special case when V consists of a single copy of the standard module for the orthogonal group, it is exactly the Laplace operator. This case is discussed in greater detail in [35]. It essentially reproduces the theory of spherical harmonics, with the added observation that r^2 and Δ, along with the modified Euler degree operator $(\Sigma_{i=1}^n x_i \frac{\partial}{\partial x_i}) + \frac{n}{2}$, span a Lie algebra, which generates the full commuting algebra in $\mathbf{W}(V)$ for the orthogonal group.) Using properties of the Weyl algebra, it is not difficult to show:

Theorem 2.2.10: *For each G-isotypic component Y in $\mathbf{P}(V)$, let Y_{min} be the homogeneous subspace of minimal degree elements belonging to Y. Then*

(a) *Y is generated by $Y_{\min}$ as a module for $g'^{(2,0)}$, that is*

$$Y = U(g'^{(2,0)}) \cdot Y_{min}.$$

(Here $U(g'^{(2,0)})$ indicates the universal enveloping algebra of $g'^{(2,0)}$. The subalgebra of $\mathbf{P}(V)$ generated by $g'^{(2,0)}$ is a quotient of $U(g'^{(2,0)})$.)

(b) *The joint action of $G \times g'^{(1,1)}$ on Y_{min} is irreducible.*

(c) *Thus, each G isotypic component can be expressed as a tensor product*

$$Y = \sigma \otimes \tau(\sigma),$$

where σ is an irreducible representation of G, and $\tau(\sigma)$ is the corresponding representation of g'.

(*d*) *Moreover, the correspondence $\sigma \longrightarrow \tau$ is a bijection between the relevant sets of representations of G and of g'.*

Details of the argument are given in [36].

2.3. *Connection with unitary representation theory of classical groups*

As noted above the space $\tilde{\mathbf{W}}^{(2)}$ has the structure of the Lie algebra sp_{2n}. Since the field of scalars for the Weyl algebra is $\mathbf{C}$, the complex numbers, this will be a complex Lie algebra. However, the unitary representation of the Heisenberg group, and its extension to (the metaplectic covering of) $Sp_{2n}(\mathbf{R})$ imposes a real structure on these Lie algebras, with the real subspace consisting of those elements that get sent to skew-adjoint operators (and so exponentiate to unitary one-parameter groups).

If we look at the real form of G that is acting on $\mathbf{P}(V)$, we find that it is compact. The real form of g' is a Lie algebra with Hermitian symmetric structure [4], whose maximal compact subalgebra is the real form of ${g'}^{(1,1)}$, and with ${g'}^{(2,0)}$ and ${g'}^{(0,2)}$ being the positive and negative nilradicals for the Hermitian structure. The representations of g' that appear in the situation described above are unitary, and of the type known as *holomorphic* representations [1], [44]. Holomorphic representations were the first representations constructed by Harish-Chandra [25], [26], in his early efforts to find the discrete series of real Lie groups. He soon realized that they were not sufficient for his needs, but they remained a distinguished class of representations, and have received considerable attention [13], [41], [1], [28], [44], etc.

In the early 1970s, several papers from several sets of authors appeared that gave a construction of holomorphic representations for the three main classes of groups with Hermitian structure. The papers [17], [18], deal with G an orthogonal group and g' a symplectic Lie algebra; papers [21], [22] deal with G a symplectic group and g' an orthogonal Lie algebra; and the paper [42] treats the case when $G = U_n$, the compact form of GL_n, and $g' = gl_m$, with $u_{p,q}$ the corresponding Hermitian real form. The results above, which were stimulated by communication with Stephen Gelbart, constitute a unification of these works. They appear in [36]. Although this paper was not published until 1989, it was written in 1974, and circulated as a preprint for many years. Its composition was stimulated by the explosion of interest at that time in Lie superalgebras, and at the cost of added complication, it presents a result that applies to tensor products of Weyl algebras and Clifford algebras.

2.4. *Dual pairs; reciprocity*

To summarize, this is the situation we described above. Given a vector space V, the Weyl algebra $\mathbf{W}(V)$ is a filtered algebra, and the space of elements of degree at most 2 decomposes

$$\mathbf{W}^{(2)} \simeq sp(V \oplus V^*) \oplus \mathbf{W}^{(1)} \simeq sp(V \oplus V^*) \oplus W(V) \oplus \mathbf{C},$$

where $W(V) = V \oplus V^*$ is understood to be endowed with its canonical symplectic form extending the duality pairing between V and V^*, and $\mathbf{C}$ is the scalars.

Also, suppose that V supports a classical action of a (classical) group G. Then the action of G on $W(V)$ by conjugation is also a classical action. Let g' be the Lie subalgebra of $sp(V \oplus V^*)$ that commutes with the action of G. Then Weyl's FFT implies that g' generates the full algebra of operators in $\mathbf{W}(V)$ that commute with G. It follows almost immediately (the case of $G = O_n$ needs checking, since O_n is not connected), that the Lie algebra g of G is then the centralizer of g' in $sp(W)$.

Moreover, the joint action of G and g' on $\mathbf{P}(V)$ creates a decomposition

$$\mathbf{P}(V) \simeq \Sigma_{\rho \in \hat{G}}\, \rho \otimes \rho',$$

where ρ runs through the irreducible representations of G, and ρ' are corresponding irreducible representations for g'. (If a given ρ does not appear in $\mathbf{P}(V)$, we understand that $\rho' = \{0\}$.) The spaces $\rho \otimes \rho'$ are the isotypic components of $\mathbf{P}(V)$ for either of G or g', and the correspondence $\rho \leftrightarrow \rho'$ is a bijection. It is also the case that g' has a Hermitian symmetric structure, and the ρ' are representations of holomorphic type relative to this structure, but that is not the point of emphasis here. We focus instead on the fact that g and g' are each the centralizer of the other inside $sp(V \oplus V')$, or that, if G' is the classical subgroup of $Sp(W)$ with Lie algebra g', then G and G' are mutual centralizers in $Sp(W)$. Thus, the restriction of the oscillator representation of $\tilde{S}p(W)$ to $\tilde{G} \times \tilde{G}'$ gives rise to a bijection between certain sets of representations of $\tilde{G}$ and of $\tilde{G}'$.

Given a pair of subgroups (H, H') of a group Γ, the condition that each be the (full) centralizer of the other inside Γ is clearly a very special relationship. We call such a pair a *dual pair* of subgroups of Γ. Starting with any subgroup H_o of Γ, we can easily find a dual pair (H, H') in Γ

with $H_o \subset H$. Just let H' be the centralizer of H_o, and let $H'' = H$ be the centralizer of H'. Then it is easy to check that (H', H'') must be a dual pair in Γ.

Thus, on one hand, dual pairs in a group are common. On the other hand, they are rather special, and dual pairs in Sp_{2n} or $\tilde{Sp}_{2n}$ are readily classifiable. The ones in $Sp_{2n}(\mathbf{C})$ that satisfy a natural irreducibility condition belong to two families:

$$(GL_k(\mathbf{C}), GL_\ell(\mathbf{C})), \quad \text{with } n = k\ell, \quad \text{and}$$
$$(O_k(\mathbf{C}), Sp_{2\ell}(\mathbf{C})), \quad \text{with } n = k\ell.$$

Dual pairs inside $Sp_{2n}(\mathbf{R})$ are made from real forms of these pairs. Dual pairs in exceptional Lie groups have been classified in [57].

Having defined the general notion that describes the relationship between g and g', one is tempted to ask the more general question: does the restriction of the oscillator representation ω from $Sp_{2n}(\mathbf{R})$ to a dual pair $G \cdot G'$ of reductive subgroups (a *reductive dual pair*) define a bijection between representations of G and representations of G'? There are technical issues that need to be taken into account in formulating a precise question, but one such question was posed in [31]. This question has been answered in the affirmative, but the result involves many cases, which were proved in pieces over a period of years. The proof for when the base field is $\mathbf{R}$ is given in [37]. When the base field is a p-adic field, the most manageable cases, which form a fairly large class, were treated in [50]. See also [71]. One of the final cases is established in [16].

Remarkably, this result, and its extension to the adelic context, includes, and in some sense "explains" many of the previously known constructions of automorphic forms. This is the focus of the next section.

3. General Dual Pairs: Applications to Automorphic Forms

In 1829, Carl Gustav Jacobi published constructions of a collection of functions of a complex variable, now known as *theta functions* [38]. Among the simplest and most basic of these is the function

$$\Theta(z) = \Sigma_{n=-\infty}^{\infty} e^{2\pi i n^2 z}. \tag{3.1}$$

This series converges for z in the upper half of the complex plane. On the real axis it is a formal Fourier series, with $e^{2\pi i k z}$ appearing with non-zero coefficient when k is a perfect square. It is clear that for $\ell \geq 1$, we have

$$\Theta^\ell(z) = \Sigma_{m=-\infty}^{\infty} S_\ell(m) e^{2\pi i m z}, \tag{3.2}$$

where $S_\ell(m)$ is the number of ways to express m as a sum of ℓ perfect squares.

In 1834, Jacobi [39] published an explicit formula for $S_4(m)$ — that it is equal 8 times the sum of the divisors of m that are not themselves divisible by 4:

$$S_4(m) = 8\ \Sigma_{4\nmid k|m} k. \tag{3.3}$$

It is obvious that this sum is non-zero. This implies that every whole number is the sum of four perfect squares, which was one of the major results of 18th century number theory, proved in 1770 by J. L. Lagrange. In the late 1840s, Gotthold Eisenstein gave a rationale for the pleasant formula for the coefficients of Θ^4 by showing that it was what is now called an *Eisenstein series*, which are examples of *modular forms* — holomorphic functions on the complex upper half plane that transform in specified ways under coordinate changes given by linear fractional transformations coming from the *modular group* $SL_2(\mathbf{Z})$, or certain of its subgroups. This connected theta functions with the theory of modular forms, which have attracted continued and growing interest over the years because of the number-theoretic information they encode. The term "modular form" is usually attributed to Erich Hecke.

Gradually, modular forms came to be understood as functions on a coset space $Sl_2(\mathbf{R})/\mathbf{\Gamma}$, where Γ is some congruence subgroup of $SL_2(\mathbf{Z})$. Modular forms were further generalized to *automorphic forms*, for which $SL_2(\mathbf{R})$ is replaced by any reductive Lie group, and Γ by any arithmetic subgroup [5], [27].

Over the course of the 20th century, there appeared a succession of constructions of modular or automorphic forms. Hecke constructed analogs of theta functions for real quadratic fields. In the 1930s, Carl Ludwig Siegel [65] constructed "Siegel modular forms", which were functions on the *Siegel upper half space*, which is the space of symmetric $n \times n$ matrices, with positive definite real part. This is the hermitian symmetric space for the real symplectic group $Sp_{2n}(\mathbf{R})$.

In 1949, Hans Maass, a student of Hecke, constructed functions on $SL_2(\mathbf{R})/\Gamma$ that were real analytic rather than complex analytic, connected with real quadratic fields [46]. In 1969, Joseph Shalika and Shunichi Tanaka [64] constructed automorphic forms for $SL_2(K)$, for any algebraic number field K. Then in 1972, Goro Shimura [63] constructed a correspondence between modular forms of half-integral weight, and modular forms of even weight. Shimura's construction inspired several follow-up papers, examining aspects of it more closely, [53], [70], [72], [15]. Stephen Rallis and Gerard Schiffmann [55], [56] wrote several papers studying construction of automorphic forms using the oscillator (or in their terminology, Weil) representation.

It turns out that all of these constructions, and several others that appeared in these years, are examples of dual pair correspondences on adele groups [19], [55].

4. Applications to Harmonic Analysis and PDE

Since the Heisenberg group combines the action of $\mathbf{R}^n$ on $L^2(\mathbf{R}^n)$ by translations, with the action of $(\mathbf{R}^n)^*$ by multiplication by characters, it should be relevant to harmonic analysis on $\mathbf{R}^n$. It is indeed, in part quite directly, and in part through the symmetries provided by the symplectic group, and by an extension of $Sp_{2n}(\mathbf{R})$ that we refer to as the *oscillator semigroup.*

4.1. *Elementary facts*

A: *Plancherel Theorem and Fourier inversion*:

First consider Fourier transform. It is easy to check that it interchanges translation and multiplication by characters. Thus, it normalizes the Heisenberg group in the algebra of operators on $L^2(\mathbf{R}^n)$. *Ipso facto*, it represents an element of $Sp_{2n}(\mathbf{R})$. This immediately implies unitarity (up to multiples — the correct multiple can be checked by computing the transform of any function, selected to make things easy, such as a Gaussian.) Since repeating Fourier transform twice will restore translations and multiplications-by-characters to themselves, it must have order two, modulo the center of the metaplectic group. This is the essential fact of Fourier Inversion.

B: *Hermite functions; the quantum harmonic oscillator*:

To get a sense of the way that the CCR impose structure, suppose that we have a pair $c^{\pm}$ of operators $c^{\pm}$ such that $c^- = (c^+)^*$, and that satisfy the (slightly differently normalized) CCR:

$$[c^+, c^-] = 2. \tag{4.1.1}$$

Suppose further that we have a vector $\vec{v}_0$ such that $c^-(\vec{v}_0) = 0$. Set

$$\vec{v}_n = (c^+)^n(\vec{v}_0) = c^+(\vec{v}_{n-1}).$$

Then we can compute

$$c^-(\vec{v}_n) = c^-c^+(\vec{v}_{n-1}) = (c^+c^- - 2)(\vec{v}_{n-1}). \tag{4.1.2}$$

Assume inductively that, for $k < n$,

$$c^-(\vec{v}_k) = -2k\vec{v}_{k-1}. \tag{4.1.3}$$

Then from the equation above, we conclude

$$\begin{aligned} c^-(\vec{v}_n) &= c^-c^+(\vec{v}_{n-1}) = (c^+c^- - 2)(\vec{v}_{n-1}) \\ &= c^+(-2(n-1)\vec{v}_{n-2}) - 2\vec{v}_{n-1} = -2n\vec{v}_{n-1}. \end{aligned} \tag{4.1.4}$$

Thus the Equation (4.1.3) holds for all k. Applying c^+ to this shows that

$$c^+c^-(\vec{v}_n) = -2n\vec{v}_n. \tag{4.1.5}$$

This says that the $\vec{v}_n$ are a set of eigenfunctions with distinct eigenvalues for the self-adjoint operator c^+c^-, and thus, they form an orthogonal basis for their span, which will be invariant under c^+ and c^-, and thus is all of $L^2(\mathbf{R})$.

Note that for the specific operators $c^- = \frac{d}{dx} + x$ and $c^+ = \frac{d}{dx} - x$, which are linear combinations of Heisenberg's original p and q, we have $c^-(e^{\frac{-x^2}{2}}) = 0$, so the discussion above applies to this pair of $c^\pm$.

These functions are in fact classical special functions, the *Hermite functions* [14]. (These come in two standard normalizations; these are called the Hermite functions for physicists. This construction of them was first described by Dirac.)

Also, the operator $c^+c^- + c^-c^+ = 2(\frac{d^2}{dx^2} - x^2)$ is essentially the Hamiltonian operator for the quantum harmonic oscillator [34]. Moreover, the Hermite functions are the basis functions of the isotypic components (which are in fact irreducible for this case) for the (very smallest) dual pair $(U_1, U_1) \subset SL_2(\mathbf{R})$.

4.2. *Oscillator semigroup*

This section is taken mainly from [35]. See also [14] for a more detailed and contextual account.

Recall from Theorem 1.7.4 that the space $L^2(W)$ is sent via the representation ρ_χ of H_n to the algebra of Hilbert-Schmidt operators on $L^2(\mathbf{R}^n)$. A fairly rich family of functions in $L^2(W)$ is provided by the *generalized Gaussian functions*, which have the form $e^{-B(w,w)}$, where B is a symmetric complex bilinear form, with positive definite real part. It turns out that this set has a natural algebraic structure: it is a semigroup under composition of operators. Let $\mathcal{A} = \begin{bmatrix} A & B \\ B^T & D \end{bmatrix}$ be a $2n \times 2n$ symmetric matrix, made up of $n \times n$ complex symmetric matrices A and D with positive definite real part, and an arbitrary complex $n \times n$ matrix B, with transpose B^T. We let $\gamma_{\mathcal{A}}$ denote the associated Gaussian function on W:

$$\gamma_{\mathcal{A}}(x, y) = \exp(-\pi(x^T Ax + 2x^T By + y^T Dy)), \tag{4.2.1}$$

where x and y are in $\mathbf{R}^n$. Denote the set of such operators as Ω_n.

The main fact about the operators in Ω_n is that the twisted convolution/composition of two of them produces another one, up to multiples. Thus, the scalar multiples of the operators in Ω form a semigroup. (The same is true if these functions are considered as integral kernel operators. This is discussed in [35]. However, the formula for composition of the associated integral kernel operators is cumbersome, and will not be dealt with here.) This can be verified by a fairly direct explicit computation. The formula for the composed operator is

$$\gamma_{\mathcal{A}_1} \sharp \gamma_{\mathcal{A}_2} = \det(\mathcal{A}_1 + \mathcal{A}_2)^{-\frac{1}{2}} \gamma_{\mathcal{A}_3}, \tag{4.2.2}$$

where

$$\mathcal{A}_3 = \mathcal{A}_2 - (\mathcal{A}_2 + iJ/2)(\mathcal{A}_1 + \mathcal{A}_2)^{-1}(\mathcal{A}_2 - iJ/2). \tag{4.2.3a}$$

Here J is the matrix of the symplectic form. Later, we will want to specify also that $J^2 = -I$. By elementary manipulations, we can rewrite this formula in the forms

$$\mathcal{A}_3 + iJ/2 = (\mathcal{A}_2 + iJ/2)(\mathcal{A}_1 + \mathcal{A}_2)^{-1}(\mathcal{A}_1 + iJ/2) \tag{4.2.3b}$$

or

$$\mathcal{A}_3 - iJ/2 = (\mathcal{A}_1 - iJ/2)(\mathcal{A}_1 + \mathcal{A}_2)^{-1}(\mathcal{A}_2 - iJ/2). \tag{4.2.3c}$$

This formula may seem somewhat complicated and obscure at first sight. However, it is much simpler than the corresponding formula for the composition of two integral operators with Gaussian kernels [35]. It can be further explicated and simplified using the Cayley transform. We will do this below.

Also, note that this family of operators is closed under taking adjoint. For this, we have the simple formula

$$(\gamma_{\mathcal{A}})^t = \gamma_{\overline{\mathcal{A}^t}}. \tag{4.2.4}$$

In this formula, $(\gamma_{\mathcal{A}})^t$ indicates the operator adjoint of $\gamma_{\mathcal{A}}$, and $\mathcal{A}^t$ indicates the transpose of the matrix $\mathcal{A}$.

We note also that the semigroup of multiples of Gaussians under twisted convolution is normalized by the action of the metaplectic group $\tilde{Sp}_{2n}(\mathbf{R})$

on $L^2(\mathbf{R}^n)$ by the oscillator representation. In fact, it is preserved not only by conjugation, but by composition on the right or the left by the operators of $GL_n(\mathbf{R})$, and also by multiplication by Gaussian functions, or their limits when the real part of the quadratic form decreases to 0.

4.3. *Normalization and Cayley transform; Estimates*

The connection between our semigroup and Sp_{2n} is stronger still. We have been working with arbitrary scalar multiples of the Gaussian kernel operators, but it turns out that there is a best multiple, up to sign, of each one. For $\mathcal{A}$ in Ω_n, set

$$\gamma^0_{\mathcal{A}} = (\det(\mathcal{A} + iJ/2)^{\frac{1}{2}} \gamma_{\mathcal{A}}. \tag{4.3.1}$$

Of course, there is an ambiguity in the choice of square root. We will not try to resolve this. Rather, we want to think of Equation (4.3.1) as a definition of a pair of operators, each the negative of the other.

With the normalization of Equation (4.3.1), it is straightforward to deduce from Equations (4.2.3a), (4.2.3b), (4.2.3c) that

$$\gamma^0_{\mathcal{A}_1} \natural \gamma^0_{\mathcal{A}_2} = \gamma^0_{\mathcal{A}_3}, \tag{4.3.2}$$

with $\mathcal{A}_3$ being determined by Equation (4.2.3b). This gives us some extra motivation to elucidate the relationship of $\mathcal{A}_3$ to $\mathcal{A}_1$ and $\mathcal{A}_2$. The key to understanding this is the Cayley transform, which connects Sp_{2n} with its Lie algebra sp_{2n}. sp_{2n} is the Lie algebra of infinitesimal isometries of the form $\langle \cdot, \cdot \rangle$ on W: it is the space of operators

$$S \in End(W); \langle Sw, w' \rangle + \langle w, Sw' \rangle = 0. \tag{4.3.3}$$

In terms of the skew-symmetric matrix J that represents $\langle \cdot, \cdot$, this says $JS + S^t J = 0$, where S^t denotes the transpose of S. Since J is itself skew-adjoint this condition may also be expressed as

$$JS = -(JS)^t = -S^t J^t = -S^t J = S^t J^t = (JS)^t. \tag{4.3.4}$$

Thus, JS is a symmetric matrix. Now consider the complex symplectic group $Sp_{2n}(\mathbf{C})$. There is a well-known ([80], [20]) rational map $c : sp \leftrightarrow Sp$, known as the *Cayley transform*. It can be defined in any associative algebra, by the formula

$$c(x) = \frac{x+1}{x-1} \tag{4.3.5}$$

for the set of x such that $x - 1$ is invertible. On this set, it is an involution: $c(c(x)) = x$, as is easily verified. For the algebra of $n \times n$ matrices, it is defined on the (open, Zariski dense) set of matrices which do not have 1 as an eigenvalue. It can also be verified by an elementary, although more involved, computation, that it takes the Lie algebra sp to the group Sp; and vice versa, since it is of order 2.

If we take two elements x and y of sp for which c is defined, and apply c, and multiply the results, then apply c again to get back to sp, the result is

$$c(c(x)c(y)) = (y-1)(x+y)^{-1}(x-1)+1. \tag{4.3.6}$$

The similarity of this to the formulas (4.2.3a), (4.2.3b), (4.2.3c) above relating $\mathcal{A}_3$ to $\mathcal{A}_1$ and $\mathcal{A}_2$ is hopefully evident. It may be thought of as a "rational version" for the classical groups of the Campbell-Hausdorff formula.

We now have the following collection of structures:

- a symplectic form J on $W \simeq \mathbf{R}^{2n}$;
- the symplectic group $Sp(W) \simeq Sp_{2n}(\mathbf{R})$ of linear transformations that preserve J;
- the oscillator/metaplectic representation ω, a unitary representation of the (two-fold metaplectic cover of the) symplectic group;
- the Lie algebra sp_{2n} of $Sp_{2n}(\mathbf{R})$;
- the collection of (multiples of) Gaussian distributions e^{-wQw^t}, where Q is a symmetric $2n \times 2n$ real matrix with positive definite real part on $W \simeq \mathbf{R}^{2n}$, which form a semigroup under twisted convolution; and more particularly, the subsemigroup of normalized Gaussians, described in formula (4.3.1).

We want to clarify the relations between these structures.

Here are the key relations.

i) The condition that a matrix g in $M_{2n}(\mathbf{R})$ belong to Sp_{2n} is that

$$gJg^t = J. \tag{4.3.7a}$$

Here g^t is the matrix transpose, aka adjoint, of g.

ii) The condition that a matrix X belong to sp_{2n} is that $\exp(tX)$ belong to Sp for all t. Differentiating at $t=0$ gives the equation

$$XJ + JX^t = 0. \tag{4.3.7b}$$

Since $J^t = -J$, the above condition can be written as $XJ = -JX^t = -(XJ^t)^t = (XJ)^t$. In other words, XJ is symmetric. This lets us define a map β, from the space $\mathcal{S}_{2n}$ of complex symmetric $2n \times 2n$ matrices to the Lie algebra $sp_{2n}(\mathbf{C})$ by the formula

$$\mathcal{A} \to \beta(\mathcal{A}) = -2iJ\mathcal{A}. \tag{4.3.8}$$

This map β is a linear isomorphism from $\mathcal{S}_{2n}$ to $sp_{2n}(\mathbf{C})$.

We note that any skew-symmetric J defines such a map. The particular map, however, will of course vary with choice of J. For our purposes, we want to work in coördinates corresponding to a standard symplectic basis, so that J will be a direct sum of matrices $\begin{bmatrix} 0 & 1 \\ -1 & 0 \end{bmatrix}$, which means that $J^2 = -I$.

Now we take:

- $\mathcal{S}_{2n}^+$ = cone of $2n \times 2n$ complex symmetric matrices with positive definite real part.
- Ω_n = scalar multiples of Gaussian functions $\gamma_{\mathcal{A}}$ on W corresponding to elements $\mathcal{A}$ of $\mathcal{S}_{2n}^+$.
- Ω_n^0 = normalized multiples $\gamma_{\mathcal{A}}^0 = (\det(\mathcal{A} + iJ/2)^{\frac{1}{2}}\gamma_{\mathcal{A}}$, excluding those for which $\det(\mathcal{A} + iJ/2)$ is equal to zero.

We have shown, in formula (4.3.2) that Ω_n^0 is a subsemigroup of Ω_n.

We can map Ω_n^0 to $Sp_{2n}(\mathbf{C})$ by the recipe

$$\omega^{-1}(\gamma_{\mathcal{A}}^0) = c(-2iJ\mathcal{A}) = \frac{2iJ\mathcal{A} - 1}{2iJ\mathcal{A} + 1} = c \circ \beta \circ \gamma^{-1}(\gamma_{\mathcal{A}}^0). \tag{4.3.9}$$

Combining our various formulas above shows

Proposition 4.3.10: *The mapping*

$$\omega^{-1} : \Omega^0 \quad \to \quad Sp_{2n}(\mathbf{C})$$

is a 2 to 1 homomorphism from Ω^0 to an open subsemigroup of $Sp_{2n}(\mathbf{C})$.

Moreover, it can be shown (see [35], §§13, 15, or [14] §5.5 for details) that

Proposition 4.3.11: *The semigroup Ω^0 consists of contraction operators.*

This result provides a basis for estimating the operator norms of superpositions of translates of elements of Ω^0 by elements of the Heisenberg group. As sketched in [32] and [35], this in turn allows one to derive some of the important estimates in the PDE literature of pseudodifferential and

Fourier integral operators, including the Calderón-Vaillancourt (0, 0) estimate [10], and those of Beals [3], Hörmander [29], Unterberger [67]. A key aspect of the elements of Ω^0 in proving these estimates is the rapid decay away from the origin of the corresponding integral kernel operators. This allows to provide efficient estimates of superpositions of elements of Ω^0 translated by elements of the Heisenberg group.

4.4. *Passing to the boundary*

The elements of the normalized oscillator semigroup Ω^0 are contractions as operators, and as sketched above, they may also be considered as elements in an open subsemigroup of the complex symplectic group. We might wonder what happens if we have a sequence $\mathcal{A}_k$ of elements Ω^0, whose inverse images $\omega^{-1}(\mathcal{A}_k)$ corresponds to a sequence of elements that converge in $Sp_{2n}(\mathbf{C})$. This establishes a nice connection with the oscillator representation [35].

Theorem 4.4.1:

(*a*) *The real symplectic group* $Sp_{2n}(\mathbf{R})$ *is in the closure of* $\omega^{-1}(\Omega^0)$.
(*b*) *If* $\mathcal{A}_k$ *is a sequence in* Ω^0 *such that* $\omega^{-1}(\mathcal{A}_k)$ *converge in* $Sp_{2n}(\mathbf{C})$ *to an element* g *of* $Sp_{2n}(\mathbf{R})$, *then the sequence* $\{\mathcal{A}_k\}$ *converges in the strong operator topology to* $\omega(g)$.

Here as elsewhere operators such as $\omega(g)$ are considered to be defined only up to sign.

References

1. J. Adams, "Unitary highest weight modules", *Adv. Math.*, **63** (1987), 113–137.
2. T. Basile, E. Joung, K. Mkrtchyan, and M. Mojaza, "Dual pair correspondence in Physics: Oscillator realizations and representations", *J. High Energy Physics*, **20** (2020), article number 20. arXiv:2006.07102v2 [hep-th].
3. R. Beals, "A general calculus of pseudo-differential operators", *Duke Math. J.* **42** (1975), 1–42.
4. A. Borel, "Les espaces hermitiens symétriques", Exposé No. 62, Séminaire Bourbaki, vol. 2. (1995), 121–132.
5. A. Borel and Harish-Chandra, "Arithmetic subgroups of algebraic groups", *Annals of Math.* **75** (1962), 485–535.
6. M. Born and W. Heisenberg, "Über eine Abänderung der formalen Regeln der Quantentheorie beim Problem der anomalen Zeeman-Effekte", *Zeit. für Physik* **26** (1924), 291–307.

7. M. Born and W. Heisenberg, "Über quantentheoretische Umdeutung kinematischer und mechanischer Beziehungen", *Zeit. für Physik* **33** (1925), 879–893. [English translation in: van der Waerden 1968, "Quantum-Theoretical Re-interpretation of Kinematic and Mechanical Relations"]
8. M. Born and P. Jordan, "Zur Quantenmechanik", *Zeit. für Physik* **34** (1925), 858–888. [English translation in: van der Waerden 1968, "On Quantum Mechanics"]
9. M. Born, W. Heisenberg and P. Jordan, "Zur Quantenmechanik II", *Zeit. für Physik* **35** (1926), 557–615. [English translation in: van der Waerden 1968, 15 "On Quantum Mechanics II"]
10. A. Calderon and R. Vaillancourt, "A class of bounded pseudodifferential operators", *Proc. Nat. Acad. Sci. USA* **69** (1972), 1185–1187.
11. N. De Bruijn, "Uncertainty principles in Fourier analysis", Proc. Symp. on Inequalities, Academic Press, New York 1967, 57–71.
12. N. De Bruijn, "A theory of generalized functions, with applications to Wigner distributions and Weyl correspondences", *Nieuw Archief vor Wiskunde* **21** (1973), 205–280.
13. T. Enright, R. Howe and N. Wallach, "A classification of unitary highest weight modules", *Prog. Math.* **40**, Birkhuser Boston, Inc., Boston, MA, (1983), 97–143.
14. G. Folland, *Harmonic Analysis in Phase Space*, Ann. Math. Studies **122**, Princeton University Press, Princeton, NJ, 1989.
15. W. T. Gan and A. Ichino, "The Shimura-Waldspurger correspondence for Mp_{2n}", *Ann. of Math. (2)* **188** (2018), 965–1016.
16. W. T. Gan and B. Sun, "The Howe Duality Conjecture: Quaternionic Case", in: *Representation Theory, Number Theory, and Invariant Theory*, 175–192. Progr. Math., 323, Birkhüser/Springer, Cham, 2017.
17. S. Gelbart, "Holomorphic discrete series for the real symplectic group", *Invent. Math.*, **19** (1973), 49–58.
18. S. Gelbart, "A theory of Stiefel harmonics", *Trans. Amer. Math. Soc.*, **192** (1974), 29–50.
19. S. Gelbart, "Examples of dual reductive pairs", Proc. Symp. Pure Math., **33** (1977) Amer. Math. Soc., Providence RI, 287–296.
20. R. Goodman and N. Wallach, *Representations and Invariants of the Classical Groups*, Enc. of Math. and Apps. **68**, Cambridge University Press, Cambridge UK, 1998.
21. K. Gross and R. Kunze, "Bessel functions and representation theory. I", *J. Funct. Anal.*, **22** (1976), 73–105.
22. K. Gross and R. Kunze, Bessel functions and representation theory II, Holomorphic discrete series and metaplectic representations. *J. Funct. Anal.*, **25** (1977), 1–49.
23. S. Gurevich, R. Hadani and R. Howe, "Quadratic reciprocity and the sign of the Gauss sum via the finite Weil representation", *Int. Math. Res. Not.* **19** (2010), 3729–3745.

24. S. Gurevich and R. Howe, "Small representations of finite classical groups", in *Representation Theory, Number Theory, and Invariant Theory*, 209–234, Prog. in Math. **323**, Birkhäuser/Springer, Cham, 2017.
25. Harish-Chandra, "Representations of semisimple Lie groups, IV", *Amer. J. Math.*, **77** (1955), 743–777.
26. Harish-Chandra, "Representations of semisimple Lie groups, V", *Amer. J. Math.*, **78** (1956), 1–41.
27. Harish-Chandra, *Automorphic forms on semisimple Lie groups (Introduction)*, Lec. Notes in Math. **62**, Springer-Verlag, Berlin-New York, 1968.
28. H. Hecht, "The characters of some representations of Harish-Chandra", *Math. Ann.*, **219** (1976), 213–226.
29. L. Hörmander, "The Weyl calculus of pseudodifferential operators", *Comm. Pure App. Math.*, **23** (1979), 359–443.
30. L. Hörmander, *The analysis of linear partial differential operators III*, Grund. Math. Wiss. **274**, Springer Verlag, New York, 1984.
31. R. Howe, "θ-series and invariant theory", *Proc. Symp. Pure Math.* **33**, (1977), Amer. Math. Soc., Providence RI, 275–286.
32. R. Howe, "Quantum mechanics and partial differential equations", *J. Funct. Anal.* **38** (1980), 188–254.
33. R. Howe, "On the role of the Heisenberg group in harmonic analysis", *Bull. Amer. Math. Soc. (N.S.)* **3** (1980), 821–843.
34. R. Howe, "Dual pairs in physics: harmonic oscillators, photons, electrons, and singletons", *Lectures in App. Math.* **21**, Amer. Math. Soc., Providence, RI, (1985), 179–207.
35. R. Howe, "The oscillator semigroup", *Proc. Symp. Pure Math.* **48**, Amer. Math. Soc., Providence RI (1988), 61–132.
36. R. Howe, "Remarks on classical invariant theory", *Trans. Amer. Math. Soc.* **313** (1989), 539–570.
37. R. Howe, "Transcending classical invariant theory", *J. Amer. Math. Soc.* **2** (1989), 535–552.
38. C. G. Jacobi, *Fundamenta Nova Theoriae Functionum Ellipticarum*, Fratrum Bornträger, Regiomonti, 1829.
39. C. G. Jacobi, "De compositione numerorum e quatuor quadratis", *Jour. für Reine u. Angew. Math.*, **12** (1834), 167–172; Werke, 6 (1891), 245–251.
40. N. Jacobson, *Basic Algebra*, W. H. Freeman, San Francisco, 1974.
41. H.P. Jakobsen, "Hermitian symmetric spaces and their unitary highest weight modules", *J. Funct. Anal.* **52** (1983), 385–412.
42. M. Kashiwara and M. Vergne, "On the Segal-Shale-Weil representations and harmonic polynomials", *Invent. Math.* **44** (1978), 1–47.
43. A. Kirillov, *Elements of the Theory of Representations*, Grund. Math. Wiss. **220**, Springer Verlag, Berlin-New York, 1976.
44. A. Knapp, "Bounded symmetric domains and holomorphic discrete series", in *Symmetric spaces (Short Courses, Washington University)*, 211–246, Pure and App. Math., **8** M. Dekker, 1972.
45. J.S. Li, "On the classification of irreducible low rank representations of classical groups", *Comp. Math.* **71** (1989), 29–48.

46. H. Maass, "Über eine neue Art von nichtanalytischen automorphen Funktionen und die Bestimmung Dirichletscher Reihen durch Funktionalgleichungen", *Math. Ann.* **121** (1949), 141–183.
47. G. Mackey, "A theorem of Stone and von Neumann", *Duke Math. J.* **16** (1949), 313–326.
48. G. Mackey, "Induced representations of locally compact groups and applications", in *Functional Analysis and Related Fields*, Springer Verlag, New York, (1980), 132–166.
49. M. R. Murty (ed.), *Theta Functions: From the Classical to the Modern*, CRM Proceedings and Lecture Notes **1**, Amer. Math. Soc. Providence RI, 1993.
50. C. Mœglin, M.-F. Vignéras and J.-L. Waldspurger, *Correspondances de Howe sur un corps p-adique*, Lecture Notes in Math., **1291**, Springer-Verlag, Berlin, 1987.
51. S. Niwa, "Modular forms of half-integral weight and the integral of certain theta functions", *Nagoya Math. J.* **56**, (1975), 147–161.
52. T. Oda, "On modular forms associated with indefinite quadratic forms of signature $(2, n-2)$", *Math. Ann.* **231** (1977), 97–144.
53. S. Rallis, "On a relation between $\tilde{SL}_2$ cusp forms and automorphic forms on orthogonal groups", *Proc. Symp. Pure Math.* **33**, (1977) Amer. Math. Soc., Providence RI, 297–314.
54. S. Rallis, "On the Howe duality conjecture", *Comp. Math.* **51** (1984), 333–399.
55. S. Rallis and G. Schiffmann, "Automorphic forms constructed from the Weil representation, holomorphic case", *Amer. J. Math.* **100** (1978), 1049–1122.
56. S. Rallis and G. Schiffmann, "Discrete spectrum of the Weil representation", *Bull. Amer. Math. Soc.* **83** (1977), 267–270.
57. H. Rubenthaler, *Les Paires Duales dans les Algebres de Lie Reductives*, Asterisque **219**, Soc. Math. de France, 1994.
58. M. Saito, "Representations unitaires des groupes symplectiques", *J. Math. Soc. Japan* **24** (1972), 232–251.
59. I. Segal, "Foundations of the theory of dynamical systems of infinitely many degrees of freedom, I", *Mat.-Fys. Medd. Dansk, Vid. Selsk.* **31** (1959), no. 12.
60. I. Segal, *Mathematical Problems of Relativistic Physics*, Amer. Math. Soc., Providence RI, 1963.
61. D. Shale, "Linear symmetries of free boson fields", *Trans. Amer. Math. Soc.* (**103**) (1962), 149–167.
62. H. Shimizu, "Theta-series and automorphic forms on GL_2", *J. Math. Soc. Japan* **24** (1972), 232–251.
63. G. Shimura, "On modular forms of half-integral weight", *Ann. of Math. (2)* **97** (1973), 440–481.
64. J. Shalika and S. Tanaka, "On an explicit construction of a certain class of automorphic forms", *Amer. J. Math.* **91** (1969), 1049–1076.
65. C. L. Siegel, *Symplectic Geometry*, Academic Press, 1964.
66. M. Stone, "Linear transformations in Hilbert space III: operational methods and group theory", *Proc. Nat. Acad. Sci. USA* **16** (1930), 172–175.

67. A. Unterberger, "Oscillateur harmonique et opérateurs pseudodifférentiels", *Ann. Inst. Fourier (Grenoble)* **29** (1979), 201–221.
68. D. Vogan, "The size of infinite dimensional representations", *Japan J. Math.* **12** (2017), 175–210.
69. J. von Neuman, "Die Eindeutigkeit der Schödingerschem Operatoren", *Math. Ann.* **104**, 570–578.
70. J.-L. Waldspürger, "Correspondance de Shimura", *J. Math. Pures et Appl.* **59** (1980), 1–130.
71. J.-L. Waldspürger, "Demonstration d'une conjecture de dualité de Howe dans le cas p-adique, $p \neq 2$", Israel Math. Conf. Proc. **2**, Weizmann, Jerusalem, 1990, 267–324.
72. J.-L. Waldspürger, "Correspondance de Shimura et quaternions", *Forum Math.* **3** (1991), 219–307.
73. N. Wallach, *Real reductive groups. I*, Pure Appl. Math. **132**, Academic Press, Inc., Boston, MA, 1988. *Real reductive groups. II*, Pure Appl. Math., 132-II, Academic Press, Inc., Boston, MA, 1992.
74. A. Weil, "Sur certains groupes d'opérateurs unitaires", *Acta Math.* **111** (1964), 143–211.
75. A. Weil, *Basic number theory*, Die Grundlehren der mathematischen Wissenschaften **144**, Springer-Verlag New York, Inc., New York, 1973.
76. A. Weil, *Adeles and algebraic groups*, Birkhäuser, Boston, 1982.
77. E. J. Whittaker and J. N. Watson, *A Course of Modern Analysis*, 4th Ed., University Press, Cambridge, UK, 1938.
78. J. Wolf, "On the Classification of Hermitian Symmetric Spaces", *Indiana Univ. Math. J.*, 13 (1964), 489–495.
79. H. Weyl, *The Theory of Groups and Quantum Mechanics*, Dover Publications, New York, 1950.
80. H. Weyl, *The Classical Groups*, Princeton University Press, Princeton, NJ, 1946.

https://doi.org/10.1142/9789819825448_0009

Proper Actions and Representation Theory

Toshiyuki Kobayashi
Graduate School of Mathematical Sciences, The University of Tokyo
3-8-1, Komaba, Meguro, Tokyo, 153-8914, Japan
French-Japanese Laboratory in Mathematics and Its Interactions,
FJ-LMI CNRS, IR-2025, Tokyo, Japan
toshi@ms.u-tokyo.ac.jp

This exposition presents recent developments on proper actions, highlighting their connections to representation theory. It begins with geometric aspects, including criteria for the properness of homogeneous spaces in the setting of reductive groups. We then explore the interplay between the properness of group actions and the discrete decomposability of unitary representations realized on function spaces. Furthermore, two contrasting new approaches to quantifying proper actions are examined: one based on the notion of sharpness, which measures how strongly a given action satisfies properness; and another based on dynamical volume estimates, which measure deviations from properness. The latter quantitative estimates have proven especially fruitful in establishing temperedness criterion for regular unitary representations on G-spaces. Throughout, key concepts are illustrated with concrete geometric and representation-theoretic examples.

Contents

1. Introduction

The actions of non-compact groups on manifolds can exhibit highly non-trivial and "wild behavior". The notion of proper actions, introduced by

Palais [P61], abstracts and formalises the favorable features characteristic of actions of compact groups. A prototypical example of a proper action is the action of the fundamental group Γ of a manifold on its universal covering space via deck transformations.

On the other hand, when X is a Riemannian manifold on which a discrete group acts freely and by isometries, the action is automatically properly discontinuous (Proposition 3.11). The quotient space $X_\Gamma = \Gamma\backslash X$ inherits a natural Riemannian structure from X via the covering $X \to \Gamma\backslash X$, thereby becoming a Riemannian manifold. In this setting, one may regard Γ as governing the global structure of the quotient manifold $\Gamma\backslash X$, while the original manifold X determines its local structure.

However, in more general settings—such as when the Riemannian structure is replaced with a pseudo-Riemannian one (allowing indefinite metric)—the situation is significantly different: free actions by discrete groups of isometries often fail to be properly discontinuous (*e.g.*, Example 3.13).

In the study of local-to-global phenomena beyond the Riemannian setting, understanding proper actions (or properly discontinuous actions) is therefore crucial.

In this chapter, we examine recent progress concerning proper actions from both geometric and representation-theoretic perspectives.

The exposition begins with the topological and geometric framework related to group actions by using binary relations $\pitchfork$ and $\sim$ on the power set of G. Sections 3 and 4 present criteria for the properness of group actions on homogeneous spaces. Topics include Lipsman's conjecture for nilmanifolds (Section 3.12) and the properness criterion (Theorem 4.14) in the reductive case, and several subtle examples that illustrate these results.

In Section 5, we give a brief overview of recent developments concerning cocompact discontinuous groups for reductive homogeneous spaces G/H. Inspired by Mackey's philosophy—originally developed for unitary representations—we also mention a topological analogue involving the *tangential homogeneous space* G_θ/H_θ, which arises from their associated Cartan motion groups.

The conceptual link between *properness* in topological group actions and *discrete decomposability* in unitary representation theory—traditionally seen as unrelated domains that have been developed through different methods and perspectives—was first proposed in [K00], Sec. 3. This work introduced the previously unexplored idea that non-compact subgroups may exhibit *compact-like behaviour*. Subsequent developments, particularly those involving spectral analysis on locally pseudo-Riemannian symmetric spaces $\Gamma\backslash G/H$ (*e.g.*, [KaK25]), have further deepened this perspective. In Section 6, we investigate the interplay between the properness of group actions and the discrete decomposability of unitary representations realized on function spaces.

In Section 7, we discuss two contrasting approaches to quantifying the properness of group actions. This first based on the notion of *sharpness* ([KaK16]) measures how strongly a given action satisfies the properness condition. The second takes a dynamical perspective using volume estimates to assess how far the action deviates from being proper. This latter approach has emerged as a key idea in establishing temperedness criteria for regular representations on G-spaces in recent work [BK15, BK21, BK22, BK23]. Through this discussion, we illustrate how geometric insights can inform analytic aspects of representation theory.

2. Local to Global in Geometry

2.1. *Local to Global in Riemannian Geometry*

We consider the interplay between local and global geometric properties.

- *Local properties* include curvature, (T_1) topology, locally homogeneous structure.
- *Global properties* include compactness, Hausdorffness, characteristic classes, diameter, and the fundamental group.

The study of how local geometric properties affect global structure has been one of the central themes of differential geometry over several decades, with particularly significant progress in the Riemannian setting. In contrast, relatively little is known about global properties in non-Riemannian geometry—arising, for example, from the space time model of relativity theory—or more generally in manifolds with indefinite metrics of arbitrary signature (see [G25] and references therein). For instance, the space form problem [K01], Section 2 is a long-standing problem in non-Riemannian geometry, which includes the existence problem of a compact manifold M with constant sectional curvature for a given indefinite-metric of signature (p, q), see Conjecture 5.6 below.

We begin with a classical example of a local-to-global theorem in *Riemannian geometry*.

Example 2.1: (Bonnet–Myers). Let (M, g) be an n-dimensional complete Riemannian manifold whose Ricci curvature satisfies

$$\mathrm{Ric}(g) \geq (n-1)C$$

for some positive constant C. Then M is compact and its diameter is at most $\frac{\pi}{\sqrt{C}}$.

This theorem tells us *global* properties such as compactness and the diameter are constrained by *local* information—specifically such as the positivity of curvature—in Riemannian geometry.

What can be said about the local-to-global phenomena *beyond* the traditional Riemannian setting?

2.2. *Preliminaries: Pseudo-Riemannian Manifolds*

We review briefly some basic notions from pseudo-Riemannian geometry.

Definition 2.2: A *pseudo-Riemannian manifold* (M, g) is a smooth manifold equipped with a non-degenerate symmetric bilinear form

$$g_x \colon T_xM \times T_xM \to \mathbb{R} \quad (x \in M)$$

that depends smoothly on $x \in M$.

Let (p, q) be the signature of g_x, a non-degenerate symmetric bilinear form on a $(p+q)$-dimensional manifold M. By Sylvester's law of inertia, the signature is locally constant. We say that (M, g) is a *Riemannian manifold* if $q = 0$, and is a *Lorentzian manifold* if $q = 1$.

Just as in the Riemannian case, pseudo-Riemannian manifolds (M, g) also admit natural definitions of the Levi-Civita connection, geodesics, and curvature. For example, the curvature tensor R and the sectional curvature κ for $X, Y \in T_xM$ are given by

$$\begin{aligned} R(X,Y) &:= [\Delta_X, \Delta_Y] - \Delta_{[X,Y]}, \\ \kappa(X,Y) &:= \frac{g_x(R(X,Y)Y, X)}{g_x(X,X)g_x(Y,Y) - g_x(X,Y)^2}. \end{aligned}$$

Example 2.3: (1) (Flat case) We equip $\mathbb{R}^{p+q}$ with the pseudo-Riemannian structure

$$dx_1^2 + \cdots + dx_p^2 - dx_{p+1}^2 - \cdots - dx_{p+q}^2$$

which has the signature (p, q), and denote the resulting space by $\mathbb{R}^{p,q}$.

It is a flat space; that is, the curvature tensor satisfies $R \equiv 0$. In the case where $q = 1$, $\mathbb{R}^{p,q}$ is a Lorentzian manifold known as the *Minkowski space*.

(2) (Pseudo-Riemannian space forms) The flat pseudo-Riemannian structure on $\mathbb{R}^{p,q}$ remains non-degenerate when restricted to the hypersurfaces

$$\begin{aligned} X(p-1,q)_+ =& \{x \in \mathbb{R}^{p+q} : x_1^2 + \cdots + x_p^2 - x_{p+1}^2 - \cdots - x_{p+q}^2 = 1\}, \\ X(p,q-1)_- =& \{x \in \mathbb{R}^{p+q} : x_1^2 + \cdots + x_p^2 - x_{p+1}^2 - \cdots - x_{p+q}^2 = -1\}. \end{aligned}$$

These give rise to pseudo-Riemannian manifolds of signature $(p-1, q)$ with constant sectional curvature $\kappa \equiv 1$, and of signature $(p, q-1)$ with $\kappa \equiv -1$, respectively.

(3) (Hyperbolic space: $\mathbb{H}^n = X(n,0)_-$) The hypersurface

$$\mathbb{H}^n := \{(x_1, \cdots, x_{n+1}) : x_1^2 + \cdots + x_n^2 - x_{n+1}^2 = -1\}$$

inherits a Riemannian structure from the ambient Minkowski space $\mathbb{R}^{n,1}$, and has a constant sectional curvature $\kappa \equiv -1$.

(4) (De Sitter space: $\mathrm{dS}^n = X(n-1,1)_+$) The hypersurface of $\mathbb{R}^{n,1}$,

$$\mathrm{dS}^n := \{(x_1, \cdots, x_{n+1}) : x_1^2 + \cdots + x_n^2 - x_{n+1}^2 = 1\}$$

inherits a Lorentzian metric from the ambient Minkowski space $\mathbb{R}^{n,1}$, and has a constant sectional curvature $\kappa \equiv 1$. More generally, a complete Lorentzian manifold of constant positive sectional curvature is called a *de Sitter manifold*, or a *relativistic spherical space*, as it serves as a Lorentzian analog of sphere geometry.

(5) (Anti-de Sitter space: $\mathrm{AdS}^n = X(n-1,1)_-$) This is a special case of the preceding example. The hypersurface

$$\mathrm{AdS}^n := \{(x_1, \cdots, x_{n+1}) : x_1^2 + \cdots + x_{n-1}^2 - x_n^2 - x_{n+1}^2 = -1\}$$

inherits a Lorentzian metric from $\mathbb{R}^{n-1,2}$ and has constant sectional curvature $\kappa \equiv -1$. It is regarded as a Lorentzian analog of the hyperbolic space. More generally, a complete Lorentzian manifold with constant sectional curvature $\kappa \equiv -1$ is called an *anti-de Sitter manifold*.

Remark 2.4: In Example 2.3(2), changing the signature of the flat pseudo-Riemannian structure of the ambient space $\mathbb{R}^{p+q}$ causes the signatures of the induced pseudo-Riemannian metrics on the hypersurfaces $X(p-1,q)_+$ and $X(p,q-1)_-$ to change from $(p-1,q)$ to $(q,p-1)$, and from $(p,q-1)$ to $(q-1,p)$, respectively. Furthermore, the sectional curvature is reversed in sign.

2.3. *The Calabi–Markus Phenomenon*

In contrast to Riemannian geometry, as illustrated by the Bonnet–Myers theorem (Example 2.1) the global geometry of pseudo-Riemannian manifolds exhibits markedly different behavior:

Theorem 2.5: *(Calabi–Markus [CM62]). Every relativistic spherical space (i.e., a de Sitter manifold) is non-compact. Furthermore, if the dimension is greater than two, its fundamental group is finite.*

3. Basic Problems on Discontinuous Groups for G/H

When the *homogeneous structure* is regarded as a local property, *discontinuous groups* (Definition 3.1) govern the global geometry. The study of discontinuous groups beyond the Riemannian setting is a relatively young and rapidly evolving field in group theory interacting with topology, differential geometry, representation theory, ergodic theory, and number theory, as well as other areas of mathematics. An early exposition of this subject can be found in the lecture notes [K97], and a more recent account is provided, for instance, in [G25].

This theme was also highlighted as a new direction for future research looking ahead to the 21st century on the occasion of the World Mathematical Year 2000 by Margulis [M00] and the author [K01] with both works including collection of open problems. Over the past thirty years, there have been remarkable developments employing a variety of methods. Nevertheless, several fundamental problems remain unsolved ([K23a]).

In this section, we lay the groundwork for these problems, which will be formulated more explicitly in Sections 4–5, by illustrating the basic ideas through simple examples.

3.1. *Discontinuous Groups for Acting on Manifolds X*

Beyond the Riemannian context, it is crucial to clearly distinguish between *discrete subgroups* and *discontinuous groups*.

In many cases, a discontinuous group Γ is realized as a subgroup of G acting on a manifold. Accordingly, we shall define discontinuous groups within this framework. Nevertheless, in contexts where G plays no essential role, we may omit the ambient group G and simply take $G = \Gamma$.

Definition 3.1: (Discontinuous Group). Let G be a Lie group acting on a manifold X. A discrete subgroup Γ of G is called a *discontinuous group* for X if Γ acts properly discontinuously and freely on X. See Definition 3.2 below.

The quotient space $X_\Gamma := \Gamma\backslash X$, by a discontinuous group Γ, is a (Hausdorff) manifold. Moreover, any G-invariant local geometric structure on X descends to X_Γ via the covering map $X \to X_\Gamma$ (see Proposition 3.6).

Such quotients X_Γ are examples of complete (G, X)-*manifolds* in the sense of Ehresmann and Thurston.

3.2. *Basic Notion ⋯ Proper Action*

We extend Theorem 2.5 to a broader setting formulated in the language of groups. To this end, we briefly review some basic notions in the theory of transformation groups.

Let L be a locally compact group, and X a locally compact topological space. Suppose that L acts *continuously* on X, *i.e.*, the action map

$$L \times X \to X, \quad (g, x) \mapsto gx$$

is continuous.

For a subset $S \subset X$, we define a subset $L_S \subset L$ by

$$L_S := L_{S \to S} = \{\gamma \in L : \gamma S \cap S \neq \emptyset\}.$$

If S is a singleton $\{x\}$, then $L_{\{x\}}$ coincides with the stabilizer group L_x of the point $x \in X$. In general, L_S is merely a subset of L.

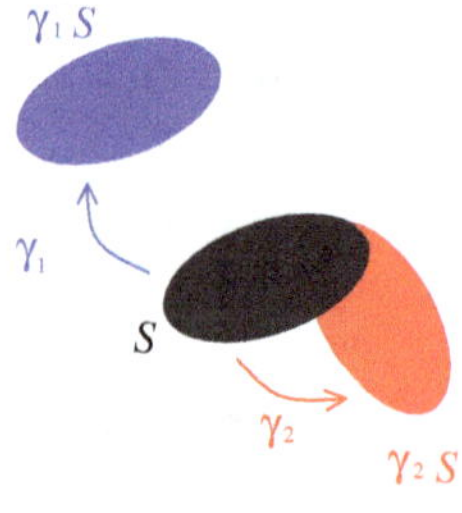

Fig. 1. $\gamma_1 \notin L_S \ni \gamma_2$.

Continuous actions possessing with the properties:

$$L_S \text{ is “small” whenever } S \text{ is “small”}$$

are precisely formulated and given the following names.

Definition 3.2: An action of L on X is called

free	if L_S is a singleton	for any singleton S;
properly discontinuous	if L_S is finite	for any compact subset S;
proper	if L_S is compact	for any compact subset S.

3.3. *Proper Maps and Proper Actions*

Let X and Y be Hausdorff, locally compact spaces. In this section, we take a closer look at some basic properties of proper actions.

Definition 3.3: A continuous map $f: X \to Y$ is called *proper* if the preimage $f^{-1}(S)$ of any compact subset $S \subset Y$ is compact.

It is worth noting that any proper map is a closed map (see *e.g.*, [B98], Chap. I, Sec. 10, Prop. 1). Indeed, let C be a closed subset of X, and let $y \notin f(C)$. Choose an open neighbourhood V of y such that its closure $\overline{V}$ is compact. Then the set $E := C \cap f^{-1}(\overline{V})$ is compact, and hence $f(E)$ is closed. It follows that the set $U := V \setminus f(E)$ is an open neighbourhood of y, disjoint from $f(C)$. Thus, $f(C)$ is closed.

For subsets S and T of X, we define

$$L_{S \to T} := \{\gamma \in L : \gamma S \cap T \neq \emptyset\}.$$

The proof of the following lemma is straightforward and is therefore omitted.

Definition–Lemma 3.4: (Proper Action). Let X be a locally compact topological space, on which a locally compact group G acts continuously. Then the following four conditions are equivalent:

(i) The action of L on X is proper in the sense of Definition 3.3.
(ii) The map $\varphi: L \times X \to X \times X$ defined by $(g, x) \mapsto (x, gx)$ is proper.
(iii) For any compact $S, T \subset X$, the set $L_{S \to T}$ is compact.
(iv) For any compact subset $S \subset X$, the set L_S $(\equiv L_{S \to S})$ is compact.

See also Lemma 7.5 for alternative characterization of proper actions from the perspective of measure theory.

3.4. *Proper + Discrete = Properly Discontinuous*

When the group L is discrete, the action of L is proper if and only if it is properly discontinuous, since a discrete set is compact if and only if it is finite.

Furthermore, the stabilizer L_x is finite for every $x \in X$ in this case. Thus, among the three properties listed in Definition 3.2, understanding of proper actions in greater depth is of particular importance.

3.5. *Discontinuous Group and Covering Transformation Group*

Suppose that X is a locally compact, Hausdorff space, on which a discrete group Γ acts continuously.

Definition 3.5: (Discontinuous Group for X). A discrete group Γ is called a *discontinuous group for X* if the action of Γ on X is properly discontinuous and free.

Let $\Gamma\backslash X$ denote the quotient space, *i.e.*, the set of Γ-orbits in X, equipped with the quotient topology induced by the natural projection $q_\Gamma\colon X \to \Gamma\backslash X$. The following result is a classical fact from general topology (see *e.g.*, [T97], Chap. 3, Sec. 3.5).

Proposition 3.6: *If Γ is a discontinuous group for a topological manifold, then the quotient space $\Gamma\backslash X$ carries a manifold structure such that the quotient map $q_\Gamma\colon X \to \Gamma\backslash X$ becomes a regular covering. Moreover, any Γ-invariant local geometric structure on X descends to $\Gamma\backslash X$ via q_Γ.*

Remark 3.7: The key condition in Definition 3.5 is that the action is properly discontinuous; freeness is of secondary importance.

There are two main reasons for this. First, suppose that Γ acts properly discontinuously on X. Then the singularities of the quotient space X_Γ are "mild", in the sense that X_Γ is locally a finite group quotient of Euclidean space, called *V-manifold* in the sense of Satake [S56] or an *orbifold* in the sense of Thurston.

Second, if Γ is a finitely generated linear group, then there exists a finite-index subgroup $\Gamma' \subset \Gamma$ such that Γ' is torsion-free by a theorem of Selberg. In particular, the Γ'-action is free and properly discontinuous, provided that the Γ-action is properly discontinuous.

In [K97], Def. 2.5, we did not require freeness in the definition of discontinuous groups, thereby allowing $X_\Gamma = \Gamma\backslash X$ to be an orbifold.

We provide some typical examples of Proposition 3.6.

Example 3.8: Suppose that M is a pseudo-Riemannian manifold. Let X be its universal covering equipped with the pull-back pseudo-Riemannian structure via the covering map $p\colon X \to M$. Let $G = \mathrm{Isom}(X)$ denote the isometry group of X, and let Γ be the fundamental group of M, based on a point $o = p(\widetilde{o}) \in M$. Then G admits the structure of a Lie group acting smoothly on X. Furthermore, Γ is regarded as a subgroup of the Lie group G, and is a discontinuous group for X with natural isomorphism $X_\Gamma \simeq M$.

As a classical example illustrating Example 3.8, we recall the uniformization of a compact Riemann surface Σ_g.

Example 3.9: (Uniformization Theorem of Klein–Poincaré–Koebe). Let Σ_g be a compact Riemann surface of genus $g \geq 2$, and let Γ denote its fundamental group $\pi_1(\Sigma_g)$, often referred to as a *surface group*. Then the universal covering space of Σ_g is biholomorphic to the Poincaré upper half plane

$$\mathbb{H} = \{z \in \mathbb{C} : \mathrm{Im}\, z > 0\}.$$

The group $PSL(2, \mathbb{R}) = SL(2, \mathbb{R})/\{\pm I_2\}$ acts holomorphically and transitively on $\mathbb{H}$ via linear fractional transformations $z \mapsto (cz + d)^{-1}(az + b)$. There is a natural diffeomorphism

$$\mathbb{H} \simeq PSL(2, \mathbb{R})/PSO(2) =: G/K.$$

The quotient $\Gamma\backslash\mathbb{H} \simeq \Gamma\backslash G/K$ can be naturally identified with the original surface Σ_g.

In this example, $G/K = PSL(2, \mathbb{R})/PSO(2)$ is a Riemannian symmetric space. More generally, the following result, which goes back to É. Cartan, provides a bridge between the geometric and group-theoretic definition of symmetric spaces.

Proposition 3.10: (Affine Locally Symmetric Space). *Any complete affine locally symmetric space is of the form $\Gamma\backslash G/H$, where G is a Lie group, H is an open subgroup of the fixed point subgroup of an involution of G, and Γ is a discrete subgroup acting properly discontinuously and freely on the symmetric space G/H.*

3.6. *Isometric Actions: Riemannian Geometry*

Let Isom(X) denote the group of isometries of a Riemannian manifold, or more generally, of a pseudo-Riemannian manifold X. Then Isom(X) is a Lie group.

In Example 3.9, $G := \mathrm{Isom}(\Sigma_g) \simeq PSL(2, \mathbb{R})$, and Γ can be regarded as a *discrete* subgroup of G.

In this subsection, we prove a converse statement, that is, any discrete subgroup Γ of Isom(X) acts properly discontinuously on X if X is a *Riemannian manifold.*

For two topological spaces X and Y, let $C(X, Y)$ denote the set of all continuous maps from X to Y. We recall that the *compact-open topology* on the set $C(X, Y)$ is a topology defined by the subbase

$$W(S, V) := \{f \in C(X, Y) : f(S) \subset V\},$$

where S is a compact subset of X and V is an open subset of Y.

The compact-open topology on $C(X, Y)$ is Hausdorff if Y is Hausdorff.

Proposition 3.11: (Isometric Transformations in Metric Spaces). *Suppose that X is a locally compact, separable, complete metric space such that X has a Heine–Borel property, that is, every bounded closed set is compact. Let Γ be a group of isometries of X endowed with compact-open topology. Then the following two conditions on Γ are equivalent:*

(i) Γ *is a discrete group.*

(ii) Γ *acts properly discontinuously on X.*

Proof: We first prove the easier direction (ii) $\Rightarrow$ (i). It suffices to show that, for any $\gamma \in \Gamma$, there exists an open set $\Gamma \subset W$ such that $\sharp(\Gamma \cap W) < \infty$, assuming (ii). Take any $x \in X$ and any open neighbourhood V of $\gamma \cdot x$ such that the closure $\overline{V}$ is compact. Let $W := W(\{x\}, V)$. Then $\gamma \in W = \Gamma_{\{x\} \to V} \subset \Gamma_{\{x\} \to \overline{V}}$. On the other hand, $\Gamma_{\{x\} \to \overline{V}}$ is finite because the Γ-action on X is properly discontinuous. Hence (ii) $\Rightarrow$ (i) is shown.

(i) $\Rightarrow$ (ii): This is a non-trivial part. The argument uses a variation of the Ascoli–Arzela theorem to the metric space (X, g). For completeness, we include a full proof below.

Suppose, on the contrary, that the action of a group Γ of isometries is not properly discontinuous. Then there exists a compact subset $S \subset X$, an infinite sequence $\{\gamma_k\} \subset \Gamma$, a sequence $\{s_k\} \subset S$ such that $\gamma_k \cdot s_k \in S$ for all $k \in \mathbb{N}$. We shall show that $\{\gamma_k\}$ cannot be discrete in the compact-open topology of Isom(X, g).

For $x \in S$, we set $M(x) \equiv M(x; S) := \max_{a \in S} d(x, a)$. For any $x \in X$, one has

$$d(x, \gamma_k \cdot x) \leq d(x, \gamma_k \cdot s) + d(\gamma_k \cdot s, \gamma_k \cdot x) \leq 2M(x).$$

Since every bounded closed set is compact, $\{\gamma_k \cdot x\}$ has a convergent subsequence in X.

We take a countable and dense subset $\{x_j\}_{j \in \mathbb{N}}$ in X. By Cantor's diagonal argument, there exist a subsequence of positive integers $k_1 < k_2 < \cdots$ such that $\gamma_{k_\ell} x_j$ converges as ℓ tends to infinity for every $j \in \mathbb{N}$. For simplicity, we continue to denote the subsequence γ_{k_ℓ} by γ_k.

We claim that the sequence of maps $\gamma_k|_C$ converges uniformly on any compact subset C in X. To see this, let $\varepsilon > 0$. Since C is compact, one can take $N > 0$ such that for any $x \in C$ there exists $j \equiv j(x) \in \{1, 2, \ldots, N\}$ with $d(x, x_j) < \frac{\varepsilon}{3}$. We take $T > 0$ such that

$$d(\gamma_k \cdot x_i, \gamma_{k'} \cdot x_i) < \frac{\varepsilon}{3}$$

for any $k, k' \geq T$ and for any $1 \leq i \leq N$. Then for any $x \in C$

$$\begin{aligned} d(\gamma_k \cdot x, \gamma_{k'} \cdot x) &\leq d(\gamma_k \cdot x, \gamma_k \cdot x_j) + d(\gamma_k \cdot x_j, \gamma_{k'} \cdot x_j) + d(\gamma_{k'} \cdot x_j, \gamma_{k'} \cdot x) \\ &< \frac{\varepsilon}{3} + \frac{\varepsilon}{3} + \frac{\varepsilon}{3} = \varepsilon, \end{aligned}$$

because γ_k is an isometry. Hence $\gamma_k \cdot x$ converges to an element, say $\gamma_C \cdot x$ in X. By taking a sequence $C_1 \subset C_2 \subset \cdots$ of compact subsets in X with $X = \bigcup_{i=1}^{\infty} C_i$, one sees that the map $\gamma_{C_j}|_{C_i}$ coincides with γ_{C_i} whenever $C_i \subset C_j$ for $i \leq j$, because of the uniqueness of the limit. Hence $\gamma \colon X \to X$ is defined as the inductive limit of γ_{C_i}.

We claim that the limiting map γ lies in $\mathrm{Isom}(X, g)$. In fact, for any $x, x' \in X$, one has

$$\begin{aligned} d(\gamma \cdot x, \gamma \cdot x') &= d(\lim_{k \to \infty} \gamma_k \cdot x, \lim_{k \to \infty} \gamma_k \cdot x') \\ &= \lim_{k \to \infty} d(\gamma_k \cdot x, \gamma_k \cdot x') = d(x, x'). \end{aligned}$$

Hence γ is an isometry. Moreover, the sequence $\{\gamma_k^{-1}\}$ yields γ^{-1} as its limit, showing that the isometry $\gamma \colon X \to X$ is a surjective map. Since γ_k converges to γ with respect to the compact-open topology, Γ is not closed in $\mathrm{Isom}(X, g)$. Since a discrete group is closed (see *e.g.*, (5.10) in [HR63]), the reverse implication (i) $\Rightarrow$ (ii) is proved. □

3.7. *Isometric Actions in Pseudo-Riemannian Geometry*

The group of isometries of any pseudo-Riemannian manifold is a Lie group. However, the proof of Proposition 3.11 relies heavily on the *positive-definiteness* of the metric on X. This leads to the following question:

Question 3.12: (Action of Isometric Discrete Group). Does the equivalence (i) $\Leftrightarrow$ (ii) Proposition 3.11 still hold in the pseudo-Riemannian setting?

Unfortunately, an analogue of Proposition 3.11 fails in pseudo-Riemannian geometry.

Let X be a pseudo-Riemannian manifold. Let $\mathrm{Isom}(X)$ denote the group of isometries, and let Γ be a subgroup of $\mathrm{Isom}(X)$. Then the implication (ii) $\Rightarrow$ (i) in Proposition 3.11 remains true, but the converse implication (i) $\Rightarrow$ (ii) does not, as demonstrated in the following example.

Example 3.13: (Isometric but Non-Proper Action). Let $\Gamma := \mathbb{Z}$ act on $X := \mathbb{R}^2$ by

$$(x, y) \mapsto (e^n x, e^{-n} y) \qquad \text{for } n \in \mathbb{Z}.$$

We first observe that there does not exist a metric d on $X = \mathbb{R}^2$ with respect to which L acts isometrically. In fact, suppose such metric d existed. Let $o := (0, 0)$ and $p := (0, 1)$. Then for $t \in \mathbb{Z}$, we compute

$$d(o, p) = d(t \cdot o, t \cdot p) = d(o, (o, e^{-t})).$$

Taking the limit as $t \to \infty$, we find $d(o, p) \to 0$, hence $d(o, p) = 0$, which contradicts the positive definiteness of d.

While no Γ-invariant *Riemannian* metric exists on X, there does exist a Γ-invariant *Lorentzian structure* on X. Indeed, consider the two-dimensional Minkowski space $\mathbb{R}^{1,1}$ with coordinates

$$(x, y) := (x_1 + x_2, x_1 - x_2),$$

where the Lorentzian metric tensor is given by $dxdy = dx_1^2 - dx_2^2$. Then the Γ-action preserves the Lorentzian structure. Thus, Γ forms a discrete group of isometries of a Lorentzian manifold, but the action is not properly discontinuous, since the origin o is fixed by all elements of Γ.

This example will be revisited from different perspectives throughout the chapter. For instance, it will appear in Example 3.20 from a group-theoretic point of view, and again, in Example 7.14 in the context of dynamical volume estimates.

3.8. *A Large Isometry Group*

As mentioned earlier, the isometry group of any pseudo-Riemannian manifold is a Lie group. Here, we present a representative class of pseudo-Riemannian manifolds whose isometry group act transitively.

Proposition 3.14: *Let $G \supset H$ be a pair of real reductive Lie groups, and let $X := G/H$. Then the homogeneous space X admits a pseudo-Riemannian structure with respect to which G acts isometrically.*

Proof: By a theorem of Mostow, there exists a Cartan involution θ of G such that $\theta H = H$. Let $\mathfrak{g} = \mathfrak{k}+\mathfrak{p}$ be the corresponding Cartan decomposition of the Lie algebra. Take an $\mathrm{Ad}(G)$-invariant, non-degenerate symmetric bilinear form B on $\mathfrak{g}$ such that $B|_{\mathfrak{k}\times\mathfrak{k}}$ is negative definite, $B|_{\mathfrak{p}\times\mathfrak{p}}$ is positive definite, and $B|_{\mathfrak{k}\times\mathfrak{p}} \equiv 0$ (*e.g.*, the Killing form if $\mathfrak{g}$ is semisimple).

Then, B induces an H-invariant, non-degenerate symmetric bilinear form $\overline{B}$ on the quotient space

$$\mathfrak{g}/\mathfrak{h} = \mathfrak{k}/(\mathfrak{h} \cap \mathfrak{k}) \oplus \mathfrak{p}/(\mathfrak{h} \cap \mathfrak{p}),$$

of signature $(d(X), e(X))$, where

$$d(X) := \dim \mathfrak{p}/(\mathfrak{h} \cap \mathfrak{p}) \quad \text{and} \quad e(X) := \dim \mathfrak{k}/(\mathfrak{h} \cap \mathfrak{k}). \tag{3.1}$$

Identifying $\mathfrak{g}/\mathfrak{h}$ with the tangent space T_oX at $o := eH \in X$, we extend this bilinear form $\overline{B}$ to each $T_{g\cdot o}X$ for $g \in G$ via the left translation map $dL_g \colon T_oX \to T_{g\cdot o}X$. This extension is well-defined because the bilinear form $\overline{B}$ is H-invariant.

Consequently, X carries a pseudo-Riemannian structure of signature $(d(X), e(X))$, on which G acts isometrically by construction. □

The numbers $d(X)$ and $e(X)$ also have natural geometric interpretations: the homogeneous space $X = G/H$ admits a K-equivariant smooth vector bundle structure

$$\mathbb{R}^{d(X)} \to X \to Y,$$

where the base space Y is the compact manifold $K/(H \cap K)$ of dimension $e(X)$, see [K89], Appendix.

Here are some classical examples:

Example 3.15: (Riemannian Symmetric Space). Let $H = K$, a maximal compact subgroup of G. Then $d(X) = \dim \mathfrak{p}$ and $e(X) = 0$. Hence the pseudo-Riemannian structure on $X = G/K$ is positive definite. The resulting Riemannian manifold G/K is called a *Riemannian symmetric space.*

Example 3.16: (Pseudo-Riemannian Space Form). Let $(G, H) = (O(p, q), O(p-1, q))$, and $X = G/H$. By a straightforward computation, we have $d(X) = q$, $e(X) = p - 1$. Thus, the pseudo-Riemannian manifold X is of signature $(q, p-1)$, and can be identified with the hypersurface

$$X(p-1, q)_+ = \{x \in \mathbb{R}^{p+q} : x_1^2 + \cdots + x_p^2 - x_{p+1}^2 - \cdots - x_{p+q}^2 = 1\}$$

in $\mathbb{R}^{p,q}$. The manifold $X(p-1, q)_+$ is diffeomorphic to a vector bundle over the sphere S^{p-1} with fiber $\mathbb{R}^q$. Note that the signature $(d(X), e(X))$ is opposite to the convention used in Example 2.3. This sign reversal is explained in Remark 2.4.

The de Sitter space $\mathrm{dS}^n = X(n-1, 1)_+$ is a special case of Example 3.16. The Calabi–Markus theorem (Theorem 2.5) can be reformulated in group-theoretic terms as follows:

Theorem 2.5′: ***(Calabi–Markus [CM62]).*** *Let $(G, H) = (O(n, 1), O(n-1, 1))$. If a discrete subgroup $\Gamma \subset G$ acts properly discontinuously on G/H, then Γ must be finite.*

3.9. *Elementary Consequences of Proper Actions*

We begin by discussing some elementary consequences of proper actions in the general setting where a locally compact group acts continuously on a locally compact Hausdorff space.

Proposition 3.17: *Suppose that a locally compact group L acts properly on a locally compact Hausdorff space. Then the following hold:*

(1) *The quotient space $L\backslash X$ is Hausdorff in the quotient topology;*
(2) *Each orbit $L \cdot x$ is closed in X for all $x \in X$;*
(3) *Each isotropy subgroup L_x is compact for all $x \in X$.*

The condition (2) is equivalent to the statement that the quotient space $L\backslash X$ satisfies the (T_1) separation axiom. Thus, the implication (1) $\Rightarrow$ (2) in Proposition 3.17 is immediate. We note that the Hausdorff property is global in nature, whereas the (T_1) property is local.

Definition 3.18: A continuous action is said to have the (CI) *property* if the condition (3) in Proposition 3.17 is satisfied.

The (CI) property is an abbreviation introduced by the author [K90], standing for "Compact Isotropy", which refers to the condition that all isotropy subgroup are compact.

Let $\varphi\colon L \times X \to X \times X$, $(g, x) \mapsto (x, gx)$ be the action map, as used in Definition-Lemma 3.4(ii). If the L-action on X is proper, then φ is a closed map.

We now provide a proof of Proposition 3.17.

Proof: (1) Let $\overline{X} := L\backslash X$ denote the quotient space, and let $\pi\colon X \to \overline{X}$ be the quotient map. To show that $\overline{X}$ is Hausdorff, it suffices to prove that the complement

$$\overline{X} \times \overline{X} \setminus \operatorname{diag}(\overline{X})$$

is open. Equivalently, it suffices to show that the preimage of the diagonal under $\pi \times \pi$, *i.e.*,

$$(\pi \times \pi)^{-1}(\operatorname{diag}(\overline{X})) = \operatorname{Image}\varphi$$

is closed in $X \times X$. Since the action is proper, φ is a proper map and hence closed, which implies that $\operatorname{Image}(\varphi)$ is closed.

(2) Again, since φ is a closed map, $\varphi(L \times \{x\}) = \{x\} \times L \cdot x$ is closed.

(3) Since φ is proper, $L_x = L_{\{x\} \to \{x\}}$ is compact. □

3.10. *Subtle Examples (Hausdorff $\neq$ (T_1))*

One may naturally ask whether the converse of the statements of Proposition 3.17 also holds. In particular, we consider whether the following implications are generally valid:

(A) free action $\overset{?}{\Longrightarrow}$ proper action,

(B) any orbit is closed $\overset{?}{\Longrightarrow}$ $L\backslash X$ is Hausdorff.

However, neither of these statements hold in the setting where X is a locally compact topological space endowed with a continuous action of a locally compact group L.

Example 3.19: Let $L := \mathbb{R}$, the additive group, act on

$$X := \mathbb{R}^2 \setminus \{(0,0)\} \quad \text{by } (x, y) \mapsto (e^t x, e^{-t} y) \quad \text{for } t \in \mathbb{R}.$$

Then the action of L on X is free, and each L-orbit is closed. However, the action is not proper. To see this, consider the compact subset of X, defined by

$$S := \{(x, y) : x^2 + y^2 = 1\}.$$

Then $L_S = L$, showing that the L-action fails to be proper.

Moreover, the two points $(0, 1)$ and $(1, 0)$ define different points in the quotient space $L\backslash X$, however, these two points cannot be separated by open sets in the quotient topology. Hence, $L\backslash X$ is not Hausdorff.

In the next section, we show how the setting of Example 3.20 naturally arises from the framework of triples $L \subset G \supset H$ of Lie groups.

3.11. *Group Theoretic Viewpoint: Properness for Triples (L, G, H)*

Let G be a locally compact group, and consider a triple of locally compact groups

$$L \subset G \supset H,$$

where L and H are closed subgroups. We consider the natural action of the subgroup L on the homogeneous space $X := G/H$.

Example 3.20: Let $G := SL(2, \mathbb{R})$. We define two subgroups of G by

$$A := \left\{ \begin{pmatrix} e^t & 0 \\ 0 & e^{-t} \end{pmatrix} : t \in \mathbb{R} \right\},$$

$$N := \left\{ \begin{pmatrix} 1 & n \\ 0 & 1 \end{pmatrix} : n \in \mathbb{R} \right\}.$$

There is a natural diffeomorphism

$$G/N \simeq \mathbb{R}^2 \setminus \{(0, 0)\}, \quad gN \mapsto g \begin{pmatrix} 1 \\ 0 \end{pmatrix},$$

under which the A-action on G/N coincides with that described in Example 3.19. Hence, the A-action on G/N is not proper. By duality—as stated in Proposition 4.10 below—the N-action on G/A is likewise not proper.

3.12. *Lipsman's Conjecture (1995)*

Suppose that $L \subset G \supset H$ is a triple of *real reductive Lie groups*. In 1989, the author obtained a criterion for the properness of the L-action on G/H. The result, originally proved in [K89], can be reformulated below, see also Theorem 4.17:

Theorem 3.21: (see Example 5(1) in [K90]). *The following two conditions are equivalent:*

(i) *the L-action on G/H is proper;*

(ii) *the L-action on G/H has the (CI) property (Definition 3.18).*

In 1995, Lipsman [L95] raised a question of whether the equivalence in Theorem 3.21 remains valid for triples of *nilpotent Lie groups*.

Conjecture 3.22: (Lipsman's Conjecture [L95]). Let G be a connected and simply connected nilpotent Lie group, and let L, H be two connected closed subgroups. Are the following two conditions equivalent?

(i) the L-action on G/H is proper;

(ii) the L-action on G/H has the (CI) property.

The implication (i) $\Rightarrow$ (ii) holds in general (see Proposition 3.17(2)). On the other hand, in the simply-connected nilpotent setting, condition (ii) is equivalent to:

(ii)$'$ *the L-action on G/H is free.*

Lipsman's conjecture has been proved affirmatively when the nilpotent Lie group G is at most 3-step; that is, when

$$[\mathfrak{g}, [\mathfrak{g}, [\mathfrak{g}, \mathfrak{g}]]] = \{0\},$$

but it fails in general. The status is summarized as follows:

True : G: 2-step nilpotent Lie groups (Nasrin [N01]),
G: 3-step nilpotent Lie groups (Baklouti–Khlif [BaKh05], Yoshino [Y04]),

False : G: 4-step nilpotent Lie groups (Yoshino [Y05]).

A counterexample discovered by Yoshino is given by a triple of simply connected nilpotent Lie groups $L \subset G \supset H$ such that

$L \simeq \mathbb{R}^2$ (abelian subgroup),
$X = G/H$ is a 5-dimensional nilmanifold $\simeq \mathbb{R}^5$,

with the following properties:

— the L-action on $X = G/H$ is free;

— all L-orbits are closed;

— the orbit space $L\backslash X$ is a (T_1) space but not Hausdorff;

— the L-action on X is not proper.

4. Properness Criterion

In this section, we present criteria for the properness of the L-action on the homogeneous space G/H, where L and H are closed subgroups of a Lie group G.

As shown in Section 3.12, where G is a simply-connected 3-step nilpotent Lie group, the (CI) property provides a convenient necessary condition for the properness of the action.

Here, we focus on the case where G is reductive.

A perspective put forward in [K89] concerning the properness criterion for the L-action on G/H emphasizes that L and H should be regarded symmetrically within the group G, rather than relying on the geometric features of the homogeneous space G/H, as in some prior approaches.

To further articulate this idea, in Section 4.2, we recall the binary relations $\pitchfork$ and $\sim$ on the power set of a locally compact group G, which were introduced in [K96] as a conceptual framework for understanding "the geometry at infinity" in the group G itself, rather than attempting to understand "the geometry at infinity" of the homogeneous space G/H.

4.1. *Expanding a Subset H of a Group G by Compact Set S*

Let H be a subset of a locally compact group G, let $S \subset G$ be a compact subset. We define the expansion of H by S through group multiplication as follows.

$$\begin{aligned} SH :=& \{bx : x \in H, b \in S\}. \\ HS :=& \{xb : x \in H, b \in S\}, \\ SHS :=& \{axb : x \in H, a, b \in S\}. \end{aligned}$$

When G is abelian, the subsets SH and HS may be thought of as tubular neighbourhoods of H, as well as the subset $SHS = (SS)H = H(SS)$. In contrast, when G is highly non-commutative—such as in the case of $SL(2,\mathbb{R})$—the set SHS can become significant "larger" in a nontrivial way. A deeper understanding of the structure of SHS allows us to reformulate the problem of properness of the L-action on G/H as a question internal to the group G itself.

Here is a straightforward and fundamental observation:

Lemma 4.1: *Suppose that both L and H are closed subgroups of G. Then the following two conditions on the pair (L, H) are equivalent:*

(i) *The action of L on G/H is proper.*

(ii) *For every compact subset $S \subset G$, the intersection $L \cap SHS$ is compact.*

Proof: Let S be a compact subset of G, and let $\overline{S} := SH/H \subset G/H$. Then $\overline{S}$ is compact. Conversely, every compact subset of G/H can be expressed in this form for some compact subset $S \subset G$. By the definition of proper actions, condition (i) is equivalent to the compactness of the set

$$L_{\overline{S}} := \{\ell \in L : \ell \cdot \overline{S} \cap \overline{S} \neq \emptyset\}$$

for every compact subset $S \subset G$.

Without loss of generality, we assume that S is symmetric, *i.e.*, $S = S^{-1}$. Under this assumption, one has

$$L_{\overline{S}} = L \cap SHS^{-1} = L \cap SHS,$$

which shows the equivalence (i) $\Leftrightarrow$ (ii). □

4.2. *Relations $\pitchfork$ and $\sim$ for Subsets in Locally Compact Groups*

Let $\mathcal{P}(G)$ denote the power set of a locally compact group G. We define the binary relations $\pitchfork$ and $\sim$ on the power set $\mathcal{P}(G)$ of the group G.

Definition 4.2: ([K96], Def. 2.1.1). For two subsets L and H of G, we define the following two binary relations $\pitchfork$ and $\sim$:
(1) $L \pitchfork H$ if for every compact subset $S \subset G$, the intersection $L \cap SHS$ is relatively compact, *i.e.*, its closure is compact;
(2) $L \sim H$ if there exists a compact subset $S \subset G$ such that both $L \subset SHS$ and $H \subset SLS$.

We illustrate these definitions with simple examples:

Example 4.3: (Abelian Case). Let G be a vector space $\mathbb{R}^n$, and let L, H be subspaces of G.

(1) $L \pitchfork H$ if and only if $L \cap H = \{0\}$.
(2) $L \sim H$ if and only if $L = H$.

Example 4.4: Let $G = SL(2, \mathbb{R})$. Up to conjugation, there are six connected subgroups of G:

$$\{e\}, K, A, N, AN, \text{ and } G,$$

where A and N are as defined in Example 3.20, and $K = SO(2)$. We then have

$$\{e\} \sim K,$$
$$A \sim N \sim AN \sim G.$$

This SL_2 example can be generalized in two directions as follows:

Example 4.5: Let G be a real reductive linear group. Then the following decompositions hold:

$$\begin{aligned} G =& KAK \quad && \text{Cartan decomposition, see (4.2),} \\ G =& KAN \quad && \text{Iwasawa decomposition,} \\ G =& KNK. && \end{aligned}$$

Hence, we have

$$\{e\} \sim K,$$
$$A \sim N \sim AN \sim G.$$

Example 4.6: Let G be a real reductive Lie group of split rank one. It follows that for any closed subgroup L, not necessarily connected, either $L \sim \{e\}$ or $L \sim G$ holds (*cf.* [K93]).

4.3. *Meaning of the Binary Relations* $\pitchfork$ *and* $\sim$

In Definition 4.2, L and H are allowed to be *subsets* of G, without assuming that they are closed subgroups.

The following two lemmas can be verified directly from Definition 4.2.

Lemma 4.7: *The relation* $\sim$ *defines an equivalence relation on* $\mathcal{P}(G)$*; that is, for any* $L_1, L_2, L_3 \in \mathcal{P}(G)$*, the following properties hold in addition to the obvious reflexivity:*

(1) (Symmetry) *$L_1 \sim L_2$ if and only if $L_2 \sim L_1$.*
(2) (Transitivity) *If $L_1 \sim L_2$ and $L_2 \sim L_3$, then $L_1 \sim L_3$.*

Lemma 4.8: *Let H, H', and L be subsets of a locally compact group G.*

(1) *$L \pitchfork H$ if and only if $H \pitchfork L$.*
(2) *If $H \sim H'$, then the following equivalence holds for any L:*

$$H \pitchfork L \iff H' \pitchfork L.$$

As an immediate consequence of Lemma 4.1 and the definition of $\pitchfork$ in Definition 4.2, we have the following proposition.

Proposition 4.9: *Let L and H be closed subgroups of a locally compact group G. Then the relation $L \pitchfork H$ holds if and only if the action of L on G/H is proper.*

Lemma 4.8(1) clarifies the symmetry between the closed subgroups L and H in G with respect to proper actions:

Proposition 4.10: *Let L and H be closed subgroups of a locally compact group G. Then, the action of L on G/H is proper if and only if the action of H on G/L is proper.*

In light of Lemma 4.8(2), we can formulate the following fundamental problem concerning a criterion for properness in the general framework, as follows:

Problem 4.11: (Properness Criterion). *Find a criterion for two subsets $L, H \subset G$ to satisfy*

$$L \pitchfork H,$$

modulo the equivalence relation $\sim$.

As we shall see in Theorem 4.12 below, the equivalence relation $\sim$ on $\mathcal{P}(G)$ is the coarsest equivalence relation that preserves the properness condition $\pitchfork$.

4.4. *Discontinuous Duality Theorem*

For a subset H of a locally compact group G, we define its "discontinuous dual" by

$$\pitchfork (H : G) := \{L \in \mathcal{P}(G) : L \pitchfork H\}.$$

The discontinuous dual $\pitchfork (H : G)$ depends solely on the equivalence class of H under the relation $\sim$, as stated in Lemma 4.7. Inspired by the Pontrjagin–Tannaka–Tatsuuma duality theorem [Ta67], which roughly states that a locally compact group G can be recovered from its unitary dual $\widehat{G}$, the present author suggested in Theorem 5.6 of [K96] a "discontinuous duality theorem" formulated as follows.

Theorem 4.12: (Discontinuous Duality Theorem). *Let G be a separable, locally compact topological group. Then any subset $H \subset G$ is determined, up to the equivalence relation $\sim$, by its discontinuous dual $\pitchfork (H : G)$.*

Theorem 4.12 was first proved for real reductive Lie groups G in [K96], and was later extended to general locally compact groups by Yoshino [Y07].

4.5. *Properness Criterion for Reductive Groups*

It is worth emphasizing that by the term "criterion", we mean an *explicit and effective* method for determining whether the relation $L \pitchfork H$ holds—not merely a theoretically correct but practically intractable reformulation. In this context, Problem 4.11 remains open for general Lie groups. However, in the case where G is a real reductive Lie group, the problem has been resolved, as reviewed in Theorem 4.14 below.

Let G be a real reductive group, $\mathfrak{g} = \mathfrak{k} + \mathfrak{p}$ a Cartan decomposition of its Lie algebra, and K the maximal compact subgroup of G with Lie algebra $\mathfrak{k}$. We take a maximal abelian subspace $\mathfrak{a}$ in $\mathfrak{p}$.

For $\alpha \in \mathfrak{a}^*$, we define the root space and the set of (restricted) roots by

$$\mathfrak{g}_\alpha := \{X \in \mathfrak{g} : \operatorname{ad}(H)X = \alpha(H) \text{ for all } H \in \mathfrak{a}\},$$
$$\Sigma(\mathfrak{g}, \mathfrak{a}) := \{\alpha \in \mathfrak{a}^* : \mathfrak{g}_\alpha \neq \{0\}\}.$$

Then the finite group

$$N_G(\mathfrak{a})/Z_G(\mathfrak{a}) = \{g \in G : \operatorname{Ad}(g)\mathfrak{a} = \mathfrak{a}\}/\{g \in G : \operatorname{Ad}(g)|_{\mathfrak{a}} = \operatorname{id}\}$$

is isomorphic to the Weyl group, to be denoted by W, of the restricted root system $\Sigma(\mathfrak{g}, \mathfrak{a})$.

We fix a set of positive roots $\Sigma^+(\mathfrak{g}, \mathfrak{a})$, and define the (closed) dominant Weyl chamber $\overline{\mathfrak{a}_+}$ by

$$\overline{\mathfrak{a}_+} := \{H \in \mathfrak{a} : \alpha(H) \geq 0 \text{ for all } \alpha \in \Sigma^+(\mathfrak{g}, \mathfrak{a})\}.$$

Then we have a natural bijection

$$\overline{\mathfrak{a}_+} \simeq \mathfrak{a}/W. \tag{4.1}$$

We set $A := \exp(\mathfrak{a})$ and $\overline{A_+} := \exp(\overline{\mathfrak{a}_+})$.

In analogy with polar coordinates in the Euclidean space $\mathbb{R}^n$, there exists a notion of polar coordinates on the Riemannian symmetric space G/K. In group-theoretic terms, this corresponds to the fact that a real reductive Lie group G admits a Cartan decomposition:

$$G = KAK = K\overline{A_+}K. \tag{4.2}$$

In this decomposition, every $g \in G$ can be written as $g \in K \exp(\mu(g))K$, where $\mu(g) \in \mathfrak{a}$ is unique up to conjugation by the Weyl group W.

The Cartan decomposition (4.2) defines the *Cartan projection*:

$$\mu\colon G \to \mathfrak{a}/W \simeq \overline{\mathfrak{a}_+}, \quad g \mapsto H \mod W, \tag{4.3}$$

characterized by the condition that $g \in K\exp(H)K$.

Example 4.13: Let $G = GL(n, \mathbb{R})$ and $K = O(n)$. Then we can identify

$$\begin{aligned} &\mathfrak{a} \simeq \mathbb{R}^n, \\ &W \simeq \mathfrak{S}_n, \\ &\overline{\mathfrak{a}_+} \simeq \mathbb{R}^n_{\geq} := \{(H_1, \ldots, H_n) : H_1 \geq \cdots \geq H_n\}. \end{aligned}$$

If $g = k_1 \operatorname{diag}(e^{H_1}, \ldots, e^{H_n})k_2$ for some $k_1, k_2 \in O(n)$ and $H_1, \ldots, H_n \in \mathbb{R}$, then

$${}^t gg = {}^t k_2 \operatorname{diag}(e^{2H_1}, \ldots, e^{2H_n})k_2.$$

Hence, the Cartan projection

$$\mu\colon GL(n, \mathbb{R}) \to \mathbb{R}^n/\mathfrak{S}_n \simeq \mathbb{R}^n_{\geq}$$

is given by

$$g \mapsto \tfrac{1}{2}(\log \lambda_1, \cdots, \log \lambda_n),$$

where $\lambda_1 \geq \cdots \geq \lambda_n \,(> 0)$ are the eigenvalues of ${}^t gg$.

The following properness criterion was established by Benoist ([B96], Thm. 5.2) and the present author ([K96], Thm. 1.1) extending the criterion given in [K89] for the special case where L and H are reductive subgroups (see Theorem 4.17).

Theorem 4.14: (Properness Criterion). *Let G be a reductive Lie group, and let H, L be subsets of G. Then the following equivalences hold:*

(1) $L \sim H$ *in* $G \iff \mu(L) \sim \mu(H)$ *in* $\mathfrak{a}$.
(2) $L \pitchfork H$ *in* $G \iff \mu(L) \pitchfork \mu(H)$ *in* $\mathfrak{a}$.

Example 4.15: Let $G = SL(2, \mathbb{R})$, $L := A$. Then the Cartan projection $\mu\colon \mathfrak{g} \to \mathfrak{a}/W \simeq \overline{\mathfrak{a}_+}$ gives

$$\mu(A) = \mu(N) = \overline{\mathfrak{a}_+}.$$

Hence, $A \not\pitchfork N$. We have seen this directly in Example 3.20, which describes the $\mathbb{R}$-action on $\mathbb{R}^2 \setminus \{(0, 0)\}$.

It is worth noting that, in the equivalences in Theorem 4.14, the left-hand sides are formulated in the non-commutative group G, whereas the right-hand sides are described in the abelian space $\mathfrak{a}$. As a consequence, the conditions $\pitchfork$ and $\sim$ can be verified on the right-hand sides, once the Cartan projections $\mu(L)$ and $\mu(H)$ are known.

Special cases of Theorem 4.14 include the following:

$\Rightarrow$ in (1): Uniform error estimates of eigenvalues under matrix perturbation.

$\Leftrightarrow$ in (2): A criterion for proper actions of groups.

Moreover, in connection with the criterion for $\pitchfork$ in (2), we discuss in Section 7 two approaches for quantifying them.

Remark 4.16: ($\sim$ and $\pitchfork$ in $\mathfrak{a}/W \simeq \overline{\mathfrak{a}_+}$). The Cartan projections $\mu(L)$ and $\mu(H)$, given in (4.3), can be interpreted in two ways: as subsets of $\overline{\mathfrak{a}_+}$ or as W-invariant subsets of $\mathfrak{a}$. In either interpretation, the relations $\sim$ and $\pitchfork$ between $\mu(L)$ and $\mu(H)$ retain the same meaning. In fact, for two subsets $S, T \subset \overline{\mathfrak{a}_+}$, define their W-invariant extensions of $\mathfrak{a}$ by

$$\widetilde{S} := W \cdot S, \quad \widetilde{T} := W \cdot T.$$

Then it is readily seen from the definitions that the following equivalences hold:

$$S \sim T \Longleftrightarrow \widetilde{S} \sim \widetilde{T},$$
$$S \pitchfork T \Longleftrightarrow \widetilde{S} \pitchfork \widetilde{T}.$$

4.6. *Properness Criterion—Special Case (Reductive Subgroups)*

In this section, we illustrate the idea behind the proof of the properness criterion (Theorem 4.14) in the special case where L and H are reductive subgroups of G.

Since the properness criterion is invariant under conjugation of L and H, we may, without loss of generality, assume that both are stable under a Cartan involution θ of G.

To treat L and H in a uniform manner, we introduce a θ-stable subgroup G', and set up the corresponding notation.

Let $\mathfrak{g} = \mathfrak{k} + \mathfrak{p}$ be the Cartan decomposition corresponding to the Cartan involution θ. Since G' is θ-stable, its Lie algebra $\mathfrak{g}'$ admits a compatible Cartan decomposition $\mathfrak{g}' = \mathfrak{k}' + \mathfrak{p}'$ with $\mathfrak{k}' \subset \mathfrak{k}$ and $\mathfrak{p}' \subset \mathfrak{p}$. Let $\mathfrak{a}_{G'}$ be a maximal abelian subspace in $\mathfrak{p}'$, which we extend to a maximal abelian subspace $\mathfrak{a}$ in $\mathfrak{p}$. Then $\mathfrak{a}_{G'} = \mathfrak{a} \cap \mathfrak{g}'$. We summarize these subspaces as follows.

$$\begin{array}{ccccccccc} \mathfrak{g} = & \mathfrak{k} & + & \mathfrak{p} & \supset & \mathfrak{p} & \underset{\text{maximal abelian}}{\supset} & \mathfrak{a} & \\ \cup & \cup & & \cup & & \cup & & \cup & \\ \mathfrak{g}' = & \mathfrak{k}' & + & \mathfrak{p}' & \supset & \mathfrak{p}' & \underset{\text{maximal abelian}}{\supset} & \mathfrak{a}_{G'} := \mathfrak{a} \cap \mathfrak{g}'. & \end{array} \tag{4.4}$$

Let $A_{G'} := \exp(\mathfrak{a}_{G'})$. By the Cartan decomposition, we have $G' \sim A_{G'}$ in G' (see Example 4.5), and hence also $G' \sim A_{G'}$ in G. The Cartan projection of G' takes the form $\mu(G') = W \cdot \mathfrak{a}_{G'}$ in $\mathfrak{a}$.

Applying the above notation to $G' = L$, we have $\mathfrak{a}_L = \mathfrak{a} \cap \mathfrak{l}$. By conjugating H by an element of K, we may assume $\mathfrak{a}_H := \mathfrak{a} \cap \mathfrak{h}$ is also a maximal abelian subspace of $\mathfrak{h} \cap \mathfrak{p}$. Then, the condition $\mu(L) \pitchfork \mu(H)$ in $\mathfrak{a}$, as stated in Theorem 4.14(2), is equivalent to the condition

$$\mathfrak{a}_H \cap W \cdot \mathfrak{a}_L = \{0\}.$$

Hence, the special case of Theorem 4.14 yields the following result:

Theorem 4.17: (Properness Criterion for Reductive Subgroups [K89]). *Let G be a real reductive Lie group, and let H and L be two reductive subgroups of G. Then the following conditions on the pair (L, H) are equivalent:*

(i) *the action of L on G/H is proper;*
(i)′ *the action of H on G/L is proper;*
(ii) $\mathfrak{a}_H \cap W \cdot \mathfrak{a}_L = \{0\}$ *in* $\mathfrak{a}$.

4.7. *Reduction of Properness Criterion to Abelian subgroups*

We now outline the key idea from [K89] that underlies the proof of Theorem 4.17. Although the proof of Theorem 4.14 in the general case is more involved, it is still based on the same principle.

First, we observe that $L \sim A_L$ and $H \sim A_H$ in G. Hence, the properness criterion in Theorem 4.17 in the reductive case can be reduced to the abelian case, where $L, H \subset A$. This reflects the fact that the ambient group G itself is highly non-commutative. This reduction is formulated as Lemma 4.18 below.

Lemma 4.18: *Suppose that L and H are connected subgroups of the split abelian subgroup A. Let $\mathfrak{l}, \mathfrak{h} \subset \mathfrak{a}$ be the (abelian) Lie algebras of L and H, respectively. Then the following are equivalent:*

(i) *The action of L on G/H is proper.*

(ii) $\mathfrak{l} \cap W\mathfrak{h} = \{0\}$.

The implication (i) $\Longrightarrow$ (ii) is the easier direction, that is, properness implies the (CI) property, as seen in Proposition 3.17(3).

The converse implication (ii) $\Longrightarrow$ (i) is more involved.

In the next section, we will give an overview of the proof.

4.8. *Proof of Lemma 4.18 for Abelian Subgroups H, $L \subset G$*

Suppose that both $\mathfrak{l}$ and $\mathfrak{h}$ are subspaces of $\mathfrak{a}$. We aim to prove that if $L \not\pitchfork H$, then $\mathfrak{l} \cap W\mathfrak{h} \neq \{0\}$.

If $L \not\pitchfork H$, then there exists a compact subset $S \subset G$ such that the intersection $L \cap SHS$ is non-compact. Hence, one can find sequences t_n, $t'_n \in \mathbb{R}$, $Y_n \in \mathfrak{l}$, $Z_n \in \mathfrak{h}$ with $\|Y_n\| = \|Z_n\| = 1$, and $c_n, d_n \in S$ such that

$$\exp(t_n Y_n) = c_n \exp(t'_n Z_n) d_n \quad \text{in } G,$$
$$\lim_{n\to\infty} t_n = \infty.$$

By passing to a subsequence, we may assume that the sequences c_n, d_n, Y_n, and Z_n converge as n tends to infinity, say,

$$c_n \to c,\ d_n \to d \text{ in } S$$
$$Y_n \to Y \neq 0 \in \mathfrak{l}, \quad Z_n \to Z \neq 0 \in \mathfrak{h}.$$

Step 1. We show that the sequence $\frac{t'_n}{t_n}$ is bounded away from 0.

Once this boundedness is established, we may again pass to a subsequence and assume that $\frac{t'_n}{t_n}$ converges. By replacing (S, t'_n, Z_n, c_n, d_n) with $(KSK, t_n, \frac{t'_n}{t_n} Z_n, c_n k, k^{-1} d_n)$ for some $k \in N_K(\mathfrak{a})$, we may further assume that $t'_n = t_n$ and $Y, Z \in \overline{\mathfrak{a}_+}$. Thus, we obtain sequences such that $\lim_{n\to\infty} t_n = \infty$ and:

$$c_n = \exp(t_n Y_n) d_n^{-1} \exp(-t_n Z_n), \tag{4.5}$$
$$c_n \to c,\ d_n \to d \text{ in } G,$$
$$Y_n \to Y \in \mathfrak{l} \cap \overline{\mathfrak{a}_+},\ Z_n \to Z \in \mathfrak{h} \cap \overline{\mathfrak{a}_+}.$$

Step 2. We now derive that $Y = Z$ from (4.5).

Both Steps 1 and 2 deal with the behavior of sequences "at infinity" in the group G. To analyze the "geometry at infinity" of the group G, we localize the analysis by examining the dynamics in terms of the root space decomposition

$$\mathfrak{g} = \bigoplus_{\alpha \in \Sigma(\mathfrak{g};\mathfrak{a}) \cup \{0\}} \mathfrak{g}_\alpha$$

via the adjoint representation.

To illustrate the method applied in both steps, we consider the following identity, which plays a crucial role in establishing $Y = Z$ in Step 2:

$$\operatorname{Ad}(c)\mathfrak{g}_\alpha = \sum_{\beta(Y) \geq \alpha(Z)} \mathfrak{g}_\beta \tag{4.6}$$

for any $\alpha \in \Sigma(\mathfrak{g}, \mathfrak{a})$. To verify (4.6), let $\operatorname{pr}_\alpha \colon \mathfrak{g} \to \mathfrak{g}_\alpha$ denote the projection associated with the root space decomposition. From (4.5), it follows that for any $\alpha \in \Sigma(\mathfrak{g}, \mathfrak{a})$,

$$\begin{aligned} \operatorname{Ad}(c)\mathfrak{g}_\alpha &= \lim_{n \to \infty} \bigoplus_{\beta \in \Sigma(\mathfrak{g},\mathfrak{a}) \cup \{0\}} e^{t_n(\beta(Y_n) - \alpha(Z_n))} \operatorname{pr}_\beta(\operatorname{Ad}(d_n^{-1})\mathfrak{g}_\alpha) \\ &\subset \bigoplus_{\beta(Y) \geq \alpha(Z)} \mathfrak{g}_\beta. \end{aligned}$$

The opposite inclusion follows similarly, and thus (4.6) holds. This argument implies $Y = Z$, hence $\mathfrak{l} \cap \mathfrak{h} \neq \{0\}$. See [K89] for further details. □

4.9. *Criterion for the Calabi–Markus Phenomenon*

The Calabi–Markus phenomenon (Theorem 2.5), originally discovered in [CM62] in the context of the de Sitter space, can be formulated in a more general setting as follows.

Corollary 4.19: (Criterion for the Calabi–Markus Phenomenon [K89]). *Let $G \supset H$ be a pair of real reductive Lie groups. Then the following four conditions (i)–(iv) are equivalent:*

(i) *G/H admits a discontinuous group $\Gamma \simeq \mathbb{Z}$.*
(ii) *G/H admits an infinite discontinuous group Γ.*
(iii) *$G \not\pitchfork H$.*
(iv) *$\operatorname{rank}_\mathbb{R} G > \operatorname{rank}_\mathbb{R} H$.*

The original result by Calabi and Markus in [CM62] shows that condition (ii) fails to hold when $(G, H) = (O(n, 1), O(n-1, 1))$. In this case, we observe $\operatorname{rank}_\mathbb{R} G = \operatorname{rank}_\mathbb{R} H = 1$, and consequently condition (iv) fails as well.

Proof: The implication (i) $\Rightarrow$ (ii) is immediate.
(ii) $\Rightarrow$ (iii). Suppose that (ii) holds. Then $\Gamma \pitchfork H$. However, since $\Gamma \not\pitchfork G$, Lemma 4.8(2) implies that $H \not\sim G$. Thus, (ii) $\Rightarrow$ (iii) is verified.
(iii) $\Rightarrow$ (iv). Without loss of generality, assume that $\mathfrak{a}_H \subset \mathfrak{a}$ as in (4.4). Since $\operatorname{rank}_{\mathbb{R}} H = \dim \mathfrak{a}_H$, the assumption $\operatorname{rank}_{\mathbb{R}} G = \operatorname{rank}_{\mathbb{R}} H$ implies $\mu(H) = \mathfrak{a}$. Hence, by the easier direction of Theorem 4.17, we have $G \sim H$. This completes the proof of (iii) $\Rightarrow$ (iv).
(iv) $\Rightarrow$ (i). This is the main step. If $\operatorname{rank}_{\mathbb{R}} G > \operatorname{rank}_{\mathbb{R}} H$, *i.e.*, if $\mathfrak{a}_H \subsetneqq \mathfrak{a}$, then there exists a one-dimensional subspace $\mathfrak{l}$ in $\mathfrak{a}$ such that $W \cdot \mathfrak{a}_H \cap \mathfrak{l} = \{0\}$. By the properness criterion in Theorem 4.17, the subgroup $L := \exp \mathfrak{l}$ acts properly on G/H. In particular, any lattice in L, isomorphic to $\mathbb{Z}$ acts properly discontinuously on G/H. □

4.10. *Proper Actions of $SL(2, \mathbb{R})$*

In the previous section, we discussed proper actions of *commutative* subgroups on homogeneous spaces. Here, we turn to the case of *non-commutative* subgroups, illustrating the discussion with the example of $SL(2, \mathbb{R})$ or $PSL(2, \mathbb{R})$.

Proposition 4.20: *Let G be a real reductive linear Lie group, and let H be a closed subgroup (possibly non-reductive, e.g., a discrete subgroup). Consider the following five conditions:*

(i) *G/H admits a discontinuous group $\Gamma \simeq \mathbb{Z}$ generated by a unipotent element.*

(ii) *G/H admits a proper action of a subgroup L which is locally isomorphic to $SL(2, \mathbb{R})$.*

(iii) *For any $g \geq 2$, G/H admits a discontinuous group Γ isomorphic to $\pi_1(\Sigma_g)$, where Σ_g is a closed oriented surface of genus g.*

(iv) *For some $g \geq 2$, G/H admits a discontinuous group $\Gamma \simeq \pi_1(\Sigma_g)$.*

(v) *G/H admits a discontinuous group Γ of infinite order, which is not virtually abelian, i.e., Γ does not contain an abelian subgroup of finite index.*

Then the following implications and equivalences hold:

$$(i) \Leftrightarrow (ii) \Rightarrow (iii) \Rightarrow (iv) \Rightarrow (v).$$

Proof: The equivalence (i) $\Leftrightarrow$ (ii) (*cf.* [K93], Lem. 3.2) follows from the Jacobson–Morozov theorem.

Since any surface group can be embedded as a discrete subgroup of $PSL(2, \mathbb{R})$, and also of $SL(2, \mathbb{R})$, the implication (ii) $\Rightarrow$ (iii) follows.

The remaining implications (iii) $\Rightarrow$ (iv) $\Rightarrow$ (v) are straightforward. □

4.11. *An Example: Actions of* $SL(2,\mathbb{R})$ *on* $SL(n,\mathbb{R})/SL(m,\mathbb{R})$

In the previous section, we discussed general properties of non-abelian groups such as surface groups and $SL(2,\mathbb{R})$ on homogeneous spaces. In this section, we examine properness of $SL(2,\mathbb{R})$-actions more concretely through an explicit example. Specifically, we consider the action of $SL(2,\mathbb{R})$ on the homogeneous space G/H via a group homomorphism

$$\varphi\colon SL(2,\mathbb{R}) \to G.$$

There are, in fact, many such homomorphisms φ, and the properness of the induced action generally depends one the choice of φ. We illustrate this dependence with the case, where $G = SL(n,\mathbb{R})$ and $H = SL(m,\mathbb{R})$ is the subgroup embedded block-diagonally in G with $m < n$.

Question 4.21: Suppose that $\varphi_n\colon SL(2,\mathbb{R}) \to SL(n,\mathbb{R})$ is an irreducible representation. Is the action of $SL(2,\mathbb{R})$ on G/H via φ_n proper?

See also Example 5.8(1) below for a related discussion on the existence problem of cocompact discontinuous groups for the same homogeneous space G/H.

Let $L := \varphi_n(SL(2,\mathbb{R}))$. To apply Theorem 4.17, we compute $\mu(H)$ and $\mu(L)$. We define a maximal abelian subspace $\mathfrak{a}$ by the diagonal embedding

$$\mathfrak{a} := \left\{ (a_1, \cdots, a_n) : \sum_{j=1}^{n} a_j = 0 \right\} \underset{\text{diag}}{\hookrightarrow} \mathfrak{g} = \mathfrak{sl}(n,\mathbb{R}).$$

Then the Cartan projection is given by $\mu\colon G \to \mathfrak{a}/\mathfrak{S}_n$ for $G = SL(n,\mathbb{R})$.

For $H = SL(m,\mathbb{R})$ in $G = SL(n,\mathbb{R})$ $(m < n)$,

$$\mu(H) = \mathfrak{S}_n \cdot \mathfrak{a}_H = \mathfrak{S}_n \cdot \left\{ (b_1, \cdots, b_m, 0, \cdots, 0) : \sum_{j=1}^{m} b_j = 0 \right\}.$$

On the other hand, for the irreducible representation φ_n, we have

$$\mu(L) = \mathfrak{S}_n \cdot \mathfrak{a}_L = \mathfrak{S}_n \cdot \mathbb{R}(n-1, n-3, \cdots, 1-n).$$

By the properness criterion in Theorem 4.17, we have the equivalences.

$$\begin{aligned} L \text{ acts properly on } G/H \quad &\iff \mu(L) \cap \mu(H) = \{0\} \\ &\iff n \text{ is even or } n - m \geq 2. \end{aligned}$$

More generally, one may ask the following question:

Question 4.22: Given a homomorphism $\varphi\colon SL(2,\mathbb{R}) \to SL(n,\mathbb{R})$, determine whether the induced action of $SL(2,\mathbb{R})$ on $SL(n,\mathbb{R})/SL(m,\mathbb{R})$, is proper.

According to the Dynkin–Kostant theory, the set of conjugacy classes $\mathrm{Hom}(SL(2,\mathbb{R}),G)/G$ is finite for any reductive Lie group G. For $G = SL(n,\mathbb{R})$, there exists a one-to-one correspondence between these conjugacy classes and the set $\mathbb{P}(n)$ of all partitions of n:

$$\mathrm{Hom}(SL(2,\mathbb{R}),G)/G \simeq \mathbb{P}(n). \tag{4.7}$$

More explicitly, any homomorphism $\varphi\colon SL(2,\mathbb{R}) \to G$ is conjugate to a direct sum of the form

$$\bigoplus_{j=1}^{n}(\overbrace{\varphi_j \oplus \cdots \oplus \varphi_j}^{m_j}),$$

where φ_j denotes the irreducible j-dimensional real representation of $SL(2,\mathbb{R})$, and m_j ($\in \mathbb{N}$) is the (possibly zero) multiplicity, satisfying

$$\sum_{j=1}^{n} j\, m_j = n.$$

Let $L := \varphi(SL(2,\mathbb{R}))$. After conjugating L if necessary, and using the convention of (4.4), we obtain

$$\mathfrak{a}_L = \mathbb{R}(\bigoplus_{j=1}^{n}(\overbrace{v_j \oplus \cdots \oplus v_j}^{m_j})),$$

where $v_j := (j-1, j-3, \ldots, 1-j) \in \mathbb{Z}^j$. Applying the properness criterion in Theorem 4.17 to the pair

$$(L,H) = (\varphi(SL(2,\mathbb{R})), SL(m,\mathbb{R})),$$

we conclude that the action of $SL(2,\mathbb{R})$ on G/H is proper if and only if:

$$\sum_{j:\mathrm{odd}} j m_j < n - m.$$

4.12. *Properly Discontinuous Actions of Surface Groups*

The previous example $G/H = SL(n, \mathbb{R})/SL(m, \mathbb{R})$ is a non-symmetric homogeneous space. When G/H is a *reductive symmetric space*, Okuda [Ok13] provided a complete classification of such spaces that admit proper actions of $SL(2, \mathbb{R})$ via a homomorphism $\varphi\colon SL(2, \mathbb{R}) \to G$. His classification relies on the properness criterion (Theorem 4.17) along with the Dynkin–Kostant theory of nilpotent orbits, as given in (4.7). Using this classification, he further established the following result:

Theorem 4.23: (Okuda [Ok13]). *Let G/H be a reductive symmetric space. Then the five conditions (i)–(v) in Proposition 4.20 are equivalent.*

For a pair of real reductive Lie groups $G \supset H$ that does *not* form a symmetric pair, the implication (v) $\Rightarrow$ (ii) in Proposition 4.20 does not necessarily hold.

4.13. *Solvable Case*

So far, we have primarily discussed proper actions (or properly discontinuous actions) on homogeneous spaces G/H in the setting where G is a reductive Lie group. In contrast, when G/H is a simply connected *solvable* homogeneous space, the Calabi–Markus phenomenon does *not* occur. In fact, the following theorem is based on a structural result on solvable Lie groups due to Chevalley [C41].

Theorem 4.24: ([K93], Thm. 2.2). *Suppose that G is a solvable Lie group and H is a proper closed subgroup of G. Then there exists a discrete subgroup Γ of G that acts properly discontinuously and freely on G/H, such that the fundamental group $\pi_1(\Gamma\backslash G/H)$ is infinite.*

5. Cocompact Discontinuous Groups

One of the central and challenging problems concerning discontinuous groups acting on non-Riemannian homogeneous spaces G/H is the following:

Problem 5.1: ([K01], Problem B). *Determine all pairs (G, H) for which G/H admits* cocompact *discontinuous groups.*

Problem 5.1 is a long-standing open problem, and it remains unsolved even when G/H is a symmetric space of rank one, as exemplified by the space form conjecture (Conjecture 5.6).

From now on, we focus on the case where G is a real reductive linear Lie group and H is a reductive subgroup. We recall, as seen in Proposition 3.14, that the homogeneous space G/H admits a pseudo-Riemannian structure with respect to which G acts as a group of isometries.

In the classical case where H is compact, a theorem of Borel [Bo63] affirms Problem 5.1 by establishing the existence of cocompact arithmetic discrete subgroups in G.

When G is noncompact, cocompact discontinuous groups for G/H are much smaller than cocompact lattices in G. For example, their cohomological dimensions are strictly smaller [K89], Cor. 5.5. A simple approach to Problem 5.1 is to consider a "continuous analog" of discontinuous groups Γ, thereby leading to the notion of the *standard quotient*, as described below.

5.1. *Standard Quotient* $\Gamma\backslash G/H$

We continue to work under the standing assumption that G is a real reductive linear Lie group and that $H \subset G$ is a reductive subgroup.

Definition 5.2: (Standard Quotient [KaK16], Def. 1.4). Suppose L is a reductive subgroup of G such that the action of L on G/H is proper. Then any torsion-free discrete subgroup Γ of L is a discontinuous group for G/H; that is, the Γ-action on G/H is properly discontinuous and free. The quotient space $\Gamma\backslash G/H$ is referred to as a *standard quotient* of G/H.

The properness criterion stated in Theorem 4.17 provides a convenient method for checking whether a given reductive subgroup $L \subset G$ satisfies the condition required in Definition 5.2.

5.2. *Finding Cocompact Discontinuous Groups*

If a subgroup L as in Definition 5.2 acts cocompactly on G/H, then G/H admits a cocompact discontinuous group Γ, obtained by taking Γ to be a torsion-free cocompact discrete subgroup of L, where existence is guaranteed by Borel's theorem. A necessary and sufficient condition for such a subgroup L, which acts properly on G/H, to act cocompactly is

$$d(L) + d(H) = d(G), \tag{5.1}$$

as established in [K89], Thm. 4.7, where $d(G) := \dim \mathfrak{p} = \dim G/K$.

A list of reductive homogeneous spaces G/H that admit proper and cocompact actions of reductive subgroups may be found in [KnK25], summarizing earlier lists including [K89, K97]. A particularly important subclass consists of irreducible symmetric spaces, which are the main focus of [KY05]. These works, in particular, provide examples of compact pseudo-Riemannian locally homogeneous spaces $\Gamma\backslash G/H$ realized as standard quotients of G/H.

The following conjecture was proposed by the author in [K01].

Conjecture 5.3: ([K01], Conj. 4.3). The homogeneous space G/H of reductive type admits a cocompact properly discontinuous group if and only if G/H admits a compact standard quotient.

If Conjecture 5.3 were proved to be true, then Problem 5.1 would reduce to the following one:

Problem 5.4: *Classify all pairs (G, H) such that G/H admits a compact standard quotient.*

This problem is expected to be tractable, as it reduces to checking a finite number of representation-theoretic conditions for each G/H in order to verify the properness criterion and the cocompactness criterion in [K89], Thms. 4.1 and 4.7.

Tojo [To19] showed that the list of irreducible symmetric spaces G/H in [KY05] admitting proper and cocompact actions of reductive subgroups L is, in fact, complete up to compact factors in the case where G is a simple Lie group. This result provides a solution to Problem 5.4 in the case where G is a symmetric space with G simple.

Furthermore, Bocheński [Bo22] studied the case where G is the direct product of two absolutely simple groups. A more recent preprint of Bocheński–Tralle [BoT24] shows that, under the assumption that G is absolutely simple, the list in [KY05] contain all the homogeneous spaces G/H that admit proper and cocompact actions of reductive subgroups L, up to compact factors and switching L and H, thereby yielding further progress on Problem 5.4.

Remark 5.5: (1) Conjecture 5.3 does not assert that all cocompact discontinuous groups are standard. Indeed, there exist reductive homogeneous spaces G/H that admit *non-standard* compact quotients; that is, there exist triples (G, H, Γ) such that Γ is a cocompact discontinuous group for G/H, while the Zariski closure of Γ fails to act properly on G/H; see [Ka12, KnK25, K98a].
(2) An analogue of Conjecture 5.3 was established by Okuda [Ok13] for semisimple symmetric spaces G/H. It is worth noting that this result replaces the key assumption of cocompactness in the original conjecture with the requirement that Γ is a surface group $\pi_1(\Sigma_g)$, as stated in Theorem 4.23.

Special cases of Conjecture 5.3 include the following:

Conjecture 5.6: (Space Form Conjecture [K01], Conj. 2.6). There exists a compact, complete, pseudo-Riemannian manifold of signature (p, q) with constant sectional curvature 1 if and only if (p, q) lies in the following list:

p	$\mathbb{N}$	0	1	3	7
q	0	$\mathbb{N}$	$2\mathbb{N}$	$4\mathbb{N}$	8

See also Example 5.10(6) below for the *tangential analogue* in the context of Cartan motion groups.

Conjecture 5.7: For any non-trivial homomorphism $\psi\colon SL(m,\mathbb{R}) \to SL(n,\mathbb{R})$ with $m < n$, the homogeneous space $SL(n,\mathbb{R})/\psi(SL(m,\mathbb{R}))$ does not admit a cocompact discontinuous group.

The following are notable special cases of Conjecture 5.7, corresponding to specific choices of ψ. For a concise overview of these methods, including a discussion of their limitations and applications, see [K01, KT24].

Example 5.8:

(1) (ψ is the identity map.) For the standard representation ψ, the homogeneous space $SL(n,\mathbb{R})/\psi(SL(m,\mathbb{R}))$ does not admit a cocompact discontinuous group.
(2) (ψ is an irreducible representation.) For any *irreducible* representation $\psi\colon SL(2,\mathbb{R}) \to SL(n,\mathbb{R})$ with $n \geq 5$, the homogeneous space $SL(n,\mathbb{R})/\psi(SL(2,\mathbb{R}))$ does not admit a cocompact discontinuous group.

The first statement for $SL(n,\mathbb{R})/SL(m,\mathbb{R})$ in Example 5.8 has been studied over 35 years with affirmative results obtained for "generic parameters". A complete solution was recently announced by Kassel, Morita, and Tholozan [KT24, KMTpre]. Earlier contributions include [LZ95, LMZ95, S00, Th15, K90, K92, Z94, B96, M17], which employed a variety of approaches from different areas.

The second statement in Example 5.8 was proved by Margulis ([M97]) based on the notion of *tempered subgroups*, defined by the asymptotic behaviour of matrix coefficients of unitary representations under the restriction from G to its subgroup H, symbolically written as $G \downarrow H$. In contrast to this notion, we will explore the notion of *tempered homogeneous spaces*

G/H in Section 7 on the regular unitary representation on $L^2(G/H)$, symbolically written as $H \uparrow G$ (see Definition 7.17).

Whereas the idea of standard quotients $\Gamma\backslash G/H$ is to replace a discrete subgroup Γ with a connected subgroup L (Definition 5.2), one may instead consider an "approximation" of Problem 5.1, by replacing the homogeneous space $X = G/H$ with the *tangential homogeneous space*

$$X_\theta := G_\theta/H_\theta,$$

where $G_\theta := K \ltimes \mathfrak{p}$ is the *Cartan motion group* associated with the real reductive group $G = K \exp \mathfrak{p}$ and similarly for H_θ. If G/H admits a compact standard quotient, then the tangential homogeneous space G_θ/H_θ admits a cocompact discontinuous group. The group G_θ is a compact extension of the abelian group $\mathfrak{p}$, and has a much simpler structure.

We consider the following *tangential* question related to Problem 5.1:

Problem 5.9: ([KY05]). *For which pairs (G, H) of real reductive Lie groups, does the tangential homogeneous space G_θ/H_θ admit a cocompact discontinuous group?*

This problem is expected to be significantly simpler than the original one, yet it remains unsolved even in the case of symmetric spaces. Nevertheless, a complete answer is available for *tangential* pseudo-Riemannian space forms, using a theorem of Adams [A62] on the maximal number of pointwise linearly independent continuous vector fields on spheres; see Example 5.10(6) below.

At the end of this section, we briefly review these problems and conjectures, taking the pseudo-Riemannian space form $X(p,q)_+$ as a representative example. We also highlight recent developments in the field (see, *e.g.*, [KT24, KnK25, K23a] and references therein).

Example 5.10: Let $(G, H) = (O(p+1, q), O(p, q))$, and let $X = X(p,q)_+ = G/H$ denote the pseudo-Riemannian space form of signature (p, q) as in Example 2.3.

(1) ([CM62, Ku81, K89]) $X(p,q)$ admits a discontinuous group of infinite order if and only if $p < q$.

(2) ([Ku81, Ok13]) $X(p,q)$ admits a discontinuous group isomorphic to a surface group if and only if $p+1 < q$ or $p+1 = q \in 2\mathbb{N}$.

(3) ([Ku81, KO90, Th15, M19]) If $X(p,q)$ admits a cocompact discontinuous group, then $pq = 0$ or $p < q$ with $q \in 2\mathbb{N}$.

(4) ([Ku81,K01]) $X(p,q)$ admits a cocompact discontinuous group if (p,q) is in the list, as stated in Conjecture 5.6.
(5) ([Ka12,KnK25]) If $(p,q) = (0,2)$, $(1,2)$, or $(3,4)$, then $X(p,q)$ admits a cocompact discontinuous group that can be continuously deformed into a Zariski dense subgroup of G, while preserving proper discontinuity of the action. Moreover, for $(p,q) = (1,2n)$ $(n \geq 2)$, the anti-de Sitter space $X(1,2n)$ admits a compact quotient which has a non-trivial continuous deformation within the class of standard quotients.
(6) ([KY05]) The tangential homogeneous space G_θ/H_θ admits a cocompact discontinuous group if and only if $p < \rho(q)$ where $\rho(q)$ is the *Radon–Hurwitz number*. Equivalently, this condition holds if and only if (p,q) appears in the following list:

p	$\mathbb{N}$	0	1	2	3	4	5	6	7	8	9	10	$\cdots$
q	0	$\mathbb{N}$	$2\mathbb{N}$	$2\mathbb{N}$	$4\mathbb{N}$	$8\mathbb{N}$	$8\mathbb{N}$	$8\mathbb{N}$	$8\mathbb{N}$	$16\mathbb{N}$	$32\mathbb{N}$	$64\mathbb{N}$	$\cdots$

6. Proper Maps and Unitary Representation

This section explores the relationship between the properness of group actions and representation theory, particularly in the context of discretely decomposable unitary representations.

6.1. *Compact-Like Actions and Compact-Like Unitary Representations*

Every continuous action of a compact group is proper (see Definition-Lemma 3.4). In this sense, a proper action may be viewed as a *compact-like action*.

Every unitary representation of a compact group decomposes discretely into a direct sum of irreducible representations. Thus, discretely decomposable unitary representations may be viewed as *compact-like representations*.

A proposal to connect two seemingly different areas—*proper actions* in topology and *discrete decomposability* in representation theory—by observing how non-compact subgroups can exhibit compact-like behaviour within infinite-dimensional automorphism groups was first articulated in the 2000 paper [K00], Sec. 3.

In this section, we review the foundational concepts and give an overview of some developments in this direction since then.

6.2. *Discrete Decomposable Unitary Representations*

Let $\widehat{G}$ denote the *unitary dual* of a locally compact group G; that is, the set of equivalence classes of irreducible unitary representations of G, endowed with the Fell topology.

By a theorem of Mautner [Mt50], Chap. VIII, Sec. 41, every unitary representation of the group decomposes into a direct integral of irreducible unitary representations.

Let G' be a subgroup of G. Suppose that $\pi \in \widehat{G}$. Then, the restriction $\pi|_{G'}$, as a unitary representation of the subgroup G', can be decomposed into a direct integral of irreducible unitary representations:

$$\pi|_{G'} \simeq \int_{\widehat{G'}}^{\oplus} m_\pi(\tau)\tau d\mu(\tau), \tag{6.1}$$

where μ is a Borel measure on the unitary dual $\widehat{G'}$, and

$$m_\pi \colon \widehat{G'} \to \mathbb{N} \cup \{\infty\}$$

is a measurable function called the *multiplicity function* for the direct integral (6.1). This irreducible decomposition is known as *the branching law*. Typically, it involves continuous spectrum when G' is non-compact.

The concept of *G'-admissible restrictions* was introduced in [K94], Sec. 1 in a general setting that includes the case where G' is a *non-compact* subgroup.

Definition 6.1: The restriction $\Pi|_{G'}$ is said to be *G'-admissible* if it can be decomposed discretely into a direct sum of irreducible unitary representations π of G':

$$\Pi|_{G'} \simeq \sum_{\pi \in \widehat{G'}}^{\oplus} m_\pi \pi \quad \text{(discrete sum)}$$

where the multiplicity $m_\pi := [\Pi|_{G'} : \pi]$ is finite for every $\pi \in \widehat{G'}$.

We refer to [K94, K98b] for the criterion of G'-admissibility for the restriction of an irreducible unitary representation of a reductive Lie group G to its reductive subgroup G'. See also Kitagawa [Ki25] for some recent developments.

Discretely decomposable restrictions may be regarded as *compact-like representations*. We examine the discrete decomposability of representations from two perspectives: one based on the properness of the moment map (Section 6.3), and the other based on *proper actions* of groups (Section 6.4 and Theorem 6.6).

6.3. *Coadjoint Orbits and Proper Maps*

Let G be a Lie group, and $\mathfrak{g}^*$ the dual of the Lie algebra $\mathfrak{g}$. The *orbit method* initiated by Kirillov, and developed by Kostant, Duflo, and Vogan among others, is a philosophy that seeks to understand the unitary dual $\widehat{G}$ through the coadjoint representation $\mathrm{Ad}^*\colon G \to GL_{\mathbb{R}}(\mathfrak{g}^*)$.

For $\lambda \in \mathfrak{g}^*$, $\mathcal{O}_\lambda := \mathrm{Ad}^*(G)\lambda$ is called a *coadjoint orbit*. The quotient space $\mathfrak{g}^*/\,\mathrm{Ad}^*(G)$ parametrizes coadjoint orbits. Loosely speaking, the orbit method suggests the existence of a "natural correspondence" between a subset of the set $\mathfrak{g}^*/\,\mathrm{Ad}^*(G)$ and the unitary dual $\widehat{G}$. Indeed, there exists a natural bijection

$$Q\colon \mathfrak{g}^*/\,\mathrm{Ad}^*(G) \xrightarrow{\sim} \widehat{G} \tag{6.2}$$

when G is a simply connected nilpotent Lie group, as Kirillov established in his 1962 celebrated paper [Ki62]. For reductive Lie groups G, there is no such natural bijection as in (6.2), however, one still expects that the orbit method provides insight into unitary representations of G via a deep relationship between $\mathfrak{g}^*/\,\mathrm{Ad}^*(G)$ and $\widehat{G}$.

For any $\lambda \in \mathfrak{g}^*$, the skew-symmetric bilinear map

$$\lambda\colon \mathfrak{g} \times \mathfrak{g} \to \mathbb{R}, \quad [X,Y] \mapsto \lambda([X,Y])$$

induces a G-invariant symplectic form on the coadjoint orbit

$$\mathcal{O}_\lambda := \mathrm{Ad}^*(G)\lambda \simeq G/G_\lambda,$$

which is known as the *Kostant–Kirillov–Souriau symplectic form*. The momentum map of the G-action on $\mathcal{O}_\lambda$ is precisely the canonical injection $\mathcal{O}_\lambda \hookrightarrow \mathfrak{g}^*$, and hence $\mathcal{O}_\lambda$ is a G-Hamiltonian manifold.

From this perspective, if one can associate an irreducible unitary representation $\Pi_\lambda := Q(\mathcal{O}_\lambda)$ naturally to a coadjoint orbit $\mathcal{O}_\lambda$, we may regard Π_λ as a geometric quantization of $\mathcal{O}_\lambda$.

Let H be a subgroup of G. By the branching problem we aim to understand the restriction $\Pi|_H$ of a representation Π of G to the subgroup H. Suppose that Π is an irreducible unitary representation that corresponds to a coadjoint orbit $\mathcal{O}$ in $\mathfrak{g}^*$. We observe that the canonical projection

$$\mathrm{pr}\colon \mathfrak{g}^* \to \mathfrak{h}^*$$

for the dual of the Lie algebras $\mathfrak{h} \hookrightarrow \mathfrak{g}$ is equivariant with respect to the coadjoint action of H. In the spirit of the orbit method, the restriction $\Pi|_H$ might be interpreted in terms of the image $\mathrm{pr}(\mathcal{O})$ as a union of H-coadjoint orbits, suggesting how the restriction $\Pi|_H$ decomposes under H.

The expected correspondences may be illustrated as follows:

$$\begin{array}{rcl} \text{unitary dual} \quad \widehat{G} \ni \Pi & \overset{\text{orbit method}}{\leftarrow \cdots \rightarrow} & \mathcal{O} \subset \mathfrak{g}^* \quad \text{(coadjoint orbit)} \\ \text{subgroup} \quad H \subset G & \leftarrow \cdots \rightarrow & \mathrm{pr}\colon \mathfrak{g}^* \to \mathfrak{h}^* \quad \text{projection} \\ \pi|_H \text{ is } H\text{-admissible} & \leftarrow \overset{?}{\cdots} \rightarrow & \mathcal{O}_\pi \hookrightarrow \mathfrak{g}^* \overset{\mathrm{pr}}{\to} \mathfrak{h}^* \text{ is proper.} \end{array}$$

Here, the question mark indicates a conjectural equivalence between H-admissibility of the restriction $\Pi|_H$ and the properness of the moment map on the coadjoint orbit $\mathcal{O}$.

Question 6.2: Suppose that $\Pi \in \widehat{G}$ is a "geometric quantization" of a coadjoint orbit $\mathcal{O} \subset \mathfrak{g}^*$ in the sense of the orbit method. Let H be a reductive subgroup of G. Is the following equivalence (i) $\Leftrightarrow$ (ii) valid?

(i) The restriction $\Pi|_H$ is H-admissible.

(ii) The projection $\mathrm{pr}\colon \mathfrak{g}^* \to \mathfrak{h}^*$ is a proper map when restricted to the coadjoint orbit $\mathcal{O}$.

See [DV10, KN03, KN18, P15] for some affirmative cases and related discussions.

Although beyond the scope of this chapter, we note that for non-reductive subgroups, Duflo introduced the notions of a *weakly proper map*, which relaxes the properness condition appearing in condition (ii) of Question 6.2. See [LOY23] for example.

6.4. *Proper Action and Discrete Decomposability*

We recall some basic notions from the theory of infinite-dimensional representations of Lie groups, not necessarily unitary. Let Π be a continuous representation of a Lie group G on a complete, locally convex topological vector space V (*e.g.*, a Banach space), and let V^∞ denote the space of smooth vectors. Then V^∞ is dense in V, and carries a natural topology. The representation Π induces a continuous representation Π^∞ on V^∞, and a dual representation $\Pi^{-\infty}$ on the continuous dual space $V^{-\infty}$ of V^∞.

Now suppose that G is a real reductive Lie group. Let $\mathcal{M}(G)$ denote the category of smooth admissible representations of finite length with moderate growth, which are defined on Fréchet topological vector spaces [W92], Chap. 11. Let $\mathrm{Irr}(G)$ denote the set of irreducible objects in $\mathcal{M}(G)$.

While we do not go into the precise definition of the category $\mathcal{M}(G)$ here, it is helpful to keep in mind that $\mathrm{Irr}(G)$ contains the smooth representations Π^∞ of irreducible unitary representations Π of G. This gives a natural injection:

$$\widehat{G} \hookrightarrow \mathrm{Irr}(G), \quad \Pi \mapsto \Pi^{\infty}. \tag{6.3}$$

Let H be a closed subgroup of a Lie group G.

Definition 6.3: We say that $\Pi \in \mathrm{Irr}(G)$ is an *H-distinguished representation* of G, if $(\Pi^{-\infty})^H \neq \{0\}$, or equivalently, by Frobenius reciprocity,

$$\mathrm{Hom}_G(\Pi, C^{\infty}(G/H)) \neq \{0\}.$$

Let $\mathrm{Irr}(G)_H$ denote the subset of $\mathrm{Irr}(G)$ consisting of H-distinguished irreducible admissible representations, and let $\widehat{G}_H := \widehat{G} \cap \mathrm{Irr}(G)_H$ via the injection given in (6.3).

In line with the philosophy of *compact-like actions* discussed in Section 6.1, which links geometry and function spaces, the following four properties are closely related:

- the action of G' on X is proper;
- the action of G' on X is *compact-like*;
- the representation of G' on $C^{\infty}(X)$ is *compact-like*;
- for any $\Pi \in \mathrm{Irr}(G)$ occurring in $C^{\infty}(X)$, the restriction $\Pi|_{G'}$ is discretely decomposable.

This philosophy holds under the additional assumption of sphericity, as formalised in Theorem 6.6 below.

We briefly recall the definition of sphericity:

Definition 6.4: Let $X_{\mathbb{C}}$ be a connected complex manifold on which a complex reductive Lie group $G_{\mathbb{C}}$ acts holomorphically. The action of $G_{\mathbb{C}}$ is said to be *spherical* if a Borel subgroup of $G_{\mathbb{C}}$ has an open orbit in $X_{\mathbb{C}}$.

Example 6.5:

(1) The complexification $G_{\mathbb{C}}/H_{\mathbb{C}}$ of a reductive symmetric space G/H is spherical.
(2) Any flag variety is spherical.

Theorem 6.6: ([K17]). *Let $X = G/H$ be a reductive symmetric space. Suppose that G' is a reductive subgroup of G, and that its complexification $G'_{\mathbb{C}}$ acts spherically on $X_{\mathbb{C}}$. If the action of G' on X is proper, then any irreducible H-distinguished unitary representation Π of G is G'-admissible;*

in particular, it decomposes discretely upon restriction to the subgroup G'. Moreover, the multiplicities are uniformly bounded:

$$\sup_{\Pi \in \mathrm{Irr}(G)_H} \sup_{\pi \in \mathrm{Irr}(G')} [\Pi|_{G'} : \pi] < \infty.$$

We now give three examples to illustrate this result.

Example 6.7: (Standard Anti-de Sitter Manifolds). Let X be an odd-dimensional anti-de Sitter space, *i.e.*,

$$X = G/H = SO(2n, 2)/SO(2n, 1).$$

The subgroup $G' := U(n, 1)$ acts properly on X, and its complexification $G'_{\mathbb{C}} = GL(n+1, \mathbb{C})$ acts spherically on the complex manifold $X_{\mathbb{C}} = SO(2n+2, \mathbb{C})/SO(2n+1, \mathbb{C})$, which is biholomorphic to the $(2n+1)$-dimensional complex sphere $S_{\mathbb{C}}^{2n+1}$. Therefore, Theorem 6.6 applies in this case. The corresponding discretely decomposable branching laws are explicitly obtained in [K94], Thm. 6.1.

Example 6.8: (Pseudo-Riemannian Space Form of Signature $(8, 7)$). Consider a 15-dimensional manifold by

$$X = G/H = SO(8, 8)/SO(8, 7).$$

Then X is a pseudo-Riemannian manifold of signature $(8, 7)$, with constant negative sectional curvature. The subgroup $G' = \mathrm{Spin}(1, 8)$ of G acts properly on X, and its complexification $G'_{\mathbb{C}} = \mathrm{Spin}(8, \mathbb{C})$ acts spherically on $X_{\mathbb{C}} \simeq S_{\mathbb{C}}^{15}$. Thus, Theorem 6.6 applies here. The corresponding discretely decomposable branching laws for the restriction $SO(8, 8) \downarrow Spin(1, 8)$ are explicitly obtained in Theorem 5.5 of [K17] or [STV18].

Example 6.9: (Indefinite Kähler Manifolds). The homogeneous space

$$X = G/H = SO(2n, 2)/U(n, 1)$$

admits a natural indefinite Kähler structure. The subgroup $G' = SO(2n, 1)$ acts properly on X, and the complexified group $G'_{\mathbb{C}} = SO(2n+1, \mathbb{C})$ acts spherically on the complex manifold $X_{\mathbb{C}} = SO(2n+2, \mathbb{C})/GL(n+1, \mathbb{C})$. Hence, Theorem 6.6 applies here. A detailed account of the geometric setting and the discretely decomposable branching laws for the restriction $G \downarrow G'$ can be found in [K09], Sec. 6, specifically for the case $n = 2$.

In the setting of Theorem 6.6, let $X_\Gamma = \Gamma\backslash G/H$ be a standard locally symmetric space (Definition 5.2), where Γ is a torsion-free discrete subgroup of G'. Equipped with the pseudo-Riemannian structure inherited from the symmetric space $X = G/H$, the space X_Γ provides a natural framework for spectral analysis. In fact, Theorem 6.6 serves as a cornerstone for the analytic theory on standard locally symmetric spaces X_Γ, as developed in the monograph [KaK25].

7. Two Quantifications of Proper Actions

Two notions that may appear unrelated at first glance—originating respectively from joint works with Kassel [KaK16] and Benoist [BK15,BK22]—in fact arose from distinct and independent motivations. In this section, however, we reinterpret them from a unified perspective: as two approaches to *quantifying the properness* of group actions.

- The notion of *sharpness* provides a means of measuring how strongly a given action satisfies the condition of properness (see Section 7.1).
- The other, based on dynamical volume estimates, quantifies the extension to which an action fails to be proper (see Sections 7.2–7.7).

7.1. *Sharp Action*

As a strengthening of the properness condition for group actions, we recall the notion of *sharpness*, introduced in [KaK16].

Let G be a linear reductive Lie group. Let

$$\mu \colon G \to \overline{\mathfrak{a}_+}$$

denote the Cartan projection associated with the Cartan decomposition $G = K\overline{A_+}K$, as defined in (4.3).

Let H be a closed subgroup, and let $X := G/H$ be the associated homogeneous space.

Definition 7.1: (Strongly Proper Action: Sharpness Constants). Let Γ be a discrete subgroup of G. We say that Γ is *sharp* for X if there exist constants $c \in (0, 1]$ and $C \geq 0$ such that

$$\|\mu(\gamma) - \mu(H)\| \geq c\|\mu(\gamma)\| - C$$

holds for all $\gamma \in \Gamma$. In this case, the quotient space $X_\Gamma := \Gamma\backslash G/H$ is called a *sharp quotient* of X.

The constants (c, C) are called the *sharpness constants*.

This notion can be reformulated in terms of the *asymptotic cone* (also known as the *limit cone*), which we recall now.

Let V be a finite-dimensional vector space over $\mathbb{R}$, and let S be a subset of V.

Definition 7.2: (Asymptotic Cone, Limit Cone). The *asymptotic cone* of S, also referred to as the *limit cone*, is a closed cone in V consisting of all limit points of sequences of the form

$$\lim_{n\to\infty} \varepsilon_n x_n,$$

where $x_n \in S$ and $\varepsilon_n > 0$ is a sequence converging to 0. We denote this cone by $S\infty$.

The following lemma is an immediate consequence of Definition 4.2 of the relation $\pitchfork$.

Lemma 7.3: *Let S and T be subsets of the vector space in V. If the asymptotic cones satisfy $S\infty \cap T\infty = \{0\}$, then $S \pitchfork T$ in V, where V is regarded as an additive group.*

We next restate Definition 7.1 in an equivalent form.

Definition 7.4: (Sharp Action). Let Γ be a discrete subgroup of G. The action of Γ on X is called *sharp* if

$$\mu(\Gamma)\infty \cap \mu(H)\infty = \{0\}.$$

If the action of Γ on $X = G/H$ is sharp, then it follows from Lemma 7.3 that

$$\mu(\Gamma) \pitchfork \mu(H) \text{ in } \mathfrak{a}.$$

Hence, by the properness criterion, as stated in Theorem 4.14, the Γ-action on X is proper.

The converse implication

$$\text{proper action} \Rightarrow \text{sharp action}$$

does not hold in general. However, there are many interesting examples in which sharpness does follow:

— When H is reductive, any standard quotient (Definition 5.2) X_Γ is sharp.

— Remarkably, Kassel and Tholozan have announced in a recent preprint [KT24] an affirmative solution to the Sharpness Conjecture

[KaK16], Conj. 4.12, which asserts that any *cocompact* discontinuous group for G/H is sharp.

An advantage of the notion of sharpness is that it becomes particularly effective in the study of *deformations* of discontinuous groups.

In contrast to the Selberg–Weil rigidity theorem for the Riemannian symmetric space G/K, irreducible pseudo-Riemannian symmetric spaces may admit cocompact discontinuous groups that are *not locally rigid*, even in arbitrarily high dimensions. This phenomenon was first observed in the early 90s (see [K93, Remarks 2 and 3]) for the group manifold G, viewed as a homogeneous space $(G \times G)/\operatorname{diag} G$.

A major difficulty in studying deformations of discontinuous groups lies in the fact that, when H is noncompact, small deformations of a discrete subgroup can easily destroy the properness of the action. In the context of 3-dimensional compact anti-de Sitter manifolds, Goldman [G85] conjectured that any small deformation of a standard cocompact discontinuous group preserves proper discontinuity. This conjecture was proved by the present author [K98a], based on the properness criterion, as stated in Theorem 4.14.

The idea introduced in [K98a], further developed by Kassel [Ka12] and related works, exploits the fact that the limit cone $\mu(\Gamma)\infty$ remains well-controlled under small deformations of Γ. Consequently, proper discontinuity is maintained through small deformation—under a mild condition—provided that the initial group is a *sharp* discontinuous group.

The notion of sharpness also plays a significant role in other problems, such as the orbit counting problem for properly discontinuous actions of Γ on pseudo-Riemannian symmetric spaces X. This is exemplified in the construction of the *stable spectrum* for $\Gamma\backslash X$ in [KaK16]. On the other hand, sharpness also proves useful in addressing the existence problem of cocompact discontinuous groups, as seen in [KT24].

7.2. *Measure-Theoretic Approach to Proper Actions*

Whereas the previous section discussed the notion of sharp actions as a quantitative strengthening of properness, the present section takes the opposite perspective: it introduces a quantitative method to measure the extent to which a group action fails to be proper.

We begin with a reformulation of the definition of proper actions (Definition-Lemma 3.4), using *measure-theoretic conditions* in lieu of the original topological definition.

Let G be a locally compact group, and let X be a locally compact space equipped with a continuous G-action. Suppose further that X carries a Radon measure μ. Then, for every compact subset $S \subset X$, the function

$$G \to \mathbb{R}, \quad g \mapsto \mathrm{vol}(S \cap gS) := \mu(S \cap gS)$$

is continuous with respect to the topology on G.

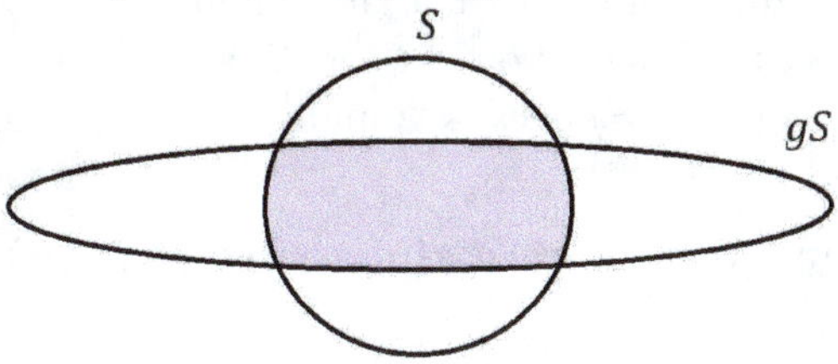

Lemma 7.5: *The following two conditions are equivalent:*

(i) *The action of G on X is proper;*

(ii) *For every compact subset $S \subset X$, the function* $\mathrm{vol}(S \cap gS)$ *has compact support on G.*

Proof: (i) $\Rightarrow$ (ii). The function $g \mapsto \mathrm{vol}(S \cap gS)$ is continuous, and its support is contained in

$$\mathrm{Supp}\,\mathrm{vol}(S \cap gS) \subset \{g \in G : S \cap gS \neq \emptyset\} =: G_S.$$

Hence, if the G-action on X is proper, (*i.e.*, G_S is compact for all compact $S \subset X$), then the function has compact support.
(ii) $\Rightarrow$ (i). Conversely, suppose that the action of G on X is not proper. Then there exists a compact subset $S \subset X$ such that G_S is not compact.

Choose an open, relatively compact subset $V \subset X$ with $S \subset V$, and let S' be the closure of V, which is compact. For each $g \in G_S$, we have

$$\emptyset \neq S \cap gS \subset V \cap gV \subset S' \cap gS'.$$

Since $S \cap gS$ is open and has positive measure (as μ is a Radon measure), it follows that $\mu(V \cap gV) > 0$. Hence,

$$\mathrm{Supp}(\mathrm{vol}(S' \cap gS')) \supset G_S,$$

which is not compact. Thus, by contraposition, (ii) implies (i). □

We now focus on the case when the action is *not* proper.

By the preceding lemma, there exists a compact subset $S \subset X$ such that the volume function

$$g \mapsto \mathrm{vol}(S \cap gS)$$

does not have compact support.

To quantitatively assess the degree of non-properness quantitatively, we examine how this function behaves at the "infinity" in G.

We may expect that the action of G on X is *close to being proper* if the volume function $\mathrm{vol}(gS \cap S)$ decays "rapidly as $g \in G$ tends to infinity".

7.3. *An Example of Volume Estimate:* vol($S \cap gS$)

To illustrate this principle, consider a simple yet instructive example showing the asymptotic behavior of $\mathrm{vol}(S \cap gS)$.

Let $G := \mathbb{R}$ act on $\mathbb{R}^2 \setminus \{(0,0)\}$ by

$$(x, y) \mapsto (e^t x, e^{-t} y), \quad \text{where } t \in \mathbb{R}.$$

As observed in Example 3.19, this action is *free* and *all orbits are closed*, but it is not *proper*. In particular, the G-action on the entire space $X := \mathbb{R}^2$ is not proper. From a measure-theoretic point of view, X and $X \setminus \{(0,0)\}$ are equivalent. To understand failure of properness quantitatively, consider the asymptotic behavior of the function

$$t \mapsto \mathrm{vol}(S \cap t \cdot S),$$

where the translate of a compact subset $S \subset G$ by $t \in \mathbb{R}$ is defined by:

$$t \cdot S := \{(e^t x, e^{-t} y) : (x, y) \in S\}.$$

Claim 7.6: If the origin $o = (0,0)$ is an interior point of a compact subset $S \subset \mathbb{R}^2$, then there exist constants C_1, $C_2 > 0$ such that

$$C_1 e^{-|t|} \le \mathrm{vol}(S \cap t \cdot S) \le C_2 e^{-|t|}$$

for all $t \in \mathbb{R}$.

Proof: We begin with the case where S is the square

$$D_R := \{(x, y) \in \mathbb{R}^2 : |x| \le R, |y| \le R\}.$$

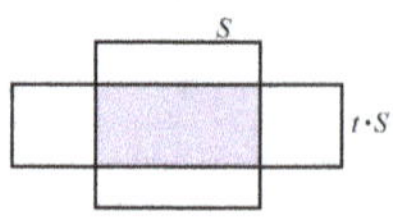

A direct computation shows that

$$\mathrm{vol}(t \cdot D_R \cap D_R) = 4R^2 e^{-|t|} = \mathrm{vol}(D_R) e^{-|t|}.$$

Now, suppose S is a compact set containing the origin as an interior point. Then there exist constants $0 < r < R$ such that $D_r \subset S \subset D_R$. It follows that

$$\mathrm{vol}(D_r \cap t \cdot D_r) \leq \mathrm{vol}(S \cap t \cdot S) \leq \mathrm{vol}(D_R \cap t \cdot D_R).$$

Using the earlier formula,

$$\mathrm{vol}(D_r)e^{-|t|} \leq \mathrm{vol}(t \cdot S \cap S) \leq \mathrm{vol}(D_R)e^{-|t|}.$$

This completes the proof. □

7.4. *Function ρ_V and Constant p_V*

The previous example extends naturally to higher dimensions. Before formulating this generalization, we recall the function ρ_V, which is associated with a finite-dimensional representation of a Lie algebra on a vector space, introduced in [BK15, BK22].

Let $\mathfrak{h}$ be a Lie algebra, and suppose

$$\tau \colon \mathfrak{h} \to \mathrm{End}_{\mathbb{R}}(V)$$

s a representation of $\mathfrak{h}$ on a finite-dimensional real vector space V. We define a non-negative function

$$\rho_V \colon \mathfrak{h} \to \mathbb{R}_{\geq 0} \tag{7.1}$$

as follows: For each $Y \in \mathfrak{h}$, let $\{\lambda_1, \ldots, \lambda_n\}$ be the multiset of generalized eigenvalue, of $\tau(Y)$, viewed as a complex-linear operator on $V_{\mathbb{C}} := V \otimes_{\mathbb{R}} \mathbb{C}$. Then define

$$\rho_V(Y) := \frac{1}{2} \sum_{j=1}^{n} |\operatorname{Re} \lambda_j|.$$

Now assume that $\mathfrak{h}$ is an algebraic Lie algebra, and that $\tau \colon \mathfrak{h} \to \mathrm{End}_{\mathbb{R}}(V)$ is an algebraic representation. Let $\mathfrak{a} \subset \mathfrak{h}$ be a maximally split abelian subalgebra. Then the function ρ_V is entirely determined by its restriction to $\mathfrak{a}$, and we have

$$\rho_V(Y) = \sum_{j=1}^{n} |\lambda_j(Y)| \quad \text{for } Y \in \mathfrak{a}$$

since $\tau(Y) \in \mathrm{End}_{\mathbb{R}}(V)$ is diagonalizable in this case.

Example 7.7: Let $\mathfrak{h}$ be a semisimple Lie algebra. For the adjoint representation

$$\mathrm{ad} \colon \mathfrak{h} \to \mathrm{End}(\mathfrak{h}),$$

the function ρ_{ad} coincides with twice the "usual ρ-function" on the positive Weyl chamber $\overline{\mathfrak{a}_+}$; that is,

$$\rho_{\text{ad}}(Y) = 2\rho(Y) = \sum_{\alpha \in \Sigma^+(\mathfrak{g},\mathfrak{a})} \alpha(Y) \quad \text{for } Y \in \overline{\mathfrak{a}_+}.$$

It is worth noting that while the "usual ρ-function" is linear, our function ρ_{ad} is only *piecewise linear*.

We now consider the ratio between two ρ-functions:

- one associated with a given representation (τ, V), and
- the other with the adjoint representation.

Definition 7.8: (The Invariant p_V). Let (τ, V) be an algebraic representation of $\mathfrak{h}$. We define the invariant p_V by

$$p_V := \max_{Y \in \mathfrak{h} \setminus \{0\}} \frac{\rho_{\mathfrak{h}}(Y)}{\rho_V(Y)}.$$

If $\mathfrak{a} \subset \mathfrak{h}$ is a maximally split abelian subalgebra, this simplifies to

$$p_V = \max_{Y \in \mathfrak{a} \setminus \{0\}} \frac{\rho_{\mathfrak{h}}(Y)}{\rho_V(Y)}.$$

In terms of eigenvalues, this becomes

$$p_V = \max_{Y \in \mathfrak{a} \setminus \{0\}} \frac{\sum |\text{eigenvalues of } \operatorname{ad}(Y) \in \operatorname{End}(\mathfrak{h})|}{\sum |\text{eigenvalues of } \tau(Y) \in \operatorname{End}(V)|}.$$

Example 7.9: Consider the standard representation of $\mathfrak{h} = \mathfrak{sl}(2, \mathbb{R})$ on $V = \mathbb{R}^2$. Let

$$\mathfrak{a} = \mathbb{R}H, \quad H := \begin{pmatrix} 1 & 0 \\ 0 & -1 \end{pmatrix}.$$

A straightforward computation shows

$$\rho_V(tH) = \frac{1}{2}(|t| + |-t|) = |t|,$$
$$\rho_{\text{ad}}(tH) = \frac{1}{2}(|2t| + 0 + |-2t|) = 2|t|.$$

Therefore, the invariant p_V is:

$$p_V = \max_{t \neq 0} \frac{2|t|}{|t|} = 2.$$

Before explaining the meaning of the invariant p_V, we introduce the concept of an *almost L^p function*, which helps clarify the broad picture.

7.5. *Almost L^p Function*

Let Z be a locally compact space equipped with a Radon measure.

Definition 7.10: (Almost L^p Function). A measurable function f is said to be *almost L^p* if

$$f \in \bigcap_{\varepsilon>0} L^{p+\varepsilon}(Z).$$

Example 7.11: Let D be the unit disk, equipped with the Poincaré metric

$$ds^2 = \frac{4(dx^2+dy^2)}{(1-x^2-y^2)^2},$$

and let Δ be the corresponding (positive) Laplace-Beltrami operator. We define the function $p(\lambda)$ by

$$p(\lambda) = \begin{cases} \frac{2}{1-\sqrt{1-4\lambda}} & \text{for } 0 \le \lambda \le \frac{1}{4}, \\ 2 & \text{for } \frac{1}{4} \le \lambda. \end{cases}$$

Suppose that $f \in C^\infty(D)$ is an eigenfunction of Δ, satisfying:

$$\Delta f = \lambda f$$

for some $\lambda \ge 0$, and suppose further that f is $SO(2)$-finite. Then f is almost $L^{p(\lambda)}$. Here, a smooth function f is said to be *$SO(2)$-finite* if the complex vector space $\mathrm{span}_{\mathbb{C}}\{f(k(x,y)) : k \in SO(2)\}$ is finite-dimensional.

If $p \le p'$, then clearly:

$$f \text{ is almost } L^p \Rightarrow f \text{ is almost } L^{p'}.$$

Hence, if a function f is almost L^p for some exponent p, then there exists a *minimal* (or *optimal*) exponent $q \le p$ such that f is almost L^q, in the sense that

$$\begin{aligned} q &= \inf\{p' > 0 : f \in L^{p'+\varepsilon}(Z) \text{ for all } \varepsilon > 0\} \\ &= \min\{p' > 0 : f \in L^{p'+\varepsilon}(Z) \text{ for all } \varepsilon > 0\}. \end{aligned}$$

7.6. *Optimal L^p-Exponent $q(G; X)$*

Suppose that a unimodular, locally compact group G acts continuously on a locally compact space X, equipped with a Radon measure. We now introduce the following invariant associated with this group action, denoted by $q(G; X)$, which measures the optimal decay rate of the volume function.

Definition 7.12: (Optimal L^p-Exponent $q(G;X)$). The invariant $q(G;X)$ is defined to be the optimal constant $q > 0$ such that, for every compact subset $S \subset X$, the function

$$g \mapsto \mathrm{vol}(S \cap gS)$$

is an almost L^q function on G. In other words,

$$\mathrm{vol}(S \cap gS) \in \bigcap_{\varepsilon>0} L^{q+\varepsilon}(G).$$

A general question is the following:

Problem 7.13: *Find an explicit formula of $q(G;X)$ in terms of geometric or representation-theoretic data associated with the action of G on X.*

Example 7.14: ($q(G;V)$ for $G = SL(2,\mathbb{R})$ acting on $V = \mathbb{R}^2$). Consider the standard action of $G = SL(2,\mathbb{R})$ on $V = \mathbb{R}^2$. Then $q(G;V) = 2$. Let us explain why this holds.

Recall the Cartan decomposition $G = KAK$, with $g = k(\theta_1)a(t)k(\theta_2)$, where

$$a(t) = \begin{pmatrix} e^t & 0 \\ 0 & e^{-t} \end{pmatrix} \text{ and } k(\theta) = \begin{pmatrix} \cos\theta & -\sin\theta \\ \sin\theta & \cos\theta \end{pmatrix}.$$

As seen in Example 3.20, the action of A on $\mathbb{R}^2$ is given by

$$(x,y) \mapsto (e^t x, e^{-t} y).$$

Now we take $S \subset \mathbb{R}^2$ to be a K-invariant compact subset (*i.e.*, S is rotationally invariant), and observe that under the Cartan decomposition $G = KAK$, the volume function satisfies

$$\mathrm{vol}(S \cap gS) = \mathrm{vol}(S \cap k(\theta_1)a(t)k(\theta_2)S) = \mathrm{vol}(S \cap a(t)S) \le Ce^{-|t|},$$

by Claim 7.6.

The Haar measure on $G = SL(2,\mathbb{R})$, expressed via the Cartan decomposition, is given by

$$\sinh(2t)dtd\theta_1 d\theta_2.$$

Therefore, the function $\mathrm{vol}(S \cap gS)$ belongs to $L^p(G)$ for any $p > 2$. Since:

$$\int_{\mathbb{R}} e^{-pt}\sinh(2t)dt < \infty \text{ if and only if } p > 2.$$

Thus, we have $q(X) \le 2$. Conversely, by Claim 7.6 again, there exists a compact subset $S \subset \mathbb{R}^2$ such that the opposite inequality also holds:

$$C' e^{-|t|} \le \mathrm{vol}(S \cap gS)$$

for some constant $C' > 0$, which shows that $q(X) \geq 2$. Hence, we conclude that $q(G;V) = 2$ if $(G,V) = (SL(2,\mathbb{R}), \mathbb{R}^2)$.

We observe that the value obtained in the above example coincides with $p_V = 2$ from Example 7.9. This is not a mere coincidence; rather, it reflects a more general principle, as reflected in Proposition 7.15 below.

Indeed, Example 7.14 extends naturally to any faithful representation of a reductive group. This generalization elucidates the relationship between the algebraic invariant p_V, defined in Definition 7.8, and the optimal constant $q(G;X)$ (see Definition 7.12) for which

$$\mathrm{vol}(S \cap hS) \in L^{p+\varepsilon}(G)$$

for all $\varepsilon > 0$, when the action of G on X is linear, as follows.

Proposition 7.15: ([BK15]). *Suppose that G is a real reductive linear group. Let $\tau\colon G \to SL_{\pm}(V)$ be a finite-dimensional representation on a real vector space V with compact kernel. Then the following equality holds:*

$$p_V = q(G;V).$$

Proof: [Sketch of Proof] Let $G = KAK$ be a Cartan decomposition. For $g = k_1 e^Y k_2$, and for a compact subset $S \subset V$ containing 0 as an interior point, one has

$$\mathrm{vol}(S \cap gS) \sim e^{-\rho_V(Y)},$$

as stated in Claim 7.6.

Asymptotically, the Haar measure dg on G satisfies

$$dg \sim e^{\rho_{\mathfrak{h}}(Y)} dk_1 dY dk_2 \quad \text{(away from wall)}.$$

This leads to the proof of Proposition 7.15. □

7.7. *Tempered G-Spaces*

We recall the notion of tempered unitary representations of a locally compact group G.

Definition 7.16: (Tempered Unitary Representation). A unitary representation π is said to be *tempered* if $\pi \prec L^2(G)$; that is, if π is weakly contained in the regular representation on $L^2(G)$.

Suppose that X is a locally compact space equipped with a Radon measure μ, on which a locally compact group G acts continuously and in a measure-preserving manner. Then there is a natural unitary representation of G on the Hilbert space $\mathcal{H} = L^2(X, \mu)$.

We note that the assumption of a G-invariant measure can be dropped. Nevertheless, one can still define a canonical unitary representation—*regular representation*—of G on the Hilbert space of L^2-sections of the half-density bundle over X.

Definition 7.17: (Tempered G-Spaces). We say that X is a *tempered space* if the regular representation of G on $L^2(X)$ is a tempered unitary representation.

A general question is the following:

Problem 7.18: *Given a homogeneous space G/H, determine a criterion on the pair (G, H) that ensures G/H is a tempered space.*

We explain the background of Problem 7.13 in connection with the theory of unitary representations.

Definition 7.19: (Almost L^p-Representation). For $p \geq 1$, a unitary representation π of G on a Hilbert space $\mathcal{H}$ is called *almost L^p* if there is a dense subspace $D \subset \mathcal{H}$ such that the matrix coefficient

$$(\pi(g)u, v)_{\mathcal{H}} \text{ is an almost } L^p \text{ function on } G \quad \text{for all } u, v \in D.$$

Cowling–Haagerup–Howe [CHH88] proved the following.

Theorem 7.20: *Let G be a semisimple Lie group. Then π is tempered if and only if π is almost L^2.*

It should be noted that an analogous equivalence may fail when G is not semisimple.

Example 7.21: Let $G = \mathbb{R}$, and let $\mathbf{1}$ denote the trivial one-dimensional unitary representation of G. Then the matrix coefficient is a constant function on G, which does not belong to $L^p(\mathbb{R})$ for any $p \neq \infty$, even though $\mathbf{1} \prec L^2(G)$.

For a compact subset $S \subset X$, we denote by χ_S the characteristic function of S, defined by

$$\chi_S(x) = \begin{cases} 1 & \text{if } x \in S, \\ 0 & \text{if } x \notin S. \end{cases}$$

Then the matrix coefficient for $\chi_S, \chi_T \in L^2(X)$, associated with compact subsets $S, T \subset X$ is given by

$$\begin{aligned}(\pi(g)\chi_S, \chi_T)_{L^2(X)} &= \int_X \chi_S(g^{-1}x)\chi_T(x)d\mu(x) \\ &= \mathrm{vol}(gS \cap T).\end{aligned}$$

Thus, Proposition 7.15, combined with Theorem 7.20, yields a solution to Problem 7.18 in the linear case:

Theorem 7.22: ([BK15]). *Suppose that G is a real reductive linear group. Let $\tau\colon G \to SL_{\pm}(V)$ be a finite-dimensional representation on a real vector space V with compact kernel. Then $L^2(V)$ is tempered if and only if $p_V \geq 2$.*

This result can be viewed as a basic case in the broader framework aimed at determining when the regular representation on $L^2(X)$ is tempered, for a general G-space X. In a series of papers [BK15,BK21,BK22,BK23], Benoist and the present author developed this perspective in a more general setting, focusing on homogeneous spaces of reductive groups, while uncovering new connections beyond the traditional scope of unitary representation theory. These developments lie beyond the scope of this chapter.

In the spirit of this section, one may interpret this line of thoughts as offering a way to *quantify the strength or failure of properness* of the group action.

Acknowledgments

The author is grateful to the anonymous referee for a careful reading of the manuscript and providing valuable comments. This chapter is based on the series of lectures delivered by the author as part of IMS Distinguished Visitor Lecture Series, held during the Satellite Conference of the Virtual ICM 2022, at the National University of Singapore, from July 1 to 15, 2022. The author would like to express his gratitude to Hung Yean Loke, Angela Pasquale, Tomasz Przebinda, Binyong Sun, and Chengbo Zhu, for their warm hospitality.

The author was partially supported by the JSPS under Grant-in Aid for Scientific Research (A) (JP23H00084), Institut Henri Poincaré (Paris), and by the Institut des Hautes Études Scientifiques (Bures-sur-Yvette).

References

[A62] J. F. Adams, *Vector fields on spheres*, Ann. Math. (2) **75**, (1962), 606–632.

[BaKh05] A. Baklouti, F. Khlif, Proper actions and some exponential solvable homogeneous spaces, Internat. J. Math. **16** (2005), 941–955.

[B96] Y. Benoist, Actions propres sur les espaces homogènes réductifs, Ann. of Math. (2) **144** (1996), no. 2, 315–347.

[BK15] Y. Benoist, T. Kobayashi, Tempered reductive homogeneous spaces. *J. Eur. Math. Soc.*, **17** (2015), 3015–3036.

[BK22] Y. Benoist, T. Kobayashi, Tempered homogeneous spaces II. In: D. Fisher, D. Kleinbock, G. Soifer (Eds.), Dynamics, Geometry, Number Theory: The Impact of Margulis on Modern Mathematics, The University of Chicago Press, 2022, pp. 213–245.

[BK21] Y. Benoist, T. Kobayashi, Tempered homogeneous spaces III. *J. Lie Theory*, **31** (2021) 833–869, Available also at arXiv:2009.10389.

[BK23] Y. Benoist, T. Kobayashi, *Tempered homogeneous spaces* IV, J. Inst. Math. Jussieu **22** (2023), 2879–2906. Available also at arXiv:2009.10391.

[Bo22] M. Bocheński, *On semisimple standard compact Clifford-Klein forms*, J. Lie Theory **32** (2022), no. 3, 813–819.

[BoT24] M. Bocheński, A. Tralle, *A solution of the problem of standard compact Clifford–Klein forms*, Available at arXiv:2403.10539v1, 2024.

[Bo63] A. Borel, *Compact Clifford–Klein forms of symmetric spaces*, Topology **2** (1963), 111–122.

[B98] N. Bourbaki, General Topology, Chapters 1–4, Springer-Verlag, Berlin, 1998, vii+437 pp.

[CM62] E. Calabi, L. Markus, *Relativistic space forms*, Ann. of Math. **75** (1962), 63–76.

[C41] C. Chevalley, *On the topological structure of solvable groups*, Ann. of Math. **42-3** (1941), 668–675.

[CHH88] M. Cowling, U. Haagerup, R. Howe, Almost L^2 matrix coefficients. *J. Reine Angew. Math.*, 387:97–110, 1988.

[DV10] M. Duflo, J. Vargas, *Branching laws for square integrable representations*, Proc. Japan Acad. Ser. A Math. Sci. **86** (2010), no. 3, 49–54.

[G85] W. M. Goldman, *Nonstandard Lorentz space forms*, J. Differential Geom. **21** (1985), 301–308.

[G25] W. M. Goldman, *Proper actions in pseudo-Riemanniann geometry*, Symmetry in Geometry and Analysis. Vol. 1, Progress in Mathematics **357**, (2025), 2–14.

[HR63] E. Hewitt, K. A. Ross, *Abstract Harmonic Analysis*, Springer Verlag, 1963.

[KnK25] K. Kannaka, T. Kobayashi, *Zariski-dense deformations of standard discontinuous groups for pseudo-Riemannian homogeneous spaces.* Preprint, 107 pages, arXiv:2507.03476.

[Ka12] F. Kassel, *Deformation of proper actions on reductive homogeneous spaces*, Math. Ann. **353** (2012), 599–632.

[KaK16] F. Kassel, T. Kobayashi, *Poincaré series for non-Riemannian locally symmetric spaces*, Adv. Math. **287** (2016), 123–236.

[KaK25] F. Kassel and T. Kobayashi, *Spectral Analysis on Standard Locally Homogeneous Spaces*, Lecture Notes in Mathematics **2367** (2025), xi+116 pages, Springer.

[KT24] F. Kassel, N. Tholozan, *Sharpness of proper action on reductive homogeneous spaces*, arXiv:2410.08179.

[KMTpre] F. Kassel, Y. Morita, N. Tholozan, in preparation.

[Ki62] A. A. Kirillov, *Unitary representations of nilpotent Lie groups*, uspehi Mat. Nauk **17** (1962).

[Ki25] M. Kitagawa, *Kobayashi's conjectures on the discrete decomposability*, In: Symmetry in Geometry and Analysis. Festschrift in honor of Toshiyuki Kobayashi, Vol. 3. 31–99. Progr. Math., **359**. Birkhäuser/Springer, 2025.

[K89] T. Kobayashi, *Proper action on a homogeneous space of reductive type*, Math. Ann. **285** (1989), 249–263.

[K90] T. Kobayashi, Discontinuous groups acting on homogeneous spaces of reductive type, In: Proceedings of ICM-90 satellite conference on Representation Theory of Lie Groups and Lie Algebras, Fuji-Kawaguchiko, 1990 (T. Kawazoe, T. Oshima and S. Sano eds.), World Scientific, 1992, pp. 59–75.

[K92] T. Kobayashi, *A necessary condition for the existence of compact Clifford–Klein forms of homogeneous spaces of reductive type*, Duke Math. J., **67** (1992), 653–664.

[K93] T. Kobayashi, *On discontinuous groups acting on homogeneous spaces with non-compact isotropy subgroups*, J. Geometry and Physics **12** (1993), 133–144.

[K94] T. Kobayashi, *Discrete decomposability of the restriction of $A_{\mathfrak{q}}(\lambda)$ with respect to reductive subgroups and its applications*, Invent. Math., **117** (1994), 181–205.

[K96] T. Kobayashi, Criterion for proper actions on homogeneous spaces of reductive groups, J. Lie Theory,**6** (1996), 147–163.

[K97] T. Kobayashi, *Discontinuous groups and Clifford-Klein forms of pseudo-Riemannian homogeneous manifolds.* Algebraic and Analytic methods in Representation Theory (Sønderborg, 1994), 99–165, Perspect. Math., **17**. Academic Press, San Diego, CA, 1997.

[K98a] T. Kobayashi, *Deformation of compact Clifford–Klein forms of indefinite-Riemannian homogeneous manifolds*, Math. Ann. **310** (1998), 395–409.

[K98b] T. Kobayashi, *Discrete decomposability of the restriction of $A_{\mathfrak{q}}(\lambda)$ with respect to reductive subgroups. II. Micro-local analysis and asymptotic K-support*, Ann. of Math. (2)**147** (1998), no. 3, 709–729.

[K00] T. Kobayashi, *Discretely decomposable restrictions of unitary representations of reductive Lie groups–examples and conjectures*, Advanced Studies in Pure Mathematics **26** 2000, Analysis on Homogeneous Spaces and Representation Theory of Lie Groups, pp. 99–127.

[K01] T. Kobayashi, *Discontinuous groups for non-Riemannian homogeneous spaces.* Mathematics Unlimited–2001 and Beyond, 723–747, Springer, Berlin, 2001.

[K09] T. Kobayashi, *Hidden symmetries and spectrum of the Laplacian on an indefinite Riemannian manifold*, In: Spectral Analysis in Geometry and Number Theory, 73–87. Contemp. Math.,**484**, Amer. Math. Soc. 2009.

[K17] T. Kobayashi, *Global analysis by hidden symmetry*, In: Representation Theory, Number Theory, and Invariant Theory: In Honor of Roger Howe on the Occasion of His 70th Birthday, Progress in Mathematics, vol. **323** (2017), pp. 359–397.

[K23a] T. Kobayashi, *Conjectures on reductive homogeneous spaces*, Mathematics Going Forward–collected mathematical brushstrokes, Lecture Notes in Math., vol. **2313**, Springer Cham., pp. 217–231, 2023.

[K23b] T. Kobayashi, *Bounded multiplicity branching for symmetric pairs*, Journal of Lie Theory, **33**, (2023), 305–328, Special Volume for Karl Heinrich Hofmann.

[KO90] T. Kobayashi, K. Ono, *Note on Hirzebruch's proportionality principle*, J. Fac. Sci. Univ. Tokyo Sect. IA Math. **37** (1990), no. 1, 71–87.

[KN03] T. Kobayashi, S. Nasrin, *Multiplicity one theorem in the orbit method*, Lie Groups and Symmetric Spaces: In memory of Professor F. I. Karpelevič (ed. S. Gindikin), Transl. Series 2. Amer. Math. Soc. **210**, 161–169 (2003).

[KN18] T. Kobayashi, S. Nasrin, *Geometry of coadjoint orbits and multiplicity-one branching laws for symmetric pairs*, Algebr. Represent. Theory **21** (2018), pp. 1023–1036, Special issue in honor of Alexandre Kirillov.

[KY05] T. Kobayashi, T. Yoshino, *Compact Clifford–Klein forms of symmetric spaces—revisited*, Pure and Appl. Math. Quarterly, **1**, (2005), 603–684. Special Issue: In Memory of Armand Borel.

[Ku81] R. S. Kulkarni, *Proper actions and pseudo-Riemannian space forms*, Adv. Math. **40** (1981), 10–51.

[LZ95] F. Labourie, R. J. Zimmer, *On the non-existence of cocompact lattices for $SL(n)/SL(m)$*, Math. Res. Lett. **2** (1995), 75–77.

[LMZ95] F. Labourie, S. Mozes, R. J. Zimmer, *On manifolds locally modelled on non-Riemannian homogeneous spaces*, Geom. Funct. Anal. **5** (1995), 955–965.

[L95] R. Lipsman, *Proper actions and a cocompactness condition*, J. Lie Theory **5** (1995), 25–39.

[LOY23] G. Liu, Y. Oshima, J. Yu, *Restriction of irreducible unitary representations of $Spin(N,1)$ to parabolic subgroups*, Represent. Theory **27** (2023), 887–932.

[M97] G. Margulis, Existence of compact quotients of homogeneous spaces, measurably proper actions, and decay of matrix coefficients. *Bull. Soc. Math. France*, **125** (3), 447–456, 1997.

[M00] G. Margulis, *Problems and conjectures in rigidity theory.* Mathematics: Frontiers and Perspectives, 161–174, Amer. Math. Soc., 2000.

[Mt50] F. I. Mautner, *Unitary representations of locally compact groups I, II*, Ann. of Math. (2) **51** (1950), 1–25; **52** (1950), 528–556.

[M17] Y. Morita, *A cohomological obstruction to the existence of compact Clifford-Klein forms*, Selecta Math. (N.S.) **23** (2017), no. 3, 1931–1953.

[M19] Y. Morita, *Proof of Kobayashi's rank conjecture on Clifford–Klein forms.* J. Math. Soc. Japan **71** (2019), no. 4, 1153–1171.

[N72] M. A. Naimark, *Normed algebras*, Wolters-Noordhoff Series of Monographs and Textbooks on Pure and Applied Mathematics. Wolters-Noordhoff Publishing, Groningen, russian edition, 1972.

[N01] S. Nasrin, Criterion of proper actions for 2-step nilpotent Lie groups. Tokyo J. Math. 24 (2001), 535–543.

[Ok13] T. Okuda, *Classification of semisimple symmetric spaces with proper* $SL(2,\mathbb{R})$*-actions*, J. Differential Geom. **94** (2013), 301–342.

[P61] R. Palais, *On the existence of slices for actions of non-compact groups*, Ann. Math. (2), **73** (1961), 295–323.

[P15] P.-E. Paradan, *Quantization commutes with reduction in the non-compact setting: the case of holomorphic discrete series*, J. Eur. Math. Soc. **17**(4), 955–990 (2015).

[S56] I. Satake, *On a generalization of the notion of manifold*, Proc. Natl. Acad. Sci., USA **42**, (1956), 359–363.

[STV18] H. Schlichtkrull, P. Trapa, D. Vogan, Laplacian on spheres, São Paulo, J. Math. Sci. **12** (2018), 295–358.

[S00] Y. Shalom, *Rigidity, unitary representations of semisimple groups, and fundamental groups of manifolds with rank one transformation group*, Ann. of Math. (2) **152** (2000), no. 1, 113–182.

[Ta67] N. Tatsuuma, A duality theorem for locally compact groups, J. Math. Kyoto Univ., **6** (1967), 187–293.

[Th15] N. Tholozan, *Volume and non-existence of compact Clifford–Klein forms*, arXiv:1511.09448v2, preprint.

[T97] W. Thurston, *Three-Dimensional Geometry and Topology*, Vol. 1. Princeton Math. Ser., **35**, Princeton University Press, Princeton, NJ, 1997. x+311 pp.

[To19] K. Tojo, *Classification of irreducible symmetric spaces which admit standard compact Clifford–Klein forms.* Proc. Japan Acad. Ser. A Math. Sci. 95 (2019), 11–15.

[W92] N. R. Wallach, *Real reductive groups. II*, Pure Appl. Math. **132-II**, Academic Press, Inc., Boston, MA, 1992.

[Y04] T. Yoshino, A solution to Lipsman's conjecture for $\mathbb{R}^4$, Int. Math. Res. Not. (2004), 4293–4306.

[Y05] T. Yoshino, *A counterexample to Lipsman's conjecture*, Internat. J. Math. **16** (2005), 561–566.

[Y07] T. Yoshino, Discontinuous duality theorem, Internat. J. Math. **18** (2007), 887–893.

[Z94] R. Zimmer, *Discrete groups and non-Riemannian homogeneous spaces*, J. Amer. Math. Soc. **7** (1994), 159–168.